S0-BKX-090

Die Grundlehren der mathematischen Wissenschaften

in Einzeldarstellungen
mit besonderer Berücksichtigung
der Anwendungsgebiete

Band 164

L. Sario · M. Nakai

Classification Theory of Riemann Surfaces

Springer-Verlag New York · Heidelberg · Berlin 1970

Prof. Leo Sario
University of California, Los Angeles

Prof. Mitsuru Nakai
Nagoya University

Printed in Germany. Title No. 5147

Universitätsdruckerei H. Stürtz AG, Würzburg

Preface

The purpose of the present monograph is to systematically develop a classification theory of Riemann surfaces. Some first steps will also be taken toward a classification of Riemannian spaces.

Four phases can be distinguished in the chronological background: the type problem; general classification; compactifications; and extension to higher dimensions.

The type problem evolved in the following somewhat overlapping steps: the Riemann mapping theorem, the classical type problem, and the existence of Green's functions. The Riemann mapping theorem laid the foundation to classification theory: there are only two conformal equivalence classes of (noncompact) simply connected regions. Over half a century of efforts by leading mathematicians went into giving a rigorous proof of the theorem: Riemann, Weierstrass, Schwarz, Neumann, Poincaré, Hilbert, Weyl, Courant, Osgood, Koebe, Carathéodory, Montel.

The classical type problem was to determine whether a given simply connected covering surface of the plane is conformally equivalent to the plane or the disk. The problem was in the center of interest in the thirties and early forties, with Ahlfors, Kakutani, Kobayashi, P. Myrberg, Nevanlinna, Speiser, Teichmüller and others obtaining incisive specific results. The main problem of finding necessary and sufficient conditions remains, however, unsolved.

At the end of his dissertation Riemann had already referred to the significance of the existence of the Green's function. This aspect gave rise to a generalization which chronologically ran somewhat parallel to the classical type problem: finding tests for the class O_G of parabolic surfaces characterized by the nonexistence of Green's functions. The class was studied by P. Myrberg and explicit criteria were established by Ahlfors, Nevanlinna, Laasonen, Wittich, and Le-Van.

For plane regions this generalized type problem formed the bridge to the classical theory of Szegö, Nevanlinna, Frostman, and others on capacities of point sets. In particular O_G turned out to be the class of regions whose boundaries have vanishing logarithmic capacity,

Schwarz's harmonic measure, or Fékete's transfinite diameter. Moreover $R \in O_G$ was necessary and sufficient for R to possess Evans-Selberg potentials.

The present monograph will only lightly touch upon these important early developments of classification theory. We start with the second aspect of the evolution, the general classification, which today continues at an ever increasing pace. It was inaugurated in 1948 with the introduction of the class O_{AD} of surfaces without nonconstant AD-functions, i.e. analytic functions with finite Dirichlet integrals. Such surfaces were said to have "hebbar" boundaries, in reference to their behavior as closed surfaces in significant function-theoretic respects (see Introduction and Bibliography). At the Trondheim Congress in 1949 a systematic array of null-classes, together with current notation, was introduced. Penetrating results on function-theoretic null sets related to several such classes were obtained in 1950 by AHLFORS and BEURLING. In 1954 the study of boundary components in classification theory was initiated by the introduction of their capacities.

During the two decades since the beginning of the general classification theory the subject has grown in depth and breadth into one of the major branches of function theory. The main achievements are due to AHLFORS, BEURLING, CONSTANTINESCU, CORNEA, HEINS, KAMETANI, KURAMOCHI, KURODA, KUSUNOKI, LEHTO, MARDEN, MATSUMOTO, A. MORI, S. MORI, L. MYRBERG, P. MYRBERG, NEVANLINNA, NOSHIRO, OHTSUKA, OIKAWA, OZAWA, PARREAU, PFLUGER, RAO, RODIN, ROYDEN, TÔKI, TSUJI, VIRTANEN, YÛJÔBÔ, among others. For a complete list of workers in the field we refer the reader to the Author Index and the Bibliography.

Although capacities of subboundaries are useful especially in the study of plane regions, detailed information on ideal boundaries can only be obtained by compactifying the surface. The mode of compactification depends on the class of functions under consideration. For the class HD of harmonic functions with finite Dirichlet integrals ROYDEN introduced in 1952 the compactification now bearing his name. For the class HB of bounded harmonic functions the Wiener compactification proved to be the most fruitful choice. The Royden and Wiener compactifications can also be described as corresponding to the solution of the Dirichlet problem by Dirichlet's principle or by Perron's method.

The most recent aspects of the theory of compactification started in 1962 with the discovery by KURAMOCHI of surfaces carrying distinguished minimal points on their boundaries. The work was continued in the authoritative treatise of Constantinescu-Cornea in 1963. The current compactification theory as it appears in the present monograph is of the recent vintage of 1966, much of it previously unpublished.

Classification of Riemannian spaces of higher dimensions is the latest facet of the theory. Although only in its infancy it has already produced surprising phenomena.

From the chapter and section headings the reader may obtain an over-all view of the plan of the book. Broadly speaking, regular functions are treated first, then those with logarithmic singularities. Among regular functions the analytic functions precede the harmonic functions; in each category those with finite Dirichlet integrals are discussed first. One denotes by AB and AD the classes of analytic bounded or Dirichlet finite functions, by HB and HD the corresponding classes of harmonic functions, and by O_{AD}, e.g., the class of surfaces without nonconstant AD-functions. The resulting scheme O_{AD}, O_{AB}, O_{HD}, O_{HB}, O_G roughly corresponds to the decreasing "magnitude" of the null classes. Treating O_{AD} at the beginning of the book also has the advantage of first encountering the numerous concrete properties that are associated with O_{AD}, more than with any other class. Finally, starting with O_{AD} somewhat follows the historical development of general classification theory.

A more detailed description of the book is given in brief surveys at the beginning of each chapter and section. For a first orientation of the nonexpert we also give, in the Introduction, some concrete examples from the early part of the book.

Every effort was made to develop the theory into a harmonious unity. The rather detailed Table of Contents and the Table of strict inclusion relations reveal some of the strands of the rather intricate logical network tying the chapters into a whole which we hope to be something more than the sum of its parts.

On occasion a result may seem isolated until its significance manifests itself in relationships given in later chapters. In this regard the Subject Index is essential as it gives cross-sections on specific topics, e.g. a particular null class.

Some subsections, indicated by an asterisk before the heading, are not needed for the understanding of the subsequent parts of the book.

Bibliographical references, summarized in the Author Index, are complete in that the source of every result not due to the authors is explicitly given.

The reader is not expected to have any previous knowledge of classification theory. For general prerequisites a standard Ph. D. curriculum is sufficient. In the few instances where we have made an exception, a precise reference is given to some well-known monograph.

The basic terminology we use is that adopted in Ahlfors-Sario [1].

Although some vague ideas for the book go back two decades, the actual planning, writing, and revising was carried out during the past five years, in particular while the junior author was visiting at UCLA in 1965–1967. We are deeply grateful to the U.S. Army Research Office–Durham for several grants which made our collaboration possible, and to Drs. J. Dawson and A. S. Galbraith for their magnificent cooperation through the entire course of the work.

We are grateful to Vice President A. Taylor, Dean L. Paige, and Professor E. Coddington for the favorable circumstances which were a conditio sine qua non for carrying out the vast project.

It is our pleasure to acknowledge our indebtedness to Professors M. H. Stone and H. Royden who read the manuscript and gave their valued advice.

Our sincere thanks are due to Y. Kwon, I. Lin, and S. Councilman who spared no effort in checking and preparing the manuscript for the printers and in helping us with the Bibliography and the Indices.

We feel privileged for the inclusion of our book in this distinguished series and we wish to thank Professor B. Eckmann for his interest and encouragement.

Dr. K. Peters and the technical experts of the Springer-Verlag met our endless wishes with endless patience.

Our thanks are also due to Mrs. Elaine Barth and Miss Ellen Cole whose teams of typists processed the manuscript and its countless revisions into a flawless professional product.

Los Angeles and Nagoya
February 1, 1970

Leo Sario
Mitsuru Nakai

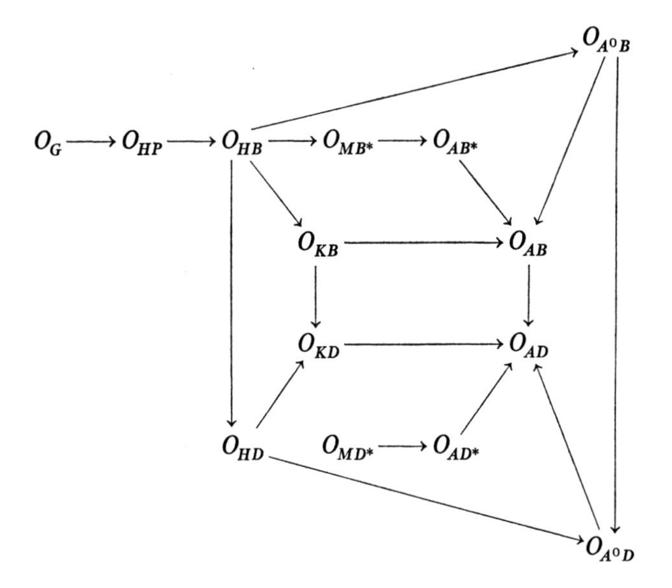

$$
\begin{array}{cccccccccccccc}
O_G < & O_{HP} & = & O_{HP}^1 & < & O_{HP}^2 & < \cdots < & O_{HP}^\infty & < & U_{HP} \cup O_G & & \\
 & \wedge & & \wedge & & \wedge & & \wedge & & \vee & & \\
 & O_{HB} & = & O_{HB}^1 & < & O_{HB}^2 & < \cdots < & O_{HB}^\infty & < & U_{HB} \cup O_G & < & O_{AB} \\
 & \wedge & & \wedge & & \wedge & & \wedge & & \wedge & & \wedge \\
 & O_{HD} & = & O_{HD}^1 & < & O_{HD}^2 & < \cdots < & O_{HD}^\infty & < & U_{\widetilde{HD}} \cup O_G & < & O_{AD}
\end{array}
$$

Table of strict inclusion relations

Contents

Chapter II · Other Classes of Analytic Functions

Chapter III · Dirichlet Finite Harmonic Functions

Chapter IV · Other Classes of Harmonic Functions

Chapter V · Functions with Logarithmic Singularities

Chapter VI · Functions with Iversen's Property

Appendix. Higher Dimensions

Introduction

The purpose of this Introduction is to give to the reader with no previous experience with classification theory some concrete examples of problems and results in the early part of the book. It is not needed for the understanding of the book proper; no proofs, bibliographical references, or rigorous definitions will be given.

The examples also serve to illustrate the ever-important role played in classification theory by plane regions, and the fascinating element of surprise offered by surfaces of infinite genus in their behavior contrary to all intuition.

An effort was made to present the examples as a logical unity.

1. Class O_{AD}. An immediate geometric property of the plane is the invariance of its infinite Euclidean area under mappings by analytic functions. The forerunner of classification theory was Riemann's mapping theorem: there are only two conformal equivalence classes of (noncompact) simply connected regions according as they do or do not have the above property.

This classification is meaningful for arbitrary Riemann surfaces R. For an analytic formulation consider an analytic mapping f of R. The area of the Riemannian image $f(R)$ is $\int_{f(R)} du\,dv = \int_R |f'(z)|^2\,dx\,dy = D(f)$, with an obvious meaning of the symbols. Because of the invariance under parametric transformations the Dirichlet integral $D(f)$ is well-defined. We denote by AD the family of analytic functions with $D(f) < \infty$ and by O_{AD} the class of Riemann surfaces on which there are no non-constant AD-functions.

2. Regions $R \notin O_{AD}$. The first problem is to characterize the plane regions R in O_{AD}. If the complement CR of R with respect to the extended plane has an interior point z_0 then $1/(z-z_0)$ gives $R \notin O_{AD}$. However interior points in CR are not necessary: the existence of a proper continuum α on ∂R already suffices. Indeed if $a, b \in \alpha$ then $((z-a)/(z-b))^{\frac{1}{2}}$ maps R into some half-plane and a subsequent mapping into a disk brings in the identity map in AD. In striking contrast with plane regions there are O_{AD}-surfaces of infinite genus with proper continua on the boundary (cf. 11 below). Similar degeneracies despite strong boundaries are encountered in quite uncomplicated higher dimensional spaces (cf. 15).

The existence of a continuum on the boundary is not necessary for $R \notin O_{AD}$. Any discrete compact plane set ∂R of positive area serves as an example. Here the λ-dimensional Hausdorff measure $m^{\lambda}(E)$ of a bounded set E is by definition $\liminf \sum r_n^{\lambda}$ where the r_n are the radii in any finite set of closed disks whose union covers E; the area is positive if and only if $m^2(E)>0$. Let $\{R_n\}_1^{\infty}$ be a regular exhaustion of R, i.e. a nested sequence of regular regions whose union is R; by a regular region we mean one that is relatively compact and bounded by a finite number of disjoint analytic Jordan curves. Choose R_1 to contain ∞ and a given z_0. Let g_n be the conformal mapping of R_n onto a region bounded by horizontal slits, with residue 1 at z_0. In the limit we obtain an image $g(R)$ of R whose boundary continues to consist of horizontal slits or points of vanishing total area (I.7 E). Thus $f(z)=g(z)-z$ cannot be constant and by the triangle inequality we have $D(f)<\infty$.

3. Modular O_{AD}-Test. On the other hand there are relatively strong discrete sets ∂R with $R \in O_{AD}$. To see this consider an exhaustion $\{R_n\}_1^{\infty}$ of an arbitrary Riemann surface R and let E_{ni} be the components of $R_{n+1}-\bar{R}_n$. Take the harmonic function u on $\bar{E}_{ni}$ with $u|\partial R_n=0, u|\partial R_{n+1}=\log \mu_{ni}$, a constant >0 such that the flux $\int_{\partial R_n} {*du} = \int_{\partial R_n} (\partial u/\partial n)\, ds = 2\pi$. The geometric meaning of the modulus is illustrated by a doubly connected E_{ni}: it is mapped by $\exp(u+i u^*)$ onto a concentric circular slit annulus with radii 1 and μ_{ni}, hence of modulus μ_{ni} in the usual sense. A general E_{ni} must first be cut along suitable level lines of (a locally defined) u^* to make the region doubly connected. Then one obtains a map to an annulus of modulus μ_{ni} with radial slits corresponding to the cuts.

Set $\mu_n = \min_i \mu_{ni}$. The modular test (I.1 D) states: if $\prod_1^{\infty} \mu_n = \infty$ for some exhaustion then $R \in O_{AD}$. The significance of the test is in that it reduces the estimation of the "conformal shape" of the entire surface R to that of surface fragments E_{ni}. To illustrate, the plane can be exhausted by regions R_n: $|z|<r_n$ whose modular product $\prod_1^m \mu_n = r_{m+1}/r_1$ diverges whereas a disk cannot be so exhausted. Intuitively speaking the divergence of the modular product assures a certain smallness of each boundary component.

We note in passing that every annulus can be subdivided into two subannuli, each having its modulus arbitrarily close to 1 (I.4 B). Starting from an exhaustion of the plane by disks we can thus form a new exhaustion such that $\prod_1^{\infty} \mu_n < 1+\varepsilon$, with $\varepsilon>0$ arbitrarily small.

4. Cantor Ternary Set as an AD-Null Set. We apply the modular test to the Cantor ternary set on the interval [0, 1]. After removing the line segment (1/3, 2/3) we encircle the segment [0, 1/3] by the annulus $1/4<|z-1/6|<1/3$ and the segment [2/3, 1] by the annulus $1/4<|z-5/6|<1/3$. By multiplying all dimensions by 1/3 and by translating the center of the configuration to 1/6 one similarly encircles the two segments of [0, 1/3] remaining after its center third has been removed. The two segments remaining from [2/3, 1] are encircled in the same manner. At each subsequent step of the construction of the Cantor set the process is repeated by again multiplying the dimensions of the two annuli by 1/3 and by suitably translating them to encircle newly formed segments. The resulting annuli are all disjoint, each has modulus 4/3, and consequently the modular product diverges. Thus the Cantor ternary set is an AD-null set (the boundary of an O_{AD}-region) although it is known to have positive logarithmic capacity (V.11 F).

There also exist discrete sets E whose projection onto a line is an interval but whose complement is in O_{AD} (II.9 C).

Despite the existence of such strong AD-null sets E they all have the property that any two points in CE can be joined by a curve in CE whose length differs arbitrarily little from their Euclidean distance (II.9 C).

5. Relation of O_{AD} to Area of ∂R. What metric properties does an AD-null set possess? To study this we modify the Cantor ternary set by removing at the nth step a p_nth part, with $p_n>1$ not necessarily 3. The resulting set has linear measure $\prod_1^\infty(1-p_n^{-1})$. Now take the Cartesian product $E^2=E\times E$ obtained by centrally removing from the unit square a crosslike part and then repeating the process for the remaining squares at all steps. The area $\prod_1^\infty(1-p_n^{-1})^2$ of E^2 vanishes if and only if $\sum p_n^{-1}=\infty$. By a modification of the above process of encircling by doubly connected regions the modular test can again be used to show that $CE^2\in O_{AD}$ if the area of E^2 vanishes. On the other hand we know that a positive area gives for any set a complement $\notin O_{AD}$. We conclude:

A necessary and sufficient condition for a Cantor set E^2 to have an O_{AD} complement is that its area vanish.

This direct relation of O_{AD} to area is valid only in some symmetric situations. In the general case we introduce discrete plane point sets E of *absolute area zero*. These are characterized by the vanishing of the area of $Cf(CE)$ under all univalent mappings f of CE; here and in the sequel the term "univalent" is understood to mean "analytic univalent." Let O_{SE} be the class of planar regions R with ∂R of absolute area zero. Then one again has a complete characterization: $O_{AD}=O_{SE}$.

6. *AD*-Removable Sets. Does this weakness of *AD*-null sets E imply that all univalent maps of CE have in fact analytic extensions to E? The answer is in the affirmative, for the property is local in the following sense. A bounded plane point set E is called *AD-removable* if for some disk U containing E every *AD*-function on $U-E$ has an analytic extension to all of U. A necessary and sufficient condition for E to be *AD*-removable is that $CE \in O_{AD}$ (I.8 D). As a consequence R is rigid if and only if it is in O_{AD}, the rigidity meaning that all univalent maps of R with residue 1 at a given point are identical. Equivalently, all univalent maps are linear. The property extends to surfaces R of finite genus: all closed surfaces into which R can be univalently mapped are conformally equivalent if and only if $R \in O_{AD}$ (I.8 F, 8 G).

7. Abelian and Schottky Covering Surfaces. A univalent mapping of an O_{AD}-surface of genus zero is unique up to a linear transformation. An important application is to the theory of uniformization. On a closed Riemann surface R_0 consider an Abelian (harmonic) differential du. Locally it is the differential of a harmonic function u. The weakest covering surface on which u is single-valued is called the Abelian covering surface corresponding to du. A modification of the modular test gives a complete result (I.18 B): all Abelian covering surfaces are in O_{AD}.

In particular this is true of the Schottky covering surface corresponding to a du which has periods along g disjoint loop cuts, g the genus of R_0. To construct it cut R_0 along g conjugate loop cuts and map the remaining planar surface conformally onto a plane region R_1 bounded by $2g$ circles. Reflect R_1 about each circle so as to obtain a region R_2 bounded by $2g(2g-1)$ circles. Then reflect the image of R_1 about each of these circles to obtain a region R_3 with $2g(2g-1)^2$ boundary circles and repeat the process ad infinitum. The resulting region R is the Schottky covering surface of R_0. Its boundary is a perfect point set, the Schottky point set.

That $R \in O_{AD}$ can be seen directly. Each component E_{ni} of $R_{n+1}-\bar{R}_n$ is a conformal duplicate of R_1 and there are only $2g$ different μ_{ni}. Consequently $\prod \mu_n$ diverges. We conclude that the Schottky uniformization of a closed surface is unique up to a linear transformation. The Schottky set is known to have positive capacity and thus offers another example of a strong *AD*-null set.

8. O_{AB}-Regions. The above rigidity properties of O_{AD}-regions are in essence manifestations of the equivalence $O_{AD}=O_{SE}$. What happens if we drop the requirement of univalency in O_{SE}? Denote by AE the class of analytic functions which omit a set of positive area and by AB the class of bounded analytic functions. One has the striking equality

$O_{AE}=O_{AB}$, i.e. as soon as a function omits a set of positive area then there also is a function which omits an entire disk (II.5 A). This result together with its implication $O_{AB}\subset O_{AD}$ continues to hold for Riemann surfaces of arbitrary genus.

What metric properties does an AB-null set possess? That a single point is such a set is the content of Liouville's theorem. More generally if the linear measure of ∂R vanishes then $R\in O_{AB}$. To see this one only has to show that for a disk $U\supset\partial R$ every $f\in AB(\bar{U}-\partial R)$ has an analytic extension to $\bar{U}$. Let $\{R_n\}_1^\infty$ be an exhaustion of R such that the length $\int_{\partial R_n}|dz|$ of ∂R_n tends to 0. If $|f|<M$ then for $z_0\in U\cap R_1$ and a constant $d<\min_{\partial R_1}|z-z_0|$ we have

$$f(z_0)=\frac{1}{2\pi i}\int_{\partial U-\partial R_n}\frac{f(z)\,dz}{z-z_0}.$$

Here $|\int_{\partial R_n}|\leq M d^{-1}\int_{\partial R_n}|dz|\to 0$ and therefore $f(z_0)$ coincides with $(2\pi i)^{-1}\int_{\partial U} f(z)(z-z_0)^{-1}\,dz$, which is analytic on U.

Whether or not vanishing of the linear measure of ∂R also is necessary for $R\in O_{AB}$ is the classical problem of Painlevé. The solution is in the negative (II.11E): the 2-dimensional Cantor set E^2 (cf. 5) with $p_n=2$ has positive linear measure but its complement is in O_{AB}. However if the $(1+\varepsilon)$-dimensional Hausdorff measure of ∂R is positive then $R\notin O_{AB}$ (II.2 F).

For sets ∂R on a line or more generally on an analytic curve, vanishing of the linear measure is necessary and sufficient for $CE\in O_{AB}$ (II.10 A). In particular a 1-dimensional Cantor set is an AB-null set if and only if its length $\prod_1^\infty(1-p_n^{-1})$ vanishes.

An intriguing and not quite well understood phenomenon is that $\sum p_n^{-1}=\infty$ is thus at once necessary and sufficient for a 1-dimensional Cantor set to be an AB-null set and for a 2-dimensional Cantor set to be an AD-null set.

9. O_{SB}-Regions. The existence of AB-functions means that the region can be mapped into a disk. When can this be done univalently? Let SB be the class of univalent functions in AB. The added requirement brings us back to D-functions: $O_{SB}=O_{SD}$, i.e. if a region has a conformally equivalent plane region of finite area then it also has one which is bounded (II.8 F).

An interesting example is the following region which has an arbitrarily small area yet is dense in the plane. Cover the plane by a net of unit squares, numbered in some order. From the nth square remove a 2-dimensional Cantor set E_n^2 of area $\prod(1-p_n^{-1})^2>1-2^{-n}\varepsilon$ and set $E=\bigcup E_n^2$. Then the region CE has area $<\varepsilon$ although its closure is the extended plane.

10. Unstable Boundary Components. An SB-mapping of CE in the above example stretches the boundary point at infinity to a proper continuum. An important general problem is to characterize such "unstable" boundary components.

Let γ be a component of the ideal boundary of a Riemann surface R. Given an exhaustion $\{R_n\}_1^\infty$ let γ_n denote the relative boundary (on ∂R_n) of that component of $R-\bar{R}_n$ whose ideal boundary contains γ. The relative boundaries of the other components of $R-\bar{R}_n$ are denoted by β_{ni}. For $z_0 \in R_1$ take the harmonic function $p_{\gamma n}$ on $\bar{R}_n - z_0$ with singularity $\log|z-z_0|$ such that $p_{\gamma n}(z)-\log|z-z_0| \to 0$ as $z \to z_0$, $p_{\gamma n}|\gamma_n = k_n$ (const), $p_{\gamma n}|\beta_{ni} = \text{const}$, $\int_{\beta_{ni}} * dp = 0$ for all i. Then the k_n increase with n, and the capacity of γ is defined as $c_\gamma = e^{-\lim k_n}$. For a plane region we have:

A point boundary component γ can be stretched into a proper continuum under some univalent mapping if and only if $c_\gamma \neq 0$. Equivalently $O_{SB} = O_{SD} = O_\gamma$ where O_γ is the class of regions whose boundary is *absolutely disconnected*, i.e. every component γ has vanishing capacity.

The geometric meaning is that the $p_{\gamma n}$ have a limit p_γ such that $\exp(p_\gamma + i\, p_\gamma^*)$ maps R into a disk with γ going to the periphery of radius c_γ^{-1} and the other components to circular slits. This periphery is the "strongest" representation of γ under all univalent maps; if it does not reduce to the point at infinity, then γ is unstable.

By replacing the circular slits by radial ones the outer contour is pushed to its furthest position. Analytically this means that the condition $p_{\gamma n}|\beta_{ni} = \text{const}$ is replaced by $(\partial p_{\gamma n}/\partial n)|\beta_{ni} = 0$ and the resulting new capacity c'_γ may vanish even if c_γ does not. A challenging unsolved problem is whether or not $c_\gamma \neq 0$, $c'_\gamma = 0$ characterize unstable components even when the existence of a pointlike realization is not known in advance.

Closely related to the above is the following rather surprising phenomenon. Let P_φ be the univalent mapping of a planar R onto the parallel slit region with inclination φ of the slits and with residue 1 at a given point. As φ varies from 0 to π only three cases can occur for a given boundary component γ: either it is always a point or always a continuum or else a point for exactly one value of φ. Whether or not instability of γ is necessary and sufficient for the third alternative is not known.

An added advantage of the classification in terms of c_γ and c'_γ is that it is meaningful on Riemann surfaces of any genus. It offers a largely unexplored field of classification theory.

11. Boundary Continua of O_{AB}-Surfaces. Illustrative of the unexpected phenomena that may occur in the case of infinite genus are O_{AB}-surfaces with proper continua on the boundary. An example is constructed as follows (cf. I.10 B).

Take a 2-sheeted covering surface R_0 of $0<|z|<3$ with branch points accumulating only at 0 and 2. Suppose the points over $|z|=1$ constitute two disjoint circles. Remove the upper circle from R_0 and then identify each point $e^{i\theta}$ on its inner shore with $e^{i(\theta+\alpha)}$ on the outer shore, with α an irrational multiple of 2π. On the resulting surface R let $f \in AB$. For points z_1, z_2 on the upper and lower sheets of R with the same projection z, the function $g(z)=\left(f(z_1(z))-f(z_2(z))\right)^2$ is in AB on $0<|z|<1$ and on $\{1<|z|<3\}-\{2\}$ and therefore has analytic extensions to 0 and 2. Since it vanishes at the branch points it must be identically 0; hence $\hat{f}(z)=f(z_1(z))=f(z_2(z))$ is a function of z on $|z|<1$ and on $1<|z|<3$. But in view of the behavior of $f(z_2)$ on the lower circle above $|z|=1$, $\hat{f}$ can be analytically extended to $|z|=1$. The behavior of $f(z_1)$ on the upper circle gives $\hat{f}(e^{i\theta})=\hat{f}(e^{i(\theta+\alpha)})$. This implies $\hat{f}=\text{const}$ on $|z|=1$ and a fortiori on $|z|<3$. We infer that $R \in O_{AB}$ despite the existence of the two boundary continua. From $O_{AB} \subset O_{AD}$ it follows that $R \in O_{AD}$ as well.

12. Boundary Properties and Quasiconformal Invariance. A property of a Riemann surface R_0 is said to be a boundary property if every Riemann surface R having a boundary neighborhood in common with R_0 has this property. The above example serves to show that, perhaps contrary to intuition, O_{AB} and O_{AD} are not boundary properties. In fact the surfaces R_0 and R have in common the complement of the annulus in the upper sheet over $\frac{1}{2}<|z|<1$, yet R_0 has the projection function z in $AB \cap AD$.

The example also shows that O_{AB} and O_{AD} are not invariant under quasiconformal mappings. By definition such mappings take infinitesimal circles into infinitesimal ellipses with a bounded ratio of the axes. The disk over $\frac{1}{2}<|z|<1$ in the upper sheet of R_0 is mapped quasiconformally onto that of R by $r e^{i\theta} \to r e^{i\theta+i\alpha(2r-1)}$ and the remainders of R_0 and R correspond by identity.

Let O_G be the class of parabolic Riemann surfaces characterized by the nonexistence of nonconstant positive superharmonic functions, in particular Green's functions. Denote by HB and HD the classes of harmonic functions which are bounded and have finite Dirichlet integrals respectively. In interesting contrast with O_{AB} and O_{AD} the classes O_G and O_{HD} are preserved under quasiconformal mappings (III.8 B, 8 H). For O_{HB} the question is open (cf. IV.11 C). Another contrast with O_{AB} and O_{AD} is that O_G, O_{HB}, and O_{HD} express properties of the ideal boundary (III.8 I and IV.11 D).

13. Inclusion Relations. Bounded Means. Classification of plane regions with respect to functions offers no interest: $O_G=O_{HP}=O_{HB}=O_{HD}$; here we have added the class HP of positive harmonic functions. For general Riemann surfaces the inclusions in $O_G < O_{HP} < O_{HB} < O_{HD}$ are also easily established but proving their strictness was one of the most difficult problems in the theory of Riemann surfaces (III.4 H, IV.3 C, V.7 C).

A relation of unforeseen importance was later discovered: if $R \in O_{HB} - O_G$ then every boundary neighborhood is in O_{AB} (IV.7 G). The relation remains true if HB and AB are replaced by HD and AD (III.5 I). Here we also have another case of O_{AB}- and O_{AD}-surfaces with proper continua on the relative boundaries.

What can be said about functions which, although perhaps not bounded, are bounded in the mean? Let $\Omega \subset R$ be a regular region containing a given point z_0. Take the function $p_\Omega(z)$ harmonic on $\bar{\Omega} - z_0$ having singularity $\log|z-z_0|$ with $p_\Omega(z) - \log|z-z_0| \to 0$ as $z \to z_0$ and such that $p_\Omega | \partial\Omega = \text{const}$. For a constant $q > 0$ and a harmonic u on R set

$$m_q(u, p_\Omega) = \int_{\partial\Omega} |u|^q * dp_\Omega .$$

Let HM_q be the family of harmonic functions u on R with $m_q(u, p_\Omega)$ bounded for all $\Omega \subset R$. An interesting equivalent property is that $|u|^q$ has a harmonic majorant on R.

Rather unexpectedly O_{HM_q} brings in no new classes: it coincides with O_{HP} or O_{HB} according as $q \leq 1$ or $q > 1$ (Corollary IV.6 B).

14. Classes O_{KB} and O_{KD}. Between the class H and the class of real parts of functions in A we insert the class K of harmonic functions whose flux $\int * du$ vanishes across all dividing cycles. What properties, if any, do the K-functions share with the H-functions or the A-functions?

For planar surfaces the K-functions are nothing but the real parts of A-functions. For finite genus the analogy with A-functions is also intimate: $O_{KB} = O_{AB}$ and $O_{KD} = O_{AD}$ (II.15 A). For arbitrary genus the modular test applies to O_{KD} if the exhaustion is done, as is always possible, by canonical regions, i.e. regular regions whose every boundary component is a dividing cycle (II.12 C). Thus e.g. all Abelian covering surfaces of closed surfaces are even in O_{KD}.

Despite this close affiliation with A-functions the inclusion relations $O_{KB} < O_{AB}$ and $O_{KD} < O_{AD}$ are strict in the general case. This is a consequence of the fact that the complement of a closed parametric disk with respect to any Riemann surface is not in $O_{KB} \cup O_{KD}$ (II.15 C), in contrast with the behavior of A-functions on such complements for surfaces in $O_{HB} - O_G$ and $O_{HD} - O_G$ (see 13 above).

Moreover the K-classes have important properties in common with the H-classes: belonging to O_{KB} and O_{KD} is a property of the ideal boundary (II.14 D), and the class O_{KD} is quasiconformally invariant (II.14 B).

On some questions more is known about K-functions than about H-functions; e.g. not even plane O_{KB}-regions are quasiconformally invariant (II.14 C) whereas the question for O_{HB} remains open. The class O_{KD} is the natural one to which the classical Riemann-Roch theorem can be extended (II.16 I). It also permits an explicit functional-theoretic characterization (II.13 C).

15. Riemannian Spaces. Entirely new phenomena are encountered in higher dimensional Riemannian spaces. An n-ball can be endowed with a Riemannian metric which makes it parabolic, despite its strong continuum boundary (App. 3 B). Similarly an n-torus punctured at a single point can be made hyperbolic (App. 3 C). The classes O_G, O_{HD}, O_{KD} are not quasiconformally and not even conformally invariant but they are quasi-isometrically invariant (App. 4 D, 4 E).

Such phenomena seem to indicate that the classification of Riemannian spaces offers a promising new field of research in classification theory.

Chapter I

Dirichlet Finite Analytic Functions

The family of (single-valued) analytic functions with finite Dirichlet integrals on a Riemann surface R will be denoted by $AD(R)$. We also use the notation AD without reference to the surface R and we let O_{AD} signify the class of Riemann surfaces for which AD does not contain any non-constant functions. As a consequence of the maximum modulus principle every closed Riemann surface is in O_{AD}. We are therefore mainly interested in open Riemann surfaces which belong to this class.

For a general orientation reference is here made to the Introduction where several examples were given of surfaces in O_{AD} and of their properties.

The present chapter is divided into four sections. In § 1 we derive tests for an arbitrarily given abstract Riemann surface to belong to O_{AD}. These tests will be used extensively throughout the chapter. Planar surfaces are discussed in § 2, which contains the most striking properties of O_{AD}-surfaces. § 3 is devoted to ramified covering surfaces of the sphere, and § 4 to smooth covering surfaces of closed surfaces.

§ 1. Arbitrary Surfaces

In 1 a modular test is given for a Riemann surface to be in O_{AD}. To this end the surface is broken up into annular fragments whose conformal shapes are expressed in terms of moduli. The conformal shape of an annular fragment is easier to estimate than that of the entire surface. The modular test is simple: if the product of the minimal moduli diverges then the surface belongs to O_{AD}. Certain "relative classes" A_0D and A^0D of AD-functions are also considered. The modular test works for O_{A_0D} and O_{A^0D} as well.

The divergence of the modular product is related to that of an integral involving lengths of boundaries of exhausting regions in terms of a conformal metric. The corresponding O_{AD}-test is established in 2. Application to fundamental regions of automorphic functions gives a particularly concrete form to this test as the length can be measured in the Euclidean metric.

Estimation of the conformal shapes of the annuli introduced in 1 is further facilitated by covering them by regularly overlapping parametric disks. An O_{AD}-test can then be given simply in terms of the numbers of these disks. This is the content of 3.

1. Modular Test

1 A. Modulus. On a Riemann surface R consider a relatively compact region E whose relative boundary ∂E consists of a finite number (≥ 2) of Jordan curves, divided into two disjoint classes α and β. Let x be a continuous function on $\bar{E}$ which is harmonic on E with

$$x|\alpha=0, \qquad x|\beta=\log\mu \tag{1}$$

where $\mu>1$ is a constant such that

$$\int_\alpha * dx=2\pi. \tag{2}$$

Here the integral is to be understood as $\lim_{\lambda>0,\,\lambda\to 0}\int_{x=\lambda} * dx$. The number μ, a conformal invariant of the configuration (E, α, β), is called its *modulus* $\operatorname{mod}(E, \alpha, \beta)$; in analogy, $\log\mu$ shall be referred to as the *logarithmic modulus* $\log\operatorname{mod}(E, \alpha, \beta)$.

We shall call x the *modulus function* for (E, α, β).

Let $w=w(\zeta; E, \beta)$ be the *harmonic measure* of β with respect to E, i.e. the continuous function on $\bar{E}$ which is harmonic on E with

$$w|\alpha=0, \qquad w|\beta=1. \tag{3}$$

By Stokes' formula

$$\int_\alpha * dw=\int_E dw\wedge * dw=D_E(w)$$

and therefore

$$\log\mu=\frac{2\pi}{D_E(w)}. \tag{4}$$

1 B. Geometric Meaning. The geometric meaning of the modulus is illuminated by the scheme of Fig. 1. The scheme will serve heuristic purposes only, and no proofs will be supplied. Let y be the (multivalued) conjugate harmonic function of the modulus function x. Consider the analytic function

$$\zeta=e^{x+iy}$$

on E with single-valued $|\zeta|$.

If E is a doubly connected planar region then ζ maps E conformally onto the annulus with radii 1 and μ (Fig. 1a), i.e. *μ is the ratio of the radii of the image annulus.*

If E is a planar region with β consisting of more than one Jordan curve, say β_1 and β_2 (Fig. 1b), then we cut E along an arc γ_β joining β_1 with β_2 on which y, considered locally, is a constant. The function ζ maps the doubly connected region so obtained onto a radial slit annulus with radii 1 and μ.

Next consider the case where the boundary of E consists of two contours α and β but the genus of E is positive, say 1 (Fig. 1c). Cut E along a Jordan curve δ which does not divide E and on which y is, locally, constant. The resulting surface has zero genus and is mapped by ζ conformally onto a radial slit annulus with radii 1 and μ.

By using both methods of cutting, one for contours and the other for positive genus, a general region E (Fig. 1d) with suitable cuts can be mapped conformally onto a radial slit annulus with radii 1 and μ. Curves belonging to α and β on the border of E are mapped onto corresponding parts of the inner and outer circular boundary of the radial slit annulus. The cuts γ_β go to radial slits issuing from the outer circular boundary; the number of these slits is $2c-2$, with c the number of contours constituting β. The analogue is true of cuts γ_α. Each cut δ which does not divide E goes to two radial slits which do not meet the circular boundaries of the radial slit annulus. *The modulus μ of (E, α, β) is again the ratio >1 of the radii of the image annulus.*

1 C. Generalization. Suppose (E_j, α_j, β_j), $j=1, \ldots, n$, are configurations as in 1 A. If the E_j are disjoint by pairs and $\alpha_j \cap \beta_i = \emptyset$ $(i \neq j)$ then we set $E=\bigcup_1^n E_j$, $\alpha=\bigcup_1^n \alpha_j$, and $\beta=\bigcup_1^n \beta_j$. For (E, α, β) we can also define the modulus function x, the harmonic measure w, and the modulus $\operatorname{mod}(E, \alpha, \beta)$ in the same fashion as in 1 A. In particular identity (4) remains valid.

Observe that $w_j = w|E_j$ is the harmonic measure of β_j with respect to E_j. In view of this we conclude by (4) that

$$\frac{1}{\log \operatorname{mod}(E, \alpha, \beta)} = \sum_{j=1}^{n} \frac{1}{\log \operatorname{mod}(E_j, \alpha_j, \beta_j)}. \tag{5}$$

Let γ be a finite set of disjoint Jordan curves in E separating α from β and dividing E into two open sets E^1 and E^2 such that $\alpha \subset \partial E^1$ and $\beta \subset \partial E^2$. Then

$$\operatorname{mod}(E, \alpha, \beta) \geq \operatorname{mod}(E^1, \alpha, \gamma) \cdot \operatorname{mod}(E^2, \gamma, \beta). \tag{6}$$

This is often referred to as the *modulus inequality*.

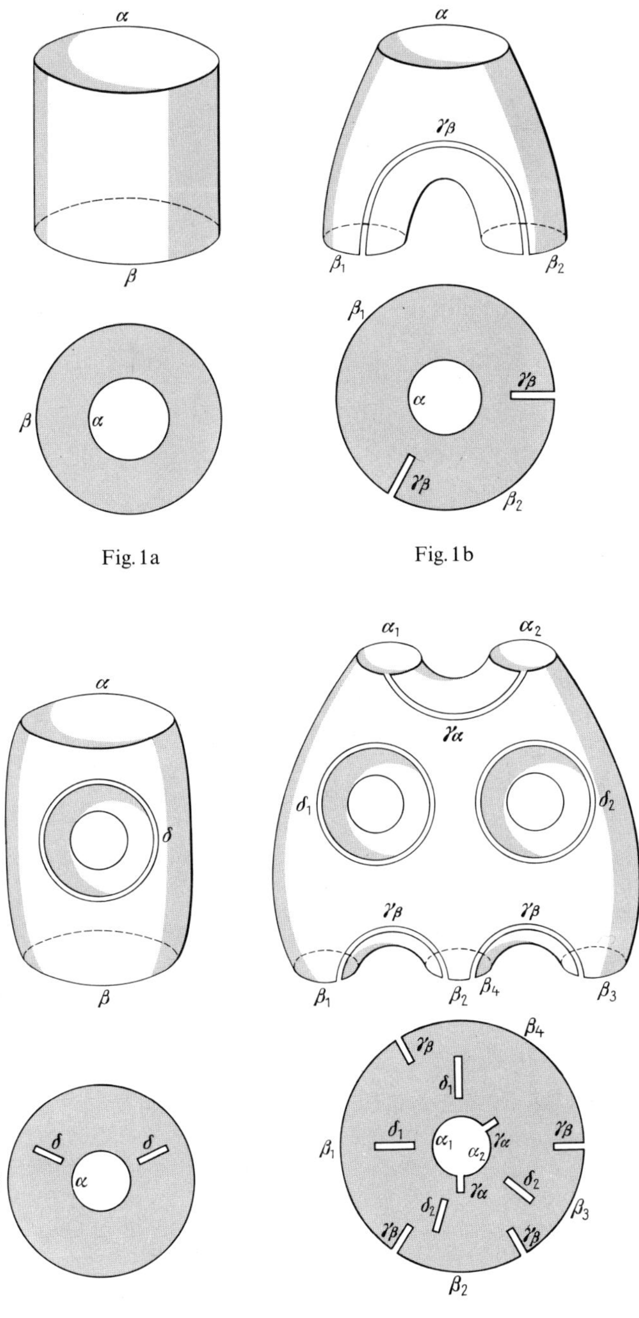

Fig. 1a

Fig. 1b

Fig. 1c

Fig. 1d

For the proof let w^1 be the harmonic measure of γ on E^1 and w^2 that of β on E^2. Define the function w^λ on $\bar{E}$ by

$$w^\lambda | \alpha \cup E^1 \cup \gamma = \lambda w^1$$

and

$$w^\lambda | \gamma \cup E^2 \cup \beta = (1-\lambda)\left(w^2 + \frac{\lambda}{1-\lambda}\right),$$

where $\lambda \in (0,1)$. By Green's formula $D_E(w^\lambda) = D_E(w) + D_E(w - w^\lambda)$ and therefore

$$D_E(w) \le D_E(w^\lambda) = \lambda^2 D_{E^1}(w^1) + (1-\lambda)^2 D_{E^2}(w^2).$$

For $\lambda_0 = D_{E^2}(w^2)\left(D_{E^1}(w^1) + D_{E^2}(w^2)\right)^{-1}$ it follows that

$$\frac{1}{D_E(w)} \ge \frac{1}{D_{E^1}(w^1)} + \frac{1}{D_{E^2}(w^2)}.$$

By (4) we see that this is the desired inequality (6).

Equality holds in (6) if and only if $w = w^{\lambda_0}$. If E is a concentric circular annulus then $w = w^{\lambda_0}$ means that γ is a concentric circle.

1 D. Modular Test. Let R be an open Riemann surface and $\{R_n\}_{n=1}^\infty$ an exhaustion of R. The open set $R_{n+1} - \bar{R}_n$ consists of a finite number of relatively compact regions E_{ni}. We denote by α_{ni} and β_{ni} the sets of contours of E_{ni} on the boundary of R_n and R_{n+1} respectively. Let μ_{ni} be the modulus of $(E_{ni}, \alpha_{ni}, \beta_{ni})$. We shall call

$$\mu_n = \min_i \mu_{ni} \tag{7}$$

the *minimum modulus* of $R_{n+1} - \bar{R}_n$. The following test will play a central role throughout this chapter.

THEOREM. *If there exists an exhaustion of an open Riemann surface R such that*

$$\prod_1^\infty \mu_n = \infty \tag{8}$$

then R belongs to O_{AD}.

Let f be an arbitrary nonconstant analytic function on R and let $u = \operatorname{Re} f$. We have to show that $D(f) = 2D(u) = \infty$. Let x_{ni} be the modulus function of $(E_{ni}, \alpha_{ni}, \beta_{ni})$ and y_{ni} the (multivalued) conjugate harmonic function of x_{ni}. For each λ with $0 \le \lambda \le \log \mu_n$ we denote by $R(n,\lambda)$ the relatively compact subregion of R bounded by the level lines $\gamma_{ni\lambda}$: $x_{ni} = \lambda$ $(0 < \lambda < \log \mu_n)$. Observe that $R_n \subset R(n,\lambda) \subset R_{n+1}$. By Green's formula

$$D_{R(n,\lambda)}(u) = \int_{\partial R(n,\lambda)} u * du = \sum_i \int_{\gamma_{ni\lambda}} (u - u_i) * du,$$

where u_i is the value of u at an arbitrarily fixed point of $\gamma_{ni\lambda}$. Since

$$|u-u_i|=\left|\int_{\gamma_{ni\lambda}}\frac{\partial u}{\partial y_{ni}}*dx_{ni}\right|\leq\left(2\pi\int_{\gamma_{ni\lambda}}\left(\frac{\partial u}{\partial y_{ni}}\right)^2*dx_{ni}\right)^{\frac{1}{2}}$$

$$\leq\left(2\pi\int_{\gamma_{ni\lambda}}|\operatorname{grad} u|^2*dx_{ni}\right)^{\frac{1}{2}}$$

on $\gamma_{ni\lambda}$ and

$$\int_{\gamma_{ni\lambda}}|*du|=\int_{\gamma_{ni\lambda}}\left|\frac{\partial u}{\partial x_{ni}}\right|*dx_{ni}\leq\left(2\pi\int_{\gamma_{ni\lambda}}\left(\frac{\partial u}{\partial x_{ni}}\right)^2*dx_{ni}\right)^{\frac{1}{2}}$$

$$\leq\left(2\pi\int_{\gamma_{ni\lambda}}|\operatorname{grad} u|^2*dx_{ni}\right)^{\frac{1}{2}},$$

we obtain

$$\left|\int_{\gamma_{ni\lambda}}(u-u_i)*du\right|\leq\int_{\gamma_{ni\lambda}}|u-u_i|\,|*du|\leq 2\pi\int_{\gamma_{ni\lambda}}|\operatorname{grad} u|^2*dx_{ni}.$$

Therefore

$$D_{R(n,\lambda)}(u)\leq 2\pi\int_{\partial R(n,\lambda)}|\operatorname{grad} u|^2*dx_{ni}=2\pi\frac{d}{d\lambda}D_{R(n,\lambda)}(u),$$

and consequently

$$\int_0^{\log\mu_n}d\lambda\leq 2\pi\int_0^{\log\mu_n}\frac{\frac{d}{d\lambda}D_{R(n,\lambda)}(u)}{D_{R(n,\lambda)}(u)}d\lambda.$$

In view of $R_n=R(n,0)\subset R(n,\log\mu_n)\subset R_{n+1}$ this implies

$$\mu_n\leq\left(\frac{D_{R_{n+1}}(u)}{D_{R_n}(u)}\right)^{2\pi}.$$

We conclude that

$$D(u)=\lim_n D_{R_n}(u)\geq D_{R_1}(u)\lim_n\left(\prod_1^n\mu_k\right)^{1/2\pi}=\infty.$$

1 E. Example. We illustrate the modular test (8) by a simple example, the complementary region R of the *Schottky point set* S in the extended plane.

Consider the complement R_1 of a finite number m (>1) of disjoint closed disks in the extended plane. Reflect R_1 about each boundary circle and denote the m images of R_1 by $E_{11},\ldots,E_{1m}$. Set

$$R_2=\left(\bigcup_{i=1}^m E_{1i}\right)\cup R_1.$$

Reflect each E_{1i} about each of its interior boundary circles so as to obtain $m-1$ images of each E_{1i}, that is, a total of $m(m-1)$ new images $E_{21}, E_{22}, \ldots, E_{2,m(m-1)}$, each conformally equivalent to R_1. Write

$$R_3 = \left(\bigcup_{i=1}^{m(m-1)} E_{2i} \right) \cup R_2 .$$

On repeating this process we obtain a sequence $\{R_n\}_{n=1}^{\infty}$ of regions such that $R_{n+1} - \bar{R}_n$ consists of $m(m-1)^{n-1}$ disjoint regions E_{ni} conformally equivalent to R_1. Let

$$R = \bigcup_{n=1}^{\infty} R_n .$$

The *Schottky point set* S with respect to the given family of circles ∂R_1 is by definition the closed complementary set of R in the extended plane.

Clearly $\{R_n\}$ is a regular exhaustion of R. Let $C_1, \ldots, C_m$ be the circles constituting ∂R_1. Since the E_{ni} are conformally equivalent to R_1 the minimum modulus μ_n of $R_{n+1} - \bar{R}_n$ is equal to the minimum $\mu > 1$ of the moduli of $(R_1, C_j, \partial R_1 - C_j)$, $j = 1, \ldots, m$. Thus

$$\prod_1^{\infty} \mu_n = \lim_k \mu^k = \infty$$

and we conclude:

THEOREM. *The complement of the Schottky set with respect to the extended plane belongs to the class O_{AD}.*

Historically the Schottky set offered a first example of an AD-null set whose capacity is positive (cf. Chapter V).

1 F. Relative Class SO_{AD}. We consider the family SO_{AD} of bordered Riemann surfaces (R, γ) with borders γ such that every AD-function f on $R \cup \gamma$ with $\operatorname{Re} f = 0$ on γ reduces to a constant. It is important to stipulate which part of the ideal boundary of R is realized as the border γ of the bordered surface (R, γ). This is illustrated by the two examples of Fig. 2. The first is the bordered surface (R, γ_1) with R: $|z| < 1$ and γ_1: $|z| = 1$ whereas in the second bordered surface (R, γ_2) we take for

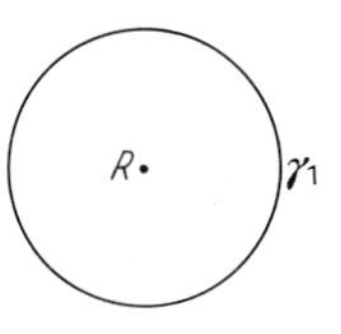

Fig. 2a

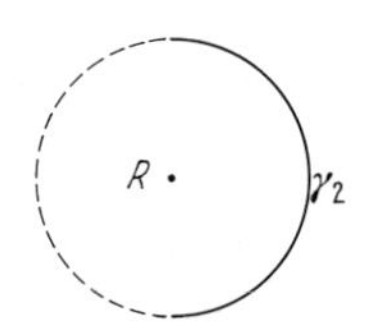

Fig. 2b

γ_2 the semicircle $\{z \mid |z|=1,\ -\pi/2<\arg z<\pi/2\}$. Clearly $(R,\gamma_1)\in SO_{AD}$, but $(R,\gamma_2)\notin SO_{AD}$.

There is an intimate relation between the relative class SO_{AD} and the absolute class O_{AD} (KURODA [3]):

THEOREM. *A bordered Riemann surface (R,γ) belongs to the class SO_{AD} if and only if the double R^γ of (R,γ) about γ belongs to O_{AD}.*

Suppose that $(R,\gamma)\in SO_{AD}$ and let $f\in AD(R^\gamma)$. Denote by φ the natural symmetric indirectly conformal self-mapping of R^γ. Consider on R^γ the functions

$$f_1(z)=f(z)-\bar{f}(\varphi(z)),$$

$$f_2(z)=-i\big(f(z)+\bar{f}(\varphi(z))\big).$$

Clearly they belong to $AD(R^\gamma)$, and $\operatorname{Re} f_1=\operatorname{Re} f_2=0$ on γ. From this and from $(R,\gamma)\in SO_{AD}$ it follows that $f_j|R=k_j$ (constant) and therefore $f_j=k_j$ on R^γ $(j=1,2)$. We conclude that $f=\frac{1}{2}(k_1+ik_2)$ on R^γ, which gives $R^\gamma\in O_{AD}$.

Next suppose that $(R,\gamma)\notin SO_{AD}$. There exists a nonconstant AD-function f on $R\cup\gamma$ with $\operatorname{Re} f=0$ on γ. Hence by the reflection principle the function F on R^γ defined by

$$F|R\cup\gamma=f,$$

$$F|\varphi(R)=-\bar{f}\circ\varphi$$

belongs to $AD(R^\gamma)$ and we have $R^\gamma\notin O_{AD}$.

1 G. Classes O_{A_0D} and O_{A^0D}. We denote by O_{A_0D} the class of Riemann surfaces R such that for every bordered subregion $\bar{R}'$ of R *with compact border $\partial R'$ and relatively compact complement $R-\bar{R}'$*, $(R',\partial R')\in SO_{AD}$. We also consider the class O_{A^0D} of Riemann surfaces R such that for every bordered subregion $\bar{R}'$ of R with compact or noncompact border $\partial R'$, $(R',\partial R')\in SO_{AD}$. From the definitions of O_{A_0D} and O_{A^0D} it follows immediately that

$$O_{A^0D}\subset O_{A_0D}. \tag{9}$$

A relation between O_{A^0D} and O_{AD} is easily obtained. Suppose there exists a nonconstant AD-function f on R. Fix a point z_0 in R and consider a component R' of the set $\{z \mid z\in R,\ \operatorname{Re} f(z)>\operatorname{Re} f(z_0)\}$. Then $f-\operatorname{Re} f(z_0)$ is a nonconstant AD-function on $\bar{R}'$ and its real part vanishes on $\partial R'$. Thus $(R',\partial R')\in SO_{AD}$, i.e. $R\in O_{A^0D}$. We have proved:

$$O_{A^0D}\subset O_{AD}. \tag{10}$$

That inclusions (9) and (10) are actually strict will be shown in 10 D.

1 H. Test for O_{A_0D} and O_{A^0D}. The modular test 1 D applies to O_{A_0D} and O_{A^0D} as well (KURODA [6]):

THEOREM. *The divergence of the modular product* (8) *is also sufficient for an open Riemann surface R to belong to O_{A^0D} and a fortiori to O_{A_0D}.*

Let f' be an analytic function on a compact or noncompact bordered subregion $\overline{R'}$ of R. Set $u' = \operatorname{Re} f'$. We have to show that if u' is nonconstant on R' and $u'=0$ on $\partial R'$ then $D_{R'}(u')=\infty$. Let $u=u'$ on R' and $u=0$ on $R-R'$. We are to prove that $D(u)=D_R(u)=\infty$.

The reasoning in 1 D can be applied with an obvious modification to the present u. The details are left to the reader.

2. Conformal Metric Test

2 A. Conformal Metric. Let $\lambda(z)$ be a strictly positive continuous function of the local parameter z such that

$$ds = \lambda(z)\,|dz|$$

is invariant under change of local parameter z of an open Riemann surface R. We call ds a *conformal metric* on R. For two points z_1 and z_2 on R we set

$$s(z_1, z_2) = \inf \int_\gamma ds,$$

where γ runs over all rectifiable curves joining z_1 and z_2. Then s is a distance function on R. Here we assume that

$$\lim_n s(z, z_n) = \infty$$

for any z in R and for any sequence $\{z_n\}$ of points in R which does not cluster in R. We may express this by saying that the distance from z to the ideal boundary of R is infinite.

Fix a point z_0 in R. Set $\Gamma(\rho) = \{z \mid z \in R, s(z_0, z) = \rho\}$ for each $\rho > 0$. If ρ is sufficiently small, then $\Gamma(\rho)$ consists of a single curve. As ρ increases, $\Gamma(\rho)$ generally breaks up at a certain ρ_1 into two or more curves. As ρ continues to increase, new decompositions may take place. If the genus of R is positive then curves will also fuse for some values of ρ. Let

$$\rho_1 < \cdots < \rho_n < \cdots \tag{11}$$

be those values of ρ for which there occur either decompositions or fusions of components of $\Gamma(\rho)$.

We impose one more requirement on the metric: the values in (11) do not accumulate anywhere. If the sequence in (11) is finite then we add to the ρ_n's an infinite sequence such that the sequence (11) thus obtained tends to infinity.

2 B. Conformal Metric Test. Let $\Lambda(\rho)$ be the length of the longest curve in $\Gamma(\rho)$ in our metric ds. Then we have:

THEOREM. *If the maximal length $\Lambda(\rho)$ of the curves grows so slowly that*

$$\int_{\varepsilon}^{\infty} \frac{d\rho}{\Lambda(\rho)} = \infty \qquad (\varepsilon > 0) \tag{12}$$

then R belongs to the class O_{AD}.

For each ρ_n in (11) let $R_n = \{z \mid z \in R, s(z_0, z) < \rho_n\}$. Then $\{R_n\}_{n=1}^{\infty}$ is an exhaustion such that $\partial R_n = \Gamma(\rho_n)$. Denote by μ_n the minimum modulus of $R_{n+1} - \bar{R}_n$. Fix n for the time being. Let E be a component of $R_{n+1} - \bar{R}_n$ with minimum modulus μ_n and set $\alpha = (\partial R_n) \cap (\partial E)$, $\beta = (\partial R_{n+1}) \cap (\partial E)$. Since the length $L(\rho)$ of $\Gamma(\rho) \cap E$ is dominated by $\Lambda(\rho)$, we have

$$\int_{\rho_n}^{\rho_{n+1}} \frac{d\rho}{\Lambda(\rho)} \leq \int_{\rho_n}^{\rho_{n+1}} \frac{d\rho}{L(\rho)}.$$

Let $w = w(z; E, \beta)$ be the harmonic measure of β with respect to E, and w^* its harmonic conjugate on E. The function $W = w + i w^*$ maps E, slit along a suitable curve γ on which w^* is constant, say 0, conformally onto the rectangle T: $0 \leq w \leq 1$, $0 \leq w^* \leq 2\pi/\log \mu_n$ in the W-plane. The Euclidean length of $\Gamma(\rho) \cap E$ in T is at least $2\pi/\log \mu_n$:

$$\frac{2\pi}{\log \mu_n} \leq \int_{\Gamma(\rho) \cap E} \left| \frac{dW}{dz} \right| |dz|.$$

By Schwarz's inequality

$$\left(\frac{2\pi}{\log \mu_n} \right)^2 \leq \int_{\Gamma(\rho) \cap E} \left| \frac{dW}{dz} \right|^2 \frac{|dz|}{\lambda} \int_{\Gamma(\rho) \cap E} \lambda |dz|.$$

Denote by $A(\rho)$ the area of that part of T which lies to the left of the image of $\Gamma(\rho) \cap E$. Since

$$\int_{\Gamma(\rho) \cap E} \left| \frac{dW}{dz} \right|^2 \frac{|dz|}{\lambda} = \frac{d}{d\rho} \int_{\rho_n}^{\rho} \int_{\Gamma(\rho) \cap E} \left| \frac{dW}{dz} \right|^2 |dz| \frac{|dz|}{d\rho} d\rho = \frac{d}{d\rho} A(\rho)$$

we have

$$\left(\frac{2\pi}{\log \mu_n} \right)^2 \leq \frac{dA}{d\rho} L(\rho),$$

which gives

$$\int_{\rho_n}^{\rho_{n+1}} \frac{d\rho}{L(\rho)} \leq \left(\frac{\log \mu_n}{2\pi} \right)^2 \int_{\rho_n}^{\rho_{n+1}} dA = \frac{1}{2\pi} \log \mu_n.$$

On summing this from $n=1$ to ∞ we obtain

$$\int_{\rho_1}^{\infty} \frac{d\rho}{L(\rho)} \leq \frac{1}{2\pi} \log \prod_{n=1}^{\infty} \mu_n .$$

Thus (12) implies the divergence of modular product (8), i.e. $R \in O_{AD}$ by Theorem 1 D.

We remark that by Theorem 1 H the *divergence of integral* (12) *also implies* $R \in O_{A^0 D} \subset O_{A_0 D}$.

***2 C. Fundamental Polygons.** As an application of the conformal metric test we consider automorphic functions. We start by recalling the definition of a fundamental polygon for a given Fuchsoid group. For details see e.g. FORD [1].

Let G be a properly discontinuous group of linear self-mappings S of the unit disc $|z|<1$, i.e. $\{z' | z' = S(z),\ S \in G\}$ has no accumulation point in $|z|<1$ for any z in $|z|<1$. The group G is called a Fuchsian group or, if it has an infinite number of generators, a *Fuchsoid group*. Clearly it is countable.

In the unit disk $|z|<1$ consider the *hyperbolic metric*

$$ds = \frac{|dz|}{1-|z|^2} .$$

The distance function s induced by this metric ds defines a non-Euclidean geometry in $|z|<1$. The non-Euclidean lines are the circular arcs in $|z|<1$ perpendicular to $|z|=1$.

Let $\{z_\nu\}_{\nu=0}^{\infty}$ be the set of points in $|z|<1$ which are equivalent to $z_0=0$, i.e. $z_\nu = S_\nu(z_0)$ with $G = \{S_\nu\}_{\nu=0}^{\infty}$. Let

$$B_\nu = \{z \,|\, |z|<1,\ s(z, z_\nu) < s(z, z_\mu)\ (\mu \neq \nu)\} .$$

We have the following easy consequences of the definition:

(a) *No two points of* B_ν *are equivalent with respect to* G.

(b) *Any point* z *in* $|z|<1$ *has an equivalent in* $B_\nu \cup \partial B_\nu$.

(c) ∂B_ν *consists of a countably infinite number of non-Euclidean lines.*

(d) *Any point* $z \in \partial B_\nu$ *has an equivalent* $z' \neq z$ *in* ∂B_ν *with* $s(z, z_\nu) = s(z', z_\nu)$.

(e) B_ν *is convex with respect to non-Euclidean segments.*

By (d), ∂B_0 consists of pairs of equivalent non-Euclidean lines $\gamma_{0k}, \gamma'_{0k}$ $(k=1, 2, \ldots)$. We call $D_\nu = B_\nu \cup \left(\bigcup_{k=1}^{\infty} S_\nu(\gamma_{0k})\right)$, where S_ν sends $z_0=0$ to z_ν and $S_\nu \in G$, a *fundamental polygon* for the Fuchsoid group G.

From (b) we see that $\{D_\nu\}$ covers the unit disk $|z|<1$, and from (a) that D_ν and D_μ do not overlap $(\nu\neq\mu)$.

The identification of equivalent points in $|z|<1$ gives rise to an open Riemann surface R. The effect of identifying equivalent sides of D_ν is the same. Conversely the D_ν are obtained by cutting R along the γ_{0k} $(k=1,2,\ldots)$.

*2 D. **Euclidean Metric Test.** The hyperbolic metric gives on R a conformal metric which satisfies all requirements of 2 A.

The arcs

$$|z|=r=\tanh\rho$$

in the fundamental polygon B_0 correspond to $\Gamma(\rho)$ of 2 A since $\Gamma(\rho)$ is the set of points in R satisfying $s(z_0,z)=\frac{1}{2}\log((1+|z|)/(1-|z|))=\rho$. These arcs consist of finitely many subsets $\Gamma_i'(r)$ which correspond to the closed curves $\Gamma_i(\rho)$ constituting $\Gamma(\rho)$. For a fixed r in $0<r<1$ let $\lambda(r)$ be the maximum of the total Euclidean lengths of the $\Gamma_i'(r)$. Then we have the following explicit test:

THEOREM. *Let G be a Fuchsoid group and B_0 its fundamental polygon containing the origin. If the maximal Euclidean length $\lambda(r)$ of the cycles on $|z|=r$ in B_0 grows so slowly that*

$$\int_\varepsilon^1 \frac{dr}{\lambda(r)}=\infty \qquad (0<\varepsilon<1), \tag{13}$$

then the Riemann surface R corresponding to G belongs to O_{AD}.

Let ρ be the hyperbolic length $s(0,r)$ corresponding to r and let $\Lambda(\rho)$ be that of the cycle $\Gamma_i'(r)$ on $|z|=r$ in B_0 which gives the maximum $\lambda(r)$. By the definition of hyperbolic metric

$$\rho=\frac{1}{2}\log\frac{1+r}{1-r}$$

and thus

$$\frac{d\rho}{dr}=\frac{1}{1-r^2}.$$

Therefore

$$\Lambda(\rho)=\int_{\Gamma_i'(r)}\frac{|dz|}{1-|z|^2}=\int_{\Gamma_i'(r)}\frac{1}{1-r^2}\,r\,d\theta$$

$$=\frac{d\rho}{dr}\int_{\Gamma_i'(r)} r\,d\theta=\frac{d\rho}{dr}\lambda(r).$$

It follows that $\Lambda(\rho)/\lambda(r)=d\rho/dr$, and clearly $\Lambda(\rho)$ gives the maximum length of the $\Gamma_i(\rho)$ in $\Gamma(\rho)$. For $\varepsilon'=\tanh^{-1}\varepsilon$

$$\int_{\varepsilon'}^{\infty}\frac{d\rho}{\Lambda(\rho)}=\int_{\varepsilon}^{1}\frac{dr}{\lambda(r)}$$

and by Theorem 2 B we conclude that $R\in O_{AD}$.

We again note that *the divergence of integral* (13) *is also sufficient for* $R\in O_{A^0D}\subset O_{A_0D}$.

Conformal metrics in parabolicity criteria (cf. Chapter V) were first used by AHLFORS [1] for simply connected surfaces and by LAASONEN [1] for arbitrary surfaces.

3. Regular Chain Test

3 A. Regular Chains. Let $\{R_n\}$ be a regular exhaustion of a given open Riemann surface R, with the boundaries $\beta_n=\partial R_n$ consisting of contours $\{\beta_{ni}\}_{i=1}^{\nu_n}$. Let C_{ni} be a doubly connected union of parametric regions D_{nij} with $\beta_{ni}\subset C_{ni}$.

The set $\{C_{ni}\}$ is called a *regular chain set* if

(α) *the* $\bar{C}_{ni}$ *are disjoint by pairs;*

(β) *there is an integer* N, *called the covering number, independent of* n, i, j, *such that no* D_{nij} *meets more than* N *regions* D_{nik};

(γ) *there exists a number* d *with* $0<d<1$, *called the covering constant, independent of* n, i, j, *such that for any two overlapping parametric regions* D_{nij} *and* D_{nik} *some point* z *in* $D_{nij}\cap D_{nik}$ *has parametric images in the parametric disks* $|z_j|<1, |z_k|<1$ *corresponding to* D_{nij}, D_{nik} *whose distances from* $|z_j|=1$ *and* $|z_k|=1$ *are greater than* d.

3 B. Regular Chain Test. Let l_{ni} be the number of regions D_{nij} in C_{ni} and set

$$\lambda_n=\max\{l_{ni};\ i=1,\dots,\nu_n\}.$$

The following result will be useful in our applications:

THEOREM. *If there exists a regular chain set such that the maximal number* λ_n *of parametric regions in the chains grows so slowly that*

$$\sum_n\frac{1}{\lambda_n}=\infty \tag{14}$$

then R *belongs to the class* O_{AD}.

For the proof let C be one of the chains in $\{C_{ni}\}$ and denote by l the number of parametric regions D_j belonging to C. Take the conformal mapping W of C onto the annulus A whose boundary ∂A consists of two concentric circles whose radii differ by 1. If we denote by r the radius of the smaller circle, then the modulus μ of A and a fortiori of C is given by

$$\mu = \frac{1+r}{r} = 1 + \frac{1}{r}.$$

We shall estimate r in terms of l. For each pair D_j and D_k in C with $D_j \cap D_k \neq \emptyset$ we choose points $z_j^k = z_k^j$ in $D_j \cap D_k$ such that their distances from ∂D_j and ∂D_k exceed d. Let z_j^0 be the center of D_j with parameter $z_j = x_j + i y_j$ and denote by $(z_j^0 z_j^k)$ the line segment joining z_j^0 and z_j^k. Clearly the total length L of the image in A of the net $\sigma = \bigcup_{j,k} (z_j^0 z_j^k)$ satisfies

$$2\pi r \le L \le \sum_{j,k} \int_{(z_j^0 z_j^k)} \left| \frac{dW}{dz_j} \right| |dz_j|.$$

Let A_j be the area of the image of D_j in A. Take $\zeta \in (z_j^0 z_j^k)$ and let Δ be the disk in D_j with center ζ and radius d. By the subharmonicity of $|dW/dz_j|^2$ in Δ

$$\left| \frac{dW}{dz_j}(\zeta) \right|^2 \le \frac{1}{\pi d^2} \int_\Delta \left| \frac{dW}{dz_j} \right|^2 dx_j \, dy_j \le \frac{A_j}{\pi d^2}.$$

Since by (β) the $(z_j^0 z_j^k)$ number at most N for each j and since the length of $(z_j^0 z_j^k)$ is <1, we have

$$2\pi r < \frac{N}{d\sqrt{\pi}} \sum_{j=1}^{l} \sqrt{A_j}.$$

Schwarz's inequality gives

$$4\pi^3 d^2 r^2 < N^2 l \sum_{j=1}^{l} A_j.$$

In view of (β) it is clear that

$$\sum_{j=1}^{l} A_j \le N\big(\pi(1+r)^2 - \pi r^2\big) = 2\pi N (r + \tfrac{1}{2}).$$

We set $c = 2\pi^2 d^2 / N^3$ and conclude that

$$\frac{1}{r} + \frac{1}{2r^2} > c \frac{1}{l}.$$

Let $\tilde{\mu}_n$ be the minimum of the moduli of the C_{ni} which is given by C_{ni_n}, say. Let $l'_n = l_{ni_n}$ and denote by r_n the corresponding inner radius as above. The inequalities

$$\frac{1}{r_n} + \frac{1}{2r_n^2} > c\,\frac{1}{l'_n} \geq c\,\frac{1}{\lambda_n}$$

show that $\sum_{n=1}^{\infty} 1/r_n = \infty$, or $\prod_{n=1}^{\infty}(1+1/r_n) = \infty$. Since $\tilde{\mu}_n = 1 + 1/r_n$ we have

$$\prod_{n=1}^{\infty} \tilde{\mu}_n = \infty.$$

We now take the exhaustion $\{R'_k\}$ of R with $\partial R'_{2n-1}$ and $\partial R'_{2n}$ consisting of the "inner" and "outer" contours of $\bigcup_i C_{ni}$. Then the corresponding modular product $\prod_k \mu'_k$ contains $\prod_n \tilde{\mu}_n$ as a subproduct and we infer that $R \in O_{AD}$.

Again we observe that by Theorem 1 H, (14) is also sufficient for $R \in O_{A^0D} \subset O_{A_0D}$.

***3 C. Second Proof.** In view of the significance of Theorem 3 B we give an alternative proof. We shall derive, directly in terms of l, an estimate of μ which is of independent interest.

Let C, l, μ and σ be as in 3 B. This time we denote by F the conformal mapping of a suitably cut C onto the rectangle $T: 0 \leq u \leq \log\mu,\ 0 \leq v \leq 2\pi$. Let L be the length of the image of σ in T. Then

$$2\pi \leq \int_{\sigma} \left|\frac{dF}{dz}\right| |dz|.$$

Let $\tilde{z}$ be a fixed point in σ. Since $F(z)$ is univalent on $|z-\tilde{z}| \leq d$ we may use Koebe's quarter theorem (e.g. Hille [2, p. 350]) to conclude that

$$\frac{d}{4}\left|\frac{dF}{dz}(\tilde{z})\right| \leq \min_{|z-\tilde{z}|=d} |F(z) - F(\tilde{z})|.$$

Geometrically it is clear that the above minimum is dominated by $\frac{1}{2}\log\mu$. Thus on α

$$\left|\frac{dF}{dz}\right| \leq \frac{2}{d}\log\mu.$$

Since $\sigma = \{(z_j^0\, z_j^k)\}$ with the length of $(z_j^0\, z_j^k) < 1$ and the number of the $(z_j^0\, z_j^k)$ less than N for each fixed j we obtain

$$2\pi \leq \frac{2}{d}\log\mu \int_{\sigma} |dz| < \frac{2Nl}{d}\log\mu.$$

Therefore

$$\log \mu > \frac{d\pi}{N} \frac{1}{l}.$$

The rest of the proof is the same as in 3 B.

3 D. Comments on Regular Chains. The purpose of conditions (β) and (γ) of 3 A is to prevent degeneracy resulting from either too numerous overlappings or too shallow intersections. As a result we were able to estimate the modulus of C_{ni} simply by the number l_{ni} of regions D_{nij} constituting C_{ni}.

It is important to observe that the theorem does not presuppose a covering of the entire surface R by a regular chain set. Only those parametric regions which belong to the chains C_{ni} are considered and no assumptions are made concerning parametric regions outside of the C_{ni}'s.

Illuminating applications of this test will be given in 11,12ff., and in particular 18 B.

3 E. Concluding Remarks. Several other explicit tests for O_{AD} can be derived from the fundamental test of Theorem 1 D. Among these we refer the reader to the deep covering test and the test by triangulation (Ahlfors-Sario [12]).

We denote by $\mathcal{M}_A$ the class of open Riemann surfaces admitting exhaustions with property (8). From (9), (10), and Theorem 1 H we obtain the following scheme:

$$\mathcal{M}_A \subset O_{A^0D} \begin{array}{l} \subset O_{A_0 D} \\ \subset O_{AD}. \end{array} \qquad (15)$$

We shall see in 10 D that the inclusion $O_{A^0D} \subset O_{AD}$ is strict: $O_{A^0D} < O_{AD}$. Therefore $\mathcal{M}_A < O_{AD}$, i.e. the modular criterion 1 D is not necessary for $R \in O_{AD}$ (see also II. 12 C). As a consequence $\mathcal{M}_A$ constitutes a new null class and it would be interesting to investigate the nature of surfaces in it. However we shall not dwell on this topic, for which reference is here made to the recent work of Accola [2].

§ 2. Plane Regions

Plane regions are particularly interesting in bringing forth the significance of the class O_{AD}.

Although the finite plane obviously belongs to O_{AD}, we shall prove that it permits an exhaustion whose modular product converges and is even arbitrarily close to 1. This shows that the convergence of the modular product of a given exhaustion has no relation to AD-degeneracy. These topics are the content of 4.

An estimate of the modulus of an arbitrary doubly connected plane region can be given in terms of a Euclidean metric quantity, the relative width. The latter leads to a useful square net test discussed in 5.

In terms of this test it will be shown in 6 that the Cartesian product of two Cantor sets has an O_{AD}-complement if its area vanishes. As a by-product of the construction we form regions of arbitrarily small area yet with a totally disconnected boundary.

That the above Cartesian product of positive area does not, conversely, have its complement in O_{AD}, follows from the characterizations of planar O_{AD}-surfaces in terms of their span, rigidity, and vanishing complementary area under all univalent mappings. Preparatory to these topics discussed in 8 we present in 7 rudiments of the theory of principal functions, one of our main tools.

In 9 we consider the class ABD of bounded AD-functions, in particular on surfaces of finite genus. It is shown that in this case $O_{AD}=O_{ABD}$, which for arbitrary genus implies $O_{AB}\subset O_{AD}$. The finite interpolation problem is solvable for surfaces of finite genus with respect to the class AD if and only if the surface is not in O_{AD}. This has the interesting consequence that for finite genus the dimension of the complex vector space AD is either 1 or infinity.

A striking difference between finite and infinite genus turns up in the problem of essential extendability discussed in 10. Using Myrberg's fascinating example showing this difference we establish the strict inclusions $O_{A^0D}<O_{A_0D}$ and $O_{A^0D}<O_{AD}$. It is also established that the O_{AD}-property is a boundary property for finite genus but not for infinite genus.

Further significant properties of plane O_{AD}-regions are illustrated by Koebe's circular mappings discussed in 11.

4. Convergent Modular Products

4 A. Estimate for Modulus. In this no. we shall use the term *annular region* for a bounded doubly connected region E of the plane. Let α be the inner boundary, β the outer boundary. Set

$$d=\sup_{\zeta\in E}\inf_{z\in\alpha\cup\beta}|z-\zeta|,$$

$$l=\inf_{\gamma} l(\gamma),$$

where γ runs over all curves in E separating α and β and $l(\gamma)$ is the length of γ. Then we have the following:

LEMMA. *The modulus μ of an annular region E has an upper bound in terms of the inequality*

$$\frac{\mu-1}{\mu+1} \leq 8\pi \frac{d}{l}. \tag{16}$$

Let E' be the annulus with radii 1 and μ, conformally equivalent to E. Consider the concentric circle γ' in E' with radius $\frac{1}{2}(\mu+1)$ and take a point w_0 on γ'. Since the conformal mapping $z=Z(w)$ of E' onto E is a univalent analytic function on $|w-w_0|<\frac{1}{2}(\mu-1)$ Koebe's quarter theorem gives

$$d \geq \tfrac{1}{4} \cdot \tfrac{1}{2}(\mu-1)\,|Z'(w_0)|.$$

Therefore

$$l \leq \int_{\gamma'} \left|\frac{dZ}{dw}\right| |dw| \leq \frac{1}{\mu-1} \cdot 8d \cdot 2\pi \cdot \tfrac{1}{2}(\mu+1)$$

and (16) follows.

4 B. Bisecting the Annulus. Given an annular region E with inner boundary α and outer boundary β let γ be a Jordan curve dividing E into two annular regions E_1 and E_2 with $\partial E_1 = \alpha \cup \gamma$ and $\partial E_2 = \gamma \cup \beta$. For the moduli μ, μ_1, and μ_2 of E, E_1, and E_2 we have by (6)

$$\mu \geq \mu_1 \mu_2.$$

Here equality occurs if and only if, under a conformal mapping of E onto a circular annulus, the image of γ is a concentric circle. This suggests that the more the image of γ deviates from the concentric circular shape the more the product $\mu_1 \mu_2$ deviates from μ. We can actually prove:

THEOREM. *Every annular region can be divided into two annular subregions such that for an arbitrarily small $\varepsilon>0$ the modulus of each subregion is less than $1+\varepsilon$.*

We have only to consider the case where E is an annulus with radii 1 and $\mu>1$. Let δ be a positive number and p an even positive integer such that

$$\frac{2\pi}{p} \leq \delta < \tfrac{1}{2}(\mu-1).$$

In E take the p radial segments

$$\left\{ r e^{i\theta} \,|\, \theta = k \cdot \frac{2\pi\mu}{p},\ 1+\delta \leq r \leq \mu-\delta \right\}, \qquad k=1,\ldots,p.$$

Join the end points of these segments by circular arcs with radii $1+\delta$ and $\mu-\delta$ alternately so as to obtain a Jordan curve dividing E into two annular regions E_1 and E_2. Let d_i, l_i, and μ_i be the quantities d, l, and μ of Lemma 4 A for E_i $(i=1,2)$. By construction $d_i<\delta$ and $l_i\geq 2\pi$. Consequently μ_i can be made arbitrarily close to 1 by choosing δ sufficiently small.

4 C. Second Proof. The foregoing proof was based on the general estimate (16). We shall now give a simpler and more elementary proof. Use will be made of the following auxiliary result (AKAZA-KURODA [3]):

LEMMA. *Let $r>0$, $a>0$, and $0<a+r<1$. Consider the annular region E_a bounded by $|z|=1$ and $|z-a|=r$ with modulus μ_a. Then*

$$\lim_{a\to 1-r} \mu_a=1.$$

In other words the modulus of an annular region bounded by two circles α and β of fixed radii approaches 1 as the distance between α and β tends to 0.

For the proof we map E_a onto the annulus E_a' with radii $\rho_a<1$ and 1 by a linear transformation

$$w=T(z)=\frac{z-t}{1-t\,z},$$

where t is a real number $|t|<1$. This is certainly possible if we can solve the equation

$$T(a+r)=-T(a-r)\ (=\rho_a).$$

On rewriting this we obtain

$$\rho_a=\frac{[(a+r)-t]+[-(a-r)+t]}{[1-t(a+r)]+[1-t(a-r)]}=\frac{r}{1-a\,t},$$

$$\rho_a=\frac{-[(a+r)-t]+[-(a-r)+t]}{-[1-t(a+r)]+[1-t(a-r)]}=\frac{t-a}{r\,t}.$$

Consequently

$$a\,t^2-(1+a^2-r^2)\,t+a=0.$$

This equation has one root between 0 and 1 and the existence of T is assured. Since

$$a\,t=\tfrac{1}{2}\left(1+a^2-r^2-(((1-a)^2-r^2)((1+a)^2-r^2))^{\frac{1}{2}}\right)$$

and $\mu_a=1/\rho_a=(1-a\,t)/r$ our assertion follows from the fact that $a\,t\to 1-r$ as $a\to 1-r$.

To complete our second proof of Theorem 4 B it again suffices to consider the case of an annulus E with radii 1 and $\mu > 1$. Let $0 < a < \frac{1}{2}(\mu - 1)$. The circle γ with center a and radius $\frac{1}{2}(\mu+1)$ divides E into annular regions E_1 and E_2. If we let $a \to \frac{1}{2}(\mu-1) = \mu - \frac{1}{2}(\mu+1)$ then by the above lemma the moduli of both E_1 and E_2 tend to 1.

Observe that in this example the dividing curve is analytic. We have proved:

THEOREM. *Every annular region can be divided by an analytic Jordan curve into two annular subregions such that the modulus of each subregion is arbitrarily close to* 1.

The theorem remains valid for the general configuration (E, α, β) not only on a Riemann surface but also in a higher dimensional Riemannian manifold (see App. 2 B).

4 D. Convergent Modular Products. In the modular O_{AD}-test we require the *existence* of an exhaustion with a divergent modular product. For a *given* exhaustion the product has no direct bearing on O_{AD}. We shall in fact show that there are always exhaustions such that (8) converges and is even as close to 1 as we wish:

THEOREM. *Every exhaustion of an open Riemann surface R has a refinement whose modular product is*

$$\prod \mu_n < 1 + \varepsilon$$

with $\varepsilon > 0$ *arbitrarily small.*

First we consider the case where R is simply connected. Let $\{R_{2n}\}$ be an exhaustion of R and let $\{\varepsilon_\nu\}_1^\infty$ be a sequence of positive numbers such that $\prod(1+\varepsilon_\nu) = 1 + \varepsilon$. By Theorem 4 C there is a regular region R_{2n+1} with $\bar{R}_{2n} \subset R_{2n+1} \subset \bar{R}_{2n+1} \subset R_{2n+2}$ such that the moduli of $R_{2n+1} - \bar{R}_{2n}$ and $R_{2n+2} - \bar{R}_{2n+1}$ are less than $1 + \varepsilon_{2n}$ and $1 + \varepsilon_{2n+1}$ respectively. Then for $\{R_\nu\}$ we have $\prod \mu_\nu < \prod(1+\varepsilon_\nu) = 1 + \varepsilon$.

Given an exhaustion of an arbitrary open Riemann surface R construct its refinement $\{R_{2n}\}$ such that the components of $R_{2n} - \bar{R}_{2n-2}$ are conformally equivalent to plane annular regions. Then an obvious modification of the above argument provides us with the proof.

5. Relative Width Test

5 A. Relative Width. Let E be an annular region with inner boundary α and outer boundary β. Denote by D its Euclidean width, i.e. the distance

$$D = \inf_{z_1 \in \alpha,\, z_2 \in \beta} |z_1 - z_2|$$

between α and β, and by L the infimum of the lengths $|\gamma|$ of curves γ in E that separate α and β and satisfy $\inf_{z_1\in\gamma, z_2\in\alpha\cup\beta}|z_1-z_2|\geq D/2$:

$$L=\inf|\gamma|.$$

The *relative width* δ of E is by definition

$$\delta=\frac{D}{L}.$$

Whereas Lemma 4 A gives an upper bound for the modulus we shall now furnish a lower bound:

LEMMA. *The modulus μ of an annular region E satisfies the inequality*

$$\delta\leq\frac{2}{\pi}(\mu-1).$$

For an arbitrarily given $\varepsilon>0$ let γ be a competing curve for L such that $|\gamma|<L+\varepsilon$. Denote by $w=W(z)$ the conformal mapping of E onto the annulus $1<|w|<\mu$. Then

$$2\pi<\int_\gamma\left|\frac{dW}{dz}\right||dz|.$$

Let z_0 be a point of γ. Since W is univalent on $|z-z_0|<D/2$, Koebe's quarter theorem gives

$$\frac{1}{4}\frac{D}{2}\left|\frac{dW}{dz}(z_0)\right|\leq\min_{|z-z_0|=D/2}|W(z)-W(z_0)|\leq\frac{\mu-1}{2}.$$

As a consequence

$$2\pi<\frac{8}{D}\cdot\frac{\mu-1}{2}\int_\gamma|dz|<\frac{4}{D}(\mu-1)(L+\varepsilon),$$

and the lemma follows.

5 B. Relative Width Test. Consider an arbitrary plane region R which contains the point at infinity. Let β be the boundary of R and take a regular exhaustion $\{R_n\}$ of R. Each $\beta_n=\partial R_n$ consists of a finite number k_n of analytic Jordan curves β_{ni}. Cover each β_{ni} by an annular region E_{ni} such that

$$E_{ni}\cap E_{mj}=\emptyset\qquad\text{for }(n,i)\neq(m,j).$$

Let δ_{ni} be the relative width of E_{ni} and set

$$\delta_n=\min_{1\leq i\leq k_n}\delta_{ni}.$$

We shall make use of the following test:

THEOREM. *If there exists a set $\{E_{ni}\}$ of annular regions in the complement R of a closed plane point set β such that the sum of the minimal relative widths δ_n diverges,*

$$\sum \delta_n = \infty, \tag{17}$$

then R belongs to O_{AD}.

Let the minimum modulus μ_n of $\{E_{ni}\}_{i=1}^{k_n}$ be attained for E_{ni} and let δ'_n be the relative width of E_{ni}. Then by Lemma 5 A

$$\sum (\mu_n - 1) > \frac{\pi}{2} \sum \delta'_n \geq \frac{\pi}{2} \sum \delta_n = \infty.$$

Therefore $\prod \mu_n = \infty$ and by Theorem 1 D, $R \in O_{AD}$.

5 C. Square Net Test. Let β be a closed bounded plane point set. Cover the plane by a net of squares of side 1. Consider the union Q_0 of those closed squares that meet β. By induction let Q_n be the union of squares of side 2^{-n}, obtained by refining the net with sides 2^{-n+1}, that intersect β. Denote by q_{ni} the number of squares in the component Q_{ni} of Q_n and set

$$q_n = \max q_{ni}.$$

We shall show:

THEOREM. *Let R be a plane region with bounded boundary. If the maximum number of squares increases so slowly that*

$$\sum \frac{1}{q_n} = \infty \tag{18}$$

then R belongs to O_{AD}.

For a fixed n the Q_{ni} have distances not less than 2^{-n} from each other. Enclose Q_{ni} by polygonal Jordan curves α_{ni} and β_{ni} whose segments are parallel to the free sides of Q_{ni} with distances $2^{-(n+1)}$ and $2^{-(n+2)}$ respectively. The set bounded by α_{ni} and β_{ni} is an annular region with Euclidean width

$$D_{ni} = \frac{1}{2^{n+2}}.$$

If Q_{ni} consists of only one square then the length L^*_{ni} of the curve α_{ni} is $16/2^{n+1}$. In the general case of q_{ni} squares

$$L^*_{ni} \leq q_{ni} \cdot 16 \cdot \frac{1}{2^{n+1}}.$$

Since the length (see 5 A) L_{ni} of this annular region is less than L_{ni}^*

$$\delta_{ni}=\frac{D_{ni}}{L_{ni}}>\frac{1}{32}\frac{1}{q_{ni}}.$$

We apply this to the annular region which gives δ_{ni} its minimum δ_n and conclude by $q_{ni}\leq q_n$ that $\sum q_n^{-1}=\infty$ implies $\sum \delta_n=\infty$. A fortiori $R\in O_{AD}$.

In the same manner as before (18) *is also sufficient for* $R\in O_{A^0D}\subset O_{A_0D}$.

6. Generalized Cantor Sets

6 A. Vanishing Linear Measure. Remove from the unit interval the p_1th part $(p_1>1)$ such that the remainder $E(p_1)$ consists of two equal intervals. Similarly remove from each interval of $E(p_1)$ the p_2th part $(p_2>1)$ such that the remainder $E(p_1 p_2)$ consists of four equal intervals. In general form $E(p_1\ldots p_n)$ by centrally removing the p_nth part from each component of $E(p_1\ldots p_{n-1})$. The *generalized Cantor set*

$$E(p_1 p_2\ldots)=\bigcap_{n=1}^{\infty} E(p_1\ldots p_n)$$

is a totally disconnected perfect set. If $p_n=3$ $(n=1,2,\ldots)$ then $E(p_1 p_2\ldots)$ is the Cantor ternary set. In the sequel we choose the p_n such that $p_n\leq p_{n+1}$ for all n.

THEOREM. *The complement of a generalized Cantor set* $E(p_1 p_2\ldots)$ *of vanishing linear measure belongs to the class* O_{AD}.

We shall later see that the theorem is an immediate consequence of the nonexistence of nonconstant bounded analytic functions on the complement of sets of vanishing linear measure. Here we give the proof as a first application of the square net test, to prepare for further extensions. This approach will allow us to replace O_{AD} by O_{A^0D} and O_{A_0D} in the theorem.

The total length of the approximating set $E(p_1\ldots p_n)$ is $\prod_{\nu=1}^{n}(1-1/p_\nu)$. Therefore

$$\prod_{\nu=1}^{\infty}\left(1-\frac{1}{p_\nu}\right)=0$$

or equivalently

$$\sum_{\nu=1}^{\infty}\frac{1}{p_\nu}=\infty.$$

Let $\{Q_n\}$ be the sequence of unions of squares for the set $E(p_1 p_2\ldots)$ as in 5 C. Choose a subsequence $\{Q^n\}$ of $\{Q_n\}$ such that the lengths s_n

of the sides of the squares in Q^n satisfy

$$\frac{1}{4p_n} h_{n-1} \le s_n < \frac{1}{2p_n} h_{n-1}$$

where

$$h_n = \frac{1}{2^n} \prod_{\nu=1}^{n} \left(1 - \frac{1}{p_\nu}\right)$$

is the length of the intervals in $E(p_1 \ldots p_n)$ and h_{n-1}/p_n is their minimal distance apart. The set Q^n is decomposed into components Q^{ni} $(i=1, \ldots, 2^n)$. The number q^{ni} of squares in Q^{ni} must satisfy

$$q^{ni} \le 2\left(\frac{h_n}{s_n} + 2\right) \le 4p_n.$$

Thus if we set $q^n = \max\{q^{ni} | i = 1, \ldots, 2^n\}$ then

$$\sum_{n=1}^{\infty} \frac{1}{q_n} \ge \sum_{n=1}^{\infty} \frac{1}{q_n} \ge \frac{1}{4} \sum_{n=1}^{\infty} \frac{1}{p_n} = \infty.$$

In view of Theorem 5 C the proof is herewith complete.

Remark. We shall later see (V. 11 F) that a Cantor set is of zero capacity, or equivalently that the complement of a Cantor set belongs to the class O_G if and only if

$$\prod_{\nu=1}^{\infty} \left(1 - \frac{1}{p_\nu}\right)^{2^{-\nu}} = 0.$$

A comparison with

$$\prod_{\nu=1}^{\infty} \left(1 - \frac{1}{p_\nu}\right) = 0$$

shows the great difference between sets of zero capacity and those with O_{AD} complements.

As an example choose $p_\nu = p > 1$ $(\nu = 1, 2, \ldots)$. We denote this particular Cantor set by $E(p^\infty)$. Then

$$\prod_{\nu=1}^{\infty} \left(1 - \frac{1}{p_\nu}\right)^{2^{-\nu}} = 1 - \frac{1}{p}$$

and

$$\prod_{\nu=1}^{\infty} \left(1 - \frac{1}{p_\nu}\right) = \lim_n \left(1 - \frac{1}{p}\right)^n = 0.$$

Hence in addition to the Schottky sets *the Cantor sets $E(p^\infty)$ are simple examples of sets of positive capacity whose complements belong to O_{AD}.*

6 B. Zero Area. Let $E^2(p_1 p_2 \ldots)$ be the Cartesian product in the plane of the Cantor set $E(p_1 p_2 \ldots)$ with itself. We maintain:

THEOREM. *The complement of $E^2(p_1 p_2 \ldots)$ belongs to O_{AD} if and only if $E^2(p_1 p_2 \ldots)$ has zero area.*

The necessity will be proved in 8 C with the aid of an extremal function constructed in 7. Here we establish the sufficiency as a consequence of the relative width test.

Around each square belonging to $E^2(p_1 \ldots p_n)$ draw squares α_{ni} and β_{ni} with sides parallel to those of the enclosed square and at distances $h_{n-1}/2p_n$ and $h_{n-1}/4p_n$ from it (cf. 6 A). The annular region bounded by α_{ni} and β_{ni} has Euclidean width

$$D_{ni} = \frac{h_{n-1}}{4p_n} = \frac{1}{2^{n+1}} \frac{1}{p_n} \prod_{\nu=1}^{n-1} \left(1 - \frac{1}{p_\nu}\right).$$

The length of α_{ni} is

$$L_{ni}^* = 4\left(h_n + \frac{h_{n-1}}{p_n}\right) = \frac{1}{2^{n-2}} \left(1 + \frac{1}{p_n}\right) \prod_{\nu=1}^{n-1} \left(1 - \frac{1}{p_\nu}\right).$$

The length L_{ni} of the annular region used in defining the relative width is dominated by L_{ni}^*. Hence the relative width of this region is

$$\delta_n = \delta_{ni} = \frac{D_{ni}}{L_{ni}} > \frac{1}{8(1+p_n)}.$$

As a consequence

$$\sum_{n=1}^{\infty} \delta_n > \frac{1}{16} \sum_{n=1}^{\infty} \frac{1}{p_n} = \infty,$$

since $\prod_{\nu=1}^{\infty} (1 - 1/p_\nu)^2 = 0$ by the assumption that the area of $E^2(p_1 p_2 \ldots)$ vanishes. The lemma follows by Theorem 5 B.

We shall return to Cantor sets in connection with AD-null sets in II. 11.

6 C. Regions of Area ε. We digress to point out a by-product due to P. MYRBERG (see SARIO [1, p. 77]):

THEOREM. *There exist regions of arbitrarily small area yet with a totally disconnected boundary.*

Let ε be an arbitrarily small positive number. Cover the plane with a net of squares of side 1. In one of these squares we construct a Cantor set $E^2(p_1^0 p_2^0 \ldots)$ of area

$$\prod_{\nu=1}^{\infty} \left(1 - \frac{1}{p_\nu^0}\right)^2 = 1 - \frac{\varepsilon}{2}.$$

To do this we only have to choose for example $p_\nu^0 = 1/(1-(1-\varepsilon/2)^{2^{-\nu-1}})$. Then the area of the complement of $E^2(p_1^0 p_2^0 \ldots)$ with respect to the square is $\varepsilon/2$. In each of the eight neighboring squares (first generation) construct the Cantor set $E^2(p_1^1 p_2^1 \ldots)$ whose complementary area with respect to the square is $8^{-1} \cdot 2^{-1} \cdot \varepsilon/2$. In general construct in each square belonging to the nth generation of $8n$ squares the Cantor set $E^2(p_1^n p_2^n \ldots)$ whose complementary area with respect to the square is $(8n)^{-1} \cdot 2^{-n} \cdot \varepsilon/2$. Then the region

$$R = \{|z| < \infty\} - \bigcup_{n=0}^{\infty} E^2(p_1^n p_2^n \ldots)$$

has the required properties.

7. Extremal Functions and Conformal Mappings

7 A. Principal Functions. Let R_1 be the complement of a compact bordered subregion of an arbitrary Riemann surface R, with border α_1 of R_1 oriented positively with respect to R_1. A *normal operator* L associated with R_1 is a linear operator assigning to a bounded continuous function f on α_1 a continuous function Lf on $\bar{R}_1$, harmonic on R_1, and such that

$$Lf|\alpha_1 = f, \tag{19}$$

$$\min f \leq Lf \leq \max f, \tag{20}$$

$$\int_{\alpha_1} * dLf = 0. \tag{21}$$

In condition (21), $*dLf$ may not exist on α_1 but we understand this expression as the limit of $\int_\alpha * dLf$ as $\alpha \to \alpha_1$, where α is an analytic curve in R_1 homologous to α_1. Since Lf is harmonic on R_1, $\int_\alpha * dLf$ is independent of α and $\int_{\alpha_1} * dLf$ is well-defined.

The following existence theorem will play a fundamental role in this section:

THEOREM. *Let s be a finitely continuous function on $\bar{R}_1$, harmonic on R_1. The condition*

$$\int_\alpha * ds = 0 \tag{22}$$

is necessary and sufficient for the existence of a harmonic function p on R such that

$$(p-s)|\bar{R}_1 = L((p-s)|\alpha_1). \tag{23}$$

The function p is unique up to an additive constant and it reduces to a constant if and only if $s = L(s|\alpha_1)$.

The function p is called the *principal function* corresponding to (s, L, R_1), and s will be referred to as a *singularity function.*

7 B. Proof. The necessity of (22) is clear. We shall prove its sufficiency. Let R_0 be a regular region with $R-R_1 \subset R_0$ and denote by α_0 the boundary of R_0 oriented positively with respect to R_0.

Let T be the linear operator assigning to each bounded continuous function f on α_0 a continuous function Tf on $\bar{R}_0$, harmonic on R_0 and with $Tf|\alpha_0 = f$. Set $Kf = L(Tf)$ or more precisely $Kf = L(Tf|\alpha_1)$, and

$$s_0 = s - Ls.$$

Then $K^n s_0 = L(TK^{n-1} s_0)$ $(n = 1, 2, \ldots)$.

Denote by ω the continuous function on $\overline{R_0 \cap R_1}$, harmonic on $R_0 \cap R_1$ such that $\omega = 0$ on α_1 and $\omega = 1$ on α_0. We shall show that

$$\int_{\alpha_1} TK^n s_0 * d\omega = 0 \tag{24}$$

for all $n = 0, 1, \ldots$. Let $u = TK^n s_0 - K^n s_0$. By Green's formula

$$\int_{\alpha_0} (u * d\omega - \omega * du) = -\int_{\alpha_1} (u * d\omega - \omega * du).$$

Since $u = 0$ on α_0, $\int_{\alpha_0} * du = 0$ by (21) and (22), and $\omega = 0$ on α_1, we obtain

$$\int_{\alpha_1} u * d\omega = 0.$$

By $s_0 = 0$ on α_1 the substitution of $Ts_0 - s_0$ for u gives

$$\int_{\alpha_1} Ts_0 * d\omega = \int_{\alpha_1} s_0 * d\omega = 0.$$

Again on taking $u = TK^n s_0 - K^n s_0$ we conclude by $K^n s_0 = L(TK^{n-1} s_0) = TK^{n-1} s_0$ on α_1 that

$$\int_{\alpha_1} TK^n s_0 * d\omega = \int_{\alpha_1} TK^{n-1} s_0 * d\omega.$$

Statement (24) follows.

Let $\mathscr{F}$ be the class of bounded continuous functions v on $\bar{R}_0$, harmonic on R_0 and with $\int_{\alpha_1} v * d\omega = 0$. In particular $TK^n s_0 \in \mathscr{F}$ by (24). We shall show that there exists a constant $q = q_{\mathscr{F}}$ with $0 < q < 1$ such that

$$\max_{\alpha_1} |v| \le q \max_{\alpha_0} |v|$$

for any v in $\mathscr{F}$. It suffices to consider functions $v \not\equiv 0$. Contrary to the assertion assume the existence of a sequence $\{v_n\} \subset \mathscr{F}$ such that

$\max_{\alpha_1}|v_n|/\max_{\alpha_0}|v_n|\to 1$ as $n\to\infty$. The function $u_n=v_n/\max_{\alpha_0}|v_n|$ again belongs to $\mathscr{F}$, and $|u_n|\leq 1$ on $\bar{R}_0$. On choosing a subsequence if necessary we may assume that $\{u_n\}$ converges to a function u harmonic on R_0. Since

$$\int_{\alpha_1} u*d\omega=\lim_n \int_{\alpha_1} u_n*d\omega=0$$

and $*d\omega<0$ on α_1, u is not of constant sign. On the other hand

$$\max_{\alpha_1}|u|=\lim_n \max_{\alpha_1}|u_n|=1.$$

By the maximum principle $|u|=1$ identically on R_0, and u is constant. This contradicts the fact that $u\in\mathscr{F}$.

In particular

$$\max_{\alpha_1}|TK^{n-1}s_0|\leq q\max_{\alpha_0}|TK^{n-1}s_0|=q\max_{\alpha_0}|K^{n-1}s_0|$$

for any $n=1,2,\ldots$. By (20)

$$\max_{\alpha_0}|K^n s_0|=\max_{\alpha_0}|L(TK^{n-1}s_0)|\leq\max_{\alpha_1}|TK^{n-1}s_0|.$$

Therefore

$$\max_{\alpha_0}|K^n s_0|\leq q\max_{\alpha_0}|K^{n-1}s_0|.$$

By repetition we finally obtain

$$\max_{\alpha_0}|K^n s_0|\leq q^n\max_{\alpha_0}|s_0|. \tag{25}$$

Consider the function φ defined on α_0 by

$$\varphi=\sum_{n=0}^{\infty}K^n s_0.$$

It is bounded and continuous on α_0 by (25). Again by (25), $K\varphi=\varphi-s_0$. Hence

$$\varphi-K\varphi=s-Ls$$

on α_0. On setting $f=T\varphi$ on α_1 we have $\varphi-Lf=s-Ls$ on α_0, i.e.

$$\varphi-s=L(f-s) \quad \text{on} \quad \alpha_0,$$
$$f=T\varphi \quad \text{on} \quad \alpha_1.$$

We define p on R by

$$p|R_0=T\varphi,$$
$$p|R_1=Lf+s-Ls.$$

This is well-defined on R, because $T\varphi-(Lf+s-Ls)=(\varphi-s)-L(f-s)=0$ on α_0 and $T\varphi-(Lf+s-Ls)=f-f-s+s=0$ on α_1 and thus $T\varphi=Lf+s-Ls$ on $R_0\cap R_1$. We conclude that p is harmonic on R and

$$L(p-s)=L(Lf-Ls)=L(f-s)=p-s$$

on R_1, i.e. p satisfies (23).

Assume that p and p' are solutions of (23). Then $L(p-p')=p-p'$ shows that $p''=p-p'$ takes its maximum on α_1 and hence p'' is constant on R. Similarly $Ls=s$ implies $Lp=p$ and thus p must be constant.

7 C. Operators L_0 and L_1. Let R_1 be as above and let $\Omega\subset R_1$ be a relatively compact regularly imbedded region in R with boundary $\alpha_1\cup\beta_\Omega$. Given f on α_1 as above we designate by $u_{0\Omega}$ and $u_{1\Omega}$ the harmonic functions on Ω with boundary values f on α_1 and with

$$*du_{0\Omega}=0, \qquad u_{1\Omega}=c_j=\text{const}$$

on $\beta_{j\Omega}$ such that $\int_{\beta_{j\Omega}}*du_{1\Omega}=0$, where the $\beta_{j\Omega}$ are the borders on β_Ω of the components of $R_1-\Omega$. For convenience we set

$$L_{i\Omega}f=u_{i\Omega}$$

for $i=0,1$. By Green's formula we have for $\bar{\Omega}\subset\Omega'$ and $f\in C^1(\alpha_1)$

$$\begin{aligned} D_\Omega(L_{0\Omega'}f-L_{0\Omega}f)&=D_\Omega(L_{0\Omega'}f)-D_\Omega(L_{0\Omega}f)\\ &\le D_{\Omega'}(L_{0\Omega'}f)-D_\Omega(L_{0\Omega}f) \end{aligned} \tag{26}$$

and

$$D_\Omega(L_{1\Omega}f-L_{1\Omega'}f)\le D_\Omega(L_{1\Omega}f)-D_{\Omega'}(L_{1\Omega'}f). \tag{27}$$

From (27) we see that $u_1=\lim_{\Omega\to R_1}L_{1\Omega}f$ exists, $D_{R_1}(u_1)<\infty$, and $\lim_{\Omega\to R_1}D_\Omega(L_{1\Omega}f-u_1)=0$.

Again by Green's formula

$$D_\Omega(L_{1\Omega}f-L_{0\Omega}f)=D_\Omega(L_{1\Omega}f)-D_\Omega(L_{0\Omega}f). \tag{28}$$

This with (26) gives the existence of $u_0=\lim_{\Omega\to R_1}L_{0\Omega}f$ with $D_{R_1}(u_0)<\infty$ and $\lim_{\Omega\to R_1}D_\Omega(L_{0\Omega}f-u_0)=0$. Since $C^1(\alpha)$ is uniformly dense in $C(\alpha)$, we can define

$$L_i f=\lim_{\Omega\to R_1}L_{i\Omega}f \tag{29}$$

for all $f\in C(\alpha_1)$ and $i=0,1$.

If $f\in C^1(\alpha)$ then

$$D_{R_1}(L_i f)<\infty \tag{30}$$

and

$$\lim_{\Omega \to R_1} D_\Omega(L_i f - L_{i\Omega} f) = 0. \tag{31}$$

We can easily see that the operator L_i $(i=0, 1)$ thus defined is normal.

7 D. Functions with Singularities. For $\zeta \in R$ fix a punctured disk $R_{1\zeta}$ about ζ and a local parameter z on $R_{1\zeta} \cup \zeta$. Let $R_{1\beta}$ be the interior of the complement of a regular region of R such that $\bar{R}_{1\zeta} \cap \bar{R}_{1\beta} = \emptyset$. Set $R_1 = R_{1\zeta} \cup R_{1\beta}$. Define a harmonic function s^θ on $\bar{R}_1$ by

$$s^\theta | \bar{R}_{1\zeta} = \operatorname{Re} \frac{e^{i\theta}}{z-\zeta},$$

$$s^\theta | \bar{R}_{1\beta} = 0.$$

Then s^θ satisfies (22). Consider L_j $(j=0, 1)$ on R_1. On $R_{1\zeta}$ every normal operator coincides with the Dirichlet operator giving the solution to the boundary value problem on $R_{1\zeta} \cup \zeta$.

Let p_j^θ be the principal function with respect to (s^θ, L_j, R_1). It is a harmonic function on $R - \zeta$ with singularity $\operatorname{Re}(e^{i\theta}/(z-\zeta))$ at ζ and with L_j-behavior at the ideal boundary β of R. Here we assume that $p_j^\theta - \operatorname{Re}(e^{i\theta}/(z-\zeta))$ vanishes at ζ.

For real numbers h, k let $\{p\}_{h+k}^\theta$ be the class of harmonic functions p on $R - \zeta$ with the singularity $(h+k)\operatorname{Re}(e^{i\theta}/(z-\zeta))$ at ζ. Let

$$p = \operatorname{Re}\left\{(h+k)\frac{e^{i\theta}}{z-\zeta} + \sum_{n=1}^{\infty} a_n (z-\zeta)^n\right\}$$

at ζ and set

$$\alpha^\theta = \operatorname{Re}(a_1 e^{i\theta}).$$

For $p = p_j^\theta$ write $\alpha^\theta = \alpha_j^\theta$ $(j=0, 1)$ and consider the function

$$p_{hk}^\theta = h\, p_0^\theta + k\, p_1^\theta$$

where the expression $\int_\beta p * dp$ means $\lim_{\Omega \to R} \int_{\partial\Omega} p * dp$.

We shall establish the following deviation formula:

THEOREM. *The identity*

$$\int_\beta p * dp + 2\pi(k-h)\,\alpha^\theta = 2\pi(k^2 \alpha_1^\theta - h^2 \alpha_0^\theta) + D(p - p_{hk}^\theta) \tag{32}$$

is valid for every p in the class $\{p\}_{h+k}^\theta$.

Let Ω be a regular region in R such that $\Omega \supset R_{1\zeta}$ and $\partial\Omega \subset R_{1\beta}$. Denote by $p_{hk\Omega}^\theta$ the function p_{hk}^θ constructed on Ω instead of R. By 7 C it is seen that $p_{hk\Omega}^\theta - p_{hk}^\theta \to 0$ on R and $D_\Omega(p_{hk\Omega}^\theta - p_{hk}^\theta) \to 0$ as $\Omega \to R$. Hence we have only to prove (32) in the case where R is a compact bordered surface with border β and p is harmonic on $R\cup\beta-\zeta$. By Green's formula

$$D(p-p_{hk}^{\theta})=\int_{\beta} p*dp+\int_{\beta} p_{hk}^{\theta}*dp_{hk}^{\theta}-\int_{\beta} p*dp_{hk}^{\theta}-\int_{\beta} p_{hk}^{\theta}*dp. \tag{33}$$

Let γ be the boundary of $R_{1\zeta}\cup\zeta$. The second integral of the right-hand side becomes

$$\int_{\beta} p_{hk}^{\theta}*dp_{hk}^{\theta}=hk\int_{\gamma}(p_0^{\theta}*dp_1^{\theta}-p_1^{\theta}*dp_0^{\theta}).$$

Let $Q_j=e^{i\theta}/(z-\zeta)+a_1^{(j)}(z-\zeta)+\cdots$ with $\operatorname{Re} Q_j=p_j^{\theta}$ $(j=0,1)$ on $R_{1\zeta}\cup\zeta$. Then

$$\begin{aligned}\int_{\beta} p_{hk}^{\theta}*dp_{hk}^{\theta}&=hk\operatorname{Im}\int_{\gamma} Q_0\,dQ_1\\&=hk\operatorname{Im}\int_{\gamma}\left(\frac{e^{i\theta}}{z-\zeta}+a_1^{(0)}(z-\zeta)+\cdots\right)\left(\frac{-e^{i\theta}}{(z-\zeta)^2}+a_1^{(1)}+\cdots\right)dz\\&=hk\operatorname{Im}\left[\int_{\gamma}\frac{-a_1^{(0)}e^{i\theta}}{z-\zeta}dz+\int_{\gamma}\frac{a_1^{(1)}e^{i\theta}}{z-\zeta}dz\right]\\&=hk\operatorname{Im}(2\pi i\,a_1^{(1)}e^{i\theta}-2\pi i\,a_1^{(0)}e^{i\theta})=2\pi hk(\alpha_1^{\theta}-\alpha_0^{\theta}).\end{aligned}$$

The third and last integrals on the right of (33) become in the same manner as above

$$-\int_{\beta} p*dp_{hk}^{\theta}=-k\int_{\beta} p*dp_1^{\theta}=-k\int_{\beta}(p*dp_1^{\theta}-p_1^{\theta}*dp)=-2\pi k\big((h+k)\,\alpha_1^{\theta}-\alpha^{\theta}\big)$$

and

$$-\int_{\beta} p_{hk}^{\theta}*dp=-h\int_{\beta} p_0^{\theta}*dp=-h\int_{\beta}(p_0^{\theta}*dp-p*dp_0^{\theta})=-2\pi h\big(\alpha^{\theta}-(h+k)\,\alpha_0^{\theta}\big).$$

Substitution in (33) gives

$$\begin{aligned}D(p-p_{hk}^{\theta})=\int_{\beta} p*dp+2\pi hk(\alpha_1^{\theta}-\alpha_0^{\theta})-2\pi k\big((h+k)\,\alpha_1^{\theta}-\alpha^{\theta}\big)\\-2\pi h\big(\alpha^{\theta}-(h+k)\,\alpha_0^{\theta}\big).\end{aligned}$$

On rewriting this we obtain (32).

7 E. Conformal Mappings. Hereafter in 7 we always assume that R is a *planar Riemann surface*, i.e. it can be viewed as a subregion of the extended plane. We write $p_j=p_j^0$ $(j=0,1)$.

Recall that principal functions p_j $(j=0,1)$ introduced in 7 D have vanishing flux across every dividing cycle in $R-\zeta$. However since R is planar every cycle is dividing. Therefore there exists a unique single-valued analytic function P_j on $R-\zeta$ such that

$$p_j = \operatorname{Re} P_j, \tag{34}$$

$$P_j(z) = \frac{1}{z-\zeta} + a_1^{(j)}(z-\zeta) + a_2^{(j)}(z-\zeta)^2 + \cdots \tag{35}$$

at ζ. We also set

$$P_{hk} = h P_0 + k P_1$$

for real h, k.

In order to investigate O_{AD}-subregions of the plane we need the following mapping theorem, which is by now classical. In view of its basic importance for our characterization of O_{AD}-subregions we shall give a complete proof. (For references and further generalizations see the monographs Rodin-Sario [3] and Sario-Oikawa [27].)

Theorem. *Each of the functions P_{hk} for $(h, k) = (1, 0)$, $(0, 1)$, and $(1, 1)$ gives a univalent conformal mapping of R onto a region R_{hk} in the extended plane.*

The image region R_{10} (resp. R_{01}) is a horizontal (resp. vertical) slit region with vanishing complementary area.

After a preliminary discussion in 7 F – 7 I the univalency of $P_0 = P_{10}$ and $P_1 = P_{01}$ together with the above properties of R_{10} and R_{01} will be established in 7 J. The univalency proof for $P_0 + P_1 = P_{11}$ will be given in 7 K.

7 F. Principal Functions P_j^θ. In addition to the functions P_j it is convenient to introduce functions P_j^θ defined as follows. As in 7 E there exist single-valued analytic functions P_j^θ on $R-\zeta$ such that

$$p_j^\theta = \operatorname{Re} P_j^\theta,$$

$$P_j^\theta(z) = \frac{e^{i\theta}}{z-\zeta} + a_1^{(j)}(z-\zeta) + a_2^{(j)}(z-\zeta)^2 + \cdots$$

at ζ $(j=0, 1)$.

We shall show that $P_0^\theta = -i P_1^{\theta+\pi/2}$. If R is a compact bordered surface then $q_0^\theta + p_1^{\theta+\pi/2}$ is harmonic on R where $q_0^\theta = \operatorname{Im} P_0^\theta$. On each contour it is constant and has vanishing flux; therefore it reduces to a constant. It is easily seen that these functions on an arbitrary R are the limits of the corresponding functions on compact bordered subregions (cf. 7 C). Thus $q_0^\theta + p_1^{\theta+\pi/2}$ is constant on an arbitrary R. This implies that $P_0^\theta + i P_1^{\theta+\pi/2}$ is a constant which must be zero by the definition of P_j^θ. We may therefore restrict our attention to the function P_0^θ.

Observe that $P_0 = P_0^0$ and $P_1 = P_1^0$. In the same manner as above it can be seen that

$$P_0^\theta = P_0 \cos\theta + i P_1 \sin\theta. \tag{36}$$

7 G. Univalency of P_0^θ. First suppose that R is the interior of a compact bordered surface. Since P_0^θ extends analytically onto each contour and $\operatorname{Im} P_0^\theta = \text{const}$ there, $\operatorname{Re} P_0^\theta$ varies between a finite maximum and a finite minimum, and the image of each contour is a horizontal slit. By the argument principle P_0^θ is thus univalent.

For an arbitrary R the univalency of P_0^θ is a consequence of the Hurwitz theorem which gives the univalency of a nondegenerate limit of univalent functions.

7 H. An Extremal Property of P_0^θ. Let P be a univalent analytic function on $R - \zeta$ with $p = \operatorname{Re} P \in \{p\}_1^\theta$. Denote by A_P the area of the complement X_R of $P(R)$ with respect to the plane.

Take a regular subregion Ω of R containing ζ and let X_Ω be the complement of $P(\Omega)$ with respect to the plane. If we choose a positive orientation of $\partial\Omega$ then $P(\partial\Omega)$ is the negatively oriented boundary of X_Ω. Set $P = u + i v$. Since

$$\int_{\partial\Omega} p * dp = \int_{P(\partial\Omega)} u\, dv = -\int_{X_\Omega} du\, dv$$

and $X_\Omega \to X_R$ as $\Omega \to R$ we conclude that

$$-A_P = \int_\beta p * dp, \tag{37}$$

where again $\int_\beta p * dp = \lim_{\Omega \to R} \int_{\partial\Omega} p * dp$.

This together with (32) for $h = 1$ and $k = 0$ implies

$$A_P + 2\pi\, \alpha^\theta = 2\pi\, \alpha_0^\theta - D(p - p_0^\theta) \tag{38}$$

for every admissible P.

From (38) it follows that for every admissible P

$$\alpha^\theta \leq \alpha_0^\theta \tag{39}$$

where equality holds if and only if $P = P_0^\theta$. In particular

$$\alpha_0^\theta \geq 0 \tag{40}$$

since $P = e^{i\theta}/(z - \zeta)$ is admissible.

Another consequence of (38) is that

$$A_P + 2\pi\, \alpha^\theta \leq 2\pi\, \alpha_0^\theta \tag{41}$$

for every admissible P, with equality only for $P=P_0^\theta$. In particular

$$A_{P_0^\theta}=0. \tag{42}$$

7 I. Horizontal Slits. The complement X_R of $P_0^\theta(R)$ has been seen by (42) to have vanishing area. Let E be a component of X_R. We shall next show that E either consists of a single point or is a horizontal slit.

Suppose that E were not a point. By Riemann's mapping theorem the complement S of E with respect to the plane is conformally equivalent to the unit disk, and the function P_0^0 for the unit disk maps it conformally onto a horizontal slit region. Therefore the function P_0^0 for S, which we call φ, maps S onto a horizontal slit region. Here we suppose that φ has the expansion

$$\varphi(w)=w+b\,w^{-1}+\cdots$$

at ∞. By (40), $\operatorname{Re} b\geq 0$ and equality holds if and only if $\varphi(w)\equiv w$ since the identity function w is admissible.

We return to our original P_0^θ for R and observe that

$$(\varphi\circ P_0^\theta)(z)=\frac{e^{i\theta}}{z-\zeta}+(a_1^{(0)}+b\,e^{-i\theta})(z-\zeta)+\cdots.$$

Since $P=\varphi\circ P_0^\theta$ is admissible (39) reads $\operatorname{Re}(a_1^{(0)}e^{i\theta})\geq\operatorname{Re}(a_1^{(0)}e^{i\theta}+b)$, which in turn implies $\operatorname{Re} b=0$. Therefore $\varphi(w)\equiv w$, and E must be a horizontal slit.

7 J. Mappings P_0 and P_1. Theorem 7 E follows for P_0 and P_1 from the properties of P_0^θ and $P_0^\theta(R)$ established in 7 F – 7 I and from the relations

$$P_0=P_0^0,\qquad P_1=-i\,P_0^{\pi/2}$$

which are special cases of (36).

7 K. Mapping P_0+P_1. We next prove the univalency of $P_{11}=P_0+P_1$. It suffices to consider the case where R is the interior of a compact bordered region (cf. 7 G).

Let β_k be a contour of R and denote by σ_k the horizontal slit $P_0^\theta(\beta_k)$. For a point w in the plane consider $(2\pi i)^{-1}\int_{\beta_k} f^{-1}\,df$ where $f=P_0^\theta-w$. By the extended residue theorem the principal value of this integral is $n(w)-n(\infty)$ where $n(w)$ is the number of times P_0^θ takes the value w in $\bar{R}$. Multiplicities are counted as usual for w-points in R and with half their ordinary values if the points lie on the boundary. We have seen that $\int_{\beta_k} f^{-1}\,df=0$ if $w\notin\sigma_k$. If $w\in\sigma_k$ the integral reduces to $\int_{\beta_k} d\log(P_0^\theta-w)$. Thus it is zero in all cases and $n(w)=n(\infty)=1$. It follows that f has either a simple zero in R, two simple zeros on some β_k, or a double zero on some β_k. Consequently $p_0^\theta|\beta_k$ takes on every value between its maximum and minimum at two distinct points.

On β_k we have by (36), $p_0^\theta = \operatorname{Re} P_0^\theta = \operatorname{Re}[(P_0+P_1)\,e^{i\theta}] + \text{const}$, which shows that the line $\operatorname{Re}(w\,e^{i\theta}) = \text{const}$ has at most two points of intersection with the curve $w=(P_0+P_1)|\beta_k$. Thus the image γ_k of β_k under P_0+P_1 is convex.

We shall next show that each γ_k is an analytic curve, i.e. $(P_0+P_1)' \neq 0$ on β_k. We have seen that both P_0' and P_1' have two simple zeros on β_k. These zeros must be located at distinct points. In fact suppose that $P_0' = P_1' = 0$ at a point b of β_k. By (36) this would imply that the first two derivatives of P_0^θ vanish at b for an appropriate choice of θ. Actually P_0'' with respect to the curve parameter is real and P_1'' is imaginary and hence there is a θ such that $P_0''/P_1'' = -i\tan\theta$ at b. On the other hand P_0^θ takes on every value with multiplicity at most 2, and thus P_0' and P_1' cannot vanish simultaneously. Since P_0' is real and P_1' is imaginary along β_k, $(P_0+P_1)'$ cannot vanish on β_k.

Let us compute the winding number of γ_k about a point inside it. The tangent to γ_k has the slope $\operatorname{Im} F$ with $F = P_1'/P_0'$. Observe that $F \neq 0, \infty$ on R and that $F(\zeta)=1$. On each β_k, F is imaginary except for two simple poles. At the two points where P_0^θ has vanishing derivative we see from (36) that $F = i\cot\theta$. Thus F takes on every imaginary value at least twice. For a fixed θ apply the extended residue theorem to $F - i\cot\theta$. There are two poles and at least two zeros on each β_k but no poles in R. Hence there are exactly two zeros on each β_k, and none in R, and therefore $\operatorname{Re} F \neq 0$ on R. Since $\operatorname{Re} F(\zeta) = 1$ we conclude that $\operatorname{Re} F > 0$ on R and consequently $\operatorname{Im} F$ decreases as β_k is traced in the positive direction. Therefore each γ_k is an analytic convex curve traced exactly once in the direction of decreasing slope.

Take a point w encircled by m curves γ_k. The winding number of $(P_0+P_1)(\partial R)$ about w is $-m$. This integer is the value of $n(w)-n(\infty)$, and therefore $m=1$, i.e. P_0+P_1 is univalent.

This completes the proof of Theorem 7 E.

8. Characterization of O_{AD}-Regions

8 A. The Analytic Span. Let $\{P\}$ be the class of univalent analytic functions on a *planar region* R less a point ζ at which

$$P(z) = \frac{1}{z-\zeta} + a_1(z-\zeta) + \cdots. \tag{43}$$

Observe that $p = \operatorname{Re} P \in \{p\}_1^0$ (cf. 7 D). In terms of A_P of 7 H, (32) now reads

$$A_P + 2\pi(h-k)\,\alpha = 2\pi(h^2\alpha_0 - k^2\alpha_1) - D(p - p_{hk}) \tag{44}$$

where $p_{hk} = \operatorname{Re} P_{hk}$, $\alpha = \operatorname{Re} a_1$, $\alpha_j = \operatorname{Re} a_1^{(j)}$ (cf. 7 D, 7 E). Using the choices $(h, k) = (1, 0)$ and $(0, 1)$ we see that

$$\alpha_1 \leq \alpha \leq \alpha_0, \tag{45}$$

with P_0 (resp. P_1) the unique function maximizing (minimizing) α in $\{P\}$.

Following SCHIFFER [1] we consider the quantity

$$S_A = S_A(R, \zeta) = \alpha_0 - \alpha_1 \tag{46}$$

where the use of the symbol ζ is understood to include the choice of the local parameter at ζ: The quantity is called the *span* or more precisely the *analytic span* or the *A-span* of R. It depends not only on R but also on $\zeta \in R$ and thus it is not a genuine conformal invariant. However the vanishing of S_A will be seen to be independent of ζ and therefore conformally invariant. It will also be shown that $a_1^{(0)} - a_1^{(1)}$ is in fact real and thus equal to S_A.

On setting $(h, k) = (\frac{1}{2}, \frac{1}{2})$ in (44) we obtain

$$A_P = \tfrac{1}{2}\pi S_A - D(p - p_{\frac{1}{2}\frac{1}{2}}). \tag{47}$$

In other words

$$\tfrac{1}{2}\pi S_A = \max_{\{P\}} A_P = A_{\frac{1}{2}(P_0 + P_1)}. \tag{48}$$

8 B. Regular Functions. If U is in the class $A(R)$ of analytic functions on R then clearly $u = \operatorname{Re}(U - U(\zeta)) \in \{p\}_0^0$. Choose $(h, k) = (1, -1)$ in (32). Since $\int_\beta u * du = D(u)$ we obtain

$$D(u) - 4\pi \operatorname{Re} U'(\zeta) = -2\pi S_A + D(u - (p_0 - p_1)). \tag{49}$$

In particular if $U \equiv 0$ then

$$S_A = \frac{1}{2\pi} D(p_0 - p_1). \tag{50}$$

Consider the class $\{U\}_1 \subset A(R)$ with $U'(\zeta) = 1$. Obviously $D_U = D(u)$ is the area of the Riemannian image $U(R)$. If $S_A \neq 0$ then $(P_0 - P_1)/S_A$ belongs to $\{U\}_1$, and (32) with

$$(h, k) = \left(\frac{1}{S_A}, -\frac{1}{S_A}\right) \tag{51}$$

implies that

$$\frac{2\pi}{S_A} = \min_{\{U\}_1} D_U = D_{(P_0 - P_1)/S_A}. \tag{52}$$

If $S_A = 0$, (52) degenerates to $\infty = \infty$.

From (48) and (52) it follows that

$$\max_{\{P\}} A_P \cdot \min_{\{U\}_1} D_U = \pi^2. \tag{53}$$

8 C. Characterizations. After this somewhat lengthy preparation we are now able to give a complete characterization of O_{AD}-regions:

THEOREM. *The following ten properties for plane regions are equivalent:*

(a) $R \in O_{AD}$,
(b) $S_A = 0$ *for some* $\zeta \in R$,
(c) $S_A = 0$ *for every* $\zeta \in R$,
(d) $P_0 = P_1$ *for some* $\zeta \in R$,
(e) $P_0 = P_1$ *for every* $\zeta \in R$,
(f) $\{P\}$ *consists of only* P_0 *for some* $\zeta \in R$,
(g) $\{P\}$ *consists of only* P_0 *for every* $\zeta \in R$,
(h) *all univalent maps of R are linearly dependent,*
(i) *all univalent maps of R are linear transformations,*
(j) *the complementary area vanishes under all univalent maps of R.*

By (52) we see immediately that (a) and (c) are equivalent. That (b) is equivalent to (d) and that (c) is equivalent to (e) follows from (50). By (45) it is seen that (d) (resp. (e)) is equivalent to (f) (resp. (g)).

Let F be an arbitrary univalent map of R. Note that

$$c + \frac{F'(\zeta)}{F(z) - F(\zeta)} = \frac{1}{z - \zeta} + a_1(z - \zeta) + \cdots$$

belongs to $\{P\}$, with c a suitable constant. If (g) is valid then the above function must coincide with P_0. Therefore $F(z)$ is a linear function of P_0 and this fact in turn implies (h). That (h) implies (g) is trivial and consequently (g) and (h) are equivalent.

Suppose (h) is valid. All univalent maps are linearly dependent on the map $F(z) \equiv z$, i.e. (i) holds. The converse implication is again trivial.

If (j) is true then by (48), (c) must be valid, i.e. $R \in O_{AD}$. If $R \in O_{AD}$ then again by (48), $A_P = 0$ for every $P \in \{P\}$. Since (i) also holds we obtain (j).

Clearly (c) implies (b). Finally suppose (b) is valid. By (45) and (50), $\{P\}$ consists of a single element $P_0 = P_1$, which is clearly $1/(z - \zeta)$ or z according as $\zeta \neq \infty$ or $\zeta = \infty$. Hence (i) and a fortiori (c) is true.

This completes the proof.

Property (j) is often stated by saying that the boundary of R has *absolute area zero.*

In passing we remark that the necessity in Theorem 6 B follows from (j).

A number of further metric and capacitary properties of sets with O_{AD}-complements will be discussed in connection with bounded functions in II.9–11.

8 D. Removable Sets. A compact set E of the plane is called *AD-removable* or *AD-null* if for some disk U with $E \subset U$ all AD-functions on $\bar{U}-E$ have analytic extensions to $\bar{U}$.

Let $A_0 D(\bar{U}-E)$ be the class of AD-functions on $\bar{U}-E$ whose real parts vanish on the boundary α of U.

LEMMA. *A compact point set E is AD-removable if and only if the class $A_0 D$ consists of constants.*

The necessity is obvious. For the sufficiency let f be an AD-function on $\bar{U}-E$ and set $\varphi = \operatorname{Re} f$ on α. Let u be a harmonic function on U with boundary values φ on α. Clearly $u \in HD(\bar{U})$. Let v be the conjugate of u on $\bar{U}$ and set $F = u + iv$ on $\bar{U}$. Then $F \in AD(\bar{U})$ and $f - F \in A_0 D(\bar{U}-E)$. By assumption, $f - F = c$, a constant on $\bar{U}-E$. Thus $F + c$ is the analytic extension of f to U.

By the above lemma we can show:

THEOREM. *A compact point set E is AD-removable if and only if the complement R of E belongs to O_{AD}.*

The necessity is trivial. For the sufficiency take a disk U with boundary α such that $E \subset U$. Given $F \in A_0 D(\bar{U}-E)$ and $\operatorname{Re} F = s$ on $\bar{U}-E$ let L_0 be a normal operator for $U-E$ with respect to R and p the principal function for $(s, L_0, U-E)$; the existence of the latter is assured by $\int_\alpha * ds = 0$. Since $D(p-s) < \infty$ on $\bar{U}-E$ by (30), p has a finite Dirichlet integral on $\bar{U}-E$ and thus on R. Similarly $p-s$ and hence p has vanishing periods along all cycles. Thus there exists a function P in $AD(R)$ such that $\operatorname{Re} P = p$. By $R \in O_{AD}$, p must be constant, i.e. $L_0 s = s$ on $\bar{U}-E$. In view of $s=0$ on α we have $s \equiv 0$ on $\bar{U}-E$. Therefore F is constant and by the above lemma E is AD-removable.

REMARK. *If $E_1, \ldots, E_n$ are compact and AD-removable, then so is $\bigcup_1^n E_j$.*

It suffices to show that $E = E_1 \cup E_2$ is AD-removable. Observe that E is compact and of Lebesgue measure zero. Therefore if $f \in AD$ in the complement of E and if f is analytically extendable to the complement of a compact set $K \subset E$ then $D_{\{|z|<\infty\}-E}(f) = D_{\{|z|<\infty\}-K}(f)$. Let $\{R_n\}$ be an exhaustion of the complement R of E and denote by γ_n the set of those contours of R_n which encircle points in $F = E_1 \cap E_2$. Let R'_n be the unbounded region with boundary γ_n. Every $f \in AD(R)$ is in $AD(R_n)$ for every n. Since $R'_n \cap E$ consists of two disjoint compact subsets of E_1

and E_2 respectively, f is analytically extendable to R'_n. On repeating the argument we see that f is analytically extendable to R', the complement of F. Since $D_R(f)=D_{R'}(f)$ we have $f\in AD(R')$. The fact that F is AD-removable as a compact subset of an AD-removable set E_1 implies that f is analytically extendable to all of $|z|\leq\infty$. Thus f is constant, i.e. $R\in O_{AD}$ and E is AD-removable.

The above proof is due to KAMETANI [1]. See also VI.1 L.

8 E. Surfaces of Finite Genus. Given a *Riemann surface R of finite genus* there exists a regular subregion R_0 with connected boundary α_0 such that $R-\bar{R}_0$ is planar. Map $R-R_0$ conformally into the unit disk $|z|\leq 1$ so that α_0 corresponds to $|z|=1$. The inner boundary of the image R_1 of $R-\bar{R}_0$ is a point set E which can be considered as a realization of the ideal boundary β of R. We say that R has an *AD-removable boundary* or *AD-null boundary* β if E is AD-removable in the sense of 8 D. We have only to consider the case where E is discrete.

We shall prove:

THEOREM. *A Riemann surface R of finite genus g has an AD-removable boundary if and only if R belongs to O_{AD}.*

The necessity is again clear. In order to prove the sufficiency, assume that β is not AD-removable. We can choose a regular region R_2 in R such that $R_2\supset\bar{R}_0$ and $R-\bar{R}_2$ consists of $2g+1$ regions $D_1,\ldots,D_{2g+1}$ with the property that each relative boundary α_j of D_j is connected and the set $E_j\subset E$ encircled by α_j is not AD-removable. On each D_j we can find a nonconstant AD-function F_j with vanishing real part on α_j. Let G_j be a Riemann surface such that $R\subset G_j$ and $G_j-\bar{R}_0$ is $\{|z|<1\}-E_j$. Denote by p_j the harmonic function on G_j which is the principal function for $(\operatorname{Re} F_j, L_0, D_j)$. Since $\operatorname{Re} F_j=0$ on α_j and $\operatorname{Re} F_j\neq\text{const}$ on D_j, p_j is not constant on G_j. The function

$$p=\sum_{j=1}^{2g+1} c_j\,p_j$$

is harmonic with a finite Dirichlet integral on R and not constant unless all $c_j=0$ $(j=1,\ldots,2g+1)$. Moreover p has vanishing periods along all cycles in $R-\bar{R}_0$. Let γ_k $(k=1,\ldots,2g)$ be $2g$ nondividing canonical cycles of R and consider the equations

$$\int_{\gamma_k} *dp=\sum_{j=1}^{2g+1}\Big(\int_{\gamma_k} *dp_j\Big)\,c_j=0 \qquad (k=1,\ldots,2g).$$

Since the vectors $\big(\int_{\gamma_1} *dp_j,\ldots,\int_{\gamma_{2g}} *dp_j\big)$ $(j=1,\ldots,2g+1)$ are linearly dependent there exists a solution $(c_1,\ldots,c_{2g+1})\neq(0,\ldots,0)$. The resulting harmonic function p has vanishing periods along all cycles in R and thus there exists a nonconstant AD-function P on R such that $\operatorname{Re} P = p$.

8 F. Closed Extensions. A closed Riemann surface R^* will be called a closed extension of a given Riemann surface R if R is conformally equivalent to a subregion R' of R^*.

THEOREM. *Let R be a Riemann surface of finite genus belonging to the class O_{AD}. Then all closed extensions of R are conformally equivalent.*

Take R_0 in R as in 8 E. Denote by R^* and R^{**} closed extensions of R. Let φ^* and φ^{**} be conformal imbeddings of R into R^* and R^{**} respectively. Set $\varphi^*(R) = R'$, $\varphi^*(R_0) = R'_0$, $\varphi^{**}(R) = R''$, and $\varphi^{**}(R_0) = R''_0$. Consider the univalent conformal mapping $\varphi = \varphi^{**} \circ \varphi^{*-1}$ of R' onto R'' with $\varphi(R'_0) = R''_0$. By $R' \in O_{AD}$, $R^* - R'$ must be totally disconnected in R^*. Therefore $R^* - R'_0$ is conformally equivalent to $\{|z| < 1\}$ and $R' - R'_0$ is conformally equivalent to $\{|z| < 1\} - E'$, with E' some totally disconnected set in $\{|z| < 1\}$. Similarly let $R^{**} - R''_0$ and $R'' - R''_0$ be conformally equivalent to $\{|w| < 1\}$ and $\{|w| < 1\} - E''$ respectively. Then φ gives a conformal map of $\{|z| \leq 1\} - E'$ onto $\{|w| \leq 1\} - E''$. Since

$$\varphi \in AD(\{|z| \leq 1\} - E')$$

and E' is AD-removable by Theorem 8 E, φ can be analytically extended to a mapping of all of $\{|z| \leq 1\}$ into $\{|w| \leq 1\}$. We recall Darboux's theorem: if φ maps a Jordan region D_1 onto another D_2 and if φ maps ∂D_1 univalently onto ∂D_2, then φ is univalent on $\bar{D}_1$. By this theorem or just by a simple application of the argument principle φ is a univalent conformal mapping of $\{|z| \leq 1\}$ onto $\{|w| \leq 1\}$. This shows that φ can be extended to a univalent conformal mapping of R^* onto R^{**}.

COROLLARY. *The surfaces of finite genus in O_{AD} can be classified according to their closed extensions.*

8 G. Closed Extensions (continued). The converse of Theorem 8 F holds in the following sense. Let R be a Riemann surface of finite genus, R^* and R^{**} closed extensions of R, and φ^* and φ^{**} imbeddings of R into R^* and R^{**}. Then $\varphi = \varphi^{**} \circ \varphi^{*-1}$ is a conformal map of a part $\varphi^*(R)$ of R^* into R^{**}. Assume that φ can always be extended to a conformal map of R^* onto R^{**} for all closed extensions R^* and R^{**} of R. In this case we say that R has a *unique closed extension.* Theorem 8 F also states that if $R \in O_{AD}$, then R has a unique closed extension. The converse reads (A. MORI [4]):

THEOREM. *Let R be a Riemann surface of finite genus. If R has a unique closed extension then R belongs to the class O_{AD}.*

Let g be the genus of R and let $\gamma_1, \dots, \gamma_g$ be g disjoint cycles in R such that $R_\gamma = R - \bigcup_{j=1}^{g} \gamma_j$ is planar. Here R_γ can be viewed as a plane region containing ∞ such that R_γ is bounded by $2g$ disjoint analytic Jordan curves $C_1, \dots, C_g, C'_1, \dots, C'_g$ and a closed set E. Let ψ_j be an indirectly conformal mapping of C_j onto C'_j such that the identification of C_j with C'_j by ψ_j $(j=1, \dots, g)$ creates R from R_γ.

We have to show that E is AD-removable. Contrary to the assertion assume that E is not AD-removable. Then the complement G of E does not belong to O_{AD}. By Theorem 8 C there exists a conformal transformation φ of G into the complex plane such that $\varphi(\infty) = \infty$ and φ has a nonremovable singularity at some point in E. Let $\tilde{R}_\gamma = \varphi(R_\gamma)$, $\tilde{C}_j = \varphi(C_j)$, $\tilde{C}'_j = \varphi(C'_j)$, and $\tilde{\psi}_j = \varphi \circ \psi_j \circ \varphi^{-1}$. Then an identification of $\tilde{C}_j$ and $\tilde{C}'_j$ by $\tilde{\psi}_j$ gives $\tilde{R}$ from $\tilde{R}_\gamma$, and $\varphi | R_\gamma$ can be extended to a conformal map of R onto $\tilde{R}$.

By the above identification of C_j with C'_j and of $\tilde{C}_j$ with $\tilde{C}'_j$, $R_\gamma \cup E$ and $\tilde{R}_\gamma \cup (\{|z| \leq \infty\} - \varphi(G))$ become closed Riemann surfaces R^* and R^{**} respectively, with $R \subset R^*$ and $\tilde{R} \subset R^{**}$. Clearly φ is not analytically extendable to all of R^*, since φ has a nonremovable singularity at E. This contradicts the uniqueness of a closed extension of R.

Remark. Oikawa [7] obtained the following important sharpening of the above theorem: If all closed extensions of R are conformally equivalent then R belongs to O_{AD}.

9. Class *ABD* on Surfaces of Finite Genus

9 A. $O_{AD} = O_{ABD}$ for Finite Genus. We denote by $ABD(R)$ the class of analytic functions on R which are bounded and have finite Dirichlet integrals. Trivially the class O_{ABD} of surfaces R on which $ABD(R)$ consists of constants only contains O_{AD}.

For surfaces of finite genus the converse is also true. To prove this we shall make use of the following auxiliary result:

Lemma. *If R is a plane region not belonging to O_{AD} then $P_0 - P_1$ is a nonconstant ABD-function on R.*

In fact by Theorem 8 C and (52), $P_0 - P_1$ is in $AD(R)$ and is not constant. Since P_0 and P_1 are univalent on R and send ζ to ∞ their absolute values are bounded on R outside of a disk about ζ. Therefore $|P_0 - P_1|$ is bounded on R.

Theorem. *Let R be a Riemann surface of finite genus not belonging to the class O_{AD}. Then there exists a nonconstant ABD-function on R.*

Take a closed extension R^* of R. It is readily seen that there exist $2g+1$ disjoint closed sets $E_1, \dots, E_{2g+1}$ in $R^* - R$ none of which is

AD-removable. If there exists a nonconstant ABD-function on $R^*-\bigcup_1^{2g+1} E_j$ then the same is true on R. Therefore we may suppose that $R^*-R=\bigcup_1^{2g+1} E_j$. Let R_2 be a regular subregion of R such that $R-\bar{R}_2$ consists of $2g+1$ regions $D_1, \ldots, D_{2g+1}$ with the property that the $D_j \cup E_j$ $(j=1, \ldots, 2g+1)$ are conformally disks.

By the above lemma there exists an ABD-function F_j on $\bar{D}_j - E_j$ which has a nonremovable singularity at some point of E_j. Let p_j be the principal function on R^*-E_j with respect to $(\operatorname{Re} F_j, L_0, D_j)$ where L_0 is the Dirichlet operator acting from ∂D_j into $D_j \cup E_j$. Clearly p_j is bounded and not constant.

In the same manner as in the proof of Theorem 8 E we can find a $(2g+1)$-tuple $(c_1, \ldots, c_{2g+1}) \neq (0, \ldots, 0)$ such that there exists an $F \in AD(R)$ with $\operatorname{Re} F = \sum_1^{2g+1} c_j p_j$. We observe that $c=\sup_R |\operatorname{Re} F|+1<\infty$ and conclude that $1/(c+F)$ is a nonconstant ABD-function on R.

We have proved:

$$O_{AD}=O_{ABD} \tag{54}$$

for finite genus.

As an application we can establish the following general inclusion relation (ROYDEN [1]):

$$O_{AB} \subset O_{AD} \tag{55}$$

for arbitrary genus. Here $AB(R)$ stands for the class of bounded analytic functions on R and O_{AB} is the family of Riemann surfaces R for which $AB(R)$ reduces to the complex number field.

For the proof suppose $R \notin O_{AD}$ and let $F \in AD(R)$ be nonconstant. Since the plane region $F(R)$ has finite area Theorem 8 C gives $F(R) \notin O_{AD}$. Thus there exists a nonconstant ABD-function φ on $F(R)$, and $\varphi \circ F$ is a nonconstant AB-function on R.

***9 B. Finite *ABD*-Interpolation.** For $R \in O_{AD}$ the dimension of the complex vector space $AD(R)$ is 1. If $R \notin O_{AD}$ then $\dim AD(R)>1$. The question arises: are there surfaces $R \notin O_{AD}$ with $\dim AD(R)<\infty$? We shall see that such an R, if it exists, must be of infinite genus. In other words $\dim AD(R)=\infty$ for $R \notin O_{AD}$ with finite genus. This is a consequence of the following more general fact:

THEOREM. *Let R be an open Riemann surface of finite genus. Given a finite number of distinct points $\zeta_1, \ldots, \zeta_n$ in R, local parameters $z_1, \ldots, z_n$ with $z_k(\zeta_k)=0$ $(k=1, \ldots, n)$, and complex numbers $\alpha_{\nu k}$ $(\nu=0, \ldots, m;\ k=1, \ldots, n)$, there exists an ABD-function f on R such that*

$$\frac{d^\nu f}{d z_k^\nu}(\zeta_k)=\alpha_{\nu k} \qquad (\nu=0, \ldots, m;\ k=1, \ldots, n) \tag{56}$$

if and only if R does not belong to the class O_{AD}.

The necessity is clear. Conversely assume that $R \notin O_{AD}$. By Theorem 9 A there exists a nonconstant ABD-function F on R. Choose a point ζ_0 in $R - \{\zeta_1, \ldots, \zeta_n\}$ with $F(\zeta_0) \neq F(\zeta_k)$ for $k = 1, \ldots, n$. On a closed extension R^* of R there exists a meromorphic function $r_k(z)$ regular on $R^* - \{\zeta_0, \zeta_k\}$ with a simple pole at ζ_k and a pole of order n_k, say, at ζ_0. This is a direct consequence of the Riemann-Roch theorem (see II. 16; also cf. AHLFORS-SARIO [12, p. 325]). Let m_k be the order of the zero of $\prod_{j=1}^{n}(F(z) - F(\zeta_j))^{m+1}$ at ζ_k and set $s = \max(m_k n_k; k = 1, \ldots, n)$. Then the function $(r_k(z))^{m_k - \nu} H(z)$ with

$$H(z) = (F(z) - F(\zeta_0))^s \prod_{j=1}^{n} (F(z) - F(\zeta_j))^{m+1}$$

has a zero of order ν at ζ_k, is regular at ζ_0, and belongs to $ABD(R)$. The normalized function

$$H_{\nu k}(z) = \left[(\nu!) \frac{1}{m_k!} \cdot \frac{d^{m_k} H}{d z_k^{m_k}} (\zeta_k) \cdot ((z_k\, r_k)(\zeta_k))^{m_k - \nu}\right]^{-1} (r_k(z))^{m_k - \nu} \cdot H(z)$$

has the property $(d^\nu H_{\nu k}/d z_k^\nu)(\zeta_k) = 1$.

Define $\{P_\nu(z)\}_{\nu=0}^{m}$ inductively by $P_0(z) = \sum_{k=1}^{n} \alpha_{0k} H_{0k}(z)$ and

$$P_\nu(z) = P_{\nu-1}(z) + \sum_{k=1}^{n} \left(\alpha_{\nu k} - \frac{d^\nu P_{\nu-1}}{d z_k^\nu}(\zeta_k)\right) H_{\nu k}(z) \qquad (\nu = 1, \ldots, m).$$

Then $P_\nu^{(\mu)}(\zeta_k) = P_{\nu-1}^{(\mu)}(\zeta_k)$ for $\mu < \nu$ and $P_\nu^{(\nu)}(\zeta_k) = P_{\nu-1}^{(\nu)}(\zeta_k) + (\alpha_{\nu k} - P_{\nu-1}^{(\nu)}(\zeta_k)) = \alpha_{\nu k}$. Therefore the $ABD(R)$-function $f(z) = P_m(z)$ satisfies (1).

***9 C. Finite *AD*-Interpolation with Minimum Norm.** For an open Riemann surface R of finite genus we denote by $\mathscr{F}(R) = \mathscr{F}(R; (\zeta_k), (z_k), (\alpha_{\nu k}))$ the class of AD-functions f on R with property (56). We then have:

THEOREM. *The class $\mathscr{F}(R) \neq \emptyset$ if and only if $R \notin O_{AD}$, and in this case there exists a unique function f_0 in $\mathscr{F}(R)$ such that $D(f_0) = \min_{f \in \mathscr{F}} D(f)$, or more precisely $D(f) = D(f_0) + D(f - f_0)$ for every f in $\mathscr{F}(R)$.*

Take a closed parametric disk U_k of z_k for each $k = 1, \ldots, n$ and an arbitrary parametric disk U with local parameter z such that $\zeta_k \in U$ ($k = 1, \ldots, n$). By the local subharmonicity of $|f'|^2$ for $f \in \mathscr{F}$ there exists a constant C_U such that

$$|f_1(z) - f_2(z)|^2 + \sum_{k=1}^{n} \sum_{\nu=0}^{m} \left| \frac{d^\nu f_1}{d z_k^\nu}(z_k) - \frac{d^\nu f_2}{d z_k^\nu}(z_k) \right|^2 \leq C_U D(f_1 - f_2) \quad (57)$$

for every $z \in U$, $z_k \in U_k$, and $f_1, f_2 \in \mathscr{F}$.

Take $\{f_n\} \subset \mathscr{F}$ with $D(f_n) \to d = \inf_{f \in \mathscr{F}} D(f)$. Since $\frac{1}{2}(f_1 + f_2) \in \mathscr{F}$, $D(f_n - f_{n+p}) = 2\big(D(f_n) + D(f_{n+p})\big) - 4D\big(\frac{1}{2}(f_n + f_{n+p})\big) \le 2\big(D(f_n) + D(f_{n+p})\big) - 4d$. Hence $D(f_n - f_{n+p}) \to 0$ as $n \to \infty$. Thus by (57) there exists an $f_0 \in \mathscr{F}$ such that $D(f_0) = d$. For any $f \in \mathscr{F}$ and any complex number λ, $f_0 + \lambda(f - f_0) \in \mathscr{F}$ and $D\big(f_0 + \lambda(f - f_0)\big) \ge D(f_0)$. It follows that $D(f_0, f - f_0) = 0$ and therefore $D(f) = D(f_0) + D(f - f_0)$.

10. Essential Extendability

10 A. Finite Genus. An arbitrary Riemann surface R of finite or infinite genus is called *essentially extendable* if it can be conformally and univalently mapped into a Riemann surface R^* such that the complement of the image R' of R has interior points. We claim:

THEOREM. *A Riemann surface R of finite genus belonging to O_{AD} is not essentially extendable.*

Contrary to the assertion assume that there exists a Riemann surface R^* such that $R^* \supset R$ and $R^* - R$ has an interior point. Let R_0 be a regular region of R such that $R - \bar{R}_0$ is a planar surface. There exists a point z on $\partial R \subset R^*$ and a parametric disk U about z such that $U \cap \bar{R}_0 = \emptyset$, $(R - \bar{R}_0) \cup U$ is planar, and $U \cap (R^* - R)$ has an interior point. Map $(R - \bar{R}_0) \cup U$ conformally into $|z| < 1$ such that ∂R_0 corresponds to $|z| = 1$. Then the image G of $R - \bar{R}_0$ is the disk $|z| < 1$ less a closed set E. Since E contains an interior point, which may be assumed to be 0, we conclude that $z - 1/z \in A_0 D(\bar{G})$. By virtue of Lemma 8 D and Theorem 8 E this contradicts the hypothesis $R \in O_{AD}$.

10 B. Infinite Genus. In striking contrast with the above theorem we have:

THEOREM. *There are surfaces of infinite genus belonging to O_{AD} that can be essentially extended.*

The following surface, often referred to as *Myrberg's example* [6], is by now classical. Consider a 2-sheeted covering surface Φ of the finite plane $|z| < \infty$ with an infinite number of branch points above $z = a_n$ $(n = 1, 2, \ldots)$. Here $\{a_n\}$ lies on the real line and clusters only at ∞. We denote by Φ_1 and Φ_2 the upper and lower sheets of Φ. For each z let z_1 and z_2 be the points in Φ_1 and Φ_2 over z.

Let K be a closed disk in the open upper half-plane, and K_1, K_2 the parts of Φ_1, Φ_2 above K. The Riemann surface

$$R = \Phi - K_1$$

is Myrberg's example. Clearly *it is essentially extendable* to Φ.

To see that $R \in O_{AD}$ take an AB-function f on R and consider the function

$$\varphi(z) = \big(f(z_1) - f(z_2)\big)^2$$

on the annular region $\{|z| < \infty\} - K$. It is single-valued, analytic, and bounded, with $\varphi(a_n) = 0$ ($n = 1, 2, \ldots$). It has an analytic extension to a neighborhood of ∞ in the extended plane, and its zeros cluster at ∞. Hence $\varphi \equiv 0$, i.e. $f(z_1) = f(z_2)$ for $z \notin K$. If we extend f to K_1 by setting $f(z_1) = f(z_2)$ we have $f \in AB(\Phi)$ and $f(z_1) = f(z_2)$. Therefore f can be viewed as an AB-function on $|z| \leq \infty$ and is a fortiori constant. For this reason $R \in O_{AB}$ and by (55), $R \in O_{AD}$.

10 C. Boundary Property. We say that a property of a Riemann surface is *a property of its ideal boundary* if any two Riemann surfaces with some identical boundary neighborhood both have or fail to have the property.

THEOREM. *Belonging to O_{AD} is a property of the ideal boundary for finite genus but not for infinite genus.*

The assertion for finite genus is an easy consequence of Theorem 8 E.

The proof for infinite genus is furnished by a modification of Myrberg's example (A. MORI [2]):

Example. Let Φ be the 2-sheeted covering surface of $|z| < \infty$ with branch points above $z = 6, 7, 8, \ldots$. Let R_1 be a Riemann surface obtained from Φ by removing a disk above $|z-2| \leq 1$ from the upper sheet of Φ and a disk above $|z+2| \leq 1$ from the lower sheet. Let R_2 be a Riemann surface obtained from Φ by removing disks above $|z+2| \leq 1$ from both upper and lower sheets.

Denote by R_1^0 a regular region of R_1 which is on the upper sheet above $\{|z| < 5, |z-2| > 2\}$, and by R_2^0 a regular region of R_2 on the upper sheet above $\{|z| < 5, |z+2| > 2\}$.

We see that $R_1 - \overline{R_1^0}$ and $R_2 - \overline{R_2^0}$ are identical (conformally equivalent) ideal boundary neighborhoods of R_1 and R_2 respectively. That $R_1 \in O_{AD}$ can be shown as in 10 B. However $R_2 \notin O_{AD}$ since $1/(z+2)$ can be considered to belong to $AD(R_2)$.

The example also shows the noninvariance of O_{AD} under quasiconformal mappings. This topic will be discussed in II.14.

Remark. There are several other modifications of Myrberg's example, among which we refer the reader to those of AHLFORS [6] and ROYDEN [7]. HEINS' [2] important modification will be discussed in IV.9.

The significance of Myrberg's example lies in that, despite its simplicity, it has rich implications. It was one of the incentives which brought the classification theory to its present form.

10 D. Relations $O_{A^0D} < O_{A_0D}$ and $O_{A^0D} < O_{AD}$. Myrberg's example in 10 B also serves to prove the strict inclusion relations

$$O_{A^0D} < O_{AD}, \quad O_{A^0D} < O_{A_0D}.$$

The improper inclusions were established in 1 G. Myrberg's surface R belongs to O_{AD} and also to O_{A_0D}. This is seen as in 10 B. However $R \notin O_{A^0D}$. To prove this take a disk U in Φ_1 the closure of whose projection lies in the upper half-plane and such that $U \supset K_1$ (cf. 10 B). Then $R' = U - K_1 \subset R$ and $(R', \partial U) \notin SO_{AD}$. Hence $R \notin O_{A^0D}$.

As for the classes O_{AD} and O_{A_0D} it is easily seen that

$$O_{A_0D} = O_{AD} \qquad \textit{for finite genus}.$$

This follows from the argument in 8 D and 8 E. The details are left to the reader.

11. Koebe's Circular Mappings

11 A. Koebe's Principle. The following principle (Kreisnormierungsprinzip) was conjectured by Koebe [1]:

Every plane region of finite or infinite connectivity can be conformally mapped onto a plane region whose every boundary component consists of a circle or a point.

For the case of finite connectivity the proof was given by Koebe [1]. Cases with a finite number of accumulation boundary components were treated by several authors. While the general case remains unsolved we shall give the proof in certain significant cases of *infinitely* many accumulation components.

11 B. Exhaustion by Noncompact Regions. Let R be an arbitrary plane region. We form the following "exhaustion" $\{R_n\}$ of R by a nested sequence of relatively compact or noncompact subregions R_n whose union is R. The boundary of R_n consists of a finite number v_n of analytic Jordan curves β_{ni} $(i = 1, \ldots, v_n)$ in R and a finite number μ_n of isolated boundary components γ_j $(j = 1, \ldots, \mu_n)$ of R. The components γ_j for $j = \mu_n + 1, \mu_n + 2, \ldots, \mu_{n+1}$ then belong to the boundary of $R_{n+1} - R_n$. Let γ^* be the set of those boundary components of R that remain in the exterior of each R_n. In particular γ^* contains all accumulation components of the boundary of R.

11 C. Regular Chain Sets. Let $\{G\}$ be a set of chains of disks that cover the β_{ni} $(n = 1, 2, \ldots; i = 1, \ldots, v_n)$ (cf. 3 A). Denote by $\{G\}_n$ the set of those chains in $\{G\}$ that cover the curves β_{ni} for a fixed n. Let g_n be the maximum number of disks in a chain in $\{G\}_n$.

Form a set $\{D\}$ of chains of disks in R each separating a boundary component γ_j from the rest of the boundary of the corresponding $R_{n+1}-R_n$. Denote by $\{D\}_n$ the set of those chains in $\{D\}$ that encircle the components γ_j $(j=1,\ldots,\mu_n)$ belonging to the boundary of R_n. Let d_n be the maximum number of disks in any chain in $\{D\}_n$.

Choose the set $\{G\}\cup\{D\}$ such that the requirements (α), (β), and (γ) listed in 3 A for a regular chain set are fulfilled.

11 D. Test for Circular Mappings. Set

$$k_n=\max(g_n, d_n).$$

We shall show:

THEOREM. *Let R be an arbitrary plane region. If there exists an exhaustion by noncompact subregions such that*

$$\sum\frac{1}{k_n}=\infty$$

then R can be conformally mapped onto a region whose boundary components are circles and points.

By Theorem 7 E we can map R conformally onto a horizontal slit region. The required properties of the images of R, $\{R_n\}$, β_{ni}, γ_j, $\{G\}$, $\{G\}_n$, $\{D\}$, $\{D\}_n$, $\{G\}\cup\{D\}$, g_n, d_n, and k_n are not changed under this mapping. We may thus assume from the outset that R itself is a horizontal slit region.

Reflect R_1 about each proper slit γ_j $(j=1,\ldots,\mu_1)$, i.e. slit not reducing to a point. The resulting surface $\tilde{R}_1$ is (μ_1+1)-sheeted. In general construct $\tilde{R}_n$ by first reflecting R_n about each proper slit γ_j $(j=1,\ldots,\mu_n)$, then the resulting surface about each free copy of γ_j, and by repeating the process until a total of n reflections has been made. We denote the limiting surface by

$$\tilde{R}=\lim_{n\to\infty}\tilde{R}_n.$$

The chain set $\{G\}\cup\{D\}$ is carried by the reflections to a chain set on $\tilde{R}$ in an obvious manner, the numbers k_n remaining unchanged. Thus by Theorem 3 B, $\tilde{R}\in O_{AD}$. Observe that $\tilde{R}$ is planar and symmetric about each proper slit γ_j.

Now map $\tilde{R}$ conformally by a univalent analytic function w_1. Let R_w be the image of the initial copy R under this mapping, and C_j the image of an isolated boundary component γ_j of R. A reflection of $\tilde{R}$ about a proper slit γ_j gives a self-mapping of $\tilde{R}$ such that each point z goes to its conjugate $\bar{z}$. The function $w_2(z)=w_1(\bar{z})$ gives an indirectly conformal mapping so that the function $\overline{w}_2(z(w_1))=\overline{w}_2(w_1)$ is analytic and hence by Theorem 8 C a linear transformation.

The reflection of $\tilde{R}$ interchanges the regions inside and outside C_j while each point of C_j remains fixed. Choose three points on C_j and let C_j' be the circle through them. The linear transformation $\overline{w}_2(w_1)$ maps C_j' onto a circle and so does therefore the reflection $w_2(w_1)$. Since the three points remain fixed C_j' is mapped onto itself and it is readily seen that $C_j=C_j'$.

We conclude that the image of every proper slit γ_j of R is a circle in the w-plane. If γ_j is an isolated point then clearly its image is a point in the w-plane.

Let γ_w^* be the set of boundary components of R_w which do not belong to the set of images of the components γ_j discussed above. Denote by R_w^* the complement of γ_w^* in the w-plane. The image $\{G_w\}$ of $\{G\}$ on R is a regular chain set on R_w^*. Since

$$\sum \frac{1}{g_n} \geq \sum \frac{1}{k_n} = \infty$$

Theorem 3 B again gives $R_w^* \in O_{AD}$. By Theorem 8 C, γ_w^* is of absolute area zero and therefore a totally disconnected point set.

We have shown that R can be mapped conformally onto a plane region whose boundary consists of circles and points.

11 E. An Application. The following is a concrete example of a region with infinitely many accumulation components of continuum boundary components. To our knowledge Theorem 11 D is in this case the only means of establishing the existence of the Koebe mapping.

EXAMPLE. *The Schottky covering surface of the surface obtained by removing an arc from a closed Riemann surface can be conformally mapped onto a region bounded by circles and points.*

Let S be a closed surface of genus g. Remove an arc α from S and set $G=S-\alpha$. Cut G along disjoint nondividing cycles $a_1, \ldots, a_g$ such that the boundary of the resulting planar surface R_0 consists of the components $a_1, a_1^{-1}, \ldots, a_g, a_g^{-1}$, and α. Along each a_i attach to R_0 a new duplicate of R_0 along a_i^{-1}; and along each a_i^{-1} a new duplicate along a_i so as to form a $(2g+1)$-sheeted planar unramified covering surface R_2 of G. In general form R_{n+1} by attaching to R_n $(n=0, 1, \ldots)$ along each a_i on ∂R_n another copy of R_0 along a_i^{-1}; and along each a_i^{-1} on ∂R_n another copy of R_0 along a_i. The planar surface $R=\bigcup_0^\infty R_n$ is the *Schottky covering surface* of G.

On G encircle α by a disk chain D and let d be the number of disks in D. Cover each a_i in G by a disk chain G_i consisting of g_i disks such that $(D, G_1, \ldots, G_g)$ satisfies (α), (β), and (γ) of 3 A. Set

$$k = \max(d, g_1, \ldots, g_g).$$

Duplicates of D encircle the copies of α on ∂R_n, and duplicates of G_i cover the cycles a_i and a_i^{-1} on ∂R_n. Thus we have a regular chain set for R relative to $\{R_n\}$ with $k_n = k$. Here $\sum 1/k_n = \infty$ and the assertion follows.

It is clear that under any univalent mapping of R the duplicates of α, which are all isolated boundary components of R, converge to an *infinite* set of accumulation components.

§ 3. Covering Surfaces of the Sphere

From the preceding discussion of the extended plane we proceed to a study of its covering surfaces.

In 12 we consider complete covering surfaces with a finite number of projections of branch points. We represent them as polyhedric surfaces and give a related O_{AD}-test.

An application of this test to line complexes of planar surfaces is given in 13 and to strip complexes of surfaces of positive genus in 14.

In 15 we discuss punctured surfaces, i.e. complete covering surfaces from which sets of points have been deleted. This gives a generalization of the results in 12. In 16 we consider general covering surfaces of the sphere which have a finite number of sheets.

12. Finite Sets of Projections of Branch Points

12 A. Surface Elements. Let R be a Riemann surface represented as a complete covering surface of the sphere. Let $\bar{P}$: $z = z_0$ be a point in the sphere, and $\bar{U}$ a (small) disk with center $\bar{P}$. The subset of R covering $\bar{U}$ consists of a finite or infinite number of disjoint "surface elements" U. We consider three kinds of such elements:

(a) *plane elements:* U is a 1-sheeted disk;

(b) *algebraic elements:* U is a multiple disk with a finite number of sheets;

(c) *logarithmic elements:* U is a multiple disk with infinitely many sheets.

The point $\bar{P}$ in the base sphere is called a *regular point* if for a sufficiently small disk $\bar{U}$ about $\bar{P}$ every surface element U over $\bar{U}$ is a plane element. In other cases we call $\bar{P}$ a *singular point*. In this section we consider open covering surfaces R whose surface elements are of categories (a), (b), and (c) and such that *the set of singular points is finite.*

12 B. Polyhedric Representation. Let $z=z_\nu$ $(\nu=1, \ldots, q)$ be the singular points. Take a piecewise analytic Jordan curve L through these points. The set of points of R over L divides R into infinitely many halfsheets which one can view as "polygons." Thus R is represented as a *polyhedric surface.*

The half-sheets are the *faces* S, the arcs above $(z_\nu z_{\nu+1})$ $(\nu=1, \ldots, q;$ $z_{q+1}=z_1)$ the *edges*, and the end points of the arcs *vertices* of the polyhedric surface. It may happen that a "vertex" lies on the "boundary" of R (logarithmic branch point), in which case the faces adjacent to this vertex form an infinite sequence. Fix an arbitrary face S_1 as the *first generation.* The faces S_{2i} $(i=1, \ldots, t_2)$ different from S_1 which have common edges with S_1 constitute the *second generation.* In general, the faces which have common edges with the $(n-1)$st generation but do not belong to the $(n-2)$nd generation form the nth *generation.*

Let T_n be (the closure of) the union of faces of at most the nth generation; we call it the nth subpolyhedron. The boundary γ_n of T_n lies over L and consists of disjoint closed curves γ_{ni} $(i=1, \ldots, \nu_n)$, each made up of a finite number κ_{ni} of edges of the polyhedric surface R.

In each γ_{ni} there are κ_{ni} vertices E_j $(j=1, \ldots, \kappa_{ni})$. At each vertex E_j there is a sequence Z_j of faces belonging to T_n with the vertex E_j in common. Let s_j be the number of faces in Z_j and set

$$s_{ni}=\sum_{j=1}^{\kappa_{ni}} s_j.$$

12 C. O_{AD}-Test. For a fixed n write

$$\sigma_n=\max\{s_{ni};\, i=1, \ldots, \nu_n\}.$$

We have the following purely enumerative O_{AD}-test:

THEOREM. *Let R be an open complete covering surface of the sphere with a finite number of singular projection points. If the maximal numbers σ_n of faces in a polyhedric representation of R grow so slowly that*

$$\sum \frac{1}{\sigma_n}=\infty \tag{58}$$

then R belongs to the class O_{AD}.

The proof will be given in 12 D – 12 G.

12 D. Construction of an Exhaustion. Without loss of generality we assume that the closed disks

$$K_\nu^0: \ |z-z_\nu|\le 1 \qquad (\nu=1,\ldots,q)$$

are disjoint. We deform L so that $L\cap K_\nu^0$ is the diameter of K_ν^0 for each ν and the curve L thus deformed is of class C^1.

We define an exhaustion $\{R_n\}$ $(n=1,2,\ldots)$ of R as follows. The set of points in T_n over K_ν^0 consists of a finite number of components. They are contained in the interior of T_n or else have a point E_j of γ_n on their boundary. Each component G_j^0 of the latter kind consists of s_j half-disks, where s_j is the number of faces in the sequence Z_j. Let G_j^s be the part of G_j^0 which lies above the disk

$$K_\nu^s: \ |z-z_\nu|\le\frac{1}{2^s},$$

with $s=s_j$. Let

$$R_n=T_n-\bigcup_j G_j^s,$$

where j extends over every E_j on γ_n. It is easy to see that $\{R_n\}$ is an exhaustion of R.

The boundary Γ_n of R_n consists of ν_n disjoint curves Γ_{ni} $(i=1,\ldots,\nu_n)$, each composed alternately of two kinds of arcs α and β. Each α lies over an arc of L. Each β lies either over the entire boundary or half of the boundary C_ν^s of a disk K_ν^s and consists of s_j semicircles δ.

12 E. Disk Chains Covering α. For a given α let z_ν and z_μ be the projections of those vertices E_j of T_n whose associated sets G_j^s are joined by α. Denote by s and t the numbers of faces associated with those sets.

Consider the projection $\bar\alpha$ of α on the sphere. Let d be the minimum of the mutual distances of those parts of L which lie outside of the closed disks K_ν^0 $(\nu=1,\ldots,q)$. Trace the curves α_1 and α_2 on each side of $\bar\alpha$ parallel to it at distance $\min(d/3,1)$, with end points on the boundaries C_ν^0 and C_μ^0 of K_ν^0 and K_μ^0. Connect these end points by suitable radii $\{\Lambda\}$ to the corresponding centers z_ν and z_μ. Let A_0 be the simply connected region bounded by the curves α_1, α_2, (segments of) the radii $\{\Lambda\}$, and the circles C_ν^1, C_μ^1 (of radius $\frac{1}{2}$).

Similarly denote by A_{-m} $(m=1,\ldots,s-1)$ the simply connected region in K_ν^0 bounded by C_ν^{m-1}, C_ν^{m+1} and the radii $\{\Lambda\}$, and by A_m $(m=1,\ldots,t-1)$ the simply connected region in K_μ^0 bounded by C_μ^{m-1}, C_μ^{m+1}, and the radii $\{\Lambda\}$. Then the (open) arc $\bar\alpha$ is covered by the sequence

$$A_{-(s-1)}, A_{-(s-2)}, \ldots, A_{-1}, A_0, A_1, \ldots, A_{t-2}, A_{t-1}.$$

In the set of components of the part of R lying above these regions there exists a minimal subset $\{A\}$ which covers α.

12 F. Disk Chains Covering β. We turn to an arc β. Let s be the number of faces belonging to the corresponding sequence. Each of the s semicircles δ constituting β has for its projection $\bar{\delta}$ on the plane a half of the circle C_ν^s. Cover $\bar{\delta}$ by the annular sector B bounded by C_ν^{s-1}, C_ν^{s+1}, and suitable radii $\{\Lambda\}$. Some component B of the part of R over B covers δ. In this manner we obtain a set $\{B\}$ of simply connected subregions of R covering the arc β.

12 G. Enumeration. We have constructed the covering of Γ_{ni} by the sets $\{A\}$ and $\{B\}$ of regions which we shall refer to as disks. We denote by C_{ni} the union of these disks. For $i=1, \ldots, v_n$ and $n=1,2,\ldots$ the set $\{C_{ni}\}$ is a regular chain set on R (cf. 3 A). In fact,

(a) the $\bar{C}_{ni}$ are disjoint by pairs,

(b) there exists an integral covering number $N\,(=2)$ in the sense of (β) in 3 A such that each disk in $\{A\}\cup\{B\}$ meets at most N other disks in $\{A\}\cup\{B\}$,

(c) since the number of neighborhood relations between (nonsimilar) disks A or B is finite there exists a covering constant d in the sense of (γ) in 3 A.

In each chain the number of A-disks between two successive A_0 is twice the number of B-disks which in turn is equal to the number s_j of faces in the associated sequence. Thus the total number l_{ni} of elements in a chain satisfies

$$l_{ni} \leq 3\, s_{ni}$$

where

$$s_{ni} = \sum_{j=1}^{\kappa_{ni}} s_j \leq \sigma_n .$$

Therefore

$$\lambda_n = \max\{l_{ni};\, i=1, \ldots, v_n\} \leq 3\,\sigma_n$$

and

$$\sum \frac{1}{\lambda_n} \geq \frac{1}{3} \sum \frac{1}{\sigma_n} = \infty .$$

By Theorem 3 B, $R \in O_{AD}$. The proof of Theorem 12 C is herewith complete.

*13. Application to Planar Surfaces

13 A. Line Complexes. If R is planar then its polyhedric representation has a dual figure called a *line complex*, *topological tree*, or *Speiser graph*. It is constructed as follows. Choose a point in the interior of each face of the polyhedric surface, connect these points by arcs, each crossing an edge, and draw the resulting figure in the plane. To the faces of the polyhedric surface correspond the nodes of the complex, the edges are represented by line segments joining the nodes, and the vertices have for their images the components of the complement of the line complex with respect to the extended plane.

To a polyhedric subsurface T_n (see 12 B) there corresponds a subcomplex T_n. With those faces of T_n which are adjacent to a boundary component γ_{ni} is associated a *sequence* γ_{ni} *of boundary nodes*. An illustration of T_4 is given in Fig. 3 where $q=4$ and T_1 is the node at the center.

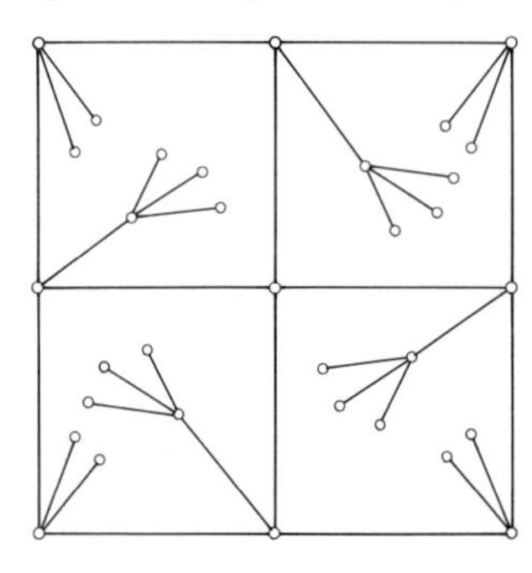

Fig. 3

The numbers of nodes in T_1, T_2, T_3, T_4 are 1, 5, 13, 33, and the corresponding numbers of boundary nodes are 1, 4, 8, 20. There are 4 sequences γ_{4i} of boundary nodes in T_4, each adjacent to one of the four regions inside the square, and each consisting of 5 nodes.

13 B. O_{AD}-Test. Set

$$k_n = \max_i k_{ni},$$

where k_{ni} is the number of nodes in the boundary sequence γ_{ni}. As an application of Theorem 12 C we see at once:

If the maximum numbers of nodes in the boundary node sequences of a line complex grow so slowly that

$$\sum \frac{1}{k_n} = \infty$$

then R belongs to the class O_{AD}.

In fact the correspondences described in 13 A give

$$k_{ni} \geq \frac{s_{ni}}{q}.$$

Therefore $k_n \geq \sigma_n/q$ and $\sum 1/\sigma_n \geq (1/q) \sum 1/k_n = \infty$.

13 C. Edge Sequences with Bounded Vertex Numbers. As in 12 B let κ_{ni} be the number of vertices on a boundary component γ_{ni}. As another application of Theorem 12 C we obtain:

If the numbers of vertices on boundary components are uniformly bounded,

$$\kappa_{ni} \leq N,$$

then R belongs to the class O_{AD}.

We again denote by σ_n the maximum of the sums of numbers s_j of faces in sequences Z_j adjacent to the γ_{ni}. By assumption there are at most N edges in γ_{ni}. In passing from T_n to T_{n+1} at most one new face is attached along each edge on the boundary of T_n. Since R is planar and therefore no uniting of different boundary curves γ_{ni} occurs as T_n increases to T_{n+1}, the increment of the sum of the numbers of faces is q for each such edge, that is, a total of qN, and this is independent of n. Therefore

$$\sigma_{n+m} \leq \sigma_n + mqN$$

for each m and

$$\sum_{m=1}^{\infty} \frac{1}{\sigma_{n+m}} \geq \sum_{m=1}^{\infty} \frac{1}{\sigma_n + mqN} = \infty.$$

By Theorem 12 C, $R \in O_{AD}$.

13 D. Finite Sets of Branch Points. The following observation is a direct consequence of 13 C:

If the number of branch points is finite then R belongs to O_{AD}.

In fact when tracing γ_{ni} we enter the boundary of a face from the boundary of another face only when we pass through a branch point. Since these are finite in number the number of faces adjacent to γ_{ni} is bounded. Every face having at most $q-1$ edges on γ_{ni}, the number of edges on γ_{ni} is bounded. Thus by 13 C, $R \in O_{AD}$.

13 E. Periodic Ends. We can also apply 13 C to the case of so-called periodic ends. Here we consider the following example. Let R be a planar surface which is represented by a line complex with M periodic ends ($M=2$ in Fig. 4). Let each period in turn consist of M periodic ends, and each of these periods again of M periodic ends and so forth.

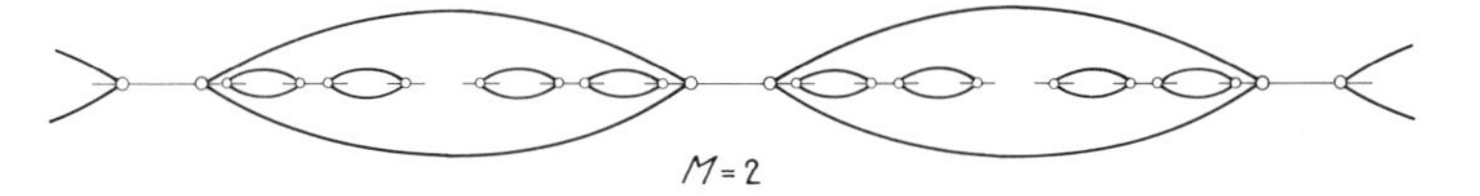

Fig. 4

Then R is an infinitely connected surface with infinitely many logarithmic branch points but the number of their projections is finite. In view of 13 C we see at once that $R \in O_{AD}$.

*14. Nonplanar Surfaces

14 A. Strip Complexes. Line complexes are usually used for planar surfaces. For surfaces of positive genus some line segments intersect

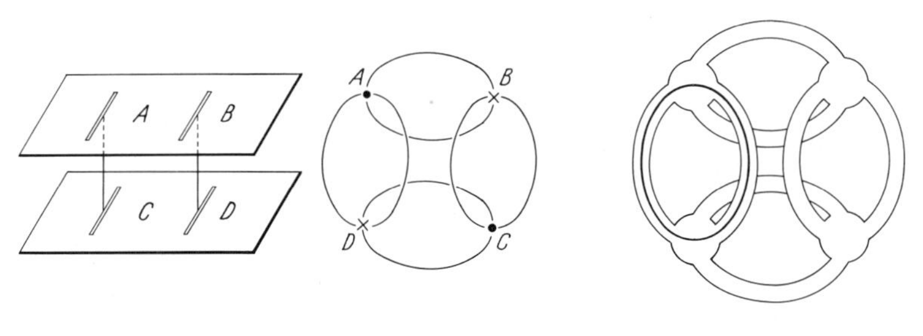

Fig. 5a Fig. 5b

each other. Nevertheless line complexes retain their illustrative significance in many cases of positive genus if the line segments are replaced by arcs which at intersections lie in different sheets. Fig. 5a shows the line complex of a torus (closed Riemann surface of genus 1). The half-sheets are denoted by A, B, C, and D.

In order to distinguish different node sequences it is advantageous to enlarge the segments to narrow strips, and the nodes to small disks. Fig. 5b gives the strip complex of the torus thus obtained. The four contours combine the nodes into four different sequences none of which is a boundary sequence.

14 B. Example. In Fig. 5c the nodes of a subcomplex T_5 are joined into two different sequences and each of these is a boundary sequence.

It is easily seen that the result in 13 B holds also for strip complexes. For example in the strip complex of Fig. 5c

$$k_{ni} = 2n-1$$

and $k_n = 2n-1$ so that

$$\sum \frac{1}{k_n} = \infty.$$

Thus the surface must belong to O_{AD}.

From the strip complex we can easily read off *the genus* q_n of the subcomplex T_n. It is the maximal number of disjoint cycles which do not

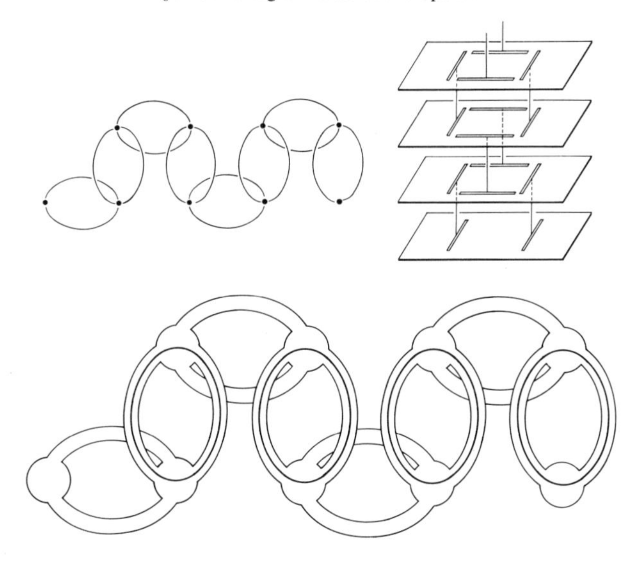

Fig. 5c

divide the strip complex T_n. In Fig. 5b the genus is 1 as indicated by one cycle. In Fig. 5c the genus g_5 of T_5 is 4, corresponding to 4 cycles.

We shall return to strip complexes in 19 G.

*15. Punctured Surfaces

15 A. Polyhedric Representation of a Punctured Surface. We turn to more general surfaces R, obtained from those considered in 12 A by removing a point set S^*. We assume that the closure S of the projection of S^* in the plane is totally disconnected and has a positive distance from the finite set of projections $z=z_\nu$ $(\nu=1, \ldots, q)$ of the branch points.

In 12 A we considered surfaces with three kinds of surface elements: (a) plane, (b) algebraic, and (c) logarithmic. We now include a fourth kind:

(d) *plane boundary elements: U* consists of a proper subset of a plane disk.

For *punctured* covering surfaces R in the above sense we also consider polyhedric representations as in 12 B. Take the piecewise analytic Jordan curve L through the z_ν such that the distance between L and S is positive. The set of points of R over L gives a polyhedric representation of R but here we must also consider faces which are *punctured* half-sheets.

In the same manner as in 12 B we arrange the faces into *generations*. Let T_n be the union of faces of at most the nth generation. We denote by

γ_n the part of the boundary of T_n which lies over L. It consists of a finite number ν_n of disjoint Jordan curves γ_{ni}. In each γ_{ni} there are κ_{ni} vertices E_j $(j=1, \ldots, \kappa_{ni})$. To each vertex E_j there corresponds a sequence Z_j of faces belonging to T_n and having E_j as a common vertex. Let s_j be the number of faces in Z_j and set

$$s_{ni} = \sum_{j=1}^{\kappa_{ni}} s_j.$$

As in 12 C we write

$$\sigma_n = \max\{s_{ni};\, i=1, \ldots, \nu_n\}$$

for each fixed n.

15 B. O_{AD}-Test. Let F be the complementary region of the set S in the sphere. We construct regular chain sets H_{ni} $(n=1, 2, \ldots; i=1, \ldots, \mu_n)$ in F (see 3). Let h_{ni} be the number of (parametric) disks belonging to the chain H_{ni} and set

$$h_n = \max\{h_{ni};\, i=1, \ldots, \mu_n\}$$

for each fixed n. We shall prove:

THEOREM. *If a punctured covering surface R satisfies the conditions*

$$\sum \frac{1}{\sigma_n} = \infty, \qquad \sum \frac{1}{h_n} = \infty$$

then it belongs to the class O_{AD}.

Let t be the local parameter and $f(t) = u(t) + i\,v(t)$ an analytic function on R with a finite Dirichlet integral $D(u)$. We have to show that u reduces to a constant. We denote by D_E the Dirichlet integral $D_E(u)$ of u taken over a subregion E of R. The proof that $D_R = 0$ will be given in 15 C – 15 F.

15 C. Chains Relative to $\{\sigma_n\}$. Without loss of generality we assume that the closed disks

$$K_\nu^0\colon |z - z_\nu| \le 1 \qquad (\nu = 1, \ldots, q)$$

are disjoint from each other and from S. We may then choose L as in 12 D, with the additional property that the regions A_0 and the set S are disjoint (see 12 E). As in 12 D we construct an exhaustion of R by removing the sets G_j^s from T_n and as in 12 E and 12 F cover the boundaries of the remaining regions by a regular chain set $\{A\} \cup \{B\}$. We denote the chains by G_{ni} $(n=1, 2, \ldots; i=1, \ldots, \nu_n)$, with g_{ni} the number of disks in G_{ni}, and set

$$g_n = \max\{g_{ni};\, i=1, \ldots, \nu_n\}.$$

The union of the chains G_{ni} for a fixed n is denoted by

$$G_n = \bigcup_{i=1}^{\nu_n} G_{ni}.$$

Let ε be an arbitrary positive number and consider the natural numbers $n = n_\nu$ $(\nu = 1, 2, \ldots)$ for which

$$\frac{1}{g_n} \leq \frac{D_{G_n}}{\varepsilon}.$$

Since $\sum_{n=1}^{\infty} D_{G_n} \leq D_R < \infty$

$$\sum_{n=n_\nu;\, \nu=1,2,\ldots} \frac{1}{g_n} < \infty.$$

From the construction of the chains G_{ni} it follows that $g_n \leq 3\sigma_n$ (see 12 G) and therefore

$$\sum_{n=1}^{\infty} \frac{1}{g_n} \geq \frac{1}{3} \sum_{n=1}^{\infty} \frac{1}{\sigma_n} = \infty.$$

If we remove from this series the terms with $n = n_\nu$ the remaining series on the left is still divergent. Thus there are infinitely many natural numbers $n \neq n_\nu$. From the sequence T_n $(n = 1, 2, \ldots)$ we remove all T_{n_ν} and rearrange the numbering so that

$$g_n D_{G_n} < \varepsilon$$

for all $n = 1, 2, \ldots$.

15 D. Chains Relative to $\{h_n\}$. We may assume that the chains H_{ni} are disjoint from the disks K_ν^0 and the regions A_0. We use the same notation H_{ni} for the chains on T_n whose "projections" are the original H_{ni} and which separate the set of boundary points of T_n over S from some fixed point of R. Let

$$H_n = \bigcup_{i=1}^{\mu_n} H_{ni}$$

and consider the natural numbers $n = n_\mu$ $(\mu = 1, 2, \ldots)$ for which

$$\frac{1}{h_n} \leq \frac{D_{H_n}}{\varepsilon}.$$

As in 15 C we conclude that there are infinitely many numbers $n \neq n_\mu$. Again we remove from $\{T_n\}$ the subsequence $\{T_{n_\mu}\}$ $(\mu = 1, 2, \ldots)$ and renumber the remaining sequence such that

$$g_n D_{G_n} < \varepsilon, \qquad h_n D_{H_n} < \varepsilon$$

for all $n = 1, 2, \ldots$.

15 E. Construction of Exhaustion. With each subpolyhedron T_n we have thus associated a chain set

$$E_n = G_n + H_n.$$

For a moment we denote by E_{ni} the chains G_{ni} and H_{ni} belonging to E_n; for varying i, E_{ni} runs over all G_{ni} and H_{ni}. Then $\{E_{ni}\}$ is a regular chain set on R. Let N be its covering number in the sense of (β) in 3 A, and d its covering constant in the sense of (γ) in 3 A. We denote by l_{ni} the number of disks in E_{ni}.

In each chain E_{ni} take a piecewise analytic Jordan curve Γ_{ni} which separates the two contours of E_{ni} and can be decomposed into arcs Λ_j ($j=1,\ldots,l_{ni}$) contained in the parametric disks D_j: $|t_j - t_j^0| < 1$ corresponding to the chain E_{ni}, at distances $>d$ from the boundaries of the disks, and having length <2.

For each fixed n the Γ_{ni} constitute the boundary Γ_n of a relatively compact subregion R_n of R. The sequence $\{R_n\}$ is an exhaustion of R.

15 F. Evaluation of D_R. Let u_i be the value of u (cf. 15 B) at some fixed point on Γ_{ni}. Then

$$D_{R_n} = \int_{\Gamma_n} u\, dv = \sum_i \int_{\Gamma_{ni}} u\, dv = \sum_i \int_{\Gamma_{ni}} (u - u_i)\, dv.$$

Observe that

$$\max |u - u_i| \le \int_{\Gamma_{ni}} \left| \frac{\partial u}{\partial s} \right| ds \le \int_{\Gamma_{ni}} \left| \frac{df}{dt} \right| ds$$

and similarly

$$\int_{\Gamma_{ni}} |dv| = \int_{\Gamma_{ni}} \left| \frac{\partial v}{\partial s} \right| ds \le \int_{\Gamma_{ni}} \left| \frac{df}{dt} \right| ds$$

where ds is the arc element of Γ_{ni}. We infer that

$$D_{R_n} \le \sum_i \left(\int_{\Gamma_{ni}} \left| \frac{df}{dt} \right| ds \right)^2 = \sum_i \left(\sum_j \int_{\Lambda_j} \left| \frac{df}{dt_j} \right| ds \right)^2.$$

We estimate $|df/dt|$ in the same manner as in the proof of Theorem 3 B and obtain by Schwarz's inequality

$$D_{R_n} \le \frac{4N}{\pi d^2} \sum_i l_{ni} D_{E_{ni}}. \tag{59}$$

In terms of the original notation G and H for the E_{ni} this inequality takes the form

$$D_{R_n} \le c\left(\sum_i g_{ni} D_{G_{ni}} + \sum_i h_{ni} D_{H_{ni}}\right)$$

where c is a constant depending only on N and d. Hence

$$D_{R_n} \leq c(g_n D_{G_n} + h_n D_{H_n}).$$

Since $g_n D_{G_n}$, $h_n D_{H_n} < \varepsilon$ we have $D_{R_n} \leq 2c \cdot \varepsilon$ for all n and conclude that $D_R = 0$.

*16. Finite Sets of Sheets

16 A. Regular and Singular Projections. Finally we consider covering surfaces R with a finite number of sheets. As before let $\bar{P}$: $z = z_0$ be a point of the plane and consider a closed disk $\bar{U}$ about $\bar{P}$. If the set of points of R over a sufficiently small $\bar{U}$ consists of a finite number μ of disjoint closed disks then we call $\bar{P}$ a *regular* projection point. Otherwise we call $\bar{P}$ *singular*.

We make no assumptions concerning the branch points or the puncturing set but assume only that *the set S of singular projection points is totally disconnected.*

Let H_{ni} $(n = 1, 2, \ldots; i = 1, \ldots, \nu_n)$ be a regular chain set on the complement F of S. Denote by h_{ni} the number of parametric disks belonging to H_{ni} and set

$$h_n = \max\{h_{ni}; i = 1, \ldots, \nu_n\}$$

for each fixed n.

16 B. O_{AD}-Test. We claim:

THEOREM. *Let R be an arbitrary covering surface of the sphere with a finite number μ of sheets and with a totally disconnected set S of singular projection points. If some exhaustion of the complement F of S in the sphere admits a regular chain set with*

$$\sum \frac{1}{h_n} = \infty$$

then R belongs to the class O_{AD}.

Let $\{F_n\}$ be the exhaustion of F associated with the chain set $\{H_{ni}\}$. The boundary γ_n of F_n consists of ν_n components γ_{ni} $(i = 1, \ldots, \nu_n)$ covered by the corresponding chains H_{ni}.

The point set in R over F_1 is composed of a finite number of disjoint regions. Let R_1 be one of them. In general let R_n be the region over F_n which contains R_{n-1}.

The boundary Γ_n of R_n is made up of a finite number of components Γ_{nj}, each over some γ_{ni}. A chain G_{nj} on R over H_{ni} covers Γ_{nj}. The number g_{nj} of disks in G_{nj}, each of which lies over the corresponding disk of H_{ni}, must satisfy $g_{nj} \le \mu h_{ni}$. If we set $g_n = \max\{g_{nj}; j\}$, then $g_n \le \mu h_n$ and consequently

$$\sum \frac{1}{g_n} \ge \frac{1}{\mu} \sum \frac{1}{h_n} = \infty.$$

Inequality (59) is valid for a general regular chain set. On setting $l_{ni} = g_{ni}$ and $E_{ni} = G_{nj}$ we therefore obtain

$$D_{R_n} \le c \cdot g_n D_{G_n}$$

where $G_n = \bigcup_j G_{nj}$ and D is the Dirichlet integral of an arbitrarily given analytic function on R. Hence

$$D_R \ge \sum_1^\infty D_{G_n} \ge \frac{1}{c} D_{R_1} \sum_1^\infty \frac{1}{g_n},$$

which shows that if the analytic function is not constant then $D_R = \infty$, i.e. $R \in O_{AD}$.

§ 4. Covering Surfaces of Riemann Surfaces

From ramified covering surfaces of the sphere we now turn to smooth covering surfaces of arbitrary Riemann surfaces.

For complete smooth covering surfaces of closed surfaces we shall derive a general result: all Abelian covering surfaces belong to the class O_{AD}. No. 17 is preparatory and the main theorem is given in 18.

The discussions in 17 and 18 will be generalized to covering surfaces of open Riemann surfaces in 19.

17. Preliminaries

17 A. Problem. Given a Riemann surface R we shall associate with each set of nondividing cycles of R a smooth complete covering surface $\tilde{R}$ of R, called an Abelian covering surface. The class $\{\tilde{R}\}$ so constructed contains in particular the Schottky covering surface and the covering surface of integral functions. We shall first schematically describe the construction method for a torus and a double torus.

17 B. Covering Surfaces of the Torus. Let R be a closed Riemann surface of genus 1. For an arbitrary nondividing loop cut a of R we construct a smooth complete covering surface $\tilde{R}$ of R as follows. Let R' be the doubly connected planar surface obtained from R by removing the cut a. Take infinitely many copies of R'. To the first copy attach a new *generation*, i.e. along a and a^{-1} of the first R' fuse new copies of R' along their contours a^{-1} and a respectively. Denote by $\tilde{R}_1$ the resulting 3-sheeted covering surface. In general attach to $\tilde{R}_n$ a new generation, i.e. along each of the contours of $\tilde{R}_n$ a new copy of R'. Let $\tilde{R}_{n+1}$ be the resulting surface. The sequence $\tilde{R}_1 \subset \tilde{R}_2 \subset \cdots$ corresponds topologically to a nested sequence of concentric circular annuli. We call

$$\tilde{R}(a) = \lim_{n\to\infty} \tilde{R}_n$$

the covering surface of R corresponding to the cut a. It is a doubly connected planar surface.

17 C. Pair of Cuts. Let b be a nondividing loop cut conjugate to a. The cuts a and b constitute *a pair $a+b$ of cuts* which form a homology base for R. We shall always assume in the sequel that for any pair $a+b$ of conjugate cuts, *$a \cap b$ consists of only one point.*

We associate a smooth complete covering surface of R with the pair $a+b$ as follows. Let R' be the surface obtained from R by removing $a+b$. It is a simply connected planar region with boundary $aba^{-1}b^{-1}$. For a topological model of R' we choose a rectangle in the plane. Take infinitely many copies of R'. To the first copy R' attach a new generation of 8 copies of R' in such a way that the boundary of the first R' is interior to the resulting 9-sheeted surface $\tilde{R}_1$. In general attach a new generation of $8(n+1)$ copies of R' to $\tilde{R}_n$ so that the boundary of $\tilde{R}_n$ is interior to the resulting $\tilde{R}_{n+1}$ $(n=1, 2, \ldots)$. On the boundary of $\tilde{R}_n$ the total number s_n of copies of a, b, a^{-1}, and b^{-1} is obviously

$$s_n = 4(1+2n).$$

The limiting surface

$$\tilde{R}(a+b) = \lim_{n\to\infty} \tilde{R}_n$$

is a simply connected planar surface, which we call *the covering surface of R corresponding to the pair $a+b$.*

17 D. Covering Surfaces of the Double Torus. Now let R be a closed Riemann surface of genus 2 and a_1+b_1 and a_2+b_2 two disjoint pairs of cuts of R which form a homology base for R. With the various combinations of a_j and b_j $(j=1, 2)$ we associate smooth unbounded covering surfaces of R described in 17 E – 17 I.

17 E. Covering Surface $\tilde{R}(a_1)$. The surface R' obtained from R by cutting along a_1 has two contours and one handle. A construction analogous to that in 17 B gives the covering surface $\tilde{R}(a_1)$ corresponding to the cut a_1.

17 F. Covering Surface $\tilde{R}(a_1+b_1)$. The surface R' obtained from R by cutting along a_1+b_1 has only one contour and carries one handle. The covering surface $\tilde{R}(a_1+b_1)$ corresponding to the pair a_1+b_1 is constructed in the same manner as the surface in 17 C.

17 G. Covering Surface $\tilde{R}(a_1, a_2)$. Let R' be the surface of connectivity 4 obtained from R by cutting along the disjoint cycles a_1 and a_2. To each contour of the first copy R' attach a new copy of R' in such a manner that a_j and a_j^{-1} are identified ($j=1, 2$). By continuing the process indefinitely we obtain in the limit the covering surface $\tilde{R}(a_1, a_2)$ of R corresponding to the cuts a_1 and a_2.

17 H. Covering Surface $\tilde{R}(a_1+b_1, a_2)$. The surface R' obtained from R by cutting along a_1, b_1 and a_2 is triply connected. In the same manner as in 17 C we attach copies of R' along $a_1, a_1^{-1}, b_1, b_1^{-1}$ and at the "vertices." We also attach copies of R' along a_2 and a_2^{-1}. Thus the first generation adjacent to the first copy contains 10 copies. On repeating the process we obtain in the limit the covering surface $\tilde{R}(a_1+b_1, a_2)$ of R corresponding to the pair a_1+b_1 and the cut a_2.

17 I. Covering Surface $\tilde{R}(a_1+b_1, a_2+b_2)$. The surface R' obtained from R by cutting along a_1+b_1 and a_2+b_2 is doubly connected. The first generation of copies to be attached contains 16 copies. We obtain in the limit the covering surface $\tilde{R}(a_1+b_1, a_2+b_2)$ of R corresponding to the pairs a_1+b_1 and a_2+b_2.

18. Covering Surfaces of Closed Surfaces

18 A. Definitions. Let R be an arbitrary closed Riemann surface of genus g. In the same manner as in the cases $g=1, 2$ we can take g disjoint pairs of cuts a_j+b_j ($j=1, \ldots, g$) in R which form a homology base for R, and for each subset (a, b) of these cuts we can construct the corresponding smooth complete covering surface $\tilde{R}((a, b))$ of R. For a more formal axiomatic approach cf. II.3.

Let φ be an Abelian integral of the first kind on R (e.g. WEYL [1]), with nonvanishing periods along each cut in (a, b). The integral of φ becomes single-valued if it is considered on the covering surface of R corresponding to the set of cuts conjugate to those in (a, b). Surfaces with this property are called *Abelian covering surfaces*. The surfaces in 17 exhaust all essentially distinct Abelian covering surfaces of a torus or a double torus. In particular *the covering surface of integral functions* in the sense of WEYL belongs to this category. This covering surface is by definition the weakest covering surface on which every Abelian integral of the first kind is single-valued. The surface considered in 17 I is an example. This surface is the covering surface of all covering surfaces of R under consideration. Among them it may thus be said to have the strongest ideal boundary. Note that the universal covering surface is stronger than the covering surface of integral functions but is not needed to make all Abelian integrals single-valued.

The Schottky covering surface is by definition the planar covering surface which has the weakest boundary among planar covering surfaces, i.e. every planar covering surface of R under consideration is a covering surface of the Schottky covering surface of R.

18 B. Main Theorem. For covering surfaces of closed surfaces we can now state a complete result:

THEOREM. *All Abelian covering surfaces of a closed Riemann surface belong to the class O_{AD}. In particular the covering surface of integral functions and the Schottky covering surface have this property.*

Let R be the given closed Riemann surface of genus g. Since $\tilde{R}=R$ if $g=0$ we may assume that $g \geq 1$. Let (a, b) be a set of disjoint pairs of cuts or single cuts on R and let $\tilde{R}$ be the covering surface corresponding to (a, b). Cover each a and b by a narrow chain of disks (parametric regions) such that for each pair of cuts $a+b$ the intersection of the two chains covering a and b is a disk which is a neighborhood of $a \cap b$ (single point) and that the chains belonging to disjoint pairs of cuts or single cuts are disjoint from each other. Let k_j be the number of disks in a chain, and k the maximum of all k_j.

The regions in $\tilde{R}$ over these chains cover the boundaries of the subsurfaces $\tilde{R}_n$ which exhaust $\tilde{R}$, and constitute a regular chain set on $\tilde{R}$.

The maximum number of sides in a boundary component of $\tilde{R}_n$ is

$$s_n = 4(1+2n)$$

if there exists at least one pair of cuts in (a, b) (cf. 17 C). If (a, b) consists of only single cuts then $s_n = 1$. Hence the maximum number λ_n of disks in a chain covering a contour of $\tilde{R}_n$ satisfies in every case the inequality

$$\lambda_n \leq 4k(1+2n).$$

Therefore

$$\sum \frac{1}{\lambda_n} = \infty.$$

By Theorem 3 B we conclude that $\tilde{R} \in O_{AD}$.

As an immediate consequence we have by Theorem 8 C:

COROLLARY. *A univalent conformal map of a planar Abelian covering surface of a closed Riemann surface is unique up to a linear transformation. In particular this is true of the Schottky covering surface.*

Since the regular chain test also works for O_{A^0D} we may replace O_{AD} in the conclusion of the theorem by O_{A^0D} and O_{A_0D}.

The universal covering surface $\tilde{R}^\infty$ of a closed Riemann surface R of genus $g>1$ is conformally equivalent to the open unit disk, and thus $\tilde{R}^\infty \notin O_{AD}$. The set $\mathscr{F}_R$ of all smooth complete covering surfaces of R constitutes a lattice in an obvious way and $\tilde{R}^\infty$ is the maximal element of $\mathscr{F}_R$. Let $\mathscr{F}_R'$ (resp. $\mathscr{F}_R''$) be the set of elements $\tilde{R}$ in $\mathscr{F}_R$ such that $\tilde{R} \notin O_{AD}$ (resp. $\tilde{R} \in O_{AD}$). We have seen that $\mathscr{F}_R' \neq \emptyset$ and $\mathscr{F}_R'' \neq \emptyset$. We propose the following problem:

Determine the classes $\mathscr{F}_R'$ and $\mathscr{F}_R''$. What are their extreme elements?

18 C. Schottky Point Sets. The boundary of the conformal image in the plane of the Schottky covering surface of a closed surface of genus $g>1$ is called the Schottky point set. It was introduced in 1 E in a different but equivalent manner. Again by Theorem 18 B we see that the Schottky point set has an O_{AD}-complement with respect to the sphere.

*19. Covering Surfaces of Open Surfaces

19 A. Covering Surfaces Associated with a Set of Cycles. Henceforth we choose an arbitrary open Riemann surface R as the base surface. In analogy with the discussion in 17 and 18 for closed base surfaces we construct the covering surface $\tilde{R}$ of R corresponding to a given set of cycles in R as follows.

A nondividing loop cut on R will be referred to as a cut and denoted as before by a, its conjugate by b. For a pair of cuts $a+b$ we again assume that $a \cap b$ consists of a single point.

Let (a, b) be a set of disjoint pairs of cuts or single cuts on R. Consider an exhaustion $\{R_n\}$ of R such that the set $\{\Gamma_n\}$ of relative boundaries Γ_n of R_n is disjoint from the set (a, b). Denote by $(a, b)_n$ the subset of (a, b) consisting of cuts in R_n, and by R_n' the surface obtained from R_n by removing the cuts in $(a, b)_n$.

If $(a, b)_1$ contains at least one $a+b$ then a model of R_1' is a square R_1' with holes and handles, bounded (in part) by $aba^{-1}b^{-1}$. Along the latter sequence of sides attach to R_1' new squares R_1' as in 17 C, 17 F, and 17 H. Similarly attach along each pair of cuts in $(a, b)_1$ new copies of R_1' to the first copy, and along each simple cut in (a, b) one new copy.

Thus a new generation has been formed around the original R'_1. Let $\tilde{R}_1$ be the resulting first approximation of $\tilde{R}$.

In the same manner attach around R'_2 a new generation and around the resulting surface another generation of copies of R'_2 to obtain $\tilde{R}_2$. We so choose the copies that $\tilde{R}_2$ contains $\tilde{R}_1$ as a relatively compact subsurface.

In general the attaching process starting from R'_n and extended to n generations around it gives a surface $\tilde{R}_n$. Its boundary $\tilde{\Gamma}_n$ consists of copies of Γ_n and copies of cuts in $(a,b)_n$.

The sequence $\tilde{R}_1 \subset \tilde{R}_2 \subset \tilde{R}_3 \subset \cdots$ defines the limiting surface

$$\tilde{R}((a,b)) = \lim_{n\to\infty} \tilde{R}_n,$$

which we call *the covering surface of R corresponding to the given set* (a,b) *of cuts on R.*

19 B. Abelian and Schottky Covering Surfaces. In analogy with 18 A we can also introduce the concepts of Abelian and Schottky covering surfaces of open Riemann surfaces R.

Let φ be a regular Abelian integral on R. We assume that φ has non-vanishing periods only along some nondividing cuts. Let $(\bar{a}, \bar{b})$ be the base of such cuts and (a,b) the set of conjugate cuts consisting of disjoint pairs and single cuts. Then φ is single-valued on the covering surface $\tilde{R}((a,b))$, which we shall call an *Abelian covering surface.*

If the set (a,b) consists of disjoint single cuts such that $R-(a,b)$ is planar, then we call $\tilde{R}((a,b))$ a *Schottky covering surface.*

19 C. Covering by Regular Chains. Let $\{R_n\}$ be an exhaustion of R used in constructing $\tilde{R}((a,b))$ (cf. 19 A); the relative boundaries Γ_n of R_n consist of disjoint closed curves. Let G be a chain covering one of these curves and let $\{G\}$ be a regular chain set consisting of such coverings G. We denote by $\{G\}_n$ the set of chains in $\{G\}$ covering the boundary Γ_n. The maximum number of disks in a chain belonging to $\{G\}_n$ will be denoted by g_n.

Cover each a and b in (a,b) by a chain H such that for a pair $a+b$ in (a,b) the intersection of the two chains corresponding to a and b is a disk containing $a \cap b$ and that

$$\{H\} \cup \{G\}$$

satisfies (β) and (γ) in 3 A. Moreover we require that $\bar{H} \cap \bar{G} = \emptyset$ and that $\bar{H}_1 \cap \bar{H}_2 \neq \emptyset$ if and only if H_1 and H_2 correspond to conjugate cuts in (a,b). Let $\{H\}_n$ be the set of chains in $\{H\}$ covering the curves in $(a,b)_n$. The maximum number of disks in a chain in $\{H\}_n$ will be denoted by h_n.

19 D. O_{AD}-Test. We set

$$k_n = \max(g_n, n h_n), \qquad q_n = \max(g_n, h_n)$$

for each fixed n. As a generalization of Theorem 18 B we obtain:

THEOREM. *Let R be an arbitrary open Riemann surface. Suppose that for a finite or infinite set (a,b) of cuts on R and for an exhaustion $\{R_n\}$ of R such that the boundaries Γ_n of R_n are disjoint from (a,b),*

$$\sum \frac{1}{k_n} = \infty.$$

Then the Abelian covering surface $\tilde{R}((a,b))$ corresponding to the Abelian integral which has nonvanishing periods along conjugate cuts $(\bar{a}, \bar{b})$ belongs to the class O_{AD}.

If (a,b) consists of only single cuts then

$$\sum \frac{1}{q_n} = \infty$$

suffices to assure that $\tilde{R}((a,b))$ belongs to O_{AD}. In particular the Schottky covering surfaces are in this class.

Let R'_n be as in 19 A. The boundary of R'_n consists of Γ_n and two copies of cuts in $(a,b)_n$. The former is covered by $\{G\}_n$ and the latter by $\{H\}_n$.

The boundary $\tilde{\Gamma}_n$ of $\tilde{R}_n$ consists of copies of Γ_n and copies of cuts in $(a,b)_n$. The former are covered by the set $\{\tilde{G}\}_n$ of copies of chains in $\{G\}_n$, the latter by the set $\{\tilde{H}\}_n$ of copies of chains in $\{H\}_n$. The set

$$\{\tilde{G}\} \cup \{\tilde{H}\} = \bigcup_{n=1}^{\infty} (\{\tilde{G}\}_n \cup \{\tilde{H}\}_n)$$

is a regular chain set on $\tilde{R}$.

The maximum number of disks in a chain in $\{\tilde{G}\}_n$ is equal to that in a chain in $\{G\}_n$ and thus g_n. The corresponding number in $\{H\}_n$ is h_n. The maximum number s_n of sides in a contour of $\tilde{R}_n$ which consists of copies of cuts in $(a,b)_n$ satisfies $s_n \le 4(1+2n)$ since the construction of $\tilde{R}_n$ is similar to that for closed base surfaces (see the proof of Theorem 18 B). Hence the number of disks in any chain in $\{\tilde{H}\}_n$ is at most $4h_n(1+2n)$.

The maximum number λ_n of disks in a chain in $\{\tilde{G}\}_n \cup \{\tilde{H}\}_n$ is therefore

$$\lambda_n \le \max\{g_n, 4h_n(1+2n)\}.$$

Since $4h_n(1+2n) \le 12 n h_n$ we obtain $\lambda_n \le 12 k_n$. Consequently

$$\sum \frac{1}{\lambda_n} \ge \frac{1}{12} \sum \frac{1}{k_n} = \infty$$

and Theorem 3 B implies that $\tilde{R}((a,b)) \in O_{AD}$.

If (a, b) consists of only disjoint cuts then $s_n = 1$ and thus $\lambda_n \leq q_n$. Therefore

$$\sum \frac{1}{\lambda_n} \geq \sum \frac{1}{q_n} = \infty$$

assures that $\tilde{R}((a, b)) \in O_{AD}$.

19 E. Finite Genus. Assume that the maximum numbers h_n of disks in the chains in $\{H\}_n$ are bounded:

$$h_n \leq K < \infty \qquad (n = 1, 2, \ldots). \tag{60}$$

If the base surface R has finite genus then we can choose $\{H\}_n$ so as to satisfy (60). In such cases we set

$$m_n = \max(g_n, n)$$

and can sharpen Theorem 19 D as follows:

Let R be an open Riemann surface of finite genus (or more generally, admitting (60)). If

$$\sum \frac{1}{m_n} = \infty$$

then every Abelian covering surface $\tilde{R}((a, b))$ of R belongs to O_{AD}.

If (a, b) consists of only disjoint (single) cuts then

$$\sum \frac{1}{g_n} = \infty$$

suffices to assure that $\tilde{R}((a, b))$ is in O_{AD}. In particular every Schottky covering surface is in this class.

The proof is by now almost trivial if we observe the relations $k_n \leq K\, m_n$ and $q_n \leq K\, g_n$.

19 F. Transcendental Hyperelliptic Surfaces. As an application of the result in 19 E we consider a transcendental hyperelliptic surface R represented as a 2-sheeted covering surface of the plane with branch points over $z = 0, 1, 2, \ldots$.

On removing from one sheet the circles

$$a_n\colon\ |z - (2n + \tfrac{1}{2})| = \tfrac{3}{4} \qquad (n = 1, 2, \ldots)$$

we obtain a planar surface. These cuts can obviously be covered by sets $\{H\}$ satisfying (60). This condition is also met by any set (a, b) of finitely many pairs of cuts or single cuts.

Let the circles

$$\Gamma_n: |z| = r_n = 2n + \tfrac{3}{2} \qquad (n = 1, 2, \ldots)$$

be the boundaries of relatively compact subregions R_n which exhaust R. The circles can be covered by a regular chain set $\{G\}$ with

$$g_n = c\, r_n$$

where c is a constant such that $\sum 1/g_n = \infty$. We conclude from 19 E:

The Schottky covering surface and the covering surface corresponding to any Abelian integral with finitely many nonvanishing periods on the transcendental hyperelliptic Riemann surface R belong to the class O_{AD}.

19 G. Strip Complexes. For a second application of 19 E we return to covering surfaces R of the complex plane with branch points over a finite number of points. For the set $\{G\}$ one can choose the regular chain set constructed in 12 E – 12 G, which covers the boundaries of the exhaustion $\{R_n\}$ of R formed in 12 D. If (58) in 12 C is valid then we see that $\sum 1/g_n = \infty$.

In the case of strip complexes of 14 A – 14 B the counterpart of series (58) is the series in 13 B. Moreover we can easily see that the set (a, b) of nondividing cycles in R which correspond in a natural manner to nondividing cycles in the strip complex can be covered by a chain set $\{H\}$ such that the number h of disks in a chain H is equal to the number of nodes through which the corresponding cycle passes in the strip complex. We conclude:

The results in 19 D *and* 19 E *remain valid for strip complexes if we replace the numbers* g_n *and* h_n *by the corresponding maximal numbers of nodes in boundary node sequences and in nondividing cycles of the strip complex respectively.*

For example the Schottky covering surface of the Riemann surface depicted in Fig. 5c belongs to the class O_{AD}.

Chapter II

Other Classes of Analytic Functions

In the preceding chapter we discussed tests for and properties of the class O_{AD}. We now turn to relations of O_{AD} to other null classes determined by analytic functions.

The central object of our interest is the class O_{AB} of Riemann surfaces R on which the class $AB(R)$ of bounded analytic functions reduces to the field of complex numbers. To some extent the class O_{AB} can be discussed parallel to O_{AD}. However at the present state of knowledge the metric properties of the ideal boundary of O_{AB}-surfaces are still obscure compared with those for O_{AD}.

The second most important topic is the class of what we shall call K-functions, i.e. harmonic functions whose flux vanishes across all dividing cycles. The null class O_{KD} inherits some of the properties of O_{AD} but also possesses interesting characteristics of its own.

In § 1 we establish the basic inclusion relations between null classes of Riemann surfaces related to analytic functions and give some tests for O_{AB}. This is followed in § 2 by a specialization to plane regions and a characterization of the several null classes by conformal invariants. The final § 3 is devoted to the class of K-functions.

§ 1. Inclusion Relations

We start by defining a number of significant classes of analytic functions. As an immediate consequence of the extremum theorems in Chapter I we obtain the basic inclusion relations among the corresponding null classes for both planar and arbitrary Riemann surfaces.

The only new class turns out to be O_{AB}, and as counterparts of the tests for O_{AD} in Chapter I we give conformal metric tests and modular tests for O_{AB}. The important special case of Painlevé's problem is discussed in connection with the Newtonian capacity and the Hausdorff measure. A brief discussion of the relative classes A_0B and A^0B is appended.

It will be seen that, as in the case of O_{AD}, membership in the class O_{AB} is not a property of the ideal boundary. This leads us to the conclusion that both O_{AD} and O_{AB} are quasiconformally noninvariant.

The significance of modular O_{AB}-tests is illustrated by their close connection with the vanishing linear measure of the boundary of a plane region and by the O_{AB}-property of certain covering surfaces of a closed surface.

1. Basic Inclusions

1 A. Plane Regions. We retain the meaning of A and D used in Chapter I and use the following further abbreviations: B for "bounded," E for "omitting a set of positive area," and S for "univalent analytic." The null classes are denoted correspondingly. For example O_{SE} is the class of Riemann surfaces R on which the family $SE(R)$ of univalent analytic functions omitting a set of positive area is empty. We also consider the special null class $\mathcal{S}_A$ of plane regions whose analytic span S_A vanishes (cf. I.8 A).

THEOREM. *For plane regions*

$$O_{AB} \subset O_{SE} = O_{ABD} = O_{AD} = \mathcal{S}_A. \tag{1}$$

The identities $O_{SE} = O_{AD} = \mathcal{S}_A$ were proved in I.8 C. Identities (54) and (55) in I.9 A complete relations (1).

1 B. Arbitrary Surfaces. From the above theorem we easily deduce:

THEOREM. *For arbitrary Riemann surfaces*

$$O_{AE} = O_{AB} \subset O_{AD}. \tag{2}$$

The second relation was established in I.9 A. The inclusion $O_{AE} \subset O_{AB}$ is obvious. Suppose $f \in AE(R)$ is nonconstant. Then the identity function on the image R' of R under f is in $SE(R')$, and consequently by (1) there exists a nonconstant $g \in ABD(R')$. The function $g \circ f$ is nonconstant and in $AB(R)$. This gives the first relation (2).

Remark. The equality $O_{SE} = O_{AD}$ for plane regions was obtained in AHLFORS-BEURLING [5] by a different method. It will be covered in § 2 of this chapter. The above proof of $O_{AE} = O_{AB}$ in the general case is due to ROYDEN [4].

2. The Class O_{AB}

2 A. Conformal Metric Test. In view of Theorems 1 A and 1 B the only new class is O_{AB}; the strict inclusion $O_{AB} < O_{AD}$ will be established in 11. We shall first give a conformal metric test and then derive a modular test for O_{AB}. Thus the process is reversed compared with the case of O_{AD} (cf. I. 1, 2).

Consider a conformal metric $ds=\lambda(z)|dz|$ on an open Riemann surface R and the induced metric $s(z_1, z_2)=\inf\int_\gamma ds$ where γ runs over all rectifiable curves joining z_1 and z_2 (cf. I. 2 A). Let $\Gamma(\rho)=\{z\in R|s(z, z_0)=\rho\}$ for a fixed z_0 and any $\rho>0$. We assume that $\Gamma(\rho)$ consists of a finite number, say $n(\rho)$, of Jordan curves $\Gamma_1(\rho), \ldots, \Gamma_{n(\rho)}(\rho)$ and that these are piecewise smooth. Let $\Lambda(\rho)$ be the length of the longest of these curves and set $M(\rho)=\max n(\rho')$ for $0<\rho'\leq\rho$.

The following result is due to PFLUGER [3]:

THEOREM. *If $\Lambda(\rho)$ increases so slowly that*

$$\limsup_{\rho\to\infty}\left(4\pi\int_\varepsilon^\rho \frac{d\rho}{\Lambda(\rho)}-\log M(\rho)\right)=\infty \qquad (\varepsilon>0) \tag{3}$$

then R belongs to O_{AB}.

Suppose there did exist a nonconstant analytic function $w=f(z)$ on R with $|f|<1$. In the unit disk $\Delta=\{|w|<1\}$ consider the hyperbolic metric (I. 2 C). The basic tool in the proof is the well-known Schmidt isoperimetric inequality $4A(\pi+A)\leq L^2$, where A is the hyperbolic area of a Jordan region in Δ, and L is the hyperbolic length of its boundary (cf., e.g. SCHMIDT [1] and [2; p. 753, footnote 12]):

Let $R_\rho=\{z\in R|s(z_0, z)<\rho\}$. Then $\partial R_\rho=\bigcup_1^{n(\rho)}\Gamma_\nu(\rho)$. We denote by F_ρ the Riemannian image of R_ρ under f over Δ. The Riemannian images l_ν of the $\Gamma_\nu(\rho)$ under f over Δ are the boundary components of F_ρ, oriented so as to leave F_ρ to the left. The hyperbolic metric on Δ is lifted to $F_\rho\cup\bigcup_1^{n(\rho)} l_\nu$ by the projection; let $A(\rho)=A$ be the area of F_ρ, and L_ν the length of l_ν, both measured in this metric.

We denote by $N(\rho, w)$ the number of times the value $w\in\Delta$ is taken in R_ρ by the function $w=f(z)$, counted with multiplicities. Then $A(\rho)=\int_\Delta(N(\rho, w)/(1-|w|^2)^2)\,du\,dv$ where $w=u+iv$. For w not in the projection of l_ν we also denote by $N_\nu(\rho, w)$ the winding number of l_ν with respect to w: $N_\nu(\rho, w)=(1/2\pi i)\int_{l_\nu}(\omega-w)^{-1}\,d\omega$ where ω has an obvious meaning. By the residue theorem $N(\rho, w)=\sum_1^{n(\rho)} N_\nu(\rho, w)$.

For $A_\nu=\int_\Delta(N_\nu(\rho, w)/(1-|w|^2)^2)\,du\,dv$ we have $A(\rho)=\sum_1^{n(\rho)} A_\nu$. Let $S_\mu=\{w\in\Delta\,|N_\nu(\rho, w)|\geq\mu\}$, a union of finitely many disjoint regions in Δ, with $\mu=1, \ldots, m=m(\rho, \nu)$ and $S_m\neq\emptyset$, $S_{m+1}=\emptyset$. Let T_δ, $\delta=1, \ldots, k=k(\rho, \nu)$, be the Jordan regions appearing in $S_1, \ldots, S_m$. We set $\gamma_\delta=\partial T_\delta$, $\delta=1, \ldots, k$. Using the same letters T_δ, γ_δ for their areas and lengths in the hyperbolic metric we obtain

$$\sum_1^k T_\delta\geq|A_\nu|, \qquad \sum_1^k \gamma_\delta=L_\nu.$$

Observe that

$$4|A_\nu|(\pi+|A_\nu|)\leq 4\sum_1^k T_\delta\left(\pi+\sum_1^k T_\delta\right)\leq 4\pi\sum_{\delta,\delta'}^{1,\dots,k}\sqrt{T_\delta T_{\delta'}}+4\left(\sum_1^k T_\delta\right)^2$$
$$=\left(\sum_1^k\sqrt{4\pi T_\delta}\right)^2+\left(\sum_1^k 2T_\delta\right)^2\leq\left(\sum_1^k\sqrt{4\pi T_\delta+4T_\delta^2}\right)^2.$$

By Schmidt's isoperimetric inequality $\sqrt{4\pi T_\delta+4T_\delta^2}\leq\gamma_\delta$ and thus

$$4|A_\nu|(\pi+|A_\nu|)\leq\left(\sum_1^k\gamma_\delta\right)^2=L_\nu^2$$

for $\nu=1,\dots,n=n(\rho)$. Since $A\leq\sum_1^n|A_\nu|$ it follows that

$$4A\left(\pi+\frac{A}{n}\right)\leq 4\pi\sum_1^n|A_\nu|+4\sum_1^n|A_\nu|^2\leq\sum_1^n L_\nu^2.$$

On the other hand Schwarz's inequality yields

$$L_\nu^2=\left(\int_{\Gamma_\nu(\rho)}\frac{\left|\frac{df}{dz}\right|}{1-|f|^2}|dz|\right)^2=\left(\int_{\Gamma_\nu(\rho)}\frac{\left|\frac{df}{ds}\right|}{1-|f|^2}ds\right)^2$$
$$\leq\left(\int_{\Gamma_\nu(\rho)}ds\right)\left(\int_{\Gamma_\nu(\rho)}\frac{\left|\frac{df}{ds}\right|^2}{(1-|f|^2)^2}ds\right)\leq\Lambda(\rho)\int_{\Gamma_\nu(\rho)}\frac{\left|\frac{df}{ds}\right|^2}{(1-|f|^2)^2}ds$$

and hence

$$\sum_1^n L_\nu^2\leq\Lambda(\rho)\sum_1^n\int_{\Gamma_\nu(\rho)}\frac{\left|\frac{df}{ds}\right|^2}{(1-|f|^2)^2}ds=\Lambda(\rho)\frac{d}{d\rho}A(\rho).$$

Therefore

$$4A(\rho)\left(\pi+\frac{A(\rho)}{M(\rho')}\right)\leq 4A(\rho)\left(\pi+\frac{A(\rho)}{n(\rho)}\right)\leq\Lambda(\rho)\frac{d}{d\rho}A(\rho)$$

for $0<\rho\leq\rho'$.

On integrating we obtain

$$\log\left(\frac{A(\rho)}{\pi M(\rho)+A(\rho)}\cdot\frac{\pi M(\rho)+A(\varepsilon)}{A(\varepsilon)}\right)\geq 4\pi\int_\varepsilon^\rho\frac{d\rho}{\Lambda(\rho)}\qquad(\varepsilon>0)$$

and conclude that

$$\frac{A(\varepsilon)}{\pi+A(\varepsilon)}\leq M(\rho)\exp\left(-4\pi\int_{\varepsilon}^{\rho}\frac{d\rho}{\Lambda(\rho)}\right).$$

This together with (3) implies $A(\varepsilon)=0$, contrary to the assumption that f is nonconstant.

2 B. Modular Test. Let $\{R_n\}_0^\infty$ be an exhaustion of R and let $R_n-\bar{R}_{n-1}$ consist of the regions R_{nk}, $k=1,\ldots,k_n$. We may take $R_0=\{|z|<1\}$. Denote by v_n and μ_{nk} the moduli of $(R_n-\bar{R}_{n-1}, \partial R_{n-1}, \partial R_n)$ and $(R_{nk}, (\partial R_{n-1})\cap(\partial R_{nk}), (\partial R_n)\cap(\partial R_{nk}))$, and set $\mu_n=\min\mu_{nk}$ with $k=1,\ldots,k_n$ and $N(n)=\max k_j$ with $j=1,\ldots,n$.

The following reformulation of Pfluger's test is due to A. Mori [5] (cf. also Kuroda [4]); an illuminating application will be given in 2 K.

Theorem. *If there exists an exhaustion with doubly connected surface fragments R_{nk} whose minimum moduli grow so rapidly that*

$$\limsup_{n\to\infty}\left(\prod_{j=1}^{n}\mu_j\right)\Big/\sqrt{N(n)}=\infty \tag{4}$$

then R belongs to O_{AB}.

Let u_j and u_{jk} be the modulus functions corresponding to v_j and μ_{jk}. Then $u_j=c_{jk}\,u_{jk}$ on R_{jk} with $c_{jk}=(\log v_j)/(\log\mu_{jk})$. We define the function ξ on R as follows: $\xi=|z|^2$ on R_0 and $\xi=\tau_{j-1}+u_j$ on R_j-R_{j-1} with $\tau_0=1$ and $\tau_j=\sum_{t=1}^{j}\log v_t+1$. With respect to $z_0=0$ and the metric $ds=|\text{grad}\,\xi|\,|dz|$ we consider $\Gamma(\rho)$, $\Gamma_i(\rho)$, $\Lambda(\rho)$, and $M(\rho)$ as in 2 A.

For $\tau_{j-1}<\rho<\tau_j$ we have $\Gamma(\rho)\subset R_j-\bar{R}_{j-1}$. It follows that for some k with $1\leq k\leq n(\rho)$

$$\Lambda(\rho)=\int_{\Gamma_k(\rho)}ds=\int_{\Gamma_k(\rho)}*d\xi=\int_{\Gamma_k(\rho)}*du_j=c_{jk}\int_{\Gamma_k(\rho)}*du_{jk}=2\pi\,c_{jk}$$

which shows that

$$\Lambda(\rho)\leq 2\pi\frac{\log v_j}{\log\mu_j}.$$

Therefore

$$4\pi\int_{1}^{\rho}\frac{d\rho}{\Lambda(\rho)}\geq 2\sum_{1}^{j}\log\mu_t+2\frac{\log\mu_{j+1}}{\log v_{j+1}}(\rho-\tau_j) \tag{5}$$

for $\tau_j\leq\rho\leq\tau_{j+1}$. In particular

$$4\pi\int_{1}^{\tau_n}\frac{d\rho}{\Lambda(\rho)}\geq 2\sum_{1}^{n}\log\mu_j.$$

Since the R_{nk} are annuli, $M(\tau_n)=N(n)$. As a consequence

$$4\pi\int_{\varepsilon}^{\tau_n}\frac{d\rho}{\Lambda(\rho)}-\log M(\tau_n)\geq 2\log\left(\left(\prod_{j=1}^{n}\mu_j\right)\Big/\sqrt{N(n)}\right)$$

with $0<\varepsilon<1$, and we infer that (4) implies (3). The proof is complete.

Note that the above tests are more stringent than those for O_{AD}. This is in keeping with the inclusion $O_{AB}\subset O_{AD}$. Another version of these tests, such as the regular chain test, can be given as in the case of O_{AD} with obvious modifications which are left to the reader.

2 C. Relative Classes. By replacing D by B in I.1 F and I.1 G we define the classes SO_{AB}, O_{A_0B}, O_{A^0B} and prove the counterpart of Theorem I.1 F (Kuroda [4]):

Theorem. *A bordered Riemann surface (R,γ) belongs to the class SO_{AB} if and only if the double $\hat{R}^{\gamma}$ of (R,γ) about γ belongs to O_{AB}.*

Using this theorem we can show as in I.1 G that

$$O_{A^0B}<O_{A_0B},\qquad O_{A^0B}<O_{AB}.$$

Here the strictness follows as in I.10 D from Myrberg's example I.10 B.

We also have

$$SO_{AB}\subset SO_{AD}.$$

In fact if $(R,\gamma)\in SO_{AB}$ then by the above theorem $\hat{R}^{\gamma}\in O_{AB}$ and by (1), $\hat{R}^{\gamma}\in O_{AD}$. It follows in view of Theorem I.1 F that $(R,\gamma)\in SO_{AD}$. From this we conclude at once that

$$O_{A_0B}\subset O_{A_0D},\qquad O_{A^0B}\subset O_{A^0D}.$$

In 11 we shall prove that the inclusion $SO_{AB}\subset SO_{AD}$ is strict.

2 D. Plane Regions. The nature of (4) is best illustrated in the case of plane regions (2 K). It is for such regions that the O_{AB}-problem was initiated by Painlevé in the following form: given a plane compact set E, when is E a *Painlevé set*, i.e. an AB-removable set? Here E is called *AB-removable* or *AB-null* if for some disk D with $E\subset D$ all AB-functions on $\bar{D}-E$ have analytic extensions to $\bar{D}$. Clearly we have only to consider the case where E is totally disconnected.

By the same procedure as in I.8 D we can show:

Theorem. *A plane compact set E is AB-removable if and only if the complement of E with respect to the extended plane belongs to O_{AB}.*

The counterpart of I. 8 E is also easily proved by the method employed there.

Remark. It is readily seen that if $E_1, \ldots, E_n$ are compact AB-removable sets then $\bigcup_1^n E_j$ is also AB-removable (cf. I. 8 D; see also VI. 1 L).

2 E. Hausdorff Measure. Let $h(t)$ be an increasing function of t on $[0, 1]$ with $h(0)=0$. We cover a plane compact set E by at most a countable number of squares with sides $0<t_i<t\leq 1$ parallel to the coordinate axes and set

$$\tilde{h}(t)=\inf \sum_i h(t_i)$$

where the infimum is taken over all such coverings. Clearly $\tilde{h}(t)$ increases as t decreases and we may define

$$h^*(E)=\lim_{t\to 0} \tilde{h}(t).$$

It is easy to see that h^* is a Carathéodory outer measure and a measure on the Borel field. We call $h^*(E)$ the *Hausdorff h-measure* of E. If $h(t)=t^\lambda$ ($\lambda>0$) then $h^*(E)$ is the *λ-dimensional Hausdorff measure.*

Let M_λ be the set of plane regions whose complements are bounded and have vanishing λ-dimensional Hausdorff measure. We can now state (AHLFORS [3]):

THEOREM. *For plane regions*

$$M_1 \subset O_{AB} \subset M_{1+\varepsilon} \tag{6}$$

where ε is any positive number.

Let $R\in M_1$ and denote by E the complement of R. To conclude that $R\in O_{AB}$ we have to show that E is AB-removable (Theorem 2 D). Let D be a disk containing E and set $\partial D=\alpha$. Since E is compact and $R\in M_1$ we can find an "exhaustion" of $D-E$, say $\{D_n\}$, such that ∂D_n consists of α and β_n where β_n is made up of a finite number of disjoint closed polygonal lines each side of which is parallel to one of the coordinate axes, and $\int_{\beta_n} |dz|$ tends to zero as $n\to\infty$.

Take a function $f\in AB(\bar{D}-E)$. Then

$$f(z)=\frac{1}{2\pi i}\int_{\alpha\cup\beta_n}\frac{f(\zeta)}{\zeta-z}\,d\zeta$$

for $z\in D-E$ if n is sufficiently large. Clearly $\int_{\beta_n}(f(\zeta)/(\zeta-z))\,d\zeta\to 0$ as $n\to\infty$. Therefore $f(z)$ has an analytic extension $(1/2\pi i)\int_\alpha f(\zeta)/(\zeta-z)\,d\zeta$ to $\bar{D}$, i.e. E is AB-removable and $O_{AB}\supset M_1$.

2 F. Newtonian Potential. We turn to the proof of $O_{AB} \subset M_{1+\varepsilon}$. We shall actually show more:

$R \in O_{AB}$ implies the vanishing of the Newtonian capacity of E and this in turn implies that E has $(1+\varepsilon)$-dimensional measure zero.

These statements will be proved in 2 F – 2 J.

Consider the 2-dimensional Newtonian kernel $1/|z-\zeta|$ which as a function of z is subharmonic on the extended plane except for the point ζ and vanishes at ∞. The *Newtonian potential*

$$N_\mu(z) = \int \frac{d\mu(\zeta)}{|z-\zeta|}$$

of a measure μ with compact support S_μ is lower semicontinuous on the plane, subharmonic on the complement of S_μ, and vanishes at ∞.

LEMMA. *If $N_\mu(z) \leq K$ on S_μ then $N_\mu(z) \leq 2K$ on $|z| < \infty$.*

Let $z \notin S_\mu$ and $\zeta(z) \in S_\mu$ such that $|z-\zeta(z)| = \min |z-\zeta|$ for $\zeta \in S_\mu$. Then $|\zeta(z)-\zeta| \leq |z-\zeta(z)| + |z-\zeta| \leq 2|z-\zeta|$ for any $\zeta \in S_\mu$. Hence $N_\mu(z) \leq 2N_\mu(\zeta(z)) \leq 2K$.

2 G. Newtonian Capacity. For a measure μ we define the *Newtonian energy* $N(\mu)$ of μ by $\iint (1/|z-\zeta|)\, d\mu(z)\, d\mu(\zeta)$. Given a compact set E set

$$N^*(E) = 1/\inf N(\mu)$$

where μ runs over all measures μ with $S_\mu \subset E$ and $\mu(E) = 1$. The quantity $N^*(E)$ is called the *Newtonian capacity* of E.

Suppose $N^*(E) > 0$. Then we can find a sequence of admissible μ_n such that $N(\mu_n) \to 1/N^*(E)$ as $n \to \infty$. By choosing a suitable subsequence we may assume that $\{\mu_n\}$ converges to a measure μ in the sense that $\int \varphi(z)\, d\mu_n(z) \to \int \varphi(z)\, d\mu(z)$ as $n \to \infty$ for any finitely continuous function φ on $|z| < \infty$. Then $S_\mu \subset E$ and $\mu(E) = 1$. Clearly $N(\mu) \geq 1/N^*(E)$.

On the other hand since $N_n(z, \zeta) = \min(n, 1/|z-\zeta|)$ $(n = 1, 2, \ldots)$ is finitely continuous on $E \times E$

$$\begin{aligned}\iint_{E\times E} N_n(z,\zeta)\, d\mu(z)\, d\mu(\zeta) &= \lim_{m\to\infty} \iint_{E\times E} N_n(z,\zeta)\, d\mu_m(z)\, d\mu_m(\zeta)\\ &\leq \lim_{m\to\infty} N(\mu_m) = 1/N^*(E).\end{aligned}$$

Hence on letting $n \to \infty$ we conclude that $N(\mu) \leq 1/N^*(E)$ and a fortiori $N(\mu) = 1/N^*(E)$.

For the above measure μ, called the *capacitary measure* of E, we have:

LEMMA. *$N_\mu(z) \leq 2/N^*(E)$ for $|z| < \infty$.*

For the proof let $A=\{z|z\in E, N_\mu(z)<1/N^*(E)\}$ and $A_n=\{z|z\in E, N_\mu(z)\le 1/N^*(E)-1/n\}$. By the lower semicontinuity of N_μ the set A_n is compact for $n=1,2,\ldots$, and since $A=\bigcup_n A_n$, A is a Borel set. We shall first show that $\mu(A)=0$.

Assume that $\mu(A)>0$ and a fortiori $\mu(A_n)>0$ for some n. Since $\int_{S_\mu} N_\mu(z)\,d\mu(z)=1/N^*(E)$ there exists a point $\zeta_0\in S_\mu$ with $N_\mu(\zeta_0)>1/N^*(E)-1/2n$ so that ζ_0 lies outside of A_n. Hence we can find a disk $U(\zeta_0)$ such that $N_\mu(\zeta)\ge 1/N^*(E)-1/2n$ on $U(\zeta_0)$ and $\overline{U(\zeta_0)}\cap A_n=\emptyset$. Since $\zeta_0\in S_\mu$, $\mu(U(\zeta_0))=m>0$.

Let σ' be the restriction of μ to A_n and set $\sigma=m\,\mu(A_n)^{-1}\,\sigma'$. Then clearly $N(\sigma)<\infty$ and $\sigma(A_n)=m$. Let σ_1 be the signed measure defined as follows: $\sigma_1=\sigma$ on A_n, $\sigma_1=-\mu$ on $U(\zeta_0)$, and $\sigma_1=0$ for any set outside of $A_n\cup U(\zeta_0)$. Let $\mu_1=\mu+t\,\sigma_1$ $(0<t<1)$. Clearly μ_1 is a measure with $S_{\mu_1}\subset E$ and $\mu_1(E)=1$. If t is sufficiently small then

$$\begin{aligned}
N(\mu_1)-N(\mu)&=2t\int_E N_\mu\,d\sigma_1+t^2 N(\sigma_1)\\
&=2t\Big(\int_{A_n} N_\mu\,d\sigma_1+\int_{U(\zeta_0)} N_\mu\,d\sigma_1\Big)+t^2 N(\sigma_1)\\
&\le 2t\left(m\left(\frac{1}{N^*(E)}-\frac{1}{n}\right)-m\left(\frac{1}{N^*(E)}-\frac{1}{2n}\right)\right)+t^2 N(\sigma_1)\\
&=-t\left(\frac{m}{n}-t\,N(\sigma_1)\right)<0,
\end{aligned}$$

which contradicts the definition of $N(\mu)$. We have shown that $\mu(A)=0$.

Next we prove the inequality $N_\mu(z)\le 1/N^*(E)$ for $z\in S_\mu$. Suppose $N_\mu(z_0)>1/N^*(E)$ for some $z_0\in S_\mu$. We can find a disk $V(z_0)$ such that $N_\mu(z)>1/N^*(E)+\varepsilon$ on $V(z_0)$ with $\varepsilon>0$. Since $N_\mu(z)\ge 1/N^*(E)$ on $S_\mu-A$ with $\mu(A)=0$ we infer that

$$\begin{aligned}
\frac{1}{N^*(E)}&=\int_{S_\mu} N_\mu\,d\mu=\int_{S_\mu\cap V(z_0)-A} N_\mu\,d\mu+\int_{S_\mu-V(z_0)-A} N_\mu\,d\mu\\
&\ge\left(\frac{1}{N^*(E)}+\varepsilon\right)\mu(V(z_0))+\frac{1}{N^*(E)}(1-\mu(V(z_0)))\\
&=\frac{1}{N^*(E)}+\varepsilon\,\mu(V(z_0))>\frac{1}{N^*(E)}.
\end{aligned}$$

This contradiction shows that $N_\mu(z)\le 1/N^*(E)$ on S_μ, and by Lemma 2 F we conclude that $N_\mu(z)\le 2/N^*(E)$.

2 H. The Class N. We denote by N the class of plane regions R with $\infty \in R$ and such that their complements have vanishing Newtonian capacities.

First we prove (AHLFORS [3]):

THEOREM. *For plane regions $O_{AB} \subset N$.*

Suppose $R \notin N$, i.e. the complement E of R has positive Newtonian capacity $N^*(E)$. Let μ be the capacitary measure of E and consider

$$f(z) = \int_E \frac{d\mu(\zeta)}{z-\zeta}.$$

Clearly f is analytic on R and

$$|f(z)| \le \int_E \frac{d\mu(\zeta)}{|z-\zeta|} \le \frac{2}{N^*(E)} < \infty$$

on $|z| \le \infty$ and a fortiori on R, i.e. $f \in AB(R)$.

To see that f is not constant observe that

$$\operatorname{Re} f(z) = \int_E \frac{x-\xi}{(x-\xi)^2 + (y-\eta)^2}\, d\mu(\zeta)$$

where $z = x + iy$ and $\zeta = \xi + i\eta$. Since E is compact, $|\xi| \le K < \infty$ for $\zeta \in E$. Thus if $x > K$ then $\operatorname{Re} f(z) > 0$, and if $x < -K$ then $\operatorname{Re} f(z) < 0$. Therefore f cannot be constant and R does not belong to O_{AB}.

2 I. Associated Measure. Our next task is to show that

$$N \subset M_{1+\varepsilon}$$

for $\varepsilon > 0$. To this end we first prove the existence of a measure μ associated with a given function $h(t)$ used to define the Hausdorff measure.

LEMMA. *Let E be a compact set contained in the unit square J_0 with center at the origin and with $h^*(E) > 0$. There exists a measure μ such that $S_\mu \subset E$ and*

$$\mu(\bar{U}(a,t)) \le 36\, h(t)$$

where $\bar{U}(a,t)$ is the closed disk of radius t $(0 < t \le 1)$ with center at $a \in E$.

We assume that J_0 is "half open", i.e. the right and upper sides do not belong to J_0. We divide J_0 into 4^n equal squares $\{J_n^j\}_{j=1}^{4^n}$ of sides $\delta_n = 1/2^n$, which are also half open. Let B_n be the subfamily of $\{J_n^j\}_1^{4^n}$ consisting of the J_n^j with $J_n^j \cap E \ne \emptyset$.

Take the measure μ_n^1 which has a constant density on each $J_n^j \in B_n$, density zero outside of $\bigcup J_n^j$ for $J_n^j \in B_n$, and with $\mu_n^1(J_n^j) = h(\delta_n)$ for $J_n^j \in B_n$.

Next choose the measure μ_n^2 of the same kind such that if $J_{n-1}^j \in B_{n-1}$ then $\mu_n^2(J_{n-1}^j) = \min(\mu_n^1(J_{n-1}^j), h(\delta_{n-1}))$ for fixed n, j.

Again starting with μ_n^2 we construct μ_n^3. Repeating this process we reach μ_n^n, which we denote by μ_n. By the construction $S_{\mu_n} \subset J_0$ and

$$\mu_n(J_i^j) \le h(\delta_i)$$

for $i = 1, \ldots, n$ and $J_i^j \in B_i$. We also see easily that for each fixed $n \ge 1$ every $a \in E$ is contained in a $J_i^j \in B_i$ with $\mu_n(J_i^j) = h(\delta_i)$. There may exist several such J_i^j, and we choose the largest one, say $J_{i(a)}^{j(a)}$. The family $\{J_{i(a)}^{j(a)} | a \in E\}$ constitutes a disjoint finite set $\{J_{i_\nu}^{j_\nu}\}_{\nu=1}^m$ such that $J_0 = \bigcup_{\nu=1}^m J_{i_\nu}^{j_\nu}$ μ_n-a.e. Clearly

$$h(1) \ge \mu_n(J_0) = \sum_{\nu=1}^m \mu_n(J_{i_\nu}^{j_\nu}) = \sum_{\nu=1}^m h(\delta_{i_\nu}) \ge h^*(E) > 0.$$

As a consequence we can find a subsequence $\{\mu_{n_k}\}$ of $\{\mu_n\}$ and a measure μ such that $\int \varphi \, d\mu_{n_k} \to \int \varphi \, d\mu$ as $k \to \infty$ for any finitely continuous function φ. Obviously $S_\mu \subset E$ and $h(1) \ge \mu(E) \ge h^*(E)$.

Let $a \in E$ and $0 < t < 1$. Choose δ_n and t' such that $\delta_n \le t < t' < \delta_{n-1}$. Then $\bar{U}(a, t')$ meets at most 9 squares in B_{n-1} and thus at most $4 \cdot 9 = 36$ squares in B_n. It follows that

$$\mu_n(\bar{U}(a, t')) \le 36 h(\delta_n) \le 36 h(t),$$

and this in turn gives

$$\mu(\bar{U}(a, t)) \le \lim_{k \to \infty} \mu_{n_k}(\bar{U}(a, t')) \le 36 h(t).$$

2 J. The Proof of $N \subset M_{1+\varepsilon}$. We now consider the special function $h(t)$ which as $t \to 0$ decreases to 0 so rapidly that

$$\int_0^1 \frac{h(t)}{t^2} \, dt < \infty.$$

The choice $h(t) = t^{1+\varepsilon}$ $(\varepsilon > 0)$ is an example.

LEMMA. *For a compact set E, $N^*(E) = 0$ implies $h^*(E) = 0$.*

For the proof we may assume that $E \subset J_0$ and that its diameter is less than 1. Suppose $h^*(E) > 0$. We take the associated measure μ of Lemma 2 I and consider

$$N_\mu(z) = \int_E \frac{d\mu(\zeta)}{|z - \zeta|}.$$

Fix a point $z \in E$ arbitrarily and set $v(t) = \mu(\bar{U}(z,t))$. By Lemma 2 I, $v(t) \leq 36\,h(t)$, and thus

$$N_\mu(z) = \int_0^1 \frac{dv(t)}{t} = \left[\frac{v(t)}{t}\right]_0^1 + \int_0^1 \frac{v(t)}{t^2}\,dt$$

$$\leq \left[\frac{v(t)}{t}\right]_0^1 + 36 \int_0^1 \frac{h(t)}{t^2}\,dt \equiv K < \infty.$$

We conclude that $N_\mu(z) \leq K$ on $S_\mu \subset E$,

$$N(\mu) = \int N_\mu\,d\mu \leq K\,\mu(S_\mu),$$

and therefore $N(\mu/\mu(S_\mu)) < \infty$. This gives $N^*(E) > 0$.

The proof of (6) is herewith complete.

2 K. O_{AB}-Test and Linear Measure. The gap between M_1 and $M_{1+\varepsilon}$ in (6) cannot be closed: we shall see in 11 E that $M_1 < O_{AB}$. For linear sets, however, it will be shown in 10 A that $M_1 = O_{AB}$.

In this context it is interesting that the test 2 B.(4) applied to plane regions "almost" implies M_1.

For preparation we state the following Golusin inequality [1]. Consider an annular region with modulus μ and areas A_1, A_2 of the simply connected regions bounded by its outer and inner contours respectively.

LEMMA. *The modulus μ of an annular region has the bound*

$$\mu^2 \leq \frac{A_1}{A_2}. \tag{7}$$

Here the equality holds if and only if the annular region is a circular annulus.

For the convenience of the reader we include here a proof of this quite well-known inequality.

Take the conformal mapping $z = \varphi(\zeta)$ of $1/\mu < |\zeta| < 1$ onto the annular region (cf. I.1 B) and denote by S_ρ the area of the simply connected region bounded by the image curve C_ρ of $|\zeta| = \rho$ $(1/\mu < \rho < 1)$.

Let the Laurent expansion of $\varphi(\zeta)$ be

$$\varphi(\zeta) = \sum_{n=-\infty}^{\infty} a_n \zeta^n.$$

Then the equation of C_ρ is

$$z = x + i y = \sum_{n=-\infty}^{\infty} a_n \rho^n e^{in\theta} \qquad (0 \leq \theta < 2\pi)$$

and therefore S_ρ is obtained as

$$S_\rho = \left| \frac{1}{2} \int_0^{2\pi} (x y' - x' y) \, d\theta \right| = \left| -\frac{1}{2} \operatorname{Im} \int_0^{2\pi} z \bar{z}' \, d\theta \right| = \pi \sum_{n=-\infty}^{\infty} n |a_n|^2 \rho^{2n}.$$

Since $S_\rho \to A_1$ as $\rho \to 1$ and $S_\rho \to A_2$ as $\rho \to 1/\mu$

$$\frac{A_1}{A_2} = \frac{\sum_{-\infty}^{\infty} n |a_n|^2}{\sum_{-\infty}^{\infty} n |a_n|^2 \mu^{-2n}}.$$

On the other hand

$$\frac{\sum n |a_n|^2}{\sum n |a_n|^2 \mu^{-2n}} - \mu^2 = \frac{\sum_{n \neq 0,1} n |a_n|^2 (1 - \mu^{-2n+2})}{\sum_{n \neq 0} n |a_n|^2 \mu^{-2n}} \geq 0,$$

which implies

$$\frac{A_1}{A_2} - \mu^2 = \frac{\sum_{n \neq 0,1} n |a_n|^2 (1 - \mu^{-2n+2})}{\sum_{n \neq 0} n |a_n|^2 \mu^{-2n}} \geq 0.$$

The equality here and a fortiori in (7) holds if and only if $a_n = 0$ $(n \neq 0, 1)$, that is $\varphi(\zeta) = a_0 + a_1 \zeta$. Then the annular region is in fact an annulus, and conversely.

2 L. Test for Linear Measure Zero. Consider a plane totally disconnected compact set E with complement R. Let $\{R_n\}_0^\infty$ be an exhaustion of R such that the resulting surface fragments R_{nk} (cf. 2 B) are annular regions. We postulate the existence of a constant $\delta > 0$ independent of n, k with the property

$$\delta \cdot d_{nk}^2 \leq A_{nk} \tag{8}$$

where d_{nk} is the diameter of the inner contour β_{nk} of R_{nk}, and A_{nk} is the area of the simply connected region bounded by β_{nk}. This condition means that the deviation of β_{nk} from the circular shape is small. We claim (OZAWA-KURODA [8]):

THEOREM. *For exhaustions of the complement R of E with* (8) *the condition* (4) *implies that $R \in M_1$.*

To see this denote by α_{nk} the outer contour of R_{nk}, and by A'_{nk} the area of the region bounded by α_{nk}. For the minimum modulus μ_n, (7) yields $\mu_n^2 \leq A'_{nk}/A_{nk}$. Thus

$$\sum_{k=1}^{k_n} \sqrt{A_{nk}} \leq \frac{1}{\mu_n} \sum_{k=1}^{k_n} \sqrt{A'_{nk}}.$$

By Schwarz's inequality and (8)

$$\sqrt{\delta} \sum_{k=1}^{k_n} d_{nk} \leq \frac{1}{\mu_n} \sum_{k=1}^{k_n} \sqrt{A'_{nk}} \leq \frac{1}{\mu_n} \sqrt{\sum A'_{nk}} \cdot \sqrt{k_n}.$$

Since

$$\sum_1^{k_n} A'_{nk} \leq \sum_1^{k_{n-1}} A_{n-1,k} \leq \frac{1}{\mu_{n-1}^2} \sum_1^{k_{n-1}} A'_{n-1,k} \leq \frac{1}{\mu_{n-1}^2} \sum_1^{k_{n-2}} A_{n-2,k}$$

$$\leq \frac{1}{\mu_{n-1}^2} \cdot \frac{1}{\mu_{n-2}^2} \sum_1^{k_{n-2}} A'_{n-2,k} \leq \cdots \leq \frac{1}{\left(\prod_1^{n-1} \mu_k\right)^2} \sum_1^{k_1} A'_{1k},$$

we conclude that

$$\sqrt{\delta} \sum_{k=1}^{k_n} d_{nk} \leq \frac{\sqrt{k_n}}{\prod_{k=1}^{n} \mu_k} \sqrt{\sum_{k=1}^{k_1} A'_{1k}}.$$

Condition (4) of 2 B implies that $\liminf_{n\to\infty} \sum_1^{k_n} d_{nk} = 0$. Let $h(t) = t$ and $t_n = \min_k d_{nk}$. Then (cf. 2 E)

$$\tilde{h}(t_n) \leq \sum_1^{k_n} d_{nk}$$

and since $t_n \to 0$

$$h^*(E) = \lim_n \tilde{h}(t_n) \leq \liminf_n \sum_1^{k_n} d_{nk} = 0.$$

The proof is complete.

Remark. Under the same assumption as in the above theorem with (4) and (8) replaced by the single condition

$$\limsup_{n\to\infty} \left(\prod_{j=1}^{n} \mu_j\right) \Big/ N(n) = \infty$$

Suita [1] proved that $R \in M_1$. This condition is closely related to an O_{A^0B}-criterion to be discussed in VI. 3 F.

2 M. Boundary Property. We close with a remark on general O_{AB}-surfaces. By the modification I. 10 C of Myrberg's example I. 10 B we also see:

THEOREM. *Membership in O_{AB} is not a property of the ideal boundary.*

In analogy with O_{AD} the general class O_{AB} is not quasiconformally invariant as will be seen in 14 A. In the case of finite genus membership in O_{AD} turned out to be a boundary property (I. 10 C) but this is not so for O_{AB} as will also be seen in 14. We have a similar situation for quasiconformal invariance: O_{AD}-surfaces of finite genus are invariant whereas this is not true for O_{AB}-surfaces of finite genus. The proofs will be furnished in 14 and 15.

**3. Covering Surfaces of Closed Surfaces

3 A. Commutative Covering Surfaces. Let R be a *closed* Riemann surface of genus $g \geq 1$. Consider a smooth complete covering surface $\tilde{R}$ of R. Take a closed curve C on R, and two curves $\tilde{C}_1$ and $\tilde{C}_2$ on $\tilde{R}$ lying over C. It may happen that $\tilde{C}_1$ is closed and $\tilde{C}_2$ is open. If such a situation never occurs then we call $\tilde{R}$ *regular.*

Let $\tilde{R}$ be a regular covering surface of R and T a topological mapping of $\tilde{R}$ onto itself such that for any $\tilde{P} \in \tilde{R}$, $T(\tilde{P})$ has the same projection on R as $\tilde{P}$. We call such a T a *covering transformation* of $\tilde{R}$. For two covering transformations T_1 and T_2, $T_1 \circ T_2$ is again a covering transformation of $\tilde{R}$. Under this operation the totality $G(\tilde{R})$ of covering transformations of $\tilde{R}$ forms a group, which shall be referred to as the *covering transformation group* of $\tilde{R}$.

Take any two distinct points $\tilde{O}_1$ and $\tilde{O}_2$ on $\tilde{R}$ with the same projection O on R. We shall show that there exists a $T \in G(\tilde{R})$ such that $T(\tilde{O}_1) = \tilde{O}_2$. This property of the group $G(\tilde{R})$ is called *transitivity.*

The proof is as follows. For any $\tilde{P}_1 \in \tilde{R}$ take a curve $\tilde{C}_1$ connecting $\tilde{O}_1$ to $\tilde{P}_1$; let P and C be the projections of $\tilde{P}_1$ and $\tilde{C}_1$. There exists one and only one curve $\tilde{C}_2$ on $\tilde{R}$ starting from $\tilde{O}_2$ and lying over C. Let $\tilde{P}_2$ be the end point of $\tilde{C}_2$ which lies over P. By the regularity of $\tilde{R}$, $\tilde{P}_2$ is uniquely determined by $\tilde{P}_1$. Let $\tilde{P}_2 = T(\tilde{P}_1)$. Then T is a covering transformation of $\tilde{R}$ with $T(\tilde{O}_1) = \tilde{O}_2$. From this proof it is also easy to see that such a T is uniquely determined by $\tilde{O}_1$ and $\tilde{O}_2$.

In the sequel we consider only regular covering surfaces $\tilde{R}$ whose covering transformation group $G(\tilde{R})$ is commutative. For convenience we shall call such an $\tilde{R}$ a *commutative covering surface.* It is also known as an Abelian covering surface but in the present book this term is used in another sense (I. 18 A).

3 B. Generators of $G(\tilde{R})$. Let $a_j, b_j, j = 1, \ldots, g$, be a homology base for R, i.e. a_j, b_j are analytic Jordan curves on R with the following three properties:

(α) the system $a_j, b_j, j = 1, \ldots, g$, does not divide R;

(β) except for the g pairs (a_j, b_j), $j=1, \ldots, g$, no two of the curves have points in common;

(γ) for each j, $a_j \cap b_j$ consists of exactly one point.

If we remove these curves a_j, b_j from R we obtain a planar surface.

Let $\tilde{O}$ be an arbitrary but then fixed point of $\tilde{R}$. For each $j=1, \ldots, g$ denote by $a_j(\tilde{O})$ and $b_j(\tilde{O})$ the end points of curves on $\tilde{R}$ starting from $\tilde{O}$ with projections on R which are closed and homotopic to the curves a_j and b_j respectively. Clearly all $a_j(\tilde{O})$ and $b_j(\tilde{O})$ have the same projection as $\tilde{O}$, and some of these points may be identical. Since $G(\tilde{R})$ is transitive and the regularity of $\tilde{R}$ is postulated there exists a unique $T \in G(\tilde{R})$ with $T(\tilde{O}) = a_j(\tilde{O})$ or $b_j(\tilde{O})$. By this correspondence, T may be identified with a_j or b_j. If T_j corresponds to $c_j \in \{a_j, b_j | j=1, \ldots, g\}$ then $T_1 \circ T_2$ corresponds to $c_1 + c_2$. Therefore $G(\tilde{R})$ may be considered as an additive group with the *system of generators* $\{a_j, b_j | j=1, \ldots, g\}$ which satisfy a number of *defining relations*

$$\sum_{j=1}^{g} \gamma_{kj} a_j + \sum_{j=1}^{g} \gamma_{k, g+j} b_j = 0 \qquad \text{(identity transformation).} \tag{9}$$

Here $k=1, \ldots, q$ $(1 \leq q \leq 2g)$ and the γ_{kj} are integers whose $q \times 2g$ matrix

$$(\gamma_{kj}) = \begin{pmatrix} \gamma_{11} & \gamma_{12} & \cdots & \gamma_{1,2g} \\ \gamma_{21} & \gamma_{22} & \cdots & \gamma_{2,2g} \\ \vdots & & & \vdots \\ \gamma_{q1} & \gamma_{q2} & \cdots & \gamma_{q,2g} \end{pmatrix} \tag{10}$$

has rank q. The number $2g - q = r$ is referred to as the *rank* of the additive group $G(\tilde{R})$.

With each element

$$\sum_{j=1}^{g} m_j a_j + \sum_{j=1}^{g} m_{g+j} b_j \in G(\tilde{R})$$

with integral coefficients m_j we associate the lattice point $(m_1, \ldots, m_{2g})$ in the $2g$-dimensional Euclidean space E^{2g}. Let N be the group of transformations of E^{2g} generated by the $2g$ translations carrying the origin to the $2g$ unit points $(1, 0, \ldots, 0), \ldots, (0, \ldots, 0, 1)$, and let $N(\tilde{R})$ be the subgroup of N generated by the q translations taking the origin to $(\gamma_{k1}, \ldots, \gamma_{k,2g})$, $k=1, \ldots, q$. Clearly two lattice points of E^{2g}, or equivalently two elements of N, represent the same element of $G(\tilde{R})$ if and only if they are equivalent with respect to $N(\tilde{R})$, or what amounts to the same, they belong to the same equivalence class of N by $N(\tilde{R})$. Therefore $G(\tilde{R})$ is isomorphic to the factor group $N/N(\tilde{R})$.

3 C. Construction of $\tilde{R}$. We have analyzed the group $G(\tilde{R})$ when a commutative covering surface $\tilde{R}$ of R is given. Clearly if two covering surfaces $\tilde{R}_1$ and $\tilde{R}_2$ of R have isomorphic groups then $\tilde{R}_1=\tilde{R}_2$. In other words the additive group G with the defining relations (9) determines at most one $\tilde{R}$ with $G(\tilde{R})=G$. Next we shall show that such an $\tilde{R}$ always exists and is of course unique.

Suppose a closed Riemann surface R and an abstract additive group with defining relations (9) are given. Fix the orientations of the curves a_j, b_j and denote the two shores of a_j, b_j by a_j^+, a_j^- and b_j^+, b_j^- so that b_j intersects a_j from a_j^+ to a_j^- and similarly a_j intersects b_j from b_j^+ to b_j^-. Cut R along each a_j and b_j. The resulting planar surface R' has g contours, each consisting of four sides a_j^+, b_j^+, a_j^-, and b_j^-.

With each residue class $(m_1, \ldots, m_{2g})$ modulo $N(G)$ where $N(G)$ is determined from (9) in the same way as $N(\tilde{R})$ was defined in 3 B we associate a replica $R'(m_1, \ldots, m_{2g})$ of R'. We identify the side b_j^+ of $R'(m_1, \ldots, m_j, \ldots, m_g, m_{g+1}, \ldots, m_{g+j}, \ldots, m_{2g})$ with the side b_j^- of $R'(m_1, \ldots, m_j+1, \ldots, m_g, m_{g+1}, \ldots, m_{g+j}, \ldots, m_{2g})$, and a_j^+ of $R'(m_1, \ldots, m_j, \ldots, m_g, m_{g+1}, \ldots, m_{g+j}, \ldots, m_{2g})$ with a_j^- of $R'(m_1, \ldots, m_j, \ldots, m_g, m_{g+1}, \ldots, m_{g+j}+1, \ldots, m_{2g})$ in such a fashion that each point on a_j^+ or b_j^+ is identified with the corresponding point on a_j^- or b_j^- respectively. By this procedure each side of each copy of R' is identified with some unique side of some other or possibly the same copy of R' and four copies of R' meet at each vertex of each copy of R': $R'(m_1, \ldots, m_j, \ldots, m_g, m_{g+1}, \ldots, m_{g+j}, \ldots, m_{2g})$, $R'(m_1, \ldots, m_j+1, \ldots, m_g, m_{g+1}, \ldots, m_{g+j}, \ldots, m_{2g})$, $R'(m_1, \ldots, m_j, \ldots, m_g, m_{g+1}, \ldots, m_{g+j}+1, \ldots, m_{2g})$ and $R'(m_1, \ldots, m_j+1, \ldots, m_g, \ldots, m_{g+j}+1, \ldots, m_{2g})$; some of these four may be identical.

Let $\tilde{R}$ be the surface thus constructed. It is not difficult to show that $\tilde{R}$ is a commutative covering surface of R with $G(\tilde{R})=G$.

3 D. Structures of $\tilde{R}$ and $G(\tilde{R})$. Assume that $\tilde{R}$ is closed and $O\in R$. Then there are only a finite number of points $\tilde{O}_1, \ldots, \tilde{O}_n$ on $\tilde{R}$ lying over O. An element $T\in G(\tilde{R})$ determines an $\tilde{O}_j$ with $T(\tilde{O}_1)=\tilde{O}_j$, and $\tilde{O}_j$ determines a $T\in G(\tilde{R})$ with $T(\tilde{O}_1)=\tilde{O}_j$. Therefore $G(\tilde{R})$ is finite. Conversely if $G(\tilde{R})$ is finite, by considerations similar to the above we can conclude that there are only a finite number of points lying over a point $O\in R$ and therefore $\tilde{R}$ is closed. In this case and in this case only the rank $r=2g-q$ of $G(\tilde{R})$ is zero. We have proved *the equivalence of the compactness of $\tilde{R}$, the finiteness of $G(\tilde{R})$, and the condition $r=0$.*

If $r>0$ then $\tilde{R}$ has a nonempty ideal boundary. In view of the construction of $\tilde{R}$ from $G(\tilde{R})$ it is easily seen that (cf. 3 E below) if $r=1$ then *the number of ideal boundary components of $\tilde{R}$ is two; if $r\geq 2$ then it is one.*

Let $\tilde{g}$ be the genus of $\tilde{R}$ and ν the order of $G(\tilde{R})$. From the construction we see that *if* $\nu<\infty$ *then* $\tilde{g}=\nu(g-1)+1$*; if* $\nu=\infty$ *and* $g=1$ *then* $\tilde{g}=0$*; if* $\nu=\infty$ *and* $g\geq 2$ *then* $\tilde{g}=\infty$.

3 E. Standard Exhaustion $\{\tilde{R}_n\}$ of $\tilde{R}$. We shall construct a convenient exhaustion of $\tilde{R}$. We consider only an open $\tilde{R}$ so that $r>0$. From the $2g$ unit basis vectors $(1,0,\ldots,0),\ldots,(0,\ldots,0,1)$ of E^{2g} we choose r vectors $A_1,\ldots,A_r$ such that $A_1,\ldots,A_r$ and the $q=2g-r$ vectors $B_1,\ldots,B_q$, with $B_k=(\gamma_{k1},\ldots,\gamma_{k,2g})$, $k=1,\ldots,q$, are linearly independent. We introduce a new coordinate system $(x_1,\ldots,x_q;y_1,\ldots,y_r)$ in E^{2g} with the same origin as before and with $B_1,\ldots,B_q;A_1,\ldots,A_r$ as basis vectors:

$$(m_1,\ldots,m_{2g})=(x_1,\ldots,x_q;y_1,\ldots,y_r)\,T. \tag{11}$$

Here T is a $2g\times 2g$ matrix obtained from (10)

$$T=\begin{pmatrix} \gamma_{11} & \cdots\cdots\cdots\cdots & \gamma_{1,2g} \\ \vdots & & \vdots \\ \gamma_{q1} & \cdots\cdots\cdots\cdots & \gamma_{q,2g} \\ 0 & \ldots\ 0\ \ 1\ \ 0\ \ldots & 0 \\ \vdots & \ddots & \vdots \\ 0 & \cdots\cdots\cdots\ 0 & 1 \end{pmatrix}$$

where the $(q+j)$th row is the unit vector whose $(q+j)$th element is 1. We set

$$T^{-1}=\begin{pmatrix} \alpha_{11} & \ldots & \alpha_{1q} & \beta_{11} & \ldots & \beta_{1r} \\ \vdots & & \vdots & \vdots & & \vdots \\ \alpha_{2g,1} & \ldots & \alpha_{2g,q} & \beta_{2g,1} & \ldots & \beta_{2g,r} \end{pmatrix}. \tag{12}$$

Two lattice points $(m_1,\ldots,m_{2g})$ and $(m_1',\ldots,m_{2g}')$ are equivalent with respect to $N(\tilde{R})$ if and only if their difference has integral x-coordinates and vanishing y-coordinates:

$$\begin{aligned} &\sum_{j=1}^{2g}\alpha_{jk}(m_j'-m_j)\equiv 0 \pmod 1, && k=1,\ldots,q, \\ &\sum_{j=1}^{2g}\beta_{jl}(m_j'-m_j)=0, && l=1,\ldots,r. \end{aligned} \tag{13}$$

The number Q of lattice points $(m_1,\ldots,m_{2g})$ contained in the $2g$-dimensional parallelepiped: $0\leq x_k<1$, $0\leq y_l<1$, $k=1,\ldots,q$, $l=1,\ldots,r$, is equal to the absolute value of the determinant $|T|$. For the proof of this well-known fact in algebra we refer e.g. to Minkowski [1].

We note that if $(x_1,\ldots,x_q;y_1,\ldots,y_l,\ldots,y_r)$ represents a lattice point in E^{2g}, i.e. if the components of the vector $(x_1,\ldots,x_q;y_1,\ldots,y_r)\,T$ are all integers then $(x_1, \ldots, x_q; y_1, \ldots, y_l \pm 1, \ldots, y_r)$ also represent lattice points and these are adjacent to the former. Actually from (11) it is seen that the $(0, \ldots, 0; 0, \ldots, \pm 1, \ldots, 0)\, T$ have the form $(0, \ldots, \pm 1, \ldots, 0)$.

Consider two copies of R' in $\tilde{R}$ adjacent to each other along some side or having some vertex in common:

$$R'(m_1, \ldots, m_i, \ldots, m_g, m_{g+1}, \ldots, m_{g+i}, \ldots, m_{2g}) = R'[x_1, \ldots, x_q; y_1, \ldots, y_r]$$

and

$$R'(m_1, \ldots, m_i + \varepsilon_1, \ldots, m_g, m_{g+1}, \ldots, m_{g+i} + \varepsilon_2, \ldots, m_{2g}) \\ = R'[x'_1, \ldots, x'_q; y'_1, \ldots, y'_r]$$

with $\varepsilon_1, \varepsilon_2 = 0, \pm 1$ and $\varepsilon_1^2 + \varepsilon_2^2 \neq 0$. By (11) and (12) we see that $y'_l - y_l = \varepsilon_1 \beta_{i,l} + \varepsilon_2 \beta_{g+i,l}$, $l = 1, \ldots, r$. Hence if we denote by M the sum of 1 and the integral part of $\max(|\beta_{il}| + |\beta_{g+i,l}|, l = 1, \ldots, r; i = 1, \ldots, g)$, then

$$|y'_l - y_l| < M, \qquad l = 1, \ldots, r. \tag{14}$$

For any $n \geq 0$ we now define $\tilde{R}_n$ as the union of all $R'(m_1, \ldots, m_{2g}) = R'[x_1, \ldots, x_q; y_1, \ldots, y_r]$ satisfying $-Mn \leq y_l < M(n+1)$, $l = 1, \ldots, r$. The number of such R' is $M^r (2n+1)^r Q$, and $\{\tilde{R}_n\}_0^\infty$ is an exhaustion of $\tilde{R}$. From (14) it also follows that any R' of $\tilde{R}$ having points in common with some R' of $\tilde{R}_n$ belongs to $\tilde{R}_{n+1}$. Hence the closure of $\tilde{R}_n$ belongs to the interior of $\tilde{R}_{n+1}$. Moreover $\tilde{R}_n$ is connected for sufficiently large n; for example $n \geq n_0 = \sum_{k,j} |\gamma_{kj}| + rM$. To see this we connect each of the $M^r Q$ duplicates of R' in $\tilde{R}_0$ to $R'[0, \ldots, 0; 0, \ldots, 0]$ by a chain of regions R' in $\tilde{R}$ and choose n_0 so large that $\tilde{R}_{n_0}$ contains these $M^r Q$ chains. Suppose $R'[x_1, \ldots, x_q; y_1, \ldots, y_r] \subset \tilde{R}_n$. By reducing the absolute values of the y_l one by one we construct a chain of regions R' in $\tilde{R}_n$ which connects $R'[x_1, \ldots, x_q; y_1, \ldots, y_r]$ to one of the $M^r Q$ regions R' of $\tilde{R}_0$. Thus $R'[x_1, \ldots, x_q; y_1, \ldots, y_r]$ can be joined to $R'[0, \ldots, 0; 0, \ldots, 0]$ in $\tilde{R}_n$ for $n \geq n_0$.

3 F. Main Theorem. As a counterpart of Theorem I. 18 B we shall show that a significant class of commutative covering surfaces of closed surfaces belongs to O_{AB}.

Let R be a closed surface and $\tilde{R}$ its commutative covering surface. Let $\{a_j, b_j\}$, $j = 1, \ldots, g$, be a system of generators of the covering transformation group $G(\tilde{R})$ of $\tilde{R}$ as described in 3 B. We consider covering surfaces with the property that for each $j = 1, \ldots, g$

$$\gamma_j a_j + \gamma_{g+j} b_j = 0 \tag{15}$$

with at least one of the integral coefficients γ_j and γ_{g+j} different from zero for each j.

We are now able to state (A. MORI [5]):

THEOREM. *A commutative covering surface $\tilde{R}$ with* (15) *belongs to the class O_{AB}.*

Such covering surfaces are best illustrated by the following example. Let R_a be a planar surface obtained from R by cutting along g disjoint nondividing loop cuts a_j, $j=1, \ldots, g$. Consider a commutative covering surface $\tilde{R}_a$ of R consisting of a finite or infinite number of replicas of R_a. Then $\tilde{R}_a \in O_{AB}$.

In fact we can construct g curves b_j such that the system $\{a_j, b_j\}$, $j=1, \ldots, g$, is a homology base as described in 3 B. Then (15) is satisfied with $\gamma_j=1$, $\gamma_{g+j}=0$.

The proof of the theorem will be given in 3 G – 3 I.

3 G. Estimation of Length of $\partial\tilde{R}_n$. Suppose that a curve C on $\tilde{R}$ consists of some arcs $a_j^+, a_j^-, b_j^+, b_j^-$, $j=1, \ldots, g$, on the boundaries of some regions R' in $\tilde{R}$. Hereafter in 3 the *length* $|C|$ of the curve will be understood to be the number of these arcs constituting C.

Our first task is to show that $L_n=|\partial\tilde{R}_n|$ satisfies

$$L_n=O(n^{r-1}). \tag{16}$$

For any r integers $t_1, \ldots, t_r$ let $Z(t_1, \ldots, t_r)$ be the part of $\tilde{R}$ which consists of the $M^r Q$ regions $R'[x_1, \ldots, x_q; y_1, \ldots, y_r]$ satisfying $M t_l \leq y_l < M(t_l+1)$, $l=1, \ldots, r$. By (14) any R' in $\tilde{R}$ adjacent to some R' of $Z(t_1, \ldots, t_r)$ belongs to one of the sets $Z(t_1+\delta_1, \ldots, t_r+\delta_r)$ with $\delta_l=0, \pm 1$ for $l=1, \ldots, r$. For δ's not all zero let $\gamma(t_1, \ldots, t_r; \delta_1, \ldots, \delta_r)$ be the part of the boundary of $Z(t_1, \ldots, t_r)$ along which $Z(t_1, \ldots, t_r)$ is adjacent to $Z(t_1+\delta_1, \ldots, t_r+\delta_r)$, and set $L(\delta_1, \ldots, \delta_r)=|\gamma(t_1, \ldots, t_r; \delta_1, \ldots, \delta_r)|$. Since every $Z(t_1, \ldots, t_r)$ is congruent to $Z(0, \ldots, 0)=\tilde{R}_0$, $L(\delta_1, \ldots, \delta_r)$ does not depend on $t_1, \ldots, t_r$ but only on $\delta_1, \ldots, \delta_r$. The sum of the $L(\delta_1, \ldots, \delta_r)$ with respect to the 3^r-1 admissible value combinations of δ is equal to the length L_0 of the boundary of $\tilde{R}_0$.

Suppose $Z(t_1, \ldots, t_r) \subset \tilde{R}_n$, i.e. $-n \leq t_l \leq n$, $l=1, \ldots, r$. The curve $\gamma(t_1, \ldots, t_r; \delta_1, \ldots, \delta_r)$ belongs to the boundary of $\tilde{R}_n$ if and only if $Z(t_1+\delta_1, \ldots, t_r+\delta_r) \not\subset \tilde{R}_n$. The number of such value combinations $(t_1, \ldots, t_r)$ for a fixed $(\delta_1, \ldots, \delta_r)$ is $(2n+1)^r-(2n)^{r'}(2n+1)^{r-r'}$ where r' is the number of nonvanishing components of $(\delta_1, \ldots, \delta_r)$. Therefore the total length of the boundary of $\tilde{R}_n$ is

$$\begin{aligned} L_n &= \sum [(2n+1)^r-(2n)^{r'}(2n+1)^{r-r'}]\, L(\delta_1, \ldots, \delta_r) \\ &\leq \sum [(2n+1)^r-(2n)^r]\, L(\delta_1, \ldots, \delta_r) \\ &= [(2n+1)^r-(2n)^r]\, L_0 \end{aligned}$$

where the summation ranges over the $3^r - 1$ value combinations of δ. We conclude that (16) holds.

3 H. Length of Components of $\partial \tilde{R}_n$. Let $\partial \tilde{R}_n$ consist of components Γ_n^k $(k=1, \ldots, k(n))$. Under the assumption (15) we shall estimate $L_n^k = |\Gamma_n^k|$ from above:

$$L_n^k = O(1). \tag{17}$$

Suppose that a boundary arc α, one of $a_1^+, a_1^-, b_1^+, b_1^-$ say, of an $R'(m_1, \ldots, m_{2g})$ of $\tilde{R}_n$ belongs to $\partial \tilde{R}_n$. At each end point of α there meet four R''s of $\tilde{R}$ not all belonging to $\tilde{R}_n$, say $R'(m_1, \ldots, m_{g+1}, \ldots, m_{2g})$, $R'(m_1 + \varepsilon_1, m_2, \ldots, m_{g+1}, \ldots, m_{2g})$, $R'(m_1, \ldots, m_g, m_{g+1} + \varepsilon_2, m_{g+2}, \ldots, m_{2g})$, and $R'(m_1 + \varepsilon_1, m_2, \ldots, m_g, m_{g+1} + \varepsilon_2, m_{g+2}, \ldots, m_{2g})$ with $\varepsilon_1, \varepsilon_2 = \pm 1$. Thus the arc of $\partial \tilde{R}_n$ which is adjacent to α at this point must be one of $a_1^+, a_1^-, b_1^+, b_1^-$ of one of the above four R'. From this we infer:

The component of $\partial \tilde{R}_n$ containing α consists of only the arcs a_1^+, a_1^-, b_1^+, b_1^- of R''s with constants $m_2, \ldots, m_g, m_{g+2}, \ldots, m_{2g}$.

Let $R'(m_1^*, \ldots, m_g^*, m_{g+1}^*, \ldots, m_{2g}^*)$ be an R' belonging to $\tilde{R}_n$. We denote by $\Delta_n'' = \Delta_n''(m_2^*, \ldots, m_g^*, m_{g+2}^*, \ldots, m_{2g}^*)$ the part of $\tilde{R}_n$ consisting of R''s of the form $R'(m_1, m_2^*, \ldots, m_g^*, m_{g+1}, m_{g+2}^*, \ldots, m_{2g}^*)$, and by $\gamma_n'' = \gamma_n''(m_2^*, \ldots, m_g^*, m_{g+2}^*, \ldots, m_{2g}^*)$ the part of the boundary of Δ_n'' consisting of the arcs $a_1^+, a_1^-, b_1^+, b_1^-$. We have:

In order to estimate from above the length of a component of $\partial \tilde{R}_n$ consisting of the arcs $a_1^+, a_1^-, b_1^+, b_1^-$, it is sufficient to estimate the length $L_n'' = L_n''(m_2^*, \ldots, m_g^*, m_{g+2}^*, \ldots, m_{2g}^*)$ of γ_n''.

The region $R'(m_1, m_2^*, \ldots, m_g^*, m_{g+1}, m_{g+2}^*, \ldots, m_{2g}^*)$ belongs to $\tilde{R}_n$ by the definition of $\tilde{R}_n$ if and only if

$$-Mn \le \beta_{1l} m_1 + \beta_{g+1,l} m_{g+1} + \sum_{j \neq 1, g+1} \beta_{jl} m_j^* < M(n+1), \quad l = 1, \ldots, r. \tag{18}$$

From (15) it follows that the g lattice points $(\gamma_1, 0, \ldots, 0, \gamma_{g+1}, 0, \ldots, 0)$, $\ldots, (0, \ldots, 0, \gamma_i, 0, \ldots, 0, \gamma_{g+i}, 0, \ldots, 0), \ldots, (0, \ldots, 0, \gamma_g, 0, \ldots, 0, \gamma_{2g})$ in E^{2g} are equivalents of $(0, \ldots, 0)$ with respect to the group $N(\tilde{R})$. Therefore on any (m_i, m_{g+i})-plane through any lattice point in E^{2g} there exist pairs of lattice points equivalent to each other. Hence the rank of the $2 \times r$ matrix

$$\begin{pmatrix} \beta_{11}, \ldots, \beta_{1r} \\ \beta_{g+1,1}, \ldots, \beta_{g+1,r} \end{pmatrix} \tag{19}$$

is one or zero.

First we consider the case in which the rank of (19) is one. Suppose for instance that $|\beta_{11}| + |\beta_{g+1,1}| \neq 0$. Then (18) defines a parallel strip region δ_n'' on the (m_1, m_{g+1})-plane, which is parallel to the line $\beta_{11} m_1 + \beta_{g+1,1} m_{g+1} = 0$. A lattice point (m_1, m_{g+1}) on the (m_1, m_{g+1})-plane is by (13) an equivalent of the lattice point (0, 0) if and only if

$$\begin{aligned} &\alpha_{1k} m_1 + \alpha_{g+1,k} m_{g+1} \equiv 0 \pmod 1, && k=1,\dots,q, \\ &\beta_{1l} m_1 + \beta_{g+1,l} m_{g+1} = 0, && l=1,\dots,r. \end{aligned} \tag{20}$$

Thus any equivalent of (0, 0) on this plane lies on the line $\beta_{11} m_1 + \beta_{g+1,1} m_{g+1} = 0$. Moreover since the elements of the matrix QT^{-1} are integers the point $(-Q^2 \beta_{g+1,1}, Q^2 \beta_{11})$ is actually a lattice point different from (0, 0) and satisfies (20).

Let $(\mu_1, \mu_{g+1}) \neq (0, 0)$ be one of the equivalents of (0, 0) closest to (0, 0). Any equivalent of (0, 0) is then represented in the form $(\nu \mu_1, \nu \mu_{g+1})$, $\nu = 0, \pm 1, \dots$. Hence two lattice points (m_1, m_{g+1}) and (m_1', m_{g+1}') are equivalent if and only if $m_1' = m_1 + \nu \mu_1$, $m_{g+1}' = m_{g+1} + \nu \mu_{g+1}$ for some integer ν.

With each lattice point (m_1, m_{g+1}) we associate a square $S(m_1, m_{g+1})$ whose center is (m_1, m_{g+1}) and whose sides are parallel to the coordinate axes and have unit length. The square $S(m_1, m_{g+1})$ may be considered as a model of the replica $R'(m_1, m_2^*, \dots, m_g^*, m_{g+1}, m_{g+2}^*, \dots, m_{2g}^*)$; the sides parallel to the m_1-axis are models of the boundary arcs a_1^+ and a_1^-, and the sides parallel to the m_{g+1}-axis are models of b_1^+ and b_1^-. The model of Δ_n'' is constructed from the sum of all $S(m_1, m_{g+1})$ with $(m_1, m_{g+1}) \in \delta_n''$ by identifying the squares $S(m_1 + \nu \mu_1, m_{g+1} + \nu \mu_{g+1})$, $\nu = 0, \pm 1, \dots$. The resulting cylinder-shaped figure has two boundary components, each of length $|\mu_1| + |\mu_{g+1}|$. Hence we have

$$L_n'' = 2(|\mu_1| + |\mu_{g+1}|) \leq 2Q^2 (|\beta_{11}| + |\beta_{g+1,1}|)$$

and conclude that (17) holds.

Next we consider the case in which the rank of (19) is zero. Then any point on the (m_1, m_{g+1})-plane satisfies (18), and the lattice points $(Q, 0)$ and $(0, Q)$ both satisfy (20), i.e. they are equivalents of (0, 0). By the well-known Minkowski procedure [1] we can find two lattice points (μ_1, μ_{g+1}) and (μ_1', μ_{g+1}') equivalent to (0, 0) such that any equivalent of (0, 0) is uniquely expressed in the form $(\nu \mu_1 + \nu' \mu_1', \nu \mu_{g+1} + \nu' \mu_{g+1}')$ with integral coefficients ν, ν'.

As in the first case we take the square $S(m_1, m_{g+1})$ as a model of $R'(m_1, m_2^*, \dots, m_g^*, m_{g+1}, m_{g+2}^*, \dots, m_{2g}^*)$ and construct the model of Δ_n'' from the entire (m_1, m_{g+1})-plane by identifying the equivalent squares $S(m_1 + \nu \mu_1 + \nu' \mu_1', m_{g+1} + \nu \mu_{g+1} + \nu' \mu_{g+1}')$ for $\nu, \nu' = 0, \pm 1, \dots$. The resulting torus-shaped figure has no boundary arcs, i.e. γ_n'' is empty. Thus $L_n'' = 0$. We again infer that (17) is true.

3 I. Membership in O_{AB}. For each $j = 1, \dots, 2g$ we construct on R a doubly connected strip region D_j containing the curve a_j ($j \leq g$) or b_{j-g}

$(j>g)$ in its interior such that any two D_j, except the g pairs D_i, D_{g+i}, $i=1, \ldots, g$, have no points in common and that for each i, $D_i \cap D_{g+i}$ is a simply connected region. We assume further that, in the process of cutting R into R' (see 3 C), D_j is bisected into two simply connected strip regions D_j^+ and D_j^- in R', respectively adjacent to the boundary arcs a_j^+ and a_j^-, or b_{j-g}^+ and b_{j-g}^- of R'. Then $D_i^+ \cap D_{i+g}^+$, $D_i^+ \cap D_{i+g}^-$, $D_i^- \cap D_{i+g}^+$, and $D_i^- \cap D_{i+g}^-$ are simply connected subregions of R with a vertex of R' on the boundary.

Let $\{\tilde{R}_n\}$ be the exhaustion of $\tilde{R}$ used in 3 E and let Γ_n^k $(k=1, \ldots, k(n))$ be a component of the boundary Γ_n of $\tilde{R}_n$. Denote by A_n^k the union of the replicas of D_j^+, D_j^-, $D_i^+ \cap D_{g+i}^+$, $D_i^+ \cap D_{g+i}^-$, $D_i^- \cap D_{g+i}^+$, and $D_i^- \cap D_{g+i}^-$ on $\tilde{R}_n$, which neighbor the arcs of Γ_n^k or have some vertex of Γ_n^k on the boundary. Clearly A_n^k is a doubly connected subregion of $\tilde{R}_n$. Because of $\partial \tilde{R}_n = \bigcup_{k=1}^{k(n)} \Gamma_n^k$ we see that A_n^k, $k=1, \ldots, k(n)$, are the totality of components of $\overline{\tilde{R}_n - \tilde{R}_{n-1}}$ for large n.

Since A_n^k consists of L_n^k replicas of D_j^+, D_j^- and at most the same number of replicas of $D_i^+ \cap D_{g+i}^+$, $D_i^+ \cap D_{g+i}^-$, $D_i^- \cap D_{g+i}^+$, or $D_i^- \cap D_{g+i}^-$ it is easily seen that

$$\log \mu_n^k \geq \frac{c}{L_n^k} \tag{21}$$

where $\mu_n^k = \operatorname{mod} A_n^k$ and c is a positive constant independent of k and n. From (17) and (21) it follows that

$$\mu_n = \min_{1 \leq k \leq k(n)} \mu_n^k \geq C > 1$$

with a constant C independent of n. On the other hand (16) implies that $k(n) = O(n^{r-1})$ and therefore

$$N(n) = \max_{1 \leq j \leq n} k(j) = O(n^{r-1}).$$

We conclude that

$$\log\left(\prod_{j=1}^{n} \mu_j / \sqrt{N(n)}\right) = \sum_{j=1}^{n} \log \mu_j - \tfrac{1}{2} \log N(n) \geq n \log C - O(\log n).$$

Thus by Theorem 2 B, $\tilde{R} \in O_{AB}$.

§ 2. Plane Regions and Conformal Invariants

We shall now give a detailed analysis of certain null classes of planar Riemann surfaces R. For analytic functions the complete string of inclusion relations is

$$O_{AB} = O_{AE} \subset O_{SE} = O_{ABD} = O_{AD} = \mathcal{S}_A \subset O_{SB} = O_{SD}.$$

Only the last two relations are new; they will also be covered in Chapter V in connection with capacity functions. In the present section these relations together with the strictness of the two inclusions above will be proved from the viewpoint of conformal invariants.

The leading idea of the method to be employed is as follows. Let $F(R)$ be a class of analytic functions with a given boundedness property on R. The family O_F of Riemann surfaces R on which F reduces to the set of constants can then be characterized by the vanishing of $M_F(z_0)=\sup_F|f'(z_0)|$ for all $z_0\in R$. We shall take $F=AB$, AD, AE, SB, SD, and SE under certain normalizations and compare the quantities M_F for different choices of F. This leads us to the desired conclusions.

Historically this was the first mode of obtaining these results, established by AHLFORS and BEURLING in their celebrated joint work [5]. Unless other reference is made all results in § 2 are due to them. We shall follow their presentation closely, including some interesting relations slightly outside of the actual scope of our main program in this section.

4. The Invariant M_F

4 A. Weak and Strong Monotonic Properties. Let F be a class of (single-valued) analytic functions f on plane regions R, the region varying with f. The subclass of functions on a given R is denoted by $F(R)$. For $z_0\in R$ set

$$M_F(z_0,R)=\sup_{f\in F(R)}|f'(z_0)|. \tag{22}$$

For short we also use M_F, $M_F(z_0)$, and $M_F(R)$ when no ambiguity arises.

The class F is called *monotonic* if $F(R)\subset F(R')$ for $R'\subset R$. Then by (22)

$$M_F(z_0,R)\leq M_F(z_0,R').$$

Let $z'=g(z)$ be a univalent conformal mapping of R *onto* a region R', and set $z_0'=g(z_0)$. By definition F is *conformally invariant* if $f(z')\in F(R')$ implies $f(g(z))\in F(R)$ for every such g. Clearly then

$$M_F(z_0,R)=M_F(z_0',R')\,|g'(z_0)|$$

or equivalently

$$M_F(z_0,R)\,|dz_0|=M_F(z_0',R')\,|dz_0'|.$$

Thus the differential $M_F(z,R)\,|dz|$ gives a conformally invariant metric on R. We shall be concerned mainly with the quantity M_F, which can be called a relative conformal invariant. Absolute invariants may be introduced for example as quotients of two relative invariants.

For a family F which is both monotonic and conformally invariant we have

$$M_F(z_0', R')|dz_0'| \le M_F(z_0, R)|dz_0| \tag{23}$$

where $z' = g(z)$ is a univalent conformal map of R *into* R'. This inequality will be called the *weak monotonic property* of M_F.

Conformal invariance is contrasted with *analytic invariance*. The latter is obtained from the former by suppressing the requirement that the analytic mapping g be univalent. This will of course affect neither the invariance of the metric $M_F(z, R)|dz|$ nor the monotonicity of F, and (23) remains valid for all analytic mappings. In this case we shall refer to it as the *strong monotonic property* of M_F.

4 B. Compact Function Classes. By definition F is *compact* if for any nested sequence $\{R_n\}$ exhausting R and for functions $f_n \in F(R_n)$ there exists a subsequence $\{f_{n_i}\}$ of $\{f_n\}$ converging to a limit function $f \in F(R)$, uniformly on every compact subset of R. In this case we can replace the supremum in (22) by the maximum.

THEOREM. *A monotonic, conformally invariant, and compact class F has the following properties:*

(a) $\lim_{n\to\infty} M_F(z_0, R_n) = M_F(z_0, R)$,

(b) *$M_F(z, R)$ is continuous in z,*

(c) *$\log M_F(z, R)$ is subharmonic or identically $-\infty$.*

Property (a) is an obvious consequence of the monotonicity and the compactness of F.

Take a function $f \in F(R)$ with $|f'(z_0)| = M_F(z_0, R)$. Let $z_0' \in R$ such that the disk $|z - z_0| < 2|z_0' - z_0|$ is contained in R. Clearly $M_F(z_0', R) \ge |f'(z_0')|$ and thus

$$\liminf_{z_0' \to z_0} M_F(z_0', R) \ge M_F(z_0, R).$$

Let R' be the subset of R consisting of all points whose distance from the boundary of R is greater than $|z_0' - z_0|$, and let R'' be obtained from R' by a parallel translation which takes z_0 to z_0'. Then $M_F(z_0', R) \le M_F(z_0', R'') = M_F(z_0, R')$. This with (a) implies

$$\limsup_{z_0' \to z_0} M_F(z_0', R) \le M_F(z_0, R),$$

and (b) follows.

Since $\log M_F(z, R) = \max\{\log|f'(z)| \,|\, f \in F(R)\}$ and $\log|f'(z)|$ is subharmonic or identically $-\infty$ we obtain (c).

4 C. Special Classes. We are particularly interested in classes F for which $M_F(z, R)$ cannot vanish at a single point unless it vanishes identically. Several important classes to be considered will be seen to enjoy this property. We choose AB, AD, AE, SB, SD, and SE for F (cf. 1 A). For convenience we use this notation under suitable normalizations or conventions as far as the invariants M_F are concerned:

(a) $|f(z)| \leq 1$ *for* $f \in AB(R)$,

(b) $D(f) \leq 2\pi$ *for* $f \in AD(R)$,

(c) $(f(z)-f(z_0))^{-1}$ *omits an area* $\geq \pi$ *for* $f \in AE(R)$, z_0 *being a fixed point in* R,

(d) *constant functions belong to the subclasses* $SB(R)$, $SD(R)$, *and* $SE(R)$ *of* $AB(R)$, $AD(R)$, *and* $AE(R)$.

The invariants $M_{AB}, M_{AD}, M_{AE}, M_{SB}, M_{SD}, M_{SE}$ are not affected if we replace AB, AD, AE, SB, SD, SE by their subclasses $AB_0, AD_0, AE_0, SB_0, SD_0, SE_0$ determined by $f(z_0)=0$. This is obvious except for the class AB. For a function $f \in AB$ we need only observe that

$$f_0(z)=(f(z)-f(z_0))/(1-\overline{f(z_0)}\, f(z)) \in AB_0 \quad \text{and} \quad |f_0'(z_0)| \geq |f'(z_0)|.$$

It is easy to see that AB, AD, AE, SB, SD and SE are monotonic and conformally invariant. The classes AB and AE are also analytically invariant. Hence M_{AB} and M_{AE} have the strong monotonic property whereas the others have only the weak monotonic property. The subclasses $AB_0, AD_0, AE_0, SB_0, SD_0$, and SE_0 are compact. Thus we see:

All statements of Theorem 4 B *hold for each invariant* $M_{AB}, M_{AD}, M_{AE}, M_{SB}, M_{SD}$, *and* M_{SE}.

5. Invariants M_{AB} and M_{AE}

5 A. Equality of Invariants. Clearly $AB_0 \subset AE_0$ and hence $M_{AB} \leq M_{AE}$. We shall show that the opposite inequality is also true:

The invariants corresponding to AB *and* AE *are equal:*

$$M_{AB}(z_0, R) = M_{AE}(z_0, R). \tag{24}$$

Take an arbitrary $f \in AE_0(R)$ and denote by Q_f the set of values which are not taken by $1/f$. Clearly Q_f is a compact set and thus by hypothesis its area $I(Q_f)$ is finite and $\geq \pi$. We associate with f the function F_f defined by

$$F_f(z) = \frac{1}{I(Q_f)} \iint_{Q_f} \frac{du\,dv}{\dfrac{1}{f(z)} - w} \qquad (w = u + iv).$$

It is clearly analytic on R. By an easy computation we see that $F_f'(z_0) = f'(z_0)$. If we can show that $|F_f(z)| \leq 1$ on R then $M_{AB} \geq M_{AE}$ will follow.

For this purpose it suffices to prove that

$$\left|\iint_{Q_f} \frac{du\,dv}{w-\alpha}\right| \leq I(Q_f)$$

for all complex numbers α. An auxiliary congruence transformation is clearly allowed and thus we may assume that $\alpha=0$ and $\iint_{Q_f} w^{-1}\,du\,dv>0$.

Let Q_f^+ be the part of Q_f in the right half-plane. Using polar coordinates $w=r\,e^{i\theta}$ we obtain

$$0<\iint_{Q_f} \frac{du\,dv}{w} = \iint_{Q_f} \cos\theta\,dr\,d\theta \leq \iint_{Q_f^+} \cos\theta\,dr\,d\theta$$

since by assumption $\iint_{Q_f} w^{-1}\,du\,dv = \operatorname{Re}\iint_{Q_f} w^{-1}\,du\,dv$. We denote by $l(r,\theta)$ the linear measure of the set $\{w \mid w\in Q_f^+, \arg w=\theta, |w|\leq r\}$, and set $l(\theta)=\lim_{r\to\infty} l(r,\theta)$. By Fubini's theorem and then Schwarz's inequality

$$\iint_{Q_f^+} \cos\theta\,dr\,d\theta = \int_{-\pi/2}^{\pi/2} l(\theta)\cos\theta\,d\theta \leq \left(\frac{\pi}{2}\int_{-\pi/2}^{\pi/2} l^2(\theta)\,d\theta\right)^{\frac{1}{2}}.$$

Since $l(r,\theta)\leq r$ and $dr>dl(r,\theta)$ we obtain on integrating with a fixed θ

$$\int_{\{w\in Q_f^+,\,\arg w=\theta\}} r\,dr \geq \int l(r,\theta)\,dl(r,\theta) = \tfrac{1}{2}\,l^2(\theta).$$

Therefore

$$I(Q_f)\geq I(Q_f^+) = \iint_{Q_f^+} r\,dr\,d\theta \geq \tfrac{1}{2}\int_{-\pi/2}^{\pi/2} l^2(\theta)\,d\theta,$$

and we conclude that

$$\iint_{Q_f} \frac{du\,dv}{w} \leq (\pi\,I(Q_f))^{\frac{1}{2}} \leq I(Q_f).$$

5 B. Vanishing of $M_{AB}=M_{AE}$. We remark here that

$$M_{AB}(z,R)=M_{AE}(z,R)$$

cannot vanish at a single point unless it vanishes identically on R.

In fact assume that $M_{AB}(z_0,R)=0$. Then every $f\in AB$ must satisfy $f'(z_0)=0$. We wish to show that $f\equiv$const. If this were not the case then f would have the form

$$f(z)=f(z_0)+c_k(z-z_0)^k+\cdots, \qquad c_k\neq 0, \qquad k\geq 2.$$

Therefore $(z-z_0)^{1-k}(f(z)-f(z_0))$ would be in AB, with a nonzero derivative at z_0, a contradiction. Thus $M_{AB}(z,R)\equiv 0$.

5 C. Meromorphic Functions. We digress to observe a by-product:

THEOREM. *The image of $R \in O_{AB}$ under a nonconstant meromorphic function is again in O_{AB}.*

This is a direct consequence of the strong monotonic property of M_{AB} and is indeed true for any strongly monotonic class. Of course this is also a direct consequence of the definition of O_{AB} but the significance of the theorem is in that it gives a characterization of omitted sets.

The theorem can be given another formulation. Let a compact set X with a connected complement R be measured by

$$m_F(X) = M_F(\infty, R)$$

where F is a strongly monotonic class of functions.

Any normalized meromorphic function $f(z) = z + c_0 + c_1/z + \cdots$ on R omits a compact set X' with $m_F(X') \leq m_F(X)$.

5 D. Painlevé Null Sets. We recall that a compact set X is a *Painlevé null set* if and only if it is AB-removable (cf. 2 D).

THEOREM. *The value of the invariant $M_{AB}(z_0, R)$ does not change if a Painlevé null set is removed.*

In fact the family of competing functions remains the same.

6. Invariants M_{AD} and M_{SE}

6 A. Equality of Invariants. We now prove the identity $M_{AD} = M_{SE}$, a special case of which again gives $O_{AD} = O_{SE}$ (Theorem 1 A). The proof is in essence the same as that for $O_{AD} = \mathcal{S}_A$ in I.8.

THEOREM. *The invariants for AD and SE are equal:*

$$M_{AD}(z_0, R) = M_{SE}(z_0, R). \tag{25}$$

We use the notation in I.8 with $\zeta = z_0$. By I.8 B.(52) we see that

$$\frac{2\pi}{S_A} |f'(z_0)|^2 \leq D_f = \tfrac{1}{2} D(f)$$

for every $f \in A(R)$ with equality only for $f = P_0 - P_1$. In view of the normalization 4 C.(b) we conclude that

$$|f'(z_0)| \leq \sqrt{\frac{S_A}{2}},$$

and the equality holds for $f=\sqrt{1/2S_A}(P_0-P_1)\in AD(R)$. Therefore

$$M_{AD}(z_0, R)=\sqrt{\frac{S_A}{2}}. \tag{26}$$

Similarly from I.8 A.(48) we deduce in view of the normalization of the class SE that

$$M_{SE}(z_0, R)=\sqrt{\frac{S_A}{2}}. \tag{27}$$

Identity (25) follows.

6 B. Inequality $M_{AD}\leq M_{AB}$. Evidently $M_{SE}\leq M_{AE}$. Hence in conjunction with (24) and (25) we obtain

$$M_{AD}(z_0, R)\leq M_{AB}(z_0, R). \tag{28}$$

As a consequence we again see that $O_{AB}\subset O_{AD}$.

As in 5 D we also conclude:

$M_{AD}(z_0, R)$ is not affected by removing from R an AD-removable set.

7. The Invariant $M_{AD}(z_1, z_2, R)$

7 A. A Characterization. We append here a treatment of the quantity

$$M_{AD}(z_1, z_2, R)=\sup_{f\in AD(R)}|f(z_1)-f(z_2)| \tag{29}$$

where z_1 and z_2 are two distinct points in R, and $AD(R)$ is subject to the normalization 4 C.(b). This invariant will be used in our later consideration in 10 D of sets on a circle.

THEOREM. *The vanishing of $M_{AD}(z_1, z_2, R)$ is equivalent to the identical vanishing of $M_{AD}(z_0, R)$.*

This is a direct consequence of the interpolation Theorem I.9 B.

7 B. Circular and Radial Slit Mappings. The quantity $M_{AD}(z_1, z_2, R)$ has a close connection with circular and radial slit mappings of R. Although the existence of these mappings is classical we give an outline of the proof from a point of view similar to that in I.7 E–7 K.

On $R-\{z_1, z_2\}$ choose a singularity function s of the form $\log|z-z_1|$, $-\log|z-z_2|$ near z_1, z_2 respectively, and identically zero near the ideal boundary of R. As in I.7 D, Theorem I.7 A gives principal functions q_0, q_1 with this singularity. The admissible functions q are then those which have the above logarithmic poles at z_1, z_2. The flux of q is zero across every cycle separating $\{z_1, z_2\}$ from the ideal boundary of R. Observe that the flux of q across any cycle is an integral multiple of 2π and

$$Q(z) = \exp \int_{z_1}^{z} (dq + i * dq)$$

is a well-defined single-valued analytic function on $R - z_2$. It extends meromorphically to all of R with a simple pole at z_2. We choose the normalization $Q'(z_1) = 1$ for q and denote by $c = c(Q)$ the residue of Q at z_2. Let Q_0, Q_1 correspond to q_0, q_1.

First consider the case where R is the interior of a compact bordered surface $\bar{R}$. Note that Q_0, Q_1 can be extended to $\bar{R}$ so that they are analytic on $\bar{R} - z_2$ and $\arg Q_0$ and $|Q_1|$ are constant on every contour of R. In view of the argument principle Q_0 (resp. Q_1) is a univalent map onto a region bounded by a finite number of radial (resp. concentric circular) slits.

To investigate the mapping properties of Q_0 and Q_1 for an arbitrary R we derive as in I.7 D the relation

$$\int_{\beta} q * dq + 2\pi(k-h)\,\alpha = 2\pi(k^2 \alpha_1 - h^2 \alpha_0) + D(q - q_{hk}).$$

Here $h + k = 1$, $q_{hk} = h q_0 + k q_1$, $q = \mathrm{Re}\left(\sum_{n=0}^{\infty} {}_i a_n (z - z_i)^n\right) + s$ near z_i, $(i = 1, 2)$, $q_j = \mathrm{Re}\left(\sum_{n=0}^{\infty} {}_i a_n^{(j)} (z - z_i)^n\right) + s$ near z_i $(j = 0, 1)$, $\alpha = \mathrm{Re}({}_2a_0 - {}_1a_0)$, and $\alpha_j = \mathrm{Re}({}_2a_0^{(j)} - {}_1a_0^{(j)})$. The normalization gives ${}_1a_0 = 0$, ${}_2a_0 = \log c(Q)$. If Q is univalent then $\int_\beta q * dq = \int_\beta \log|Q|\, d\arg Q$ is the negative of the logarithmic area $A_{\log}(Q)$ of the complement of $Q(R)$. Therefore

$$A_{\log}(Q) + 2\pi \log|c(Q)| = 2\pi \log|c(Q_0)| - D(q - q_0),$$

$$-A_{\log}(Q) + 2\pi \log|c(Q)| = 2\pi \log|c(Q_1)| + D(q - q_1)$$

for every admissible univalent Q (cf. also V.6).

As in I.7 G and 7 I we can now see, by making use of the above extremal properties of Q_0 and Q_1 instead of those in I.7 H, that Q_0 (resp. Q_1) is a radial (resp. circular) slit mapping.

For further investigation of mappings of this type we refer the reader to the monograph Sario-Oikawa [27].

7 C. Evaluation of $M_{AD}(z_1, z_2, R)$. We have obtained univalent analytic mappings Q_0 and Q_1 of R onto regions bounded by radial and concentric circular slits respectively such that $Q_0(z_1) = Q_1(z_1) = 0$ and $Q_0(z_2) = Q_1(z_2) = \infty$. They were normalized by $Q_j'(z_1) = 1$. By multiplying Q_j by $c(Q_j)^{-1}$ we renormalize so that the residue at z_2 becomes 1. Set

$$Q_0'(z_1) = \beta_0, \quad Q_1'(z_1) = \beta_1. \tag{30}$$

THEOREM. *The invariant has the value*

$$M_{AD}(z_1, z_2, R) = \sqrt{\frac{1}{2}\log\frac{\beta_1}{\beta_0}}. \tag{31}$$

By virtue of continuity properties of $M_{AD}(z_1, z_2, R)$, β_0, and β_1 with respect to R we may restrict our attention to the case where ∂R is composed of a finite number of analytic Jordan curves. It is then clear in view of the behavior of Q_0 and Q_1 on ∂R that $\log(Q_1/Q_0)$ has a single-valued branch on R. For $f \in A(\bar{R})$ we obtain by the Cauchy-Riemann relations and Cauchy's theorem

$$\frac{1}{2} D\left(f, \log\frac{Q_1}{Q_0}\right) = \frac{i}{2}\int_{\partial R} f d\,\overline{\log\frac{Q_1}{Q_0}} = -\frac{i}{2}\int_{\partial R} f d\log Q_0 Q_1$$

$$= 2\pi(f(z_1) - f(z_2))$$

and in particular

$$\frac{1}{2} D\left(\log\frac{Q_1}{Q_0}\right) = 2\pi\log\frac{\beta_1}{\beta_0}.$$

From these two equalities and Schwarz's inequality it follows that

$$|f(z_1) - f(z_2)|^2 \leq \frac{1}{4\pi}\log\frac{\beta_1}{\beta_0} D(f)$$

and, if $D(f) \leq 2\pi$,

$$|f(z_1) - f(z_2)| \leq \left(\frac{1}{2}\log\frac{\beta_1}{\beta_0}\right)^{\frac{1}{2}}$$

with equality for $f = (\log(Q_1/Q_0))/(2\log(\beta_1/\beta_0))^{\frac{1}{2}}$. Thus we obtain (31).

8. Invariants M_{SD} and M_{SB}

8 A. Extremal Length. Our next goal is to prove that $M_{SD} = M_{SB}$. This equality will result from a comparison with a third invariant defined by means of extremal length. Primitive forms of the latter notion, which is closely related to the logarithmic modulus, were already made use of in the proof of Theorem I.3 B and in I.4 A and 5 A.

Let $\{\gamma\}$ denote a family of locally rectifiable curves in a plane region R. Consider the class of nonnegative lower semicontinuous functions $\rho(z)$ on R for which the quantities

$$L_\rho\{\gamma\} = \inf_\gamma \int_\gamma \rho|dz|, \qquad A_\rho(R) = \iint_R \rho^2\, dx\, dy \tag{32}$$

are well-defined and not simultaneously 0 or ∞. The quantity

$$\lambda\{\gamma\} = \sup_{\rho} \frac{L_\rho\{\gamma\}^2}{A_\rho(R)} \tag{33}$$

is called the *extremal length* of the family $\{\gamma\}$. Although the value of $\lambda\{\gamma\}$ does not depend on the region R the family $\{\gamma\}$ will frequently be defined with reference to a specific R.

The extremal length $\lambda\{\gamma\}$ is a conformal invariant in the following sense: for any $\{\gamma'\}$ which is obtained as an image of $\{\gamma\}$ under a univalent conformal mapping of R we have $\lambda\{\gamma\} = \lambda\{\gamma'\}$.

In view of this invariance the extremal length of the family $\{\gamma\}$ of curves on a *Riemann surface* R can also be defined; in this case we replace the functions ρ by nonnegative lower semicontinuous density differentials such that the expressions $\rho(z)|dz|$ are invariant under changes of local parameter z.

8 B. Elementary Properties. We start by compiling some fundamental properties of the extremal length $\lambda\{\gamma\}$ which will be needed in this section:

(a) *If every* $\gamma \in \{\gamma\}$ *contains a* $\gamma' \in \{\gamma'\}$, *then*

$$\lambda\{\gamma\} \geq \lambda\{\gamma'\}. \tag{34}$$

This is obvious by definition.

(b) *If* $\{\gamma_1\}$ *and* $\{\gamma_2\}$ *are contained in disjoint open sets* G_1 *and* G_2 *respectively and if every* $\gamma \in \{\gamma\}$ *contains a* $\gamma_1 \in \{\gamma_1\}$ *and a* $\gamma_2 \in \{\gamma_2\}$ *then*

$$\lambda\{\gamma\} \geq \lambda\{\gamma_1\} + \lambda\{\gamma_2\}. \tag{35}$$

We may suppose that $\lambda\{\gamma_1\}, \lambda\{\gamma_2\} > 0$ since otherwise the assertion is a special case of (a). In (33) for $\{\gamma_j\}$, $j=1, 2$, we only have to take the supremum with respect to those ρ_j that satisfy $L_{\rho_j}\{\gamma_j\} = A_{\rho_j}(R)$ and $\rho_j \equiv 0$ on $R - G_j$. Let $\rho = \rho_1 + \rho_2$. Then $L_\rho\{\gamma\} \geq L_{\rho_1}\{\gamma_1\} + L_{\rho_2}\{\gamma_2\}$ and $A_\rho(R) = A_{\rho_1}(R) + A_{\rho_2}(R) = L_{\rho_1}\{\gamma_1\} + L_{\rho_2}\{\gamma_2\}$. From these it follows that

$$\lambda\{\gamma\} \geq L_{\rho_1}\{\gamma_1\} + L_{\rho_2}\{\gamma_2\} = \frac{L_{\rho_1}\{\gamma_1\}^2}{A_{\rho_1}\{\gamma_1\}} + \frac{L_{\rho_2}\{\gamma_2\}^2}{A_{\rho_2}\{\gamma_2\}}$$

and we have (35).

(c) *If* $\{\gamma_1\}$ *and* $\{\gamma_2\}$ *are contained in disjoint open sets* G_1 *and* G_2 *respectively and if every* $\gamma_1 \in \{\gamma_1\}$ *and every* $\gamma_2 \in \{\gamma_2\}$ *contains a* $\gamma \in \{\gamma\}$ *then*

$$\frac{1}{\lambda\{\gamma\}} \geq \frac{1}{\lambda\{\gamma_1\}} + \frac{1}{\lambda\{\gamma_2\}}. \tag{36}$$

It suffices to consider the case $\lambda\{\gamma\}>0$. In the determination of $\lambda\{\gamma\}$ we may take the supremum with respect to those ρ which are normalized by $L_\rho\{\gamma\}=1$. Let $\rho_j=\rho|G_j$ be extended to R as 0 on $R-G_j$. Then

$$A_\rho(R)\geq A_{\rho_1}(R)+A_{\rho_2}(R)\geq\frac{A_{\rho_1}(R)}{L_{\rho_1}\{\gamma_1\}^2}+\frac{A_{\rho_2}(R)}{L_{\rho_2}\{\gamma_2\}^2}\geq\frac{1}{\lambda\{\gamma_1\}}+\frac{1}{\lambda\{\gamma_2\}}$$

and therefore

$$\frac{A_\rho(R)}{L_\rho\{\gamma\}^2}\geq\frac{1}{\lambda\{\gamma_1\}}+\frac{1}{\lambda\{\gamma_2\}}$$

from which (36) follows.

(d) *Let $\{\gamma\}$ be the family of curves γ joining the two sides of length a in a rectangle with sides a and b. Then*

$$\lambda\{\gamma\}=\frac{b}{a}. \tag{37}$$

Let the rectangle be T: $0\leq x\leq a$, $0\leq y\leq b$ in a region R. Take ρ as 1 on T and 0 on $R-T$. Then $L_\rho\{\gamma\}=b$ and $A_\rho(R)=ab$. Therefore $\lambda\{\gamma\}\geq b^2/ab=b/a$.

For an arbitrary ρ subject to the normalization $L_\rho\{\gamma\}=1$ and with $\rho|R-T=0$, we clearly have $\int_0^b\rho(x,y)\,dy\geq1$ for each fixed x and thus $\iint_R\rho(x,y)\,dx\,dy\geq a$. Therefore by Schwarz's inequality

$$A_\rho(R)\cdot ab\geq a^2$$

and consequently $L_\rho\{\gamma\}^2/A_\rho(R)=1/A_\rho(R)\leq b/a$, i.e. $\lambda\{\gamma\}\leq b/a$.

(e) *Let $\{\gamma\}$ be the family of curves γ separating the circles $|z|=r_0$ and $|z|=r_1>r_0$. Then*

$$\lambda\{\gamma\}=\frac{2\pi}{\log\dfrac{r_1}{r_0}}. \tag{38}$$

Let T: $r_0\leq|z|\leq r_1$ be contained in a region R. Take ρ as $1/r$ on T and 0 on $R-T$. Then $\lambda\{\gamma\}\geq L_\rho\{\gamma\}^2/A_\rho(R)=2\pi/\log(r_1/r_0)$.

For an arbitrary ρ with normalization $L_\rho\{\gamma\}=1$ and $\rho|R-T=0$ we have $\int_0^{2\pi}\rho(r,\theta)\,r\,d\theta\geq1$ for each r, and by Schwarz's inequality $\int_0^{2\pi}\rho^2\,r\,d\theta\geq 1/2\pi r$. Therefore

$$A_\rho(R)\geq\frac{1}{2\pi}\log\frac{r_1}{r_0}$$

which in turn gives $L_\rho\{\gamma\}^2/A_\rho(R)=1/A_\rho(R)\leq 2\pi/\log(r_1/r_0)$, i.e. $\lambda\{\gamma\}\leq 2\pi/\log(r_1/r_0)$.

(f) *Suppose that in* (d) *and* (e) *there are a finite number of slits parallel to the sides of length b, or concentric circular slits between* $|z|=r_0$ *and* r_1 *respectively. Then* (37) *and* (38) *remain valid.*

The same is true if the number of slits is infinite provided their projections on a side of length a, or their concentric projections on a radius are of linear measure zero.

In fact in the proofs of (d) and (e) such omissions of slits have no effect on the reasoning.

Remark. Suppose in (e) we take the family $\{\gamma\}$ of curves γ joining the two circles. Then we can show as in (e) that $\lambda\{\gamma\}=(1/2\pi)\log(r_1/r_0)$. Here $\log(r_1/r_0)=\log \operatorname{mod}(E,\alpha,\beta)$ with $E=\{z\,|\,r_0<|z|<r_1\}$, $\alpha=\{z\,|\,|z|=r_0\}$, and $\beta=\{z\,|\,|z|=r_1\}$. By the conformal invariance of λ and of the modulus,

$$\lambda\{\gamma\}=\frac{1}{2\pi}\log \operatorname{mod}(E,\alpha,\beta)$$

for every doubly connected region E.

Actually this is true for the general configuration (E,α,β) of I.1 B and the family $\{\gamma\}$ of curves γ joining α and β. The identity is still valid in the extreme situation where β is the ideal boundary of the complement $E\cup\alpha$ of a regular region on a Riemann surface. For this and further properties of extremal length we refer the reader to the monographs Rodin-Sario [3] and Sario-Oikawa [27].

8 C. Perimeter of a Set. Let z_0 be a point in a plane region R, X the complement of R, and X_0 a subset of X. Consider the class $\{\gamma\}_r$ of Jordan curves γ in R separating z_0 from X_0 and maintaining a distance of at least r from z_0. We shall show that

$$\mu(z_0,X_0)=\lim_{r\to 0}\frac{1}{r}\exp\left(-\frac{2\pi}{\lambda\{\gamma\}_r}\right) \tag{39}$$

exists.

Note first that the extremal length $\lambda\{\gamma\}_r$ tends to zero as $r\to 0$. Moreover if $r'<r$ then by (36) and (38)

$$\frac{1}{\lambda\{\gamma\}_{r'}}\geq\frac{1}{\lambda\{\gamma\}_r}+\frac{1}{2\pi}\log\frac{r}{r'},$$

that is

$$-\frac{2\pi}{\lambda\{\gamma\}_r}+\log\frac{1}{r}\geq-\frac{2\pi}{\lambda\{\gamma\}_{r'}}+\log\frac{1}{r'}.$$

On taking the exponential function on both sides we obtain

$$\frac{1}{r}\exp\left(-\frac{2\pi}{\lambda\{\gamma\}_r}\right) \geq \frac{1}{r'}\exp\left(-\frac{2\pi}{\lambda\{\gamma\}_{r'}}\right), \qquad r > r' > 0$$

and conclude that (39) converges.

The differential $\mu(z_0, X_0)|dz_0|$ is invariant under univalent conformal mappings of R if the transforms of the components of X are suitably defined (cf. 8 D). The relative invariant $\mu(z_0, X_0)$ is by (34) a nondecreasing set function of X_0 and a nonincreasing function of R. The quantity $\mu(z_0, X_0)$ will be called the *perimeter* of X_0 with respect to R and the center z_0.

For an example take $R = \{z \,|\, |z - z_0| < t\}$ and $X_0 = \{z \,|\, |z - z_0| \geq t\}$. Then by (38), $\lambda\{\gamma\}_r = 2\pi/\log(t/r)$ for $t > r > 0$. Therefore $-2\pi/\lambda\{\gamma\}_r + \log(1/r) = \log(1/t)$, or $(1/r)\exp(-2\pi/\lambda\{\gamma\}_r) = 1/t$. Hence the perimeter $\mu(z_0, X_0)$ is $1/t$.

8 D. Perimeter of a Point. The invariant $\mu(z_0, X_0)$ depends only on the set of those components of X which contain points of X_0, not on the individual points within a component. We conclude that the *perimeter* $\mu(z_0, p)$ *of a point* p equals that of the component containing p.

We are now ready to state:

THEOREM. *The invariants for SD and SB coincide and*

$$M_{SD}(z_0, R) = M_{SB}(z_0, R) = \max_{p \in X} \mu(z_0, p). \tag{40}$$

The proof will be given in 8 E – 8 G.

8 E. $M_{SB} = \max \mu$ for a Regular Region. First suppose that R is bounded by a finite number of analytic curves $\Gamma_1, \ldots, \Gamma_n$. Then $\mu(z_0, p)$ has only n values, one for each component of the complement of R. Let f_k be the univalent conformal mapping of R onto a region R_k bounded by the unit circle Γ_k' which corresponds to Γ_k, and $n-1$ concentric circular slits; we assume that $f_k(z_0) = 0$.

By (38) and the example in 8 C we infer that $\mu(0, \Gamma_k') = 1$. By the conformal invariance $\mu(z_0, \Gamma_k) = \mu(0, \Gamma_k')\,|f_k'(z_0)|$ we obtain

$$\mu(z_0, \Gamma_k) = |f_k'(z_0)| \leq M_{SB}(z_0, R)$$

since $f_k \in SB_0(R)$. From this we conclude that

$$\sup_{p \in X} \mu(z_0, p) \leq M_{SB}(z_0, R).$$

Conversely assume that $f \in SB_0(R)$. By Carathéodory's theorem the outer contour C of $f(R)$ corresponds to some contour of R, say Γ_k. Since $|f| < 1$ we have $\mu(0, C) \geq \mu(0, |w| = 1) = 1$ where $\mu(0, |w| = 1)$ is taken with respect to the region $|w|<1$. We deduce by the conformal invariance $\mu(z_0, \Gamma_k)=\mu(0, C)\,|f'(z_0)|$ that

$$|f'(z_0)| \leq \mu(z_0, \Gamma_k)$$

and therefore

$$M_{SB}(z_0, R)=\max_{p\in X} \mu(z_0, p). \tag{41}$$

8 F. $M_{SD}=M_{SB}$ for a Regular Region. We remark that $M_{SD}(z_0, R)= M_{SD}(0, R_k)\,|f_k'(z_0)|$ and $M_{SB}(z_0, R)=M_{SB}(0, R_k)\,|f_k'(z_0)|$.

For $f\in SD_0(R_k)$ set

$$L(r)=\int_{|w|=r} |f'|\,|dw|$$

whenever $|w|=r$ does not contain any slit of R_k, and

$$D(r)=\iint_{|w|<r} |f'|^2\, du\,dv, \qquad w=u+iv,$$

for every $r<1$. By Schwarz's inequality

$$L(r)^2 \leq 2\pi r\, D'(r)$$

for all but a finite number of r.

Since the image of $|w|<r$ under f always has the image of $|w|=r$ as its outer contour the isoperimetric inequality implies

$$L(r)^2 \geq 4\pi\, D(r).$$

Thus we have

$$\frac{D'(r)}{D(r)} \geq \frac{2}{r}.$$

An integration from r to 1 gives

$$D(r) \leq D(1)\, r^2 \leq \pi r^2$$

or equivalently, by virtue of the subharmonicity of $|f'(w)|^2$,

$$|f'(0)|^2 \leq \frac{1}{\pi r^2} \iint_{|w|<r} |f'(w)|^2\, du\,dv \leq 1.$$

This shows that $M_{SD}(0, R_k) \leq 1$ and consequently (cf. 8 E)

$$M_{SD}(z_0, R) \leq |f_k'(z_0)| = \mu(z_0, \Gamma_k) \leq \max_{p\in X} \mu(z_0, p).$$

On the other hand $f_k \in SD_0$ and

$$M_{SD}(z_0, R) \geq |f_k'(z_0)| = \mu(z_0, \Gamma_k)$$

for all k and thus

$$M_{SD}(z_0, R) \geq \max_{p \in X} \mu(z_0, p).$$

Statement (40) now follows from (41).

8 G. The General Case. For an arbitrary R let $\{R_n\}$ be a regular exhaustion with $z_0 \in R_1$, denote by X_n the complement of R_n, and write $\mu_n(z_0, p)$ for the invariant taken with respect to R_n. We have trivially $\mu \leq \mu_n$ and hence

$$\sup_{p \in X} \mu(z_0, p) \leq \lim_{n \to \infty} \max_{p \in X_n} \mu_n(z_0, p)$$
$$= \lim_{n \to \infty} M_{SB}(z_0, R_n) = M_{SB}(z_0, R)$$

where in the last equality we have invoked 4 B.(a).

Suppose next that $w = f(z)$, $f(z_0) = 0$, maps R conformally onto a subregion R_w of $|w| < 1$. There exists a sequence of points $w_m = f(z_m)$ which tends towards the unbounded component Y of the complement of R_w. Let p be a limit point of $\{z_m\}$. Any curve γ which separates z_0 from p has an image which separates 0 from $|w| = 1$. This means that $\mu(0, q) \geq \mu(0, |w| = 1) = 1$, with $q \in Y$. From $\mu(z_0, p) = \mu(0, q)\,|f'(z_0)|$ we conclude that

$$|f'(z_0)| \leq \mu(z_0, p)$$

and a fortiori

$$M_{SB}(z_0, R) \leq \mu(z_0, p).$$

We have proved that

$$M_{SB}(z_0, R) = \sup_{p \in X} \mu(z_0, p).$$

By virtue of 4 B.(a) the identity $M_{SB}(z_0, R_n) = M_{SD}(z_0, R_n)$ remains valid for R.

9. O_{AD}-Regions and Extremal Distances

9 A. Extremal Distance. We append here two properties of O_{AD}-regions to supplement the discussion in I.6 and I.8. We shall need the concept of extremal distance. Let X_1, X_2 be disjoint compact sets in a region R or on its boundary. The *extremal distance* $\lambda_R(X_1, X_2)$ between X_1 and X_2 with respect to R is by definition the extremal length $\lambda\{\gamma\}$ of the family of curves γ joining X_1 and X_2 in R.

Let R be an open rectangle whose sides have lengths a and b, and let X_1 and X_2 be the opposite sides of length b. Then by (37)

$$\lambda_R(X_1, X_2) = \frac{a}{b}.$$

Suppose now that a totally disconnected compact set X is removed from R. By (34)

$$\lambda_{R-X}(X_1, X_2) \geq \lambda_R(X_1, X_2) = \frac{a}{b}.$$

The question is: for what X does the equality hold? The answer gives a characterization of AD-null sets:

THEOREM. *The complement of a compact set X belongs to O_{AD} if and only if $\lambda_{R-X}(X_1, X_2) = \lambda_R(X_1, X_2)$ for a rectangle $R \supset X$ and for each pair (X_1, X_2) of opposite sides of R.*

We may assume that the rectangle R lies symmetrically with respect to the coordinate axes, with the sides of length a parallel to the x-axis. Let W be the complement of X and let $\{W_n\}_1^\infty$ be its exhaustion. We denote by P_n the function $P_0(z)$ for W_n with $\zeta = \infty$ (see I.8 A). Then $\lim_n P_n(z)$ is the function $P_0(z)$ for W with $\zeta = \infty$, which in turn is the function z if $W \in O_{AD}$ (Theorem I.8 C):

$$\lim_{n \to \infty} P_n(z) = z.$$

Hence for large n, $P_n(z)$ maps ∂R onto the boundary of a (curvilinear) quadrilateral which differs very little from ∂R. We may thus find $a_n > a$ and $b_n < b$ tending to a and b such that $|\mathrm{Re}(P_n(z))| \leq a_n/2$ on the vertical sides of R and $|\mathrm{Im}(P_n(z))| \geq b_n/2$ on the horizontal sides. Let the rectangle with sides a_n and b_n be denoted by R_n. Every curve which joins the vertical sides of R_n contains the image under P_n of a curve joining the vertical sides of R within $R - X$. By (34)

$$\lambda_{R-X}(X_1, X_2) \leq \frac{a_n}{b_n},$$

and on letting $n \to \infty$ we obtain $\lambda_{R-X}(X_1, X_2) \leq a/b$. Thus if X has its complement W in O_{AD} then $\lambda_{R-X} = \lambda_R$.

Conversely assume that $\lambda_{R-X} = a/b$ for the vertical sides and $\tilde{\lambda}_{R-X} = b/a$ for the horizontal sides. Let $t = s(z)$ be an arbitrary univalent conformal mapping of W with pole at ∞. It maps ∂R onto the boundary of a Jordan region, which in turn is mapped onto a rectangle R' of sides a' and b' by a univalent conformal mapping $w = \varphi(t)$. By (34) and the conformal invariance of λ_{R-X} we conclude that $a/b \geq a'/b'$. Similarly for $\tilde{\lambda}_{R-X}$ we obtain $b/a \geq b'/a'$ and thus

$$\frac{a'}{b'} = \frac{a}{b}.$$

Choose $\rho = |dw/dz|$ in $R - X$. For every curve γ joining the vertical sides of R in $R - X$

$$\int_\gamma \rho\, |dz| \geq a'.$$

By the definition of λ_{R-X} the region $R-X$ must be mapped onto a region whose area is at least

$$\frac{a'^2}{\lambda_{R-X}}=a'b'.$$

This means that the image of $R-X$ fills out all of R' except for a set of measure zero, and since the derivative $|\varphi'(t)|$ is bounded away from zero on the image under s of $R-X$ it follows that s must map W onto a region whose complement is of measure zero. Thus by Theorem I.8 C, $W \in O_{AD}$.

9 B. General Form. On applying an auxiliary mapping we see that R in Theorem 9 A may be an arbitrary quadrilateral. Thus we can restate Theorem 9 A in the following form, with the extremal distances taken between opposite sides of curvilinear quadrilaterals:

THEOREM. *A set X is AD-removable if and only if the removal of X does not change extremal distances.*

We also remark here that if X has an O_{AD}-complement but is not necessarily contained in a rectangle R, then the removal of X, or more precisely $X \cap R$, from R does not change extremal distances.

In fact let R' be a concentric rectangle with sides $a'>a$ and $b'<b$. Since X is totally disconnected we can find a curvilinear quadrilateral R'' which is contained in the concentric rectangle with sides a', b and contains the concentric rectangle with sides a, b' such that $\partial R'' \cap X = \emptyset$. The boundary $\partial R''$ encloses a compact subset X'' of X and we have $\lambda_{R''-X''}=\lambda_{R''}$. On the other hand we obtain by two applications of (34)

$$\lambda_{R-X} \leq \lambda_{R''-X''} = \lambda_{R''} \leq \lambda_{R'} = \frac{a'}{b'}.$$

Since a'/b' can be chosen arbitrarily close to a/b we find that $\lambda_{R-X} \leq a/b$ as desired.

9 C. Projections. Linear measure plays a significant role in the O_{AD}-problem in the following manner:

THEOREM. *A set X is AD-removable if its projections in two orthogonal directions are of linear measure zero. On the other hand if X is AD-removable then any two points in the complement W of X can be joined by a curve in W whose length differs arbitrarily little from the distance between the points.*

The first part follows easily from (37) (cf. 8 B.(f)) and Theorem 9 B. If two points have a distance in W which exceeds their distance in the plane it is clear that a thin rectangle R can be constructed such that the distance between two sides of R is greater in $R-X$ than in R. This implies $\lambda_{R-X}>\lambda_R$, and $W\notin O_{AD}$ by Theorem 9 B.

Remark. Concerning the first part of the above theorem the following interesting observation was made in TAMURA-OIKAWA-YAMAZAKI [2]:

Let $e=E(3^\infty)$ be the Cantor ternary set (cf. I.6 A) and consider $E_1=e\times e$. The projections e of E_1 into the real and imaginary axes have vanishing linear measure. Therefore E_1 is AD-removable. Rotate E_1 by 45° about its center and denote by E_2 the resulting set which is also AD-removable. Since E_1 and E_2 are compact and AD-removable so is the set

$$E=E_1\cup E_2$$

(cf. I.8 D). It is not difficult to see that the projection of E on any line is an interval. From this we conclude:

There exists an AD-removable set whose projection on any line is a nondegenerate interval.

This indicates that the condition for AD-removability in the first part of the above theorem is far from being necessary.

10. Linear Sets

10 A. Linear Measure and M_{AB}. In the present no. we always choose for z_0 of (22) the point ∞. Let X be the complement of a region R containing ∞. We consider the case where the compact set X does not divide the plane. We can then view the invariants $M_{AB}(\infty, R)$, $M_{AD}(\infty, R)$, and $M_{SB}(\infty, R)$ as functions of a compact set X.

Let $f(z)=c/z+\cdots\in AB_0(R)$. By Cauchy's formula

$$|c|\leq\frac{1}{2\pi}\int_\gamma |f(z)|\,|dz|\leq\frac{1}{2\pi}\int_\gamma |dz|$$

where γ is a curve in R which separates X from ∞. If we denote by Λ the infimum of the lengths of such γ's then

$$M_{AB}\leq\frac{1}{2\pi}\Lambda. \tag{42}$$

This again gives a part of 2 E.(6): $M_1\subset O_{AB}$.

We next consider the case where X lies on a straight line, for example the real axis. If the linear measure of X is L then we have $\Lambda=2L$, and (42) implies

$$M_{AB}\leq\frac{1}{\pi}L. \tag{43}$$

An inequality in the opposite direction is obtained by considering the function

$$f(z)=\int_X \frac{dx}{z-x}=\frac{L}{z}+\cdots.$$

It is easily seen that $|\mathrm{Im}\, f|<\pi$, and the function

$$\frac{e^{f/2}-1}{e^{f/2}+1}$$

belongs to $AB_0(R)$, with the first coefficient $L/4$. It follows that

$$M_{AB}\geq\tfrac{1}{4}L. \tag{44}$$

Once again we return to Painlevé sets. In 2 E we obtained $M_1\subset O_{AB}\subset M_{1+\varepsilon}$ but there was no way of closing the gap. However for linear sets X, (43) and (44) show that $M_1=O_{AB}$. More generally:

THEOREM. *A set X on an analytic curve has an O_{AB}-complement R if and only if it is of length zero.*

Let R' be a bounded open neighborhood of X. What is to be proved is that every AB-function on $R'-X$ is analytic on R' if and only if X is of linear measure zero. But it is clearly sufficient to prove the corresponding local statement, which follows from the fact that every point on the analytic curve γ on which X lies has a neighborhood which can be mapped conformally so that γ will correspond to a segment of the real axis. Here it is of course necessary to choose a neighborhood whose boundary does not intersect X. If X is totally disconnected this is always possible, and if X contains an arc neither M_{AB} nor the linear measure can be zero.

Remark. It was also shown in AHLFORS-BEURLING [5] that $M_{AD}\leq L/4$ for a linear set X of length L.

10 B. Invariants M_{SB} and M_{AD}. The relation $M_{SB}=M_{SD}\leq M_{AD}$ can be supplemented by the following property:

THEOREM. *The invariants M_{SB} and M_{AD} for linear sets X are both positive or both zero.*

Assume $M_{SB}=0$. By (40) the perimeter μ vanishes for every boundary point of R, the complement of X. This property is obviously invariant under univalent mappings of R onto R' and a fortiori the complement of R' is always totally disconnected. Thus in particular the linearity of X implies that both slit mappings P_0 and P_1 must degenerate. It follows that $a_1^{(0)}=a_1^{(1)}=0$ (cf. I.8) and $M_{AD}=0$.

10 C. Sets of Capacity Zero. Let X be the complement of a region R containing ∞. Consider the class $\widetilde{AB}(R)$ of all multivalued analytic functions f with $|f|$ single-valued and ≤ 1. We set

$$\operatorname{cap}(X) = M_{\widetilde{AB}}(\infty, R).$$

For a general X we define $\operatorname{cap}(X)$ as the supremum of $\operatorname{cap}(Y)$ for compact $Y \subset X$. The quantity $\operatorname{cap}(X)$ is called the *capacity* of X. Thus

$$M_{SB} = M_{SD} \leq M_{AD} = M_{SE} \leq M_{AE} = M_{AB} \leq \operatorname{cap}(X) \tag{45}$$

and the vanishing of the capacity of X implies the degeneracy of X with respect to all six classes.

Consider the special case where there exists a harmonic function u on $R - \infty$ such that $u = c$ on ∂R and $u(z) + \log|z| \to 0$ as $z \to \infty$. Such a function u is called the *capacity function* of X. Then

$$\operatorname{cap}(X) = e^{-c}.$$

In fact take $f_0 \in \widetilde{AB}(R)$ such that $|f_0(z)| = e^{u(z) - c}$ or $f_0(z) = e^{-c}(z^{-1} + \cdots)$. Then $|f_0'(\infty)| = e^{-c}$. For $f \in \widetilde{AB}_0(R)$ the maximum principle and the fact that $|f_0(z)| = 1$ on ∂R give $|f(z)/f_0(z)| \leq 1$ on R and in particular $|f'(\infty)| \leq |f_0'(\infty)|$. Thus $M_{\widetilde{AB}}(\infty, R) = e^{-c}$.

We shall return to a more systematic treatment of capacity functions and the capacity in Chapter V.

10 D. Sets on a Circle. We now consider linear sets X with O_{AD}-complements. It will be convenient to choose the sets on the unit circle.

Let X be a compact set on $|z| = 1$ and denote by X' the complement of X with respect to $|z| = 1$. We shall first prove:

$$M_{AD}(0, \infty, R) = \sqrt{2 \log \frac{1}{\operatorname{cap}(X')}} \tag{46}$$

where R is the complement of X (cf. 7 A.(29)).

We begin the proof by supposing that X consists of a finite number of closed arcs Y_i, and thus X' consists of open arcs Y_i'. As in 7 C, $M_{AD}(0, \infty, R)$ is determined by mappings Q_0 and Q_1. Clearly $Q_1(z) \equiv z$ in the present case and thus

$$\beta_1 = Q_1'(0) = 1.$$

Again in the present situation $1/\overline{Q_0(1/\bar{z})}$ is a radial slit mapping whose residue at ∞ is $1/\bar{\beta}_0$. Therefore we must have

$$\overline{Q_0\left(\frac{1}{\bar{z}}\right)}\, Q_0(z) = \bar{\beta}_0.$$

From this relation we conclude that

$$|Q_0(z)| = \sqrt{\beta_0}$$

on each arc Y_i'. On the other hand we know that $\partial \log|Q_0(z)|/\partial n = 0$ on Y_i. For this reason $(1/\sqrt{\beta_0})\,Q_0$ can be continued from $|z|>1$ across the Y_i to a function Q defined and single-valued outside of the arcs Y_i' which satisfies

$$\overline{Q\left(\frac{1}{\bar{z}}\right)} = Q(z).$$

Now we see that

$$u(z) = \tfrac{1}{2}\left(-\log|z\,Q(z)| - \tfrac{1}{2}\log\beta_0\right)$$

is harmonic outside of X' except for a logarithmic pole at ∞ such that $u(z)+\log|z|$ vanishes for $z=\infty$. Thus u is the capacity function of X' and $u = -4^{-1}\log\beta_0$ on X'. As a consequence

$$\operatorname{cap}(X') = \sqrt[4]{\beta_0}.$$

On the other hand $Q_1'(0) = \beta_1 = 1$, and $Q_0'(0) = \beta_0$ with 7 C.(31) gives

$$M_{AD}(0,\infty,R) = \sqrt{\frac{1}{2}\log\frac{1}{\beta_0}} = \sqrt{2\log\frac{1}{\sqrt[4]{\beta_0}}}$$

as desired.

From the above special case we obtain (46) in the general case by an inner approximation of X' and a routine limiting process.

10 E. Circular Sets with O_{AD}-Complements. As a consequence of (46) and Theorem 7 A we conclude at once:

THEOREM. *A closed set X on the unit circle is AD-removable if and only if the capacity of its complement with respect to the circle is* 1.

This does not of course imply the vanishing of the capacity of X (cf. 11 D). The same is true if we replace in the above theorem a circular set X by a set X on an analytic arc.

11. Counterexamples

11 A. General Relations. Once more we make use of the inequalities

$$M_{SB} = M_{SD} \leq M_{AD} = M_{SE} \leq M_{AE} = M_{AB}.$$

In terms of the corresponding null classes we have

$$O_{SB} = O_{SD} \supset O_{AD} = O_{SE} \supset O_{AE} = O_{AB}.$$

We now complete our discussion of these relations by giving examples to prove the strictness of the two inclusions:

$$O_{SD} > O_{AD}, \qquad O_{SE} > O_{AE}.$$

These examples are mostly generalized Cantor sets and will be discussed in connection with the class M_λ of regions R whose complements X have a vanishing λ-dimensional Hausdorff measure. Explicitly we shall show that $O_{SB} - M_1 \neq \emptyset$, $O_{SB} - M_2 \neq \emptyset$, and $O_{AD} - M_1 \neq \emptyset$.

11 B. M_1 and O_{SB}. We start with $O_{SB} - M_1 \neq \emptyset$:

THEOREM. *There exists a linear Cantor set whose complement belongs to $O_{SB} - M_1$.*

Let $\{q_i\}_1^\infty$ be an increasing sequence of real numbers $0 < q_i < 1$ and $\{n_i\}_1^\infty$ an increasing sequence of positive integers. We shall construct a generalized linear *Cantor set* $E(q_i; n_i)$ as a closed subset of $E_0 = \{t \mid 0 \leq t \leq 1\}$; this set is more general than those discussed in I.6. The first step is to divide E_0 into $2n_1 + 1$ subintervals, the odd ones of length $a_1 = q_1/(n_1+1)$ and the even ones of length $b_1 = (1-q_1)/n_1$. For short they will be referred to as the a_1-intervals and b_1-intervals, and the union of the closed a_1-intervals is denoted by E_1. In the next step each a_1-interval is subdivided into $2n_2+1$ alternating a_2- and b_2-intervals of lengths

$$a_2 = \frac{q_2}{n_2+1} a_1, \qquad b_2 = \frac{1-q_2}{n_2} a_1.$$

The union of all closed a_2-intervals is denoted by E_2. The process is repeated and we obtain a nested sequence $E_1 \supset E_2 \supset \cdots$. The intersection $E = \bigcap_1^\infty E_k$ is the desired Cantor set $E(q_i; n_i)$. The length of $\bigcap_1^n E_k$ is $\prod_1^n q_k$ and thus the length of $E(q_i; n_i)$ is $\prod_1^\infty q_k$. Hence the length is positive if and only if

$$\sum_1^\infty (1 - q_k) < \infty. \tag{47}$$

Under this condition the complement R of $E = E(q_i; n_i)$ does not belong to M_1.

Next we seek a condition on $\{q_i\}$ and $\{n_i\}$ which will ensure that R belongs to O_{SB} or, equivalently, that $M_{SB}(\infty, R) = \sup_{p \in E} \mu(\infty, p)$ vanishes (cf. Theorem 8 D). By (36) and (38) this will be the case if each point of E can be surrounded by a sequence of disjoint annuli σ_ν which do not meet E and whose decreasing radii r_ν'' and r_ν' satisfy the condition

$$\sum_1^\infty \log \frac{r_\nu''}{r_\nu'} = \infty. \tag{48}$$

Let us fix our attention on a point $t \in E$. This point belongs to a certain a_k-interval, say $a_k(t)$, for each k. We surround t by annuli centered at the midpoint of $a_k(t)$ which pass through the b_k-intervals contained in $a_{k-1}(t)$. Of course some of these may intersect the real axis in only one b_k-interval. We consider only those annuli whose inner radius is at least equal to $a_k + b_k$ and outer radius is at most b_{k-1}. It is clear that such annuli cannot intersect any a_{k-1}-interval other than $a_{k-1}(t)$ and hence cannot meet E. Furthermore an annulus of the $(k+1)$st generation cannot meet an annulus of the kth generation, for a common point would at once be at a distance dominating $a_k + b_k$ from the center of $a_k(t)$ and at a distance less than b_k from the center of $a_{k+1}(t)$; this is impossible since the two centers have a mutual distance less than a_k.

The smallest annulus of the kth generation which satisfies the imposed conditions has radii $(3/2)\, a_k + b_k$ and $(3/2)\, a_k + 2b_k$, and these radii are increased by $a_k + b_k$ at each step of the kth generation. The number ν_k of the last permissible annulus is therefore determined by

$$\tfrac{1}{2} a_k + b_k + \nu_k (a_k + b_k) \le b_{k-1}.$$

Since $a_k + b_k < a_{k-1}/n_k$ and $b_{k-1} > ((1-q_{k-1})/q_{k-1})\, a_{k-1} > (1-q_{k-1})\, a_{k-1}$, it is sufficient to take

$$\nu_k = [n_k(1-q_{k-1})] - 1$$

whenever this is positive; here [ ] is the greatest dominated integer. Moreover we have for the νth annulus $(1 \le \nu \le \nu_k)$ of the kth generation

$$\begin{aligned} &\log\left(\tfrac{1}{2} a_k + b_k + \nu(a_k + b_k)\right) - \log\left(\tfrac{1}{2} a_k + \nu(a_k + b_k)\right) \\ &\quad = -\log\left(1 - \frac{b_k}{\tfrac{1}{2} a_k + b_k + \nu(a_k + b_k)}\right) > \frac{b_k}{\tfrac{1}{2} a_k + b_k + \nu(a_k + b_k)} \\ &\quad > \frac{1}{\nu+1}\, \frac{b_k}{a_k + b_k} > \frac{1-q_k}{\nu+1}. \end{aligned}$$

Thus the annuli of the kth generation contribute to the sum (48) an amount greater than

$$(1-q_k)\left(\frac{1}{2} + \frac{1}{3} + \cdots + \frac{1}{\nu_k + 1}\right) > (1-q_k) \log \frac{\nu_k + 2}{2} \ge (1-q_k) \log \frac{n_k(1-q_{k-1})}{2}.$$

We conclude that $R \in O_{SB}$ if

$$\sum_{1}^{\infty} (1-q_k) \log \frac{n_k(1-q_{k-1})}{2} = \infty. \tag{49}$$

Thus by choosing $\{q_i\}$ and $\{n_i\}$ so as to satisfy (47) and (49) we obtain R in $O_{SB} - M_1$. For example take $q_k = 1 - k^{-2}$, $n_k = k^2\, 2^k$.

11 C. M_2 and O_{SB}. We have seen in Theorem I.8 C that $M_2 \supset O_{AD}$. We shall show that $O_{SB} - M_2 \neq \emptyset$. This gives the desired strict relation $O_{SB} > O_{AD}$.

THEOREM. *There exists a region R whose complement is a 2-dimensional Cantor set such that $R \in O_{SB} - M_2$ and therefore $R \in O_{SB} - O_{AD}$.*

The Cartesian product X of two identical linear Cantor sets of positive length is a 2-dimensional Cantor set of positive area. Hence the complement R of X does not belong to M_2. This R is again in O_{SB} whenever condition (49) is valid. The proof is the same as above except that it is convenient to replace the circular rings by quadratic frames. In (48) we let r_ν''/r_ν' be the ratio of the outer and inner dimensions of the frames, and it is elementary to show that the divergence of (48) is still a sufficient condition for $R \in O_{SB}$.

11 D. M_1 and O_{AD}. Next we construct a linear set X whose complement R does not belong to M_1 but belongs to O_{AD} (cf. also II.E). For linear sets, $M_1 = O_{AB}$ (cf. Theorem 10 A) and thus this serves also to prove the strict inclusion $O_{AB} < O_{AD}$.

More precisely we construct a compact set X on $|z|=1$ with positive length such that the complement X' of X with respect to $|z|=1$ has capacity 1. Then by Theorem 10 E, $R \in O_{AD}$. Observe that $\operatorname{cap}(X) > 0$.

THEOREM. *There exists a region R whose complement X is on a circle or on a line such that $R \in O_{AD} - M_1$.*

It is easy to see that the capacity of a finite number of open arcs on $|z|=1$ is equal to that of the corresponding closed arcs.

Let α be an arc on $|z|=1$ and map the complement of α conformally onto $|w|>\rho$ by a univalent function with expansion $w(z) = z + a_0 + a_1/z + \cdots$ at $z=\infty$. Such a ρ is determined uniquely by α and is referred to as the *mapping radius* of α. Clearly $\log(1/|w(z)|)$ gives the capacity function of α and a fortiori $\operatorname{cap}(\alpha) = \rho$ (cf. 10 C).

Suppose moreover that α has length $4/\lambda$. By applying elementary transformations one readily sees that $\rho = \sin(1/\lambda)$. For further information concerning relations between mapping radii and capacities we refer the reader to e.g. TSUJI [5, pp. 84–86].

Let u be the capacity function of the arc

$$X_1' = \{e^{i\theta} \,|\, |\theta - \theta_0| < 2/\lambda \pmod{2\pi}\}.$$

Then $u(z^n)/n$ is seen to be the capacity function of the set

$$X_n' = \{e^{i\theta} \,|\, |n\theta - \theta_0| < 2/\lambda \pmod{2\pi}\}$$

and we conclude that

$$\operatorname{cap}(X_n') = \sqrt[n]{\sin(1/\lambda)} = 1 - \frac{\log\lambda}{n} + O\left(\left(\frac{\log\lambda}{n}\right)^2\right),$$

and the length of X_n' is $4/n\lambda$.

Choose a sequence $\{X_n'\}_1^\infty$ of such sets of length $L_n = 4/n\lambda_n$ with $\lim_{n\to\infty}(\log\lambda_n)/n = 0$ and $\sum_1^\infty 4/n\lambda_n \le L$ for some $L < 2\pi$. Then the union X' of the sequence $\{X_n'\}_1^\infty$ has a length not greater than L, and its capacity is 1. For example the choice $\lambda_n = 8n/L$ qualifies.

As a consequence of the theorem we also obtain the strict inclusion

$$SO_{AB} < SO_{AD}$$

anticipated at the end of 2 C. For the proof take a compact set E which is AD-removable but not AB-removable and is contained in the unit disk. Let γ be the unit circle and R the unit disk less E. Then $(R,\gamma) \in SO_{AD}$ but $(R,\gamma) \notin SO_{AB}$.

11 E. Strict Inclusion $M_1 < O_{AB}$. We close this section by showing that the first inclusion relation in Theorem 2 E is strict: For plane regions

$$M_1 < O_{AB},$$

i.e. there exists a Painlevé null set X (cf. 2 D, 5 D) whose 1-dimensional Hausdorff measure (cf. 2 E) is positive. This was first indicated by VITUŠKIN [1]. The following surprisingly simple example is due to GARNETT [1].

THEOREM. *The 2-dimensional Cantor set $E^2(p_1 p_2 \ldots)$ with $p_n = 2$ for every n (cf. I. 6 B) is a Painlevé null set but of positive linear measure.*

We shall denote the above Cantor set by X. The complement R of X with respect to $|z| \le \infty$ belongs to $O_{AB} - M_1$. The proof will be given in 11 F – 11 K.

11 F. Positive Length. Denote by E_0 the unit square $[0,1]\times[0,1]$ and by E_n the Cartesian product $E^2(p_1 \ldots p_n) = E(p_1 \ldots p_n) \times E(p_1 \ldots p_n)$, $n = 1, 2, \ldots$ (cf. I. 6 A). The set E_n consists of 4^n squares E_{nj} $(j = 1, \ldots, 4^n)$. Clearly

$$X = \bigcap_{n=1}^{\infty} E_n.$$

Designate by X_{nj} the intersection $X \cap E_{nj}$.

Let Y be the projection of $e^{-\alpha i} X$ $(\alpha = \arctan \frac{1}{2})$ on the real axis. The projection of $e^{-\alpha i} E_n$ is $[0, 3/\sqrt{5}]$ for every n, and therefore $Y = [0, 3/\sqrt{5}]$.

From the definition of Hausdorff measure (cf. 2 E) it is easily seen that $h^*(e^{-\alpha i} X) \geq h^*(Y) = 3/\sqrt{5}$. We conclude that $h^*(X) > 0$, that is, X is of positive 1-dimensional Hausdorff measure.

11 G. Analytic Capacity. The conformal invariant

$$ac(X) = M_{AB}(\infty, R)$$

is often referred to as the *analytic capacity* of X. We denote by $A(X, K)$ the class of analytic functions f on R such that $|f| \leq K$ and $f(\infty) = 0$. The Taylor expansion of $f \in A(X, K)$ at ∞ has the form

$$f(z) = \frac{a(f)}{z} + \sum_{n=2}^{\infty} \frac{a_n}{z^n},$$

with $\lim_{z \to \infty} z f(z) = f'(\infty) = a(f)$. Therefore

$$ac(X) = \sup_{f \in A(X, K)} \frac{|a(f)|}{K}.$$

By the linear transformation $w = 4^{-n} z + c_{nj}$, with c_{nj} the lower left corner point of X_{nj}, X is mapped onto X_{nj} and it follows that $ac(X_{nj}) = 4^{-n} ac(X)$. We must show that $ac(X) = 0$. To this end we shall deduce auxiliary relations in 11 G – 11 J.

Take an arbitrary $f \in A(X, K_1)$ where K_1 is a finite positive constant. For $z \notin X$, let Γ_{nj}^z be a cycle in R with winding number 1 (resp. 0) with respect to points in X_{nj} (resp. $(X - X_{nj}) \cup \{z\}$). Set

$$f_{nj}(z) = -\frac{1}{2\pi i} \int_{\Gamma_{nj}^z} \frac{f(\zeta)}{\zeta - z} \, d\zeta.$$

By Cauchy's integral theorem the functions f_{nj} $(j = 1, \ldots, 4^n)$ are independent of the choice of Γ_{nj}^z and can therefore be considered to be analytic in the complement of X_{nj}. We maintain:

(a) $f(z) = \sum_{j=1}^{4^n} f_{nj}(z)$ *for* $z \notin X$;

(b) *there exists a finite constant* $K_2 = (1 + 6/\pi) K_1$ *such that* $f_{nj} \in A(X_{nj}, K_2)$ $(j = 1, \ldots, 4^n)$;

(c) $|a(f_{nj})| \leq 4^{-n} K_2 \, ac(X)$ $(j = 1, \ldots, 4^n)$.

In fact, let $\Gamma^z = \sum_{j=1}^{4^n} \Gamma_{nj}^z \subset R$. Cauchy's integral formula yields $f(z) = -(1/2\pi i) \int_{\Gamma^z} f(\zeta)(\zeta - z)^{-1} d\zeta$, from which (a) follows.

Take the open square S concentric with E_{nj} and with side $3/4^n$. We can choose $\Gamma^z - \Gamma_{nj}^z$ so that $\Gamma^z - \Gamma_{nj}^z - \partial S$ is homologous to zero with respect to R. If $z \in S - X_{nj}$, then by Cauchy's integral formula

$$f(z) - f_{nj}(z) = -\frac{1}{2\pi i} \int_{\Gamma^z - \Gamma^z_{nj}} \frac{f(\zeta)}{\zeta - z}\, d\zeta$$

$$= -\frac{1}{2\pi i} \int_{\partial S} \frac{f(\zeta)}{\zeta - z}\, d\zeta.$$

Therefore

$$|f_{nj}(z)| \leq |f(z)| + |f(z) - f_{nj}(z)|$$

$$\leq K_1 + \frac{1}{2\pi} \int_{\partial S} \frac{|f(\zeta)|}{|\zeta - z|}\, |d\zeta|$$

$$\leq K_1 \left[1 + \frac{1}{2\pi} \cdot 4 \cdot \frac{3}{4^n} \left(\inf_{\zeta \in \partial S} |\zeta - z|\right)^{-1}\right].$$

This implies that

$$\limsup_{z \notin X_{nj},\, z \to X_{nj}} |f_{nj}(z)| \leq K_1 \left(1 + \frac{6}{\pi}\right).$$

The maximum modulus principle yields (b).

To prove assertion (c) we only have to observe that

$$|a(f_{nj})| \leq \sup_{\varphi \in A(X_{nj}, K_2)} |a(\varphi)| = K_2 \sup_{\varphi \in A(X_{nj}, 1)} |a(\varphi)|$$

$$= K_2\, ac(X_{nj}) = 4^{-n} K_2\, ac(X).$$

11 H. Cauchy Potentials. The Cauchy potentials

$$h_{nj}(z) = 4^{2n}\, a(f_{nj}) \iint_{E_{nj}} \frac{d\xi\, d\eta}{\zeta - z} \qquad (\zeta = \xi + i\eta)$$

are analytic in the complements of the corresponding E_{nj}'s. We claim:

The sequence $\{h_n(0)\}_1^\infty$ *with* $h_n(z) = \sum_{j=1}^{4^n} h_{nj}(z)$ *is bounded.*

Observe that if $z \in E_{nj}$ then

$$\left|\iint_{E_{nj}} \frac{d\xi\, d\eta}{\zeta - z}\right| \leq \iint_{|\zeta - z| \leq \sqrt{2}\, 4^{-n}} \frac{d\xi\, d\eta}{|\zeta - z|} = \int_0^{2\pi} \int_0^{\sqrt{2}\, 4^{-n}} \frac{r\, dr\, d\theta}{r} = 2\sqrt{2}\, 4^{-n} \pi.$$

By Lemma 2 F we obtain

$$\left|\iint_{E_{nj}} \frac{d\xi\, d\eta}{\zeta - z}\right| \leq 4\sqrt{2}\, 4^{-n} \pi$$

for $|z|\le\infty$. In view of 11 G.(c) we infer on setting $K_3=4\sqrt{2}\,\pi K_2$ that

$$|h_{nj}(z)|\le K_3\,ac(X).$$

For $g_{nj}(z)=f_{nj}(z)+h_{nj}(z)$ and $g_n(z)=\sum_{j=1}^{4^n} g_{nj}(z)=f(z)+h_n(z)$ we have

$$|g_{nj}(z)|\le K_4 \qquad (z\notin X_{nj})$$

where $K_4=K_2+K_3\,ac(X)$. Let $g_n^*(0)=\limsup_{z\notin E_n,\,z\to 0}|g_n(z)|$ and $g_{nj}^*(0)=\limsup_{z\notin E_n,\,z\to 0}|g_{nj}(z)|$.

For an arbitrary $\varepsilon>0$ set $U_\varepsilon(z)=\{\zeta\,|\,|\zeta-z|<\varepsilon\}$. If $z\in U_\varepsilon(0)$ then

$$\begin{aligned}\left|\iint_{E_{nj}}\frac{d\xi\,d\eta}{\zeta}-\iint_{E_{nj}}\frac{d\xi\,d\eta}{\zeta-z}\right| &\le\left|\iint_{E_{nj}-U_\varepsilon(0)}\left(\frac{1}{\zeta}-\frac{1}{\zeta-z}\right)d\xi\,d\eta\right|\\ &\quad+\iint_{U_\varepsilon(0)}\frac{d\xi\,d\eta}{|\zeta|}+\iint_{U_{2\varepsilon}(z)}\frac{d\xi\,d\eta}{|\zeta-z|}\\ &=\left|\iint_{E_{nj}-U_\varepsilon(0)}\left(\frac{1}{\zeta}-\frac{1}{\zeta-z}\right)d\xi\,d\eta\right|+2\pi\varepsilon+4\pi\varepsilon.\end{aligned}$$

Therefore

$$\limsup_{z\to 0}\left|\iint_{E_{nj}}\frac{d\xi\,d\eta}{\zeta}-\iint_{E_{nj}}\frac{d\xi\,d\eta}{\zeta-z}\right|\le 6\pi\varepsilon.$$

This shows that $\lim_{z\to 0}h_n(z)=h_n(0)$.

From $|h_n(z)|\le|f(z)|+|g_n(z)|\le K_1+|g_n(z)|$ $(z\notin E_n)$, we obtain $|h_n(0)|\le K_1+g_n^*(0)$. Thus it suffices to show that $\{g_n^*(0)\}_1^\infty$ is bounded. We have

$$a(h_{nj})=h'_{nj}(\infty)=\lim_{z\to\infty}z\,h_{nj}(z)=-4^{2n}\,a(f_{nj})\iint_{E_{nj}}d\xi\,d\eta=-a(f_{nj}).$$

Therefore $a(g_{nj})=a(h_{nj}+f_{nj})=0$ and $g_{nj}(z)$ has a zero of order at least 2 at ∞. For $\zeta\in E_{nj}$ the function $(z-\zeta)^2\,g_{nj}(z)$ is analytic in the complement of E_{nj}. The maximum modulus principle yields

$$|(z-\zeta)^2\,g_{nj}(z)|\le\limsup_{z\notin E_{nj},\,z\to\partial E_{nj}}|z-\zeta|^2\,K_4\le 2\cdot 4^{-2n}\,K_4$$

for every $z\notin E_{nj}$ and $\zeta\in E_{nj}$. Hence on setting

$$d(z,E_{nj})=\inf_{\zeta\in E_{nj}}|z-\zeta|$$

we obtain for $z\notin E_{nj}$

$$|g_{nj}(z)|\le\frac{2K_4\,4^{-2n}}{d(z,E_{nj})^2}.$$

Let E_{nj_n} be the square with $0 \in E_{nj_n}$. Set

$$B_n = \sum_{j \neq j_n}^{1,\ldots,4^n} \frac{2K_4 4^{-2n}}{d(0, E_{nj})^2}.$$

In order to compare B_n with

$$B_{n-1} = \sum_{j \neq j_{n-1}}^{1,\ldots,4^{n-1}} \frac{2K_4 4^{-2(n-1)}}{d(0, E_{n-1,j})^2}$$

observe that

$$\sum_{j \neq j_n, E_{nj} \subset \{x < \frac{1}{2}, y < \frac{1}{2}\}} \frac{2K_4 4^{-2n}}{d(0, E_{nj})^2} = B_{n-1}$$

and

$$\sum_{E_{nj} \subset \{x > \frac{1}{2} \text{ or } y > \frac{1}{2}\}} \frac{2K_4 4^{-2n}}{d(0, E_{nj})^2} \le 3 \cdot 4^{n-1} \cdot \frac{2K_4 4^{-2n}}{2^{-2}}.$$

It follows that $B_n \le B_{n-1} + 6K_4 4^{-n}$ which in turn gives $B_n \le 2K_4$.

Let z be close to 0 with $z \notin E_n$. Then

$$|g_n(z)| \le \sum_{j=1}^{4^n} |g_{nj}(z)| = |g_{nj_n}(z)| + \sum_{j \neq j_n}^{1,\ldots,4^n} |g_{nj}(z)|$$

$$\le K_4 + \sum_{j \neq j_n}^{1,\ldots,4^n} \frac{2K_4 4^{-2n}}{d(z, E_{nj})^2}.$$

On taking upper limits of both extremes of the above as $z \to 0$ with $z \notin E_n$, we obtain

$$g_n^*(0) \le K_4 + B_n \le 3K_4.$$

11 I. Perturbation. We next assert:

If $a(f) > 0$, then for some n and j, $a(f_{nj}) \neq 4^{-n} a(f)$.

Suppose to the contrary that $a(f_{nj}) = 4^{-n} a(f)$ for all n and j. Set $A_n = a(f)^{-1} \operatorname{Re}(h_n(0))$. By 11 H, $\{A_n\}_1^\infty$ is a bounded sequence. Since $h_n(0) = \sum_{j=1}^{4^n} 4^{2n} a(f_{nj}) \iint_{E_{nj}} \zeta^{-1} d\xi d\eta$ and $4^n a(f_{nj}) = a(f)$

$$A_n = 4^n \sum_{j=1}^{4^n} \iint_{E_{nj}} \operatorname{Re}\left(\frac{1}{\zeta}\right) d\xi d\eta = 4^n \sum_{j=1}^{4^n} \iint_{E_{nj}} \frac{\xi d\xi d\eta}{\xi^2 + \eta^2}.$$

Observe that

$$4^n \sum_{E_{nj} \subset \{\xi > \frac{1}{2}\}} \iint_{E_{nj}} \frac{\xi d\xi d\eta}{\xi^2 + \eta^2} \ge 4^n \cdot 2 \cdot 4^{n-1} \cdot \tfrac{1}{4} \cdot 4^{-2n} = \tfrac{1}{8}.$$

On the other hand we obtain by the change of variables $(\xi,\eta)\to(4\xi,4\eta)$

$$A_{n-1}=4^{n-1}\sum_{j=1}^{4^{n-1}}\iint_{E_{n-1,j}}\frac{\xi\,d\xi\,d\eta}{\xi^2+\eta^2}=4^n\sum_{j=1}^{4^{n-1}}\iint_{\frac{1}{4}E_{n-1,j}}\frac{\xi\,d\xi\,d\eta}{\xi^2+\eta^2}$$

$$=4^n\sum_{E_{nj}\subset\{\xi<\frac{1}{2},\eta<\frac{1}{2}\}}\iint_{E_{nj}}\frac{\xi\,d\xi\,d\eta}{\xi^2+\eta^2}.$$

This implies that $A_n>A_{n-1}+\frac{1}{8}$. Thus $\{A_n\}$ cannot be bounded, a contradiction.

11 J. An Estimate. We conclude our auxiliary considerations by the following assertion:

For any two positive numbers ε and K there exists a positive constant $\delta=\delta(\varepsilon,K)$ such that for any $f\in A(X,K)$ with $|a(f)|\geq\varepsilon$ the inequality

$$\sup_{n,j}4^n|a(f_{nj})|>(1+\delta)\,|a(f)|$$

is valid.

Suppose this were not the case. Then there would exist a decreasing sequence $\{\delta_k\}$ with $\lim_k\delta_k=0$ and a sequence $\{{}^kf\}\subset A(X,K)$ with $|a({}^kf)|\geq\varepsilon$ such that

$$|a({}^kf_{nj})|\leq 4^{-n}(1+\delta_k)\,|a({}^kf)|$$

for every n and j.

On choosing a subsequence, if necessary, we may assume that $\{{}^kf\}$ converges to an $f\in A(X,K)$. Clearly $|a(f)|\geq\varepsilon$ and

$$|a(f_{nj})|\leq 4^{-n}|a(f)|$$

for every n and j. On replacing f by $e^{i\theta}f$ for a suitable θ we may again suppose that $a(f)>0$. By virtue of

$$a(f)=\sum_{j=1}^{4^n}a(f_{nj})\leq\sum_{j=1}^{4^n}|a(f_{nj})|\leq 4^n\cdot 4^{-n}a(f)=a(f)$$

we obtain $a(f_{nj})=4^{-n}a(f)$ for every n and j. This contradicts 11 I.

11 K. Completion of the Proof. Suppose $ac(X)>0$. We can then find an $f\in A(X,1)$ with $a(f)>0$. Let $K=K_2$ with $K_1=1$ (cf. 11 G.(b)) and $\varepsilon=a(f)$. Take $\delta=\delta(\varepsilon,K)>0$ with the properties described in 11 J. Since

$f \in A(X, K)$ there exist some n_1 and j_1 with

$$|a(f_{n_1 j_1})| \geq 4^{-n_1}(1+\delta)\, a(f).$$

Let $c_{n_1 j_1}$ be the lower left corner point of $E_{n_1 j_1}$ and set

$$\varphi(z) = f_{n_1 j_1}(4^{-n_1} z + c_{n_1 j_1}).$$

We have $\varphi \in A(X, K)$ (cf. 11 G.(b)) and

$$|a(\varphi)| = 4^{n_1} |a(f_{n_1 j_1})| \geq (1+\delta)\, a(f) > \varepsilon.$$

Therefore we can again find some n'_1 and j'_1 such that

$$|a(\varphi_{n'_1 j'_1})| \geq 4^{-n'_1}(1+\delta)\, |a(\varphi)|.$$

In view of the construction (cf. 11 G) it follows that

$$\begin{aligned}\varphi_{n'_1 j'_1}(z) &= \big(f_{n_1 j_1}(4^{-n_1} z + c_{n_1 j_1})\big)_{n'_1 j'_1} \\ &= f_{n_2 j_2}(4^{-n_1} z + c_{n_1 j_1})\end{aligned}$$

for $n_2 = n_1 + n'_1$ and some j_2 with $1 \leq j_2 \leq 4^{n_2}$. Because of $a(\varphi_{n'_1 j'_1}) = 4^{n_1} a(f_{n_2 j_2})$ and $4^{n_1} |a(f_{n_2 j_2})| \geq 4^{-n'_1}(1+\delta)\, |a(\varphi)|$ we obtain $|a(f_{n_2 j_2})| \geq 4^{-(n_1+n'_1)}(1+\delta) \cdot (1+\delta)\, a(f)$, that is

$$|a(f_{n_2 j_2})| \geq 4^{-n_2}(1+\delta)^2\, a(f).$$

Again on setting $\psi(z) = f_{n_2 j_2}(4^{-n_2} z + c_{n_2 j_2})$ we see that $\psi \in A(X, K)$ (cf. 11 G.(b)) and $|a(\psi)| = 4^{n_2} |a(f_{n_2 j_2})| \geq (1+\delta)^2\, a(f) > \varepsilon$. Using the same argument as above we can find n'_2 and j'_2 such that $|a(\psi_{n'_2 j'_2})| \geq 4^{-n'_2}(1+\delta)\, |a(\psi)|$ and therefore

$$|a(f_{n_3 j_3})| \geq 4^{-n_3}(1+\delta)^3\, a(f),$$

for $n_3 = n_2 + n'_2$ and some j_3 with $1 \leq j_3 \leq 4^{n_3}$. On repeating this process we obtain a sequence $\{(n_k, j_k)\}_{k=1}^{\infty}$ of pairs n_k and j_k with

$$|a(f_{n_k j_k})| \geq 4^{-n_k}(1+\delta)^k\, a(f).$$

From 11 G.(c) it follows that

$$K_2\, ac(X) \geq (1+\delta)^k\, a(f) \qquad (k = 1, 2, \ldots).$$

This is impossible since $a(f) > 0$.

The proof of Theorem 11 E is herewith complete.

§ 3. *K*-Functions

Between the class A of analytic functions considered thus far and the class H of harmonic functions to be studied later we insert here the class K: by definition a harmonic function u on a Riemann surface R is in $K(R)$ if $*du$ is semiexact, i.e. $\int_\gamma *du=0$ for every dividing cycle γ. We shall refer to such functions simply as K-functions. For a planar R, $u\in K(R)$ is the real part of an $f\in A(R)$ and conversely. We are therefore primarily interested in surfaces of positive genus.

We start with the basic inclusion relations for K, the counterparts of those for A. Tests for O_{KD} are also given in terms of modular products, conformal metrics, and regular chain sets. A functional analytic characterization of O_{KD} is then established; it will be useful in the proof of quasiconformal invariance of the class O_{KD}. We shall only lightly touch on the class O_{KB}.

The section ends with a generalized Riemann-Roch theorem on O_{KD}-surfaces. It illustrates the important role played by O_{KD} in classification theory.

12. Inclusion Relations and Tests

12 A. Basic Inclusions. Let K be the class of harmonic functions u whose flux vanishes across all dividing cycles. For plane regions, since all cycles are dividing, $K=\{u\,|\,u=\operatorname{Re} f,\, f\in A\}$.

The principal functions $p_j=p_j^0$ $(j=0,1)$ constructed in I.7 D belong to $K(R-\zeta)$, and therefore $p_0-p_1\in K(R)$. Moreover we can easily see that

$$p_0-p_1\in KBD(R).$$

On setting $\alpha_j^0=\alpha_j$ in I.7 D we define the K-*span* of R as

$$S_K=S_K(R,\zeta,z)=\alpha_0-\alpha_1.$$

It depends not only on ζ but also on the choice of z at ζ. If R is planar and z is the standard coordinate on the plane then $S_K=S_A$. On choosing $(h,k)=(1,-1)$ and $p\equiv 0\in\{p\}_0^0$ in I.7.(32) we obtain

$$2\pi S_K=D(p_0-p_1).$$

We introduced in 1 A the class $\mathscr{S}_A$ of plane regions whose analytic span S_A vanishes, and established in 1 A and 11 A the inclusion relations $O_{AB}<O_{ABD}=O_{AD}=\mathscr{S}_A$. In the case of the K-span no restriction to plane regions is necessary if we consider the class $\mathscr{S}_K$ of Riemann surfaces whose K-span vanishes for every ζ and z.

We assert:

THEOREM. *For Riemann surfaces of arbitrary genus*

$$O_{KB}<O_{KBD}=O_{KD}=\mathscr{S}_K. \tag{50}$$

On taking $(h, k)=(1, -1)$, $\theta=0$, and $\{p\}_0^0 \cap K = KD(R)$ in I.7.(32) we obtain

$$-D(u)+4\pi \frac{\partial u}{\partial x}(\zeta)=2\pi S_K(R, \zeta, z)-D\big(u-(p_0-p_1)\big), \qquad z=x+iy,$$

for $u \in KD(R)$. Relations (50) then follow in analogy with the reasoning in I.8. In particular the strictness of the inclusion is a consequence of $O_{AB}<O_{AD}$ for plane regions, for which we know that K-functions are real parts of analytic functions.

12 B. Inclusions for H. For comparison we append here the following remarks. First we modify the operator L_1 of I.7 C in such a way that $\int_{\beta_\Omega} * du_{1\Omega}=0$ and $u_{1\Omega}|\beta_\Omega=\text{const}$. Such an L_1 is sometimes called a uniform L_1. Properties (26)–(31) in I.7 C remain valid for this uniform L_1. We can also construct principal functions p_j with respect to (s, L_j, R_1) defined in I.7 D, for $\theta=0$. It can then be similarly shown that I.7 D.(32) holds.

We define the *harmonic span* or simply *H-span* of R as

$$S_H=S_H(R, \zeta, z)=\alpha_0-\alpha_1.$$

It again depends on $\zeta \in R$ and z at ζ. In analogy with $\mathscr{S}_K$ we consider the class $\mathscr{S}_H$ of surfaces R whose H-span vanishes for every ζ and z.

Let $(h, k)=(1, -1)$, $\theta=0$ in I.7 D.(32). Then

$$-D(u)+4\pi \frac{\partial u}{\partial x}(\zeta)=2\pi S_H(R, \zeta, z)-D\big(u-(p_0-p_1)\big)$$

for every $u \in HD(R)$. From this we can again show that for Riemann surfaces of arbitrary genus

$$O_{HB} \subset O_{HBD}=O_{HD}=\mathscr{S}_H. \tag{51}$$

This relation which will be discussed in more detail in Chapter III was first proved, by a different method, by Virtanen [1].

12 C. O_{KD}-Test. Let $\{R_n\}$ be an exhaustion of R such that every component of ∂R_n is a dividing cycle of R. Then the modular test for O_{AD} given in I.1 D carries over to O_{KD} without any change, i.e. $\prod^\infty \mu_n=\infty$ implies $R \in O_{KD}$. Since the conformal metric test given in I.2 B is a corollary of the modular test for the above exhaustion it also applies to O_{KD}, i.e. $\int^\infty \Lambda(\rho)^{-1}\, d\rho=\infty$ implies $R \in O_{KD}$. The same is true of the regular chain test given in I.3 B, i.e. $\sum_1^\infty 1/\lambda_n=\infty$ assures $R \in O_{KD}$.

It is not known whether or not the O_{AB}-tests given in 2 can be used for O_{KB} (see also VI).

Rodin [2] gave an O_{KD}-characterization in terms of extremal distances, in a generalization of Theorem 9 B.

13. Characterization of O_{KD}

13 A. Spaces $\mathscr{D}^\infty$ and $\mathscr{D}$. Let $\mathscr{D}^\infty = \mathscr{D}^\infty(R)$ be the space of real-valued C^∞-functions f on R with $D(f) < \infty$. Consider the norm

$$\|f\|_{\mathscr{D}} = \sup_R |f| + \sqrt{D(f)} \leq \infty$$

and denote by $\mathscr{D} = \mathscr{D}(R)$ the space of functions f on R for which there exists a sequence $\{f_n\} \subset \mathscr{D}^\infty$ with $\|f - f_n\|_{\mathscr{D}} \to 0$.

In the next chapter we shall study the space $\mathscr{D}$ more systematically.

13 B. Spaces $\mathscr{D}_K^\infty$ and $\mathscr{D}_K$. We designate by $\mathscr{D}_K^\infty = \mathscr{D}_K^\infty(R)$ the space of functions $f \in \mathscr{D}^\infty$ such that $df \equiv 0$ outside of a compact set, which of course depends on f. This condition is equivalent to the following: there exists a regular subregion of R such that f is a constant on each component of its complement. Let $\mathscr{D}_0^\infty$ be the subclass of $\mathscr{D}^\infty$ of functions with compact supports. Then clearly $\mathscr{D}^\infty \supset \mathscr{D}_K^\infty \supset \mathscr{D}_0^\infty$.

Next denote by $\mathscr{D}_K = \mathscr{D}_K(R)$ the space of functions $f \in \mathscr{D}$ for which there exists a sequence $\{f_n\} \subset \mathscr{D}_K^\infty$ with $D(f_n - f) \to 0$. We have the following direct sum decomposition of $\mathscr{D}$ (AHLFORS [11], ROYDEN [11]):

THEOREM. *Each function $f \in \mathscr{D}$ is the sum of a $u \in KD$ and a $g \in \mathscr{D}_K$ such that $D(u, g) = 0$:*

$$\mathscr{D}(R) = KD(R) + \mathscr{D}_K(R). \tag{52}$$

Take $f \in \mathscr{D}$ with the property $D(f, g) = 0$ for all $g \in \mathscr{D}_K^\infty$. In view of $\mathscr{D}_K^\infty \supset \mathscr{D}_0^\infty$ we also have $D(f, g) = 0$ for $g \in \mathscr{D}_0^\infty$. By Weyl's lemma (see e.g. AHLFORS-SARIO [12, p. 281]) this means that $f \in H(R)$, and actually $f \in HD(R)$. Let γ be a dividing cycle in R. Take a regular open set G in R such that $\partial G = \gamma \cup \gamma'$ with γ homologous to γ'. Let $g \in \mathscr{D}_K^\infty$ be such that $g = 1$ on γ, $g = 0$ on γ', and $dg = 0$ on $R - G$; its existence is easily seen. Then

$$0 = D(g, f) = \int_{\partial G} g * df = \int_\gamma * df.$$

This means that $f \in KD$ and it follows that

$$KD \supset \{f \mid f \in \mathscr{D},\ D(f, g) = 0 \text{ for all } g \in \mathscr{D}_K^\infty\}.$$

The reverse inclusion is implied by Green's formula. We have proved (52).

13 C. Characterization of O_{KD}. As an immediate consequence of the decomposition (52) we conclude (ROYDEN [11]):

THEOREM. *A Riemann surface R belongs to O_{KD} if and only if*

$$\mathscr{D} = \mathscr{D}_K. \tag{53}$$

14. Quasiconformal Mappings and Boundary Properties

14 A. Quasiconformal Mappings as Dirichlet Mappings. Consider a topological mapping T of a Riemann surface R onto another, R'. The quasiconformality of T can be defined in several equivalent ways. We shall return to this general question in the next chapter but here we adopt the following definition which is adequate for our present purpose: T is a *quasiconformal mapping* or, equivalently for Riemann surfaces, a *Dirichlet mapping* provided (a) $f \circ T \in \mathscr{D}(R)$ if and only if $f \in \mathscr{D}(R')$, and (b) there exists a constant $K \in [1, \infty)$ such that

$$K^{-1} D(f) \leq D(f \circ T) \leq K D(f) \tag{54}$$

for every $f \in \mathscr{D}(R')$ (cf. III.7).

Let Φ, R_1, and R_2 be the Riemann surfaces of Example I.10 C. It is easy to find a C^∞-diffeomorphism T of Φ onto itself such that T is the identity mapping on the part of Φ over $|z|>5$ and that $T(R_1)=R_2$. One can easily verify that T is a Dirichlet mapping on Φ and a fortiori of R_1 onto R_2. Since $R_1 \in O_{AB} \subset O_{AD}$ and $R_2 \notin O_{AB}, O_{AD}$ (cf. I.10 C) we conclude:

THEOREM. *The classes O_{AD} and O_{AB} are not quasiconformally invariant.*

14 B. Quasiconformal Invariance of O_{KD}. In contrast with quasiconformal noninvariance of O_{AD} we state (OIKAWA [2], ROYDEN [11]):

THEOREM. *The class O_{KD} is quasiconformally invariant.*

In fact suppose that there exists a Dirichlet mapping T of $R \in O_{KD}$ onto R'. By Theorem 13 C, $\mathscr{D}(R)=\mathscr{D}_K(R)$. We infer that $\mathscr{D}(R')=\mathscr{D}_K(R')$ which, again by Theorem 13 C, implies that $R' \in O_{KD}$.

14 C. Quasiconformal Noninvariance of O_{KB}. For planar surfaces $O_{AD}=O_{KD}$ and $O_{AB}=O_{KB}$. Therefore Theorem 14 B asserts:

For planar surfaces O_{AD} is quasiconformally invariant.

In the case of O_{AB} the situation is different:

THEOREM. *Not even the class of planar O_{KB}-surfaces is quasiconformally invariant.*

For the proof one uses the deep result of BEURLING-AHLFORS [3] that there exists a quasiconformal mapping T of $|z| \leq \infty$ onto $|w| \leq \infty$ with $T\{|z|=1\}=\{|w|=1\}$, and a compact set $F \subset \{|z|=1\}$ of linear measure zero whose image $T(F)$ has positive linear measure.

Let $R=\{|z| \leq \infty\}-F$ and $R'=\{|w| \leq \infty\}-T(F)$. Then by Theorem 10 A, $R \in O_{KB}$ and $R' \notin O_{KB}$. Clearly R and R' are quasiconformally equivalent and the theorem follows.

14 D. Boundary Properties. By virtue of Myrberg's example (I. 10 B), O_{AD} and O_{AB} are not properties of the ideal boundary (I. 10 C, II. 2 M). In contrast we claim (ROYDEN [11]):

THEOREM. *Membership in O_{KD} and O_{KB} is a property of the ideal boundary.*

Suppose there exists a conformal mapping φ of an ideal boundary neighborhood V of R onto an ideal boundary neighborhood V' of R'.

Assume that $R' \in O_{KD}$ and let $u \in KD(R)$. To prove that $D(u)=0$ set $u' = u \circ \varphi^{-1}$ on V'. We may assume that ∂V and $\partial V'$ are analytic, φ is analytic on ∂V and each component of $\partial V'$ is a dividing cycle. Then in a neighborhood U' of $\partial V'$, u' is harmonic and has a single-valued harmonic conjugate v'. Take C^∞-extensions of u' and v' to $R'-V'$. Clearly $dv' = *du'$ on U'. Set $\sigma' = dv'$ on $(R'-V') \cup U'$ and $\sigma' = *du'$ on V'. Then $d\sigma' = 0$ on R'.

Let $f' \in \mathcal{D}_K^\infty(R')$. Choose a regular region R_0' of R' such that $R_0' \supset \partial V'$, $R' - \bar{R}_0' = \bigcup_{j=1}^m U_j'$ (decomposition into components), and $f' = c_j$ (a constant) on U_j'. We have

$$\int_{R'} df' \wedge \sigma' = \int_{R_0'} df' \wedge \sigma' = \int_{\partial R_0'} f' * du' = -\sum_{j=1}^m c_j \int_{\partial U_j'} * du'$$
$$= -\sum_{j=1}^m c_j \int_{\partial \varphi^{-1}(U_j')} * du = 0.$$

By the continuity of $\int_{R'} df' \wedge \sigma'$ in f' with respect to $\sqrt{D(f')}$ we conclude that $\int_{R'} df' \wedge \sigma' = 0$ for every $f' \in \mathcal{D}_K(R') = \mathcal{D}(R')$. Thus in particular $\int_{R'} du' \wedge \sigma' = 0$, i.e.

$$\int_{V'} du' \wedge * du' = -\int_{R'-V'} du' \wedge \sigma' = -\int_{\partial(R-V')} u' \sigma' = \int_{\partial V'} u' * du'.$$

On using the conformal mapping φ we deduce from this that

$$\int_V du \wedge * du = \int_{\partial V} u * du = -\int_{R-V} du \wedge * du,$$

i.e. $D(u)=0$ and therefore $R \in O_{KD}$.

We also give here a shorter proof which makes use of Theorem I. 7 A. Again suppose $R' \in O_{KD}$, choose a $u \in KD(R)$, and set $u' = u \circ \varphi^{-1}$ on V'. Let p' be the principal function on R' corresponding to (u', L_0, V'). One easily sees that $p' \in KD(R')$ and consequently that it is constant. Therefore $L_0 u' = u'$ on V' and a fortiori $L_0 u = u | V$. This shows that u is constant.

This proof applies to O_{KB} as well.

15. Surfaces of Finite Genus

15 A. Identities for A and K. We shall now show that the identity of null classes relative to A and K for planar surfaces continues to hold in the case of finite genus (ROYDEN [11]):

THEOREM. *For surfaces of finite genus $O_{AB}=O_{KB}$ and $O_{AD}=O_{KD}$. Moreover surfaces of finite genus in O_{KB} and O_{KD} are complements on closed surfaces of AB- and AD-removable sets.*

Clearly $O_{AD}\supset O_{KD}$ and we have to show that $R\in O_{KD}$ if $R\in O_{AD}$ has finite genus. Recall that there exists a closed surface R^* and an AD-removable set $E\subset R^*$ such that $R=R^*-E$ (I.8 E). Let U be a parametric disk on R^* with $U\supset E$. Take an arbitrary $u\in KD(R)$. Since every cycle in $U-E$ is dividing, $*du$ has vanishing periods along all cycles in $U-E$ and thus there exists a function $f\in AD(U-E)$ with $\operatorname{Re} f=u$. The AD-removability of E implies that u can be continued to all of U so that $u\in H(U)$, i.e. $u\in H(R^*)$. Therefore u is constant.

The proof for B is almost the same: replace I.8 E by its counterpart for AB and consider e^f in addition to the function f which is not necessarily in $AB(U-E)$.

15 B. Quasiconformal Invariance of O_{AD} for Finite Genus. On combining Theorems 14 B and 15 A we deduce (ROYDEN [11]):

THEOREM. *The class O_{AD} for surfaces of finite genus is quasiconformally invariant.*

15 C. Surfaces with Holes. We have seen in Theorem 15 A that the inclusions $O_{AD}\supset O_{KD}$ and $O_{AB}\supset O_{KB}$ are not strict for surfaces of finite genus. However they are so in the case of infinite genus. To prove this we state:

THEOREM. *Given an arbitrary Riemann surface R, closed or open, and a closed disk $\overline{U}$ in R, $R-\overline{U}$ does not belong to O_{KB} or O_{KD}.*

In fact take any nonconstant real-valued bounded continuous function f on $\partial\overline{U}$. Let L_0 be the linear operator of I.7 C with respect to $R-\overline{U}$. Then $L_0 f\in KBD(R-\overline{U})$ is nonconstant.

15 D. Strict Inclusions. Theorem 15 C yields the desired result:

THEOREM. *The following strict inclusions hold among the null classes:*

$$O_{KB}\begin{smallmatrix}\angle & O_{AB} & \nwarrow \\ \nwarrow & O_{KD} & \angle\end{smallmatrix} O_{AD}. \tag{55}$$

The Myrberg surface (cf. I.10 B) $R=\Phi-K_1$ is in O_{AB} and O_{AD}. However by Theorem 15 C it is in neither O_{KB} nor O_{KD} and we have

$$O_{KB}<O_{AB},$$

$$O_{KD}<O_{AD}.$$

Since $O_{KB}=O_{AB}<O_{AD}=O_{KD}$ for plane regions the strict inclusion $O_{KB}<O_{KD}$ is established.

15 E. The Class O_{KP}. Let P stand for "positive." Property P is meaningless for analytic functions but not for K-functions. It is obvious that $O_{KP}\subset O_{KB}$. For plane regions $O_{KP}=O_{KB}=O_{AB}$ since $e^{-(u+iv)}\in AB$ for $u\in KP$ with v the conjugate of u.

*16. The Riemann-Roch Theorem

16 A. Divisors. Let $\nu(p)$ be an integer-valued function on R such that $\nu(p)=0$ except for a finite subset of R. The formal expression $\vartheta=\prod p^{\nu(p)}$ is called a *divisor* on R. From two divisors $\vartheta_k=\prod p^{\nu_k(p)}$ $(k=1,2)$ we form a new divisor $\vartheta=\prod p^{\nu_1(p)+\nu_2(p)}$, denoted by $\vartheta_1\cdot\vartheta_2$. The totality of divisors forms an Abelian group under this multiplication. A divisor $\vartheta=\prod p^{\nu(p)}$ is said to be *integral* if $\nu(p)\geq 0$ on R. The class of integral divisors forms a semigroup.

The *carrier* of a divisor $\vartheta=\prod p^{\nu(p)}$ is the carrier or support of the function ν. If the carrier of ϑ is $\{p_1,\ldots,p_n\}$ then we also use the notation $\vartheta=p_1^{\nu_1}\ldots p_n^{\nu_n}$ where $\nu_k=\nu(p_k)$. Two divisors are said to be *disjoint* if their carriers are disjoint.

Consider the class $\mathscr{M}(R)$ of all meromorphic functions f on R such that f has a finite number of poles and a finite Dirichlet integral over the complement of a neighborhood of its poles. Let $\nu^f(p)$ be the *order* of $f\in\mathscr{M}$ at p, i.e. $\nu^f(p)=0$ if p is neither a zero nor a pole, $\nu^f(p)=n$ if p is a zero of order n, and $\nu^f(p)=-n$ if p is a pole of order n. We call $\vartheta^f=\prod p^{\nu^f(p)}$ the divisor associated with f. A function $f\in\mathscr{M}(R)$ is said to be a *multiple* of a divisor ϑ if $\vartheta^f\cdot\vartheta^{-1}$ is integral.

Let $\mathscr{N}(R)$ be the class of meromorphic differentials α, i.e. invariant expressions $\alpha=\varphi(z)\,dz$ with a meromorphic φ such that α has at most a finite number of poles and $\int\alpha\wedge\overline{*\alpha}<\infty$ over the complement of a neighborhood of its poles. For a meromorphic function f on R, $f\in\mathscr{M}$ is equivalent to $df\in\mathscr{N}$. As above, $\nu^\alpha(p)$ shall denote the order of α at p, and ϑ^α the divisor associated with α; a multiple $\alpha\in\mathscr{N}(R)$ of a divisor ϑ is defined as before.

Let $\mathscr{N}_1(R)$ be the subclass of $\mathscr{N}(R)$ consisting of *semiexact* differentials $\alpha\in\mathscr{N}(R)$, i.e. such that the period $\int_\gamma\alpha$ vanishes over every dividing cycle γ which does not separate singularities of α.

16 B. Relations. Let $\vartheta = p_1^{\nu_1} \dots p_n^{\nu_n}$ be an integral divisor and $\mathscr{G}(\vartheta)$ the space of differentials which are analytic in some neighborhood of the carrier of ϑ. Here the neighborhood may depend on the differential.

Denote by $\mathscr{L}(\vartheta)$ the space of linear functionals L on $\mathscr{G}$ such that $L(\alpha)=0$ for any multiple α of ϑ. We call each element L in $\mathscr{L}(\vartheta)$ a *relation* and say that α in $\mathscr{G}$ satisfies the relation L if $L(\alpha)=0$. The carrier of ϑ shall also be referred to as that of L.

Choose fixed local parameters $\zeta_1, \dots, \zeta_n$ at the points $p_1, \dots, p_n$. Then each $\alpha \in \mathscr{G}$ has a representation $\alpha = \varphi_k(\zeta_k)\, d\zeta_k$ at p_k, where φ_k is analytic near p_k. For $L \in \mathscr{L}$, $L(\alpha)$ depends only on the values of φ_k and its first $\nu_k - 1$ derivatives at p_k. Thus we can express L in the form

$$L(\alpha) = \sum_{k=1}^{n} \sum_{j=1}^{\nu_k} \frac{a_{kj}}{(j-1)!}\, \varphi_k^{(j-1)} \tag{56}$$

where $\varphi_k^{(j)}$ is the jth derivative of φ_k with respect to ζ_k at p_k. Once we fix the local parameters, $L \in \mathscr{L}$ determines and is determined by the matrix (a_{kj}) where $a_{kj}=0$ for $j > \nu_k$.

$\mathscr{L}(\vartheta)$ is a vector space of dimension $n(\vartheta) = \sum_1^n \nu_k$.

This dimension is called the *order* of ϑ.

16 C. Relations as Principal Parts. Let f be a function which is meromorphic in a neighborhood of the carrier of the integral divisor $\vartheta = p_1^{\nu_1} \dots p_n^{\nu_n}$ and is a multiple of ϑ^{-1}. Consider

$$L_f(\alpha) = \frac{1}{2\pi i} \sum_1^n \int_{\gamma_k} f\alpha \tag{57}$$

where γ_k is the boundary of a parametric disk, with center p_k, contained in the common domain of meromorphy of f and α, and containing p_j $(j \neq k)$. By Cauchy's theorem definition (57) does not depend on the choice of the disks. Clearly L_f is linear on $\mathscr{G}$. Let α be a multiple of ϑ. Since f is a multiple of ϑ^{-1}, $f\alpha$ is analytic on the region encircled by γ_k and thus $L_f(\alpha)=0$, i.e. $L_f \in \mathscr{L}$.

Now let ζ_k be as in 16 B with $\zeta_k(p_k)=0$. Then f has the expansion

$$f = \sum_{j=1}^{\nu_k} a_{kj}\, \zeta_k^{-j} + f_0 \tag{58}$$

at p_k with f_0 analytic at p_k. By Cauchy's formula we again obtain (56) for L_f with exactly the same coefficients as in (58).

We have established a natural one-to-one correspondence between $\mathscr{L}(\vartheta)$ and the space of principal parts of functions which are meromorphic on some neighborhood of the carrier of ϑ, and multiples of ϑ^{-1}.

We may thus say that L_f gives the principal part of f. This has the advantage of giving an expression for the principal part of f which is free of the choice of local parameters. In this vein we call $L\in\mathscr{L}$ the *principal part* of f when $L=L_f$.

16 D. An Auxiliary Formula. Hereafter we restrict our attention to surfaces R in O_{KD}. Let R_0 be a regular subregion of R, take an $f\in KD(R-R_0)$, and a square integrable α on $R-\bar{R}_0$, i.e. $\int_{R-\bar{R}_0}\alpha\wedge\overline{*\alpha}<\infty$, with $\int_\gamma\alpha=0$ for any dividing cycle γ in $R-\bar{R}_0$. Moreover we assume that α is continuous in a neighborhood of ∂R_0 relative to $R-R_0$. We shall prove:

$$\int_{\partial R_0} f\alpha=-\int_{R-\bar{R}_0} df\wedge\alpha. \tag{59}$$

Extend f to all of R such that $f\in C^\infty(R)$. Clearly $f\in\mathscr{D}_K(R)$ (cf. 13 C). Since α is continuous on ∂R_0 and, as is easily verified, $\int_\gamma\alpha=0$ for the border γ of each component of $R-R_0$, we can extend α to all of R so that α is semiexact on R. In view of Stokes' formula

$$\int_{R_0} df\wedge\alpha=\int_{\partial R_0} f\alpha.$$

By adding this to (59) one sees that it suffices to prove $\int_R df\wedge\alpha=0$. First assume that $f\in\mathscr{D}_K^\infty(R)$. We can find a regular region $R_1\supset\bar{R}_0$ such that $f=c_i$ on each component R_i' of $R-\bar{R}_1$. Stokes' formula gives

$$\int_R df\wedge\alpha=\int_{R_1} df\wedge\alpha=-\sum_i c_i\int_{\partial R_i'}\alpha=0$$

and by a limiting process we conclude that $\int_R df\wedge\alpha=0$ for $f\in\mathscr{D}_K$.

16 E. Solutions of $L_f(\alpha)=0$. Let $f\in\mathscr{M}(R)$ be a multiple of ϑ^{-1} (cf. 16 A). By definition the solutions of the relation $L_f(\alpha)=0$ include all multiples α of ϑ. In this connection we maintain (Royden [10]):

Lemma. *Let ϑ_1 and ϑ_2 be disjoint integral divisors in $R\in O_{KD}$ and let $f\in\mathscr{M}(R)$ be a multiple of ϑ_1/ϑ_2. If $\alpha\in\mathscr{N}_1(R)$ is a multiple of ϑ_1^{-1}, then* $L_f(\alpha)=0$.

Take a regular region R_0 containing the carriers of ϑ_1 and ϑ_2, and let γ_k be a small circle in R_0 about $p_k\in\vartheta_2=p_1^{\nu_1}\dots p_n^{\nu_n}$. Then $f\alpha$ is analytic on R except on the carrier of ϑ_2, and in particular on $R-R_0$. Clearly $df\wedge\alpha=0$ on $R-R_0$ and hence by (59)

$$\int_{\partial R_0} f\alpha=-\int_{R-R_0} df\wedge\alpha=0.$$

Thus by Cauchy's theorem

$$L_f(\alpha)=\frac{1}{2\pi i}\sum_1^n\int_{\gamma_k} f\alpha=-\frac{1}{2\pi i}\int_{\partial R_0} f\alpha=0.$$

16 F. Characterization of Principal Parts. For $f\in\mathcal{M}(R)$, L_f was defined as the principal part of f. However an arbitrary $L\in\mathcal{L}((\vartheta^f)^{-1})$ need not be a principal part of any function in $\mathcal{M}(R)$. When does $L=L_f$ with $f\in\mathcal{M}$? This is answered by the following statement (Royden [10]):

Lemma. *Let $R\in O_{KD}$. There is an $f\in\mathcal{M}(R)$ with the principal part L if and only if $L(\alpha)=0$ for all semiexact square integrable analytic differentials α.*

The necessity of the condition follows directly from Lemma 16 E. The proof of the sufficiency will be given in 16 G – 16 H.

16 G. Neumann's Function. We need information about the structure of $\mathcal{M}(R)$, $R\in O_{KD}$. To this end we make use of the *Neumann function* $N(p,p_0;q,q_0)$ defined as follows: N is harmonic as a function of p on $R-q-q_0$, $N+\log|\zeta(p)-\zeta(q)|$ and $N-\log|\zeta(p)-\zeta(q_0)|$ are harmonic at q and q_0 respectively, with ζ the local parameter, and $N=L_0N$ on a neighborhood of the ideal boundary of R where L_0 is the normal operator of I.7 C. By Theorem I.7 A such an N exists and is determined up to an additive constant. To make it unique we require that $N(p_0,p_0;q,q_0)=0$ with $p_0\in R-q-q_0$. Therefore N is the L_0-principal function $-\log|Q_0|$ considered in 7 B except for the normalization.

By a simple application of Green's formula we obtain

$$N(p,p_0;q,q_0)=N(q,q_0;p,p_0) \tag{60}$$

and by the uniqueness of N

$$N(p,p_0;q,q_0')=N(p,p_0;q,q_0)+N(p,p_0;q_0,q_0'). \tag{61}$$

In view of the construction N has a finite Dirichlet integral over the complement of a neighborhood of $\{q,q_0\}$. If $f\in\mathcal{D}(R)$ then by Green's formula

$$f(q)-f(q_0)=\frac{1}{2\pi}\int df\wedge *d_pN(p,p_0;q,q_0). \tag{62}$$

16 H. Sufficiency. The differential $*d_pN$ is semiexact in the sense that $\int_\gamma *d_pN=0$ for any dividing cycle γ in R which does not separate q and q_0. This is clear from the property $L_0N=N$.

Given a local parameter $\zeta=\xi+i\eta$ we shall use the notation $\partial/\partial\zeta$ and $\partial/\partial\bar\zeta$ for $\frac{1}{2}(\partial/\partial\xi-i(\partial/\partial\eta))$ and $\frac{1}{2}(\partial/\partial\xi+i(\partial/\partial\eta))$. Let q be in the domain of ζ and write $N(p,p_0;q,q_0)=N(p,p_0;\zeta,q_0)$. Then by (61), $\partial^jN/\partial\zeta^j$ is independent of q_0. From this it follows that $\partial^j N/\partial\zeta^j$ is harmonic on $R-q$. It is also seen that $\partial^j N/\partial\zeta^j$ has a finite Dirichlet integral over the complement of a neighborhood of q and is semiexact. At q we have

$$\frac{\partial^j N}{\partial\zeta^j}=\frac{(j-1)!}{2}\,\frac{1}{[\zeta(p)-\zeta(q)]^j}+\text{regular terms}. \tag{63}$$

Let $\vartheta=q_1^{\nu_1}\ldots q_n^{\nu_n}$ be an integral divisor and choose a local parameter ζ_k at q_k. Given an $L\in\mathscr{L}(\vartheta)$ consider it as a principal part in (58). Then the function

$$f(p)=2\sum_{k=1}^{n}\sum_{j=1}^{\nu_k}\frac{a_{kj}}{(j-1)!}\,\frac{\partial^j N}{\partial\zeta_k^j}, \tag{64}$$

with $\partial^j N/\partial\zeta_k^j$ the derivative of $N(p,p_0;q_k,q_0)$ with respect to ζ_k, is harmonic except at the carrier of ϑ where it has expansion (58). Moreover f has a finite Dirichlet integral over the complement of a neighborhood of the carrier of ϑ, and $*df$ is semiexact. Such a harmonic function is unique up to an additive constant.

Let $\partial_q N=(\partial N/\partial\zeta)\,d\zeta$. By (60), $\partial_q N$ is an analytic differential in q except at p and p_0 where it has simple poles. It is also square integrable over the complement of any neighborhood of p and p_0. Thus $\partial_q N$ is a multiple of $p^{-1}p_0^{-1}$, and in terms of $\partial_q N$ we may write (64) as

$$f(p)=2L(\partial_q N(p,p_0;q)).$$

Observe that $f(p_0)=0$. We therefore obtain for any $f\in\mathscr{M}(R)$

$$f(p)=2L_f(\partial_q N(p,p_0;q))+f(p_0). \tag{65}$$

We have seen that there is a meromorphic function on R with the principal part L if and only if $L(\partial_q N)$ is analytic except at the carrier of L. Let z be a local parameter at p. Then $L(\partial_q N)$ is analytic if and only if

$$\frac{\partial}{\partial\bar z}L(\partial_q N)=L\left(\partial_q\frac{\partial N}{\partial\bar z}\right)\equiv 0.$$

Here $\partial_q(\partial N/\partial\bar z)$ is an everywhere regular semiexact analytic square integrable differential in q since the singularity at p is eliminated by $\partial^2\log|z-\zeta|/\partial\zeta\,\partial\bar z=0$. The proof of Lemma 16 F is complete.

16 I. Generalized Riemann-Roch Theorem. We now turn to the main assertion. The original Riemann-Roch theorem applies to a closed surface of genus g. It gives the dimension a of the space of meromorphic functions which are multiples of a divisor ϑ of order n in terms of the dimension b of the space of meromorphic differentials which are multiples of ϑ^{-1}: $a=b-n-g+1$.

For open surfaces however the dimensions a and b are always infinite and it is for this reason that we restrict our attention to the spaces $\mathscr{M}(R)$ and $\mathscr{N}(R)$. Moreover if R is closed then $\mathscr{N}(R)=\mathscr{N}_1(R)$. This will be seen to hold also for parabolic surfaces R (cf. V.7 G). The similarity between the case of $\mathscr{N}(R)$ for a closed R and the case of $\mathscr{N}_1(R)$ for an $R \in O_{KD}$ leads to the following *generalized Riemann-Roch theorem* (ROYDEN [10]; see also KUSUNOKI [3], and, for further generalizations, RODIN [1], RODIN-SARIO [3], and YOSHIDA [1]).

THEOREM. *Let R be a Riemann surface of class O_{KD} and let ϑ_1 and ϑ_2 be disjoint integral divisors on R. In order that a given $L \in \mathscr{L}(\vartheta_2)$ be the principal part of some function f in $\mathscr{M}(R)$ which is a multiple of $\vartheta=\vartheta_1/\vartheta_2$ it is necessary and sufficient that $L(\alpha)=0$ for all α in $\mathscr{N}_1(R)$ which are multiples of ϑ_1^{-1}.*

The necessity follows from Lemma 16 E.

Suppose that $L(\alpha)=0$ for all $\alpha \in \mathscr{N}_1(R)$ which are multiples of ϑ_1^{-1}. By Lemma 16 F there is an $f \in \mathscr{M}(R)$ with the principal part L. If $\vartheta_1=1$ there is nothing to prove.

If $\vartheta_1 = p_1^{\mu_1} \dots p_m^{\mu_m}$, $\mu_1 > 0$, the function $f=L(\partial_q N(p, p_1; q))$ is analytic and vanishes at p_1. Let $z_1, \dots, z_m$ be local parameters at $p_1, \dots, p_m$. Then $\partial^j f/\partial z_k^j$ is given by $L(\partial_q(\partial^j N/\partial z_k^j))$ where $\partial^j N/\partial z_k^j$ denotes the jth derivative with respect to z_k of $N(z_k, p_1; q)$ evaluated at the point p_k. For $0 \leq j \leq \mu_k - 1$ the expression $\partial_q(\partial^j N/\partial z_k^j)$ is, in its dependence on q, a differential in $\mathscr{N}_1(R)$ and a multiple of ϑ_1^{-1}. Thus by hypothesis $L(\partial_q(\partial^j N/\partial z_k^j))=0$ which implies that f has a zero of order μ_k at p_k. We infer that f is a multiple of ϑ_1/ϑ_2 as asserted.

16 J. Classical Case. For completeness we append here a proof of the classical Riemann-Roch theorem as a consequence of the above result. Let R be a closed surface of genus g. The number of linearly independent analytic differentials is g (e.g. AHLFORS-SARIO [12, p. 319]). Every meromorphic differential is determined by its principal part to within the addition of an everywhere analytic differential. On the other hand if a principal part of a meromorphic differential satisfies the condition that the sum of the residues is zero we can construct a meromorphic differential with this principal part by taking linear combinations of the $\partial_q(\partial^j N/\partial z_k^j)(p_k, p_1; q)$. If the principal part consists of a simple pole the flux of the corresponding integral function does not vanish and there can be no meromorphic function. We conclude:

For an integral divisor ϑ the dimension of the space $V(\vartheta)$ of meromorphic differentials which are multiples of ϑ^{-1} is $g+n(\vartheta)-1$ if $n(\vartheta)>0$, and g if $n(\vartheta)=0$.

Next let ϑ_1 and ϑ_2 be disjoint integral divisors. Each element in $\mathscr{L}(\vartheta_2)$ can be considered as a linear functional on $V(\vartheta_1)$. Thus we have a natural linear mapping T of $\mathscr{L}(\vartheta_2)$ into the dual space V^* of $V(\vartheta_1)$. By Theorem 16 I the null space of T consists of principal parts L of meromorphic functions which are multiples of $\vartheta=\vartheta_1/\vartheta_2$. Thus the number a of linearly independent meromorphic functions which are multiples of ϑ is the dimension of the null space of T for $\vartheta_1 \neq 1$ and the dimension plus one for $\vartheta_1=1$ since the constant functions then also qualify. The dimension of the null space of T plus the rank r of T is equal to the dimension of $\mathscr{L}(\vartheta_2)$:

$$a+r=\begin{cases} n(\vartheta_2) & (\vartheta_1 \neq 1), \\ n(\vartheta_2)+1 & (\vartheta_1=1). \end{cases} \tag{66}$$

Let b be the number of linearly independent differentials in $V(\vartheta_1)$ which annihilate the range of T. Then $b+r$ is the dimension of $V(\vartheta_1)$ and we have

$$b+r=\begin{cases} g+n(\vartheta_1)-1 & (\vartheta_1 \neq 1), \\ g & (\vartheta_1=1). \end{cases} \tag{67}$$

Let $n(\vartheta)=n(\vartheta_1)-n(\vartheta_2)$ be the order of ϑ. Subtraction of (67) from (66) gives

$$a=b-n(\vartheta)-g+1. \tag{68}$$

Finally the set of differentials in $V(\vartheta_1)$ which annihilate the range of T is the set of those differentials α in $V(\vartheta_1)$ for which $L(\alpha)=0$ for all $L\in\mathscr{L}(\vartheta_2)$; this set in turn coincides with the set of meromorphic differentials which are multiples of ϑ_2/ϑ_1. Thus we have obtained *the classical Riemann-Roch theorem:*

Let ϑ be a divisor of order n on a closed Riemann surface of genus g. Denote by a the number of linearly independent meromorphic functions which are multiples of ϑ, and by b the number of linearly independent meromorphic differentials which are multiples of ϑ^{-1}. Then

$$a=b-n-g+1. \tag{69}$$

Chapter III

Dirichlet Finite Harmonic Functions

From the class A of analytic functions we proceed to the class H of harmonic functions. The latter are in a sense more flexible than the former and thus easier to treat. In particular the solvability of the Dirichlet problem makes it possible to obtain detailed information on the causes of degeneracy. On the other hand the lack of rigidity results in a great diversity of degeneracy phenomena. To subject them to a systematic treatment it is convenient to start with the class HD of harmonic functions with finite Dirichlet integrals and the corresponding null class O_{HD}. The close connection with Dirichlet's principle makes the class O_{HD} the most significant one among degeneracy classes related to H.

In § 1 we introduce Royden's boundary of a Riemann surface and study the behavior of functions in HD at this boundary. In § 2 we consider Dirichlet's problem concerning Royden's boundary and discuss in detail the important concept of a minimal function. The section also contains a treatment of the geometric version of the principle of harmonic majoration as well as relations between O_{AD} and certain null classes concerning harmonic functions with finite Dirichlet integrals. In particular the proofs of the strict inclusion relations are furnished.

In the final § 3 some structures on a Riemann surface in addition to its own conformal structure are considered in connection with the class O_{HD}. As an application we see that O_{HD} is quasiconformally invariant and that membership in it is a boundary property.

§ 1. Royden's Compactification

Let U (resp. $\bar{U}$) be the open (resp. closed) unit disk and $\mathscr{D}^0(\bar{U})$ the space of all (not necessarily finitely) continuous functions on $\bar{U}$ with finite Dirichlet integrals over U. Consider the problem of minimizing $D(f)$ for $f \in \mathscr{D}^0(\bar{U})$ such that $f|\partial U = \varphi$, a fixed function. The *Dirichlet principle* states that there exists a unique minimizing function $u \in \mathscr{D}^0(\bar{U})$ with $u|\partial U = \varphi$ and u in the class $H\mathscr{D}^0(\bar{U})$ of those functions in $\mathscr{D}^0(\bar{U})$

which are harmonic on U. This can be restated as follows:

$$\mathscr{D}^0(\bar{U}) = H\mathscr{D}^0(\bar{U}) + \mathscr{D}_1^0(\bar{U}) \tag{1}$$

where the sum stands for orthogonal decomposition with respect to the inner product $D(\varphi, \psi)$, and $\mathscr{D}_1^0(\bar{U})$ is the space of functions in $\mathscr{D}^0(\bar{U})$ which vanish on ∂U. Hence the space $H\mathscr{D}^0(\bar{U})$ is isomorphic to the space $\mathscr{D}^0(\bar{U})|\partial U$.

The global study of the class $H\mathscr{D}^0(\bar{U})$ is thus equivalent to that of $\mathscr{D}^0(\bar{U})|\partial U$. Since $H\mathscr{D}^0(\bar{U})$ is dense in $HD(U)$ with respect to the Dirichlet norm and $\mathscr{D}^0(\bar{U})|\partial U$ is dense in the space $\tilde{C}(\partial U)$ of integrable continuous functions on ∂U with respect to the L^1-norm, information about $HD(U)$ should be obtainable by studying $\tilde{C}(\partial U)$. The advantage of this idea is in the fact that all analytic properties in $HD(U)$ are eliminated in $\tilde{C}(\partial U)$, and the treatment of $\tilde{C}(\partial U)$ is easier than that of $HD(U)$.

However an application of this approach to an arbitrary surface R is a priori confronted with the following difficulty: R has in general no natural boundary such as ∂U in the case of U. It is for this reason that we need some abstract compactification of R. As a substitute for $\mathscr{D}^0(\bar{U})$, we take the class $\tilde{M}(R)$ of continuous functions on R with finite Dirichlet integrals; we wish to make each function in $\tilde{M}(R)$ continuous "at the ideal boundary of R." We reverse the process and take the orthogonal complement $\tilde{M}_\Delta(R)$ of $HD(R)$ in $\tilde{M}(R)$:

$$\tilde{M}(R) = HD(R) + \tilde{M}_\Delta(R). \tag{2}$$

To avoid the ambiguity of additive constants we require that each $f \in \tilde{M}_\Delta(R)$ be approximated by functions with compact supports with respect to not only the Dirichlet norm but also uniform convergence on compact sets.

This choice of $\tilde{M}_\Delta(R)$ will have the effect on $\tilde{M}(R)$ that there exists a smallest compact Hausdorff space R^*, called Royden's compactification, which contains R as an open dense subspace such that $\tilde{M}(R) \subset C(R^*)$. The boundary of R is $\Gamma = R^* - R$ and we take the set $\Delta \subset \Gamma$ of points in Γ on which every function in $\tilde{M}_\Delta$ vanishes. Then Δ corresponds to ∂U and $\Gamma - \Delta$ corresponds to the set of "irregular" points; this set is empty in the case of $R = U$. The space $HD(R)$ is isomorphic to the "dense" subspace $\tilde{M}(R)|\Delta$ of the space $\tilde{C}(\Delta)$ of "integrable" continuous functions on Δ. The study of $HD(R)$ is thus reduced to that of $\tilde{C}(\Delta)$. For example $R \in O_{HD}$ means that $\tilde{C}(\Delta)$ consists only of constants, and therefore that Δ contains at most one point.

This is the program we are going to follow. In this section we study the totality $M(R)$ of bounded members of $\tilde{M}(R)$, called Royden's algebra associated with R, and discuss in detail the corresponding decomposition (2) and the topology of R^*. Not only O_{HD} but also the class O_G of parabolic surfaces will be considered from this viewpoint.

1. Royden's Algebra

1 A. Tonelli Functions. Let $f(x, y)$ be a real-valued function defined on $(a, b)\times(c, d)$. Assume that f satisfies the following conditions:

(T. 1) *f is continuous on $(a, b)\times(c, d)$,*

(T. 2) *$x\to f(x, y)$ is absolutely continuous on (a, b) for almost every fixed $y\in(c, d)$, and the same holds if x and y are interchanged.*

From these two conditions it follows that $(\partial f/\partial x)(x, y)$ and $(\partial f/\partial y)(x, y)$ exist almost everywhere in $(a, b)\times(c, d)$ and are measurable there (see Saks [1]). We assume moreover:

(T. 3) *$(\partial f/\partial x)(x, y)$ and $(\partial f/\partial y)(x, y)$ are square integrable on any compact subset of $(a, b)\times(c, d)$.*

A real-valued function f is called a *Tonelli function* on a plane region R if f satisfies (T. 1), (T. 2), and (T. 3) for every $(a, b)\times(c, d)\subset R$. A complex-valued function f on a plane region R is a Tonelli function by definition if $\operatorname{Re} f$ and $\operatorname{Im} f$ are Tonelli functions on R. We can now introduce:

A function f on a Riemann surface R is a Tonelli function if f is one on every parametric disk of R.

One of the main reasons we take Tonelli functions instead of C^m-functions $(1\leq m\leq\infty)$ is the following important and convenient property of the former:

If f and g are Tonelli functions on a Riemann surface R then so are $f\cup g$ and $f\cap g$.

Here $f\cup g=\max(f, g)$ and $f\cap g=\min(f, g)$.

For the proof take a local parameter $z=x+iy$ in $I=(a, b)\times(c, d)$ on R and a subinterval I' with $\bar{I}'\subset I$. Observe the relations

$$f\cap g=(f-g)\cap 0+g, \qquad f\cup g=-((-f)\cap(-g)),$$

$$|f(x, y)\cap 0-f(x', y)\cap 0|\leq|f(x, y)-f(x', y)|,$$

and

$$\int_{I'}\left|\frac{\partial}{\partial x}(f(x, y)\cap 0)\right|^2 dx\,dy$$
$$=\int_{I'-(f>0)}\left|\frac{\partial}{\partial x}f(x, y)\right|^2 dx\,dy\leq\int_{I'}\left|\frac{\partial}{\partial x}f(x, y)\right|^2 dx\,dy<\infty.$$

From these we deduce that $f \cup g$ and $f \cap g$ are Tonelli functions on I and thus on R.

1 B. Definition of Royden's Algebra. The real (or complex) *Royden algebra* $\mathbf{M}(R)$ associated with a Riemann surface R is the totality of real (or complex) functions f on R satisfying the following three conditions:

(M.1) *f is bounded on R,*
(M.2) *f is a Tonelli function on R,*
(M.3) *the Dirichlet integral $D_R(f)$ of f on R is finite.*

Royden's algebra $\mathbf{M}(R)$ actually forms an algebra over the real (or the complex) number field with the usual addition and multiplication of functions and scalar multiplication defined pointwise. Concerning division we remark:

Let $f \in \mathbf{M}(R)$. The function $1/f$ belongs to $\mathbf{M}(R)$ if and only if $\inf_R |f| > 0$.

Moreover it follows from 1 A that the real Royden algebra is a vector lattice, i.e.:

The functions $f \cup g$ and $f \cap g$ belong to $\mathbf{M}(R)$ along with f and g.

1 C. Completeness. We consider several topologies on $\mathbf{M}(R)$. Let $\{f_n\}$ be a sequence of functions on R and f a function on R. We use the following terminology:

(a) $\{f_n\}$ converges to f on R in *C-topology*, $f = C\text{-}\lim_n f_n$, if $\lim_n \sup_K |f_n - f| = 0$ for each compact subset K of R;

(b) $\{f_n\}$ converges to f on R in *B-topology*, $f = B\text{-}\lim_n f_n$, if $\{f_n\}$ is uniformly bounded on R and $f = C\text{-}\lim_n f_n$ on R;

(c) $\{f_n\}$ converges to f on R in *U-topology*, $f = U\text{-}\lim_n f_n$, if $\lim_n \sup_R |f_n - f| = 0$;

(d) $\{f_n\}$ converges to f on R in *D-topology*, $f = D\text{-}\lim_n f_n$, if $\lim_n D_R(f_n - f) = 0$;

(e) $\{f_n\}$ converges to f on R in *QD-topology* ($Q = C, B, U$), $f = QD\text{-}\lim_n f_n$, if $f = Q\text{-}\lim_n f_n$ on R and $f = D\text{-}\lim_n f_n$ on R.

With respect to each of the above topologies $\mathbf{M}(R)$ becomes a topological algebra. The most important are the *CD*-, *BD*-, and *UD*-topologies. In particular the *UD*-topology is given by the norm

$$\|f\| = \|f\|_R = \sup_R |f| + \sqrt{D_R(f)}. \tag{3}$$

It is easy to see that

$$\|f\| \geq 0 \quad \textit{and} \quad \|f\| = 0 \quad \textit{if and only if } f = 0 \textit{ on } R,$$
$$\|\alpha f\| = |\alpha| \, \|f\| \quad \textit{with scalar } \alpha,$$
$$\|f+g\| \leq \|f\| + \|g\|,$$
$$\|f \cdot g\| \leq \|f\| \cdot \|g\|,$$
$$\|1\| = 1.$$

Therefore $\mathrm{M}(R)$ is a *normed algebra* in the UD-topology.

Regarding the completeness of $\mathrm{M}(R)$ we maintain (KAWAMURA [1]):

THEOREM. *Let $\{f_n\}$ be a sequence of functions in $\mathrm{M}(R)$ and f a bounded function on R satisfying*

$$f = C\text{-}\lim_{n \to \infty} f_n \tag{4}$$

on R and such that

$$D_R(f_n) \leq K < \infty \tag{5}$$

for every n. Then $f \in \mathrm{M}(R)$ and

$$D(f, g) = \lim_{n \to \infty} D(f_n, g) \tag{6}$$

for any $g \in \mathrm{M}(R)$.

Let $\Gamma^2(R)$ be the set of first order differentials α with the local representation $\alpha = a(x, y)\,dx + b(x, y)\,dy$ in terms of a local parameter $z = x + iy$ such that $a(x, y)$ and $b(x, y)$ are measurable and

$$\int_R \alpha \wedge \overline{*\alpha} = \int (|a(x, y)|^2 + |b(x, y)|^2)\,dx\,dy < \infty.$$

We set $(\alpha, \beta) = \int_R \alpha \wedge \overline{*\beta}$ for α and β in $\Gamma^2(R)$. Then $\Gamma^2(R)$ is a Hilbert space. Clearly $\{df \mid f \in \mathrm{M}(R)\} \subset \Gamma^2(R)$ and $(df, df) = D_R(f)$ for f in $\mathrm{M}(R)$.

Since $\{df_n\}$ is a bounded sequence in $\Gamma^2(R)$ it is well-known that it contains a weakly convergent subsequence $\{df_{n_k}\}$. We are going to prove that $f \in \mathrm{M}(R)$ and that the weak limit of $\{df_{n_k}\}$ is df. This will complete the proof of the theorem.

Let $z = x + iy$ be a local parameter on R in $T = (s, t) \times (u, v)$ and let $C_0^\infty(T)$ be the class of C^∞-functions on R with supports in T. An integration by parts gives

$$\int_T f_{n_k}(x, y) \frac{\partial}{\partial x} \varphi(x, y)\,dx\,dy = -\int_T \left(\frac{\partial}{\partial x} f_{n_k}(x, y) \right) \varphi(x, y)\,dx\,dy \tag{7}$$

for every φ in $C_0^\infty(T)$. Let $\alpha = a\,dx + b\,dy$ be the weak limit of $\{df_{n_k}\}$. Since $\varphi(x,y)\,dx \in \Gamma^2(R)$ and

$$\int_T \left(\frac{\partial}{\partial x} f_{n_k}(x,y)\right) \varphi(x,y)\,dx\,dy = (df_{n_k}, \varphi dx) \to (\alpha, \varphi dx) = \int_T a(x,y)\,\varphi(x,y)\,dx\,dy$$

as $k \to \infty$, we obtain on passing to the limit in (7) and using (4)

$$\int_T f(x,y) \frac{\partial}{\partial x} \varphi(x,y)\,dx\,dy = -\int_T a(x,y)\,\varphi(x,y)\,dx\,dy. \tag{8}$$

Similarly

$$\int_T f(x,y) \frac{\partial}{\partial y} \varphi(x,y)\,dx\,dy = -\int_T b(x,y)\,\varphi(x,y)\,dx\,dy. \tag{9}$$

From (8) and (9) it follows that the distribution derivatives $[\partial f/\partial x]_{\text{dis}}$ and $[df/dy]_{\text{dis}}$ of $f(x,y)$ on T are $a(x,y)$ and $b(x,y)$ respectively. By a theorem of NIKODYM f satisfies (T.2), and the usual derivatives $\partial f/\partial x$ and $\partial f/\partial y$ of f coincide with a and b on T (see SCHWARTZ [1, p. 57]). Since T is arbitrary $df = \alpha$ on R and thus $D_R(f) < \infty$, i.e. $f \in \mathrm{M}(R)$.

COROLLARY 1. *The algebra* $\mathrm{M}(R)$ *is BD-complete.*

COROLLARY 2. *The algebra* $\mathrm{M}(R)$ *is a Banach algebra with respect to* (3).

1 D. Approximation. We wish to use computational rules, such as the Green's formula known for C^∞-functions, for functions in $\mathrm{M}(R)$. For this purpose the following approximation theorem is important:

THEOREM. *For any $f \in \mathrm{M}(R)$ and positive number ε there exists a function $f_\varepsilon \in C^\infty(R) \cap \mathrm{M}(R)$ such that $\|f - f_\varepsilon\| < \varepsilon$. Moreover if f has compact support in an open set $G \subset R$ then f_ε can be chosen so as to have compact support in G.*

The First Step. Let $z = x + iy$ be a local parameter of R valid in $|z| < 2$. We assume that the support of f is contained in $|z| < 1$. Let ρ_n be the function on R defined by

$$\left(\pi \int_0^{n^{-2}} e^{-t^{-1}}\,dt\right) \rho_n(z) = \begin{cases} \exp(-(n^{-2} - x^2 - y^2)^{-1}) & \text{on } |z| < 1/n, \\ 0 & \text{on } R - \{|z| < 1/n\}. \end{cases}$$

It is a nonnegative C^∞-function on R with $\int_R \rho_n(z)\,dx\,dy = 1$. Let

$$f_n(w) = \int_R \rho_n(w - z) f(z)\,dx\,dy = \int_R \rho_n(z) f(w - z)\,dx\,dy.$$

It is easy to see that $f_n \in C^\infty(R)$, $f_n = 0$ on R outside of $|z| < 2$ and $f = U\text{-}\lim_n f_n$ on R. Clearly

$$\partial f_n(w) = \int_R \rho_n(z)\, \partial f(w-z)\, dx dy$$

almost everywhere on $|z| < 2$, where $w = u + iv$ and ∂ stands for $\partial/\partial u$ or $\partial/\partial v$. We obtain by Schwarz's inequality

$$\begin{aligned} |\operatorname{grad}_w (f_n(w) - f(w))|^2 &\leq \int_R \rho_n(z)\, dx dy \int_R \rho_n(z) |\operatorname{grad}_w (f(w-z) - f(w))|^2\, dx dy \\ &= \int_R \rho_n(z) |\operatorname{grad}_w (f(w-z) - f(w))|^2\, dx dy. \end{aligned}$$

Hence we see that

$$D_R(f_n - f) \leq \int_{|z| < 1/n} \rho_n(z) \Big(\int_{|w| < 2} |\operatorname{grad}_w (f(w-z) - f(w))|^2\, du dv \Big)\, dx dy.$$

Since $|\operatorname{grad}_w f(w)|$ is square integrable on $|w| < 2$ Lebesgue's theorem gives

$$\lim_{|z| \to 0} \int_{|w| < 2} |\operatorname{grad}_w (f(w-z) - f(w))|^2\, du dv = 0.$$

Thus $f = D\text{-}\lim_n f_n$ on R and $\lim_n \|f - f_n\| = 0$.

The Second Step. Let $\{\varphi_j\}_{j=1}^\infty$ be a sequence of C^∞-functions on R such that for some parametric disk U_j: $|z| < 2$, the support of φ_j is contained in $|z| < 1$, $\{U_j\}_{j=1}^\infty$ is a locally finite covering of R, and $\sum_{j=1}^\infty \varphi_j = 1$ on R. Since $f\varphi_j$ satisfies the requirement of the first step we can find an $f_j \in C^\infty(R) \cap \mathrm{MI}(R)$ such that the support of f_j is compact in U_j and $\|f\varphi_j - f_j\| < \varepsilon/2^j$. Moreover if f has compact support in an open subset G of R then we may assume that the support of f_j is contained in G. Set $f_\varepsilon = \sum_{j=1}^\infty f_j$. Since almost every f_j vanishes at any given point of R we have $f_\varepsilon \in C^\infty(R)$. Furthermore

$$\|f - f_\varepsilon\| \leq \sum_{j=1}^\infty \|f\varphi_j - f_j\| < \varepsilon.$$

Hence $f_\varepsilon \in C^\infty(R) \cap \mathrm{MI}(R)$ and if the support of f is contained in an open set G then the support of f_ε is contained in G.

1 E. Green's Formula. As an example of an application of the approximation theorem we give the following form of Green's formula. Let G be an open set with compact closure in R and let ∂G consist of a finite number of disjoint analytic Jordan curves.

Let $f \in \mathbb{M}(R)$ and $u \in H(\bar{G})$. Then

$$D_G(f, u) = \int_{\partial G} f * du. \tag{10}$$

In fact take $f_n \in C^\infty(R) \cap \mathbb{M}(R)$ such that $\lim_n \|f_n - f\| = 0$. By the standard Green's formula $D_G(f_n, u) = \int_{\partial G} f_n * du$ and on letting $n \to \infty$ we obtain (10).

Another form of Green's formula we shall often use is the following:

Let $f \in \mathbb{M}(R)$, $u \in HD(G)$, choose a union γ_1 of some components of ∂G and set $\gamma_2 = \partial G - \gamma_1$. Assume that $f = 0$ on γ_1 and u is harmonic on γ_2. Then

$$D_G(f, u) = \int_{\partial G} f * du = \int_{\gamma_2} f * du. \tag{11}$$

Here $*du$ is meaningless in general on γ_1 but since $f|\gamma_1 = 0$ we may write $f * du$ on γ_1 with the convention $f * du = 0$ there.

For the proof take $w \in H(\bar{G})$ such that $w|\gamma_1 = 0$ and $w|\gamma_2 = 1$. Let $G_r = \{r < w < 1\}$ and $\beta_r = \partial G_r - \gamma_2$ for $0 < r < 1$. First consider the case $f|\bar{G} = w$. Clearly

$$D_{G_r}(f, u) = \int_{\partial G_r} f * du = \int_{\beta_r} f * du + \int_{\gamma_2} f * du.$$

Here $\int_{\beta_r} f * du = r \int_{\beta_r} * du = r \int_{\gamma_2} * du \to 0$ as $r \to 0$. Since $D_{G_r}(f, u) \to D_G(f, u)$ we have (11).

For a general f, $f \cup 0$ and $f \cap 0$ belong to $\mathbb{M}(R)$ and we may assume that $f \geq 0$ on G. By considering $f + w$ instead of f we may also take $f > 0$ on $G \cup \gamma_2$. Then it is clear that $D_{G \cap \{f < c\}}(f) \to 0$ as $c > 0$ tends to zero. Set

$$f_c = (f - c) \cup 0$$

for $c > 0$. It is easily seen that $f_c \in \mathbb{M}(R)$ and

$$D_G(f - f_c) = D_G(f \cap c) = D_{G \cap \{f < c\}}(f) \to 0 \tag{12}$$

as $c > 0$ tends to zero. By (10)

$$D_{G_r}(f_c, u) = \int_{\beta_r} f_c * du + \int_{\gamma_2} f_c * du.$$

If $r > 0$ is sufficiently small then we infer by $f|\gamma_1 = 0$ that $f_c|\beta_r = 0$ and thus $\int_{\beta_r} f_c * du = 0$. Therefore on letting $r \to 0$ we obtain

$$D_G(f_c, u) = \int_{\gamma_2} f_c * du. \tag{13}$$

Next if $c > 0$ is small then $f > 0$ on γ_2 implies $f_c = f - c$ on γ_2 and consequently $\int_{\gamma_2} f_c * du = \int_{\gamma_2} f * du - c \int_{\gamma_2} * du$. For $c \to 0$ this with (12) and (13) gives (11).

1 F. Dirichlet's Principle. Let G be as in 1 E. For $f \in \mathrm{M}(R)$ and $u \in HD(G)$ with boundary values $u = f$ on ∂G

$$D_G(f) = D_G(u) + D_G(f-u), \tag{14}$$

i.e. the Dirichlet principle is valid for f and G.

In fact by (11)

$$D_G(f-u, u) = \int_{\partial G} (f-u) * du = 0$$

and (14) follows at once.

In passing we remark that for the validity of (10), (11), and (14) it suffices to assume that ∂G is piecewise analytic instead of analytic. It is also sufficient that ∂G is postulated rectifiable since these equations remain valid for smooth functions.

1 G. Potential Subalgebra. We denote by $\mathrm{M}_0(R)$ the totality of functions in $\mathrm{M}(R)$ with compact supports and by $\mathrm{M}_\Delta(R)$ the BD-closure of $\mathrm{M}_0(R)$ in $\mathrm{M}(R)$. Then $\mathrm{M}_0(R) \subset \mathrm{M}_\Delta(R) \subset \mathrm{M}(R)$. Here M_0 is a subalgebra of M_Δ and also of M, and M_Δ is a subalgebra of M.

We call $\mathrm{M}_\Delta(R)$ the *potential subalgebra of* $\mathrm{M}(R)$. Clearly $\mathrm{M}_0(R) = \mathrm{M}(R)$ if and only if R is closed. On a closed R we have of course $\mathrm{M}_\Delta(R) = \mathrm{M}(R)$ but the converse is not true. The following result (KAWAMURA [1]) corresponds to Theorem 1 C:

THEOREM. *Let $\{f_n\}$ be a sequence of functions in $\mathrm{M}_\Delta(R)$ and f a bounded function on R such that*

$$f = C\text{-}\lim_{n\to\infty} f_n \tag{15}$$

on R and

$$D_R(f_n) < K < \infty \tag{16}$$

for every n. Then f belongs to $\mathrm{M}_\Delta(R)$.

By Theorem 1 C, $f \in \mathrm{M}(R)$ and $D_R(f, g) = \lim_n D_R(f_n, g)$ for any $g \in \mathrm{M}(R)$. Let $\{R_m\}$ be a regular exhaustion of R and let φ be continuous on R such that $\varphi | \bar{R}_0 = f$, $\varphi | (R - R_1) = 0$, and $\varphi \in H(R_1 - \bar{R}_0)$. Clearly $\varphi \in \mathrm{M}(R)$ and actually $\varphi \in \mathrm{M}_0(R)$. With each f_n $(n = 1, 2, \ldots)$ we similarly associate a $\varphi_n \in \mathrm{M}_0(R)$. By considering $f - \varphi$ and $f_n - \varphi_n$ instead of f and f_n we may assume that $f = f_n = 0$ on $\bar{R}_0$.

Let u_m be continuous on R with $u_m | \bar{R}_0 = 0$, $u_m | (R - R_m) = f$, and $u_m \in H(R_m - \bar{R}_0)$. Obviously $u_m \in \mathrm{M}(R)$ and $|u_m| \leq \sup_R |f|$ on R. By Green's formula (11) or Dirichlet's principle (14) we see that $D_R(u_{m+p} - u_m) = D_R(u_m) - D_R(u_{m+p})$. Thus $\{u_m\}$ is D-Cauchy on R and since $u_m = 0$ on $\bar{R}_0$ it is also B-Cauchy. Let

$$u = BD\text{-}\lim_{m\to\infty} u_m$$

on R. Then $u \in \mathbf{M}(R)$, $u=0$ on $\bar{R}_0$, and $u \in H(R-R_0)$. Since $f-u = BD-\lim_m (f-u_m)$ and $f-u_m \in \mathbf{M}_0(R)$, $f-u$ belongs to $\mathbf{M}_\Delta(R)$. By Green's formula (11)

$$D_R(f-u_m, u) = \int_{\partial(R_m-R_0)} (f-u_m) * du = 0$$

and consequently $D_R(f-u, u)=0$. This in turn gives

$$D_R(u) = D_R(f, u) = \lim_n D_R(f_n, u). \tag{17}$$

On the other hand since $f_n \in \mathbf{M}_\Delta(R)$ there exists a sequence $\{\varphi_k\} \subset \mathbf{M}_0(R)$ such that $f_n = BD\text{-}\lim_k \varphi_k$ on R. Again by Green's formula

$$D_R(\varphi_k, u) = \int_{\partial R_0} \varphi_k * du.$$

Observe that $U\text{-}\lim_k \varphi_k = f_n = 0$ on ∂R_0. On letting $k \to \infty$ we obtain $D_R(f_n, u)=0$. Thus (17) gives $D_R(u)=0$ which in turn implies $u=0$ on R since $u=0$ on $\bar{R}_0$. In conclusion, $f = f-u \in \mathbf{M}_\Delta(R)$.

1 H. Ideals. It is clear that $\mathbf{M}_0(R)$ is not only a subalgebra but also an ideal of $\mathbf{M}_\Delta(R)$ and of $\mathbf{M}(R)$ as well. Similarly $\mathbf{M}_\Delta(R)$ *is an ideal of* $\mathbf{M}(R)$.

We must show that for $f \in \mathbf{M}_\Delta(R)$ and $g \in \mathbf{M}(R)$, $fg \in \mathbf{M}_\Delta(R)$. There exists a sequence $\{f_n\} \subset \mathbf{M}_0(R)$ with $f = BD\text{-}\lim_n f_n$ on R. It is not difficult to show that $gf = BD\text{-}\lim_n g f_n$ on R. However in view of Theorem 1 G it is sufficient to note that

$$\sup_n \sqrt{D(g f_n)} \le \sup_n \|g f_n\| \le \|g\| \sup_n \|f_n\| < \infty.$$

Since the $g f_n$ are in $\mathbf{M}_\Delta(R)$ and satisfy (15) and (16) we have $g f \in \mathbf{M}_\Delta(R)$.

2. Royden's Compactification

2 A. Definition of Royden's Compactification. Let R be a Riemann surface and $\mathbf{M}(R)$ Royden's algebra associated with R. Consider the following conditions on a topological space R^*:

(R*.1) *R^* is a compact Hausdorff space,*

(R*.2) *R^* contains R as an open and dense subspace,*

(R*.3) *every function in* $\mathbf{M}(R)$ *can be continuously extended to R^*.*

We denote by $B(R^*)$ the set of (complex- or real-valued) bounded continuous functions on R^*. The continuous extension of any $f \in \mathbf{M}(R)$ to R^* is unique by (R*.2). Hence we may use the same notation f for the extension and write $\mathbf{M}(R) \subset B(R^*)$. With this convention we consider one more condition:

(R*.4) $\mathbf{M}(R)$ *separates points in R^*.*

This means the following: if p and q are two distinct points in R^* then there exists an $f \in \mathrm{M}(R)$ such that $f(p) \neq f(q)$. We say that R^* is the *Royden compactification* of R if R^* satisfies the four conditions (R*.1) to (R*.4). First we show:

THEOREM. *For any Riemann surface R the Royden compactification R^* of R exists and is uniquely determined up to a homeomorphism fixing R elementwise.*

The proof will be given in 2 B.

2 B. Characters. Let x be a multiplicative linear functional on $\mathrm{M}(R)$ such that $x(1)=1$. We call such an x a *character* on $\mathrm{M}(R)$ and denote by X the set of all characters on $\mathrm{M}(R)$. We topologize X by the weak* topology; i.e. by definition a directed net $\{x_\lambda\} \subset X$ converges to $x \in X$ if $x_\lambda(f) \to x(f)$ for every $f \in \mathrm{M}(R)$. By Tychonoff's theorem, which is a generalization of the Cantor diagonal process, one sees easily that X becomes a compact Hausdorff space.

For a point $z \in R$, $x_z(f)=f(z)$ defines a character on $\mathrm{M}(R)$ and clearly $z \to x_z$ gives a homeomorphism of R into X. If we identify z and x_z we may view $R \subset X$ as a topological subspace. Then $f \in \mathrm{M}(R)$ may be considered as defined on X by $f(x)=x(f)$.

Clearly $\mathrm{M}(R)$ becomes a subalgebra of $B(X)$ and separates points in X. By the Stone-Weierstrass theorem $\mathrm{M}(R)$ is dense in $B(X)$ with respect to the U-topology. Hence X satisfies (R*.1)–(R*.4) except perhaps for R being open in X. However this property is a consequence of the local compactness of R. Therefore X is a Royden compactification of R.

Let R^* be a Royden compactification. Each $p \in R^*$ can be viewed as a character by setting $x_p(f)=f(p)$ for $f \in \mathrm{M}(R)$. Clearly $p \to x_p$ is a homeomorphism of R^* into X which fixes R elementwise. Thus $R \subset R^* \subset X$ as topological subspaces. Since R is dense in X and both R^* and X are compact, $R^*=X$, i.e. the uniqueness of Royden's compactification is assured. The proof of Theorem 2 A is herewith complete.

From the proof we also conclude:

COROLLARY 1. $\mathrm{M}(R)$ *is dense in* $B(R^*)$ *with respect to the U-topology.*

COROLLARY 2. *For any character x on* $\mathrm{M}(R)$ *there exists a unique point* $p \in R^*$ *such that* $x(f)=f(p)$ *for every* $f \in \mathrm{M}(R)$, *and conversely.*

2 C. Urysohn's Property. We next turn to the question: as a function space on Royden's compactification R^*, how many functions does Royden's algebra $\mathrm{M}(R)$ contain? We have seen that $\mathrm{M}(R)$ is dense in

$B(R^*)$ with respect to the U-topology. Of course $\mathbf{M}(R)$ is a proper subset of $B(R^*)$ for any R. However $\mathbf{M}(R)$ enjoys the following Urysohn property:

THEOREM. *For any two nonempty disjoint compact sets K_1 and K_2 in R^* and two real numbers a_1 and a_2 there exists a real-valued function f in $\mathbf{M}(R)$ such that $f = a_j$ on K_j $(j = 1, 2)$ and $a_1 \cap a_2 \le f \le a_1 \cup a_2$ on R^*.*

We may assume that $a_1 > a_2$. By Urysohn's lemma there exists a real-valued function $g \in B(R^*)$ such that $g = a_1 + 2$ on K_1 and $g = a_2 - 2$ on K_2. Since $\mathbf{M}(R)$ is dense in $B(R^*)$ with respect to the U-topology there exists an $h \in \mathbf{M}(R)$ with $|g - h| < 1$ on R^*. By 1 B, $f = (h \cap a_1) \cup a_2 \in \mathbf{M}(R)$ and f has the required properties.

2 D. Royden's Boundary. The set $\Gamma = R^* - R$ is a compact subset of R^*. We call Γ *Royden's boundary* of R. Clearly

$$\Gamma = \{p \mid p \in R^*, f(p) = 0 \text{ for every } f \in \mathbf{M}_0(R)\}$$

and thus $\Gamma = \emptyset$ if and only if R is compact. We next give a topological condition for distinguishing points in Γ from those in R:

THEOREM. *A point $p \in R^*$ belongs to Γ if and only if p is not a G_δ-set in R^*.*

Here a point $p \in R^*$ is a G_δ-set if there exists a countable collection $\{V_n\}_1^\infty$ of open subsets V_n of R^* such that $p = \bigcap_1^\infty V_n$. Clearly each point $p \in R$ is a G_δ-set.

We have only to show that if $p \in \Gamma$ then p is not G_δ. Contrary to the assertion assume that p is G_δ. Then we can find a sequence $\{U_n\}_{n=1}^\infty$ of open neighborhoods U_n of p such that $\overline{U}_{n+1} \subset U_n$ and $p = \bigcap_1^\infty U_n$. For each n we choose two open disks B_n and B_n' in $R \cap (U_n - \overline{U}_{n+1})$ with $B_n \supset \overline{B_n'}$ such that the annulus $A_n = B_n - \overline{B_n'}$ satisfies $\log \operatorname{mod} A_n = 2^n$. Let w be continuous on R with $w|(\bigcup_1^\infty \overline{B_n'}) = 1$, $w|(R - \bigcup_1^\infty B_n) = 0$, and $w \in H(\bigcup_1^\infty A_n)$. Clearly w is a bounded Tonelli function on R, and

$$D_R(w) = \sum_1^\infty D_{A_n}(w) = \sum_1^\infty \frac{2\pi}{\log \operatorname{mod} A_n} = 2\pi$$

(cf. I.1 A). Hence w is in $\mathbf{M}(R)$ and continuous on R^*. Let $z_n \in \partial B_n$ and $z_n' \in \partial B_n'$. Then $\lim_n z_n = \lim_n z_n' = p$ on R^* but $w(z_n) = 0$ and $w(z_n') = 1$ violate the continuity of w at p.

2 E. Harmonic Boundary. We distinguish the following important part of Royden's boundary Γ of R:

$$\Delta = \{p \mid p \in R^*, f(p) = 0 \text{ for every } f \in \mathbf{M}_\Delta(R)\}.$$

This is a compact set in R^*, and since $\mathbf{M}_0(R) \subset \mathbf{M}_\Delta(R)$, Δ is a compact subset of Γ. We call Δ Royden's *harmonic boundary* of R. First we show how Δ is distributed in Γ from the topological viewpoint:

THEOREM. *The harmonic boundary Δ is topologically so small that* $\overline{\Gamma - \Delta} = \Gamma$.

We have to show that for any $p_0 \in \Gamma$ and for any open neighborhood U of p_0, $U \cap (\Gamma - \Delta) \neq \emptyset$. Let V be an open neighborhood of p_0 with $\bar{V} \subset U$ and let $\{R_n\}_1^\infty$ be an exhaustion of R with $V \cap (R_{n+1} - \bar{R}_n) \neq \emptyset$. For each n we choose two open disks B_n and B_n' in $V \cap (R_{n+1} - \bar{R}_n)$ such that the annulus $A_n = B_n - \bar{B}_n'$ satisfies $\log \operatorname{mod} A_n = 2^n$. Let w_n be continuous on R with $w_n | (R - B_n) = 0$, $w_n | \bar{B}_n' = 1$, and $w_n \in H(A_n)$. The series $w = \sum_1^\infty w_n$ is clearly B-convergent and

$$D_R\left(w - \sum_1^m w_n\right) = \sum_{m+1}^\infty D_{A_n}(w_n) = \sum_{m+1}^\infty \frac{2\pi}{\log \operatorname{mod} A_n} = \frac{\pi}{2^{m-1}}.$$

Thus $w = BD\text{-}\lim_m \sum_1^m w_n$ on R. Since $\sum_1^m w_n \in \mathbf{M}_0(R)$, $w \in \mathbf{M}_\Delta(R)$. Let $z_n \in \bar{B}_n'$ and let p be an accumulation point of $\{z_n\}_1^\infty$. Clearly $p \in \Gamma \cap \bar{V} \subset U$. Since $w(z_n) = 1$ we have $w(p) = 1$ and thus $p \in \Gamma - \Delta$. A fortiori $p \in (\Gamma - \Delta) \cap U$.

The theorem shows that Δ is a small set in Γ from a topological viewpoint. However we shall later see that if Δ is not empty then it is large in Γ in a function-theoretic sense.

2 F. Parabolic Surfaces. For an open Riemann surface R let $\{R_n\}_0^\infty$ be an exhaustion with connected $R - \bar{R}_0$. Let $\omega_n(z; R_n, R_0)$ be the continuous function on R such that $\omega_n | \bar{R}_0 = 0$, $\omega_n | (R - R_n) = 1$, and $\omega_n \in H(R_n - \bar{R}_0)$. The function ω_n is called the harmonic measure of ∂R_n with respect to $R_n - \bar{R}_0$ (cf. I. 1 A). By the maximum principle we see that

$$\omega_{n+p} \leq \omega_n$$

and thus $\lim_{n \to \infty} \omega_n(z)$ exists on R, vanishes on $\bar{R}_0$, and is harmonic on $R - \bar{R}_0$. This function will be denoted by

$$\omega(z) = \omega(z; R, R_0)$$

and referred to as the *harmonic measure* of the ideal boundary of R with respect to $R - \bar{R}_0$.

A surface R which is either closed, or open with $\omega \equiv 0$, is called *parabolic*. Otherwise it is said to be *hyperbolic*. The class of parabolic surfaces will be denoted by O_G. We shall see that $R \in O_G$ is characterized by the nonexistence of Green's functions on R (cf. III. 6 and V), hence the notation. Observe that this notation is slightly different in nature from that of other null classes such as O_{AD}, O_{AB}, O_{HD}, etc.

We are ready to give a characterization of the class O_G in terms of the harmonic boundary (ROYDEN [3]):

THEOREM. *The following four conditions for Riemann surfaces R are equivalent:*

$$R \in O_G, \tag{18}$$

$$1 \in \mathrm{M}_\Delta(R), \tag{19}$$

$$\mathrm{M}_\Delta(R) = \mathrm{M}(R), \tag{20}$$

$$\Delta = \emptyset. \tag{21}$$

This also shows that $R \in O_G$ does not depend on the choice of R_0.

The assertion being trivial for closed surfaces we assume that R is open. Since $\mathrm{M}_\Delta(R)$ is an ideal of $\mathrm{M}(R)$ the equivalence of (19) and (20) is obvious. It is also trivial that (19) implies (21). Next assume that $\Delta = \emptyset$. For each $p \in R^*$ there exists a function $f_p \in \mathrm{M}_\Delta(R)$ such that $f_p \geq 0$ on R^* and $f_p(p) > 1$. Since R^* is compact we can find a finite number of points $p_1, \ldots, p_n$ in R^* such that $\bigcup_1^n \{p \mid p \in R^*, f_{p_j}(p) > 1\} = R^*$. Let $f = \sum_1^n f_{p_j}$. Then $f \in \mathrm{M}_\Delta(R)$ and $f > 1$ on R. Hence $1/f \in \mathrm{M}(R)$ and $1 = (1/f) f \in \mathrm{M}_\Delta(R)$, i.e. (19) and (21) are equivalent.

All that remains to be proved is the equivalence of (18) and (19). Let $u_n = 1 - \omega_n$ and $u = 1 - \omega$. Then $R \in O_G$ if and only if $u \equiv 1$ on R. By Green's formula

$$D_R(u_{n+p} - u_n, u_{n+p}) = \int_{\partial(R_{n+p} - \bar{R}_0)} (u_{n+p} - u_n) * du_{n+p} = 0$$

and thus $D_R(u_{n+p} - u_n) = D_R(u_n) - D_R(u_{n+p})$, i.e. $\{u_n\}$ is D-Cauchy. But $u_n \in \mathrm{M}_0(R)$ and $u = BD\text{-}\lim_n u_n$ mean that $u \in \mathrm{M}_\Delta(R)$. For $R \in O_G$ we therefore have $1 = u \in \mathrm{M}_\Delta(R)$.

Conversely assume that $1 \in \mathrm{M}_\Delta(R)$. Take a sequence $\{f_n\} \subset \mathrm{M}_0(R)$ such that $1 = BD\text{-}\lim_n f_n$. Clearly $u = B\text{-}\lim_n u f_n$. Let K be an arbitrary compact set in R. Then

$$\begin{aligned} D_R(uf_n - u) &\leq 2 \int_R |u|^2 \,|\mathrm{grad}(f_n - 1)|^2 \,dx\,dy + 2 \int_R |f_n - 1|^2 \,|\mathrm{grad}\, u|^2 \,dx\,dy \\ &\leq a\, D_R(f_n) + 2 \sup_K |f_n - 1|^2 \, D_R(u) + b_n \, D_{R-K}(u) \end{aligned}$$

where $a = 2 \sup_R |u|^2$ and $b_n = 2 \sup_R |f_n - 1|^2$. We set $b = \limsup_n b_n$ and obtain

$$\limsup_n D_R(uf_n - u) \leq b\, D_{R-K}(u).$$

On letting $K \to R$ we conclude that $u = D\text{-}\lim_n u f_n$ and $1 - u = BD\text{-}\lim_n (1-u) f_n$ on R. By Green's formula

$$D_R((1-u) f_n, (1-u)) = \int_{\partial(R_m - R_0)} (1-u) f_n * d(1-u) = 0$$

where $R_m \supset \operatorname{supp} f_n$. It follows that

$$D_R(u) = D_R(1-u) = \lim_n D_R\big((1-u) f_n, (1-u)\big) = 0.$$

Therefore u is a constant on R and hence $u \equiv 1$, which implies $R \in O_G$.

2 G. Maximum Principle I. The role of the harmonic boundary Δ is important in the study of harmonic functions as can already be seen from Theorem 2 F. The following is a modification for a relative case (Kusunoki-Mori [4]):

Theorem. *Let G be a subregion of R with analytic relative boundary. If $\bar{G} \cap \Delta = \emptyset$, then the double $\hat{G}$ of G about ∂G belongs to O_G.*

Since $\bar{G} \cap \Delta = \emptyset$ we can find for any $p \in \bar{G}$ an $f_p \in \mathrm{M}_\Delta(R)$ such that $f_p(p) > 1$ and $f_p \geq 0$ on R. From the compactness of $\bar{G}$ it follows that there exist a finite number of points $p_1, \ldots, p_n$ in $\bar{G}$ such that

$$\bigcup_{j=1}^{n} \{p \mid p \in R^*, f_{p_j}(p) > 1\} \supset \bar{G}.$$

The function $f = \sum_1^n f_{p_j}$ belongs to $\mathrm{M}_\Delta(R)$ and $f > 1$ on $\bar{G}$. Let $\{g_m\}_1^\infty$ be a sequence in $\mathrm{M}_0(R)$ with $f = BD\text{-}\lim_m g_m$ on R. We denote by $\hat{f}$ and $\hat{g}_m$ the symmetric extensions of $f|\bar{G}$ and $g_m|\bar{G}$ to $\hat{G}$ respectively. Clearly $\hat{f} \in \mathrm{M}(\hat{G})$, $\hat{g}_m \in \mathrm{M}_0(\hat{G})$, and $\hat{f} = BD\text{-}\lim_m \hat{g}_m$ on $\hat{G}$. Hence $\hat{f} \in \mathrm{M}_\Delta(\hat{G})$ and $\hat{f} > 1$ on $\hat{G}$. Since $(1/\hat{f})\, \hat{f} \in \mathrm{M}_\Delta(\hat{G})$ we conclude that $\hat{G} \in O_G$.

It may appear that the theorem has no relevance to the maximum principle; that it in reality contains this principle is seen by means of the following two theorems.

2 H. Maximum Principle II. First we remark that

$$O_G \subset O_{HB} \tag{22}$$

where $HB = HB(R)$ is the set of bounded harmonic functions on a Riemann surface R and O_{HB} is the class of surfaces R such that $HB(R)$ consists of only constants.

Take $u \in HB(R)$ with $R \in O_G$. We may assume that $u \geq 0$. Let $a = \sup_R u$ and $b = \sup_{R_0} u$. Using the notation of 2 F and the maximum principle we see that

$$u \leq b + a\,\omega_n$$

on R. On letting $n \to \infty$ we have $u \leq b$ on R whence $a \leq b$. Thus $u = \text{const}$, i.e. $R \in O_{HB}$.

We are ready to state:

THEOREM. *Let G be a subregion of R (G may be R) and u a real-valued harmonic function on G bounded from above (resp. below). Suppose*

$$\limsup_{z\in G, z\to p} u(z) \leq m \qquad \left(resp.\ \liminf_{z\in G, z\to p} u(z) \geq m\right) \tag{23}$$

for every $p \in (\bar{G} \cap \Delta) \cup \partial G$. Then $u \leq m$ (resp. $u \geq m$) on G.

Suppose that $u(z_0) > m$ at a point $z_0 \in G$. Choose a number c such that $u(z_0) > c > m$ and $du \neq 0$ on $\{z \mid z \in G, u(z) = c\}$. Let G_0 be the component of $\{z \mid z \in G, u(z) > c\}$ which contains z_0. Then G_0 is a subregion of G and $\bar{G}_0 \cap \Delta = \emptyset$, and thus $\hat{G}_0 \in O_G$. On the other hand $u - c \in HB(G_0)$ and $u - c = 0$ on ∂G_0. Hence $u - c$ can be continued to $\hat{G}_0$ so that $u - c \in HB(\hat{G}_0)$. This contradicts $O_G \subset O_{HB}$.

2 I. Maximum Principle III. Another form of the maximum principle which will be used often is the following:

THEOREM. *Let G be a subregion of R (G may be R) and $u \in HD(G)$ and real. Suppose*

$$-\infty \leq a \leq \liminf_{z\in G, z\to p} u(z) \leq \limsup_{z\in G, z\to p} u(z) \leq b \leq \infty \tag{24}$$

for every point p in $(\bar{G} \cap \Delta) \cup \partial G$. Then $a \leq u \leq b$ on G.

We have only to show that $b \geq u$ on G. To this end we may assume that $b < \infty$. Contrary to the assertion assume that there exists a point $z_0 \in G$ with $b < u(z_0)$. Let c be a real number such that $b < c < u(z_0)$ and $du \neq 0$ on $\{z \mid z \in G, u(z) = c\}$. Let G_0 be the component of the open set $\{z \mid z \in G, u(z) > c\}$ which contains z_0. Then $\bar{G}_0 \cap \Delta = \emptyset$ and thus $1 \in \mathrm{M}_\Delta(\hat{G}_0)$. Hence we can find a sequence $\{\varphi_n\}_1^\infty \subset \mathrm{M}_0(\hat{G}_0)$ with $1 = BD\text{-}\lim_n \varphi_n$ on $\hat{G}_0$. Clearly $u = CD\text{-}\lim_n u\,\varphi_n$ on G_0. Let $v = u - c$ and let φ_n vanish on $\hat{G}_0$ outside of a regular region S. Then the relative boundary $\partial(G_0 \cap S)$ with respect to R consists of a finite number of piecewise analytic simple curves on which $v\,\varphi_n = 0$. By Green's formula

$$D_{G_0}(v\,\varphi_n, u) = \int_{\partial(G_0 \cap S)} v\,\varphi_n * du = 0.$$

It follows that

$$D_{G_0}(u) = D_{G_0}(v, u) = \lim_n D_{G_0}(v\,\varphi_n, u) = 0.$$

As a consequence $u \equiv c$ on G_0, which contradicts $u(z_0) > c$.

2 J. Duality. Recall that Δ is defined in terms of $\mathrm{M}_\Delta(R)$ by

$$\Delta = \{p \mid p \in R^*, f(p) = 0 \text{ for every } f \in \mathrm{M}_\Delta(R)\}. \tag{25}$$

THEOREM. *The following duality relation is valid between the potential subalgebra* $\mathrm{M}_\Delta(R)$ *and the harmonic boundary* Δ:

$$\mathrm{M}_\Delta(R)=\{f \mid f\in \mathrm{M}(R),\, f=0 \text{ on } \Delta\}. \tag{26}$$

If $\Delta=\emptyset$ then the assertion is trivial. Hence we may assume that $\Delta \neq \emptyset$, i.e. $R\notin O_G$. Let $f\in \mathrm{M}(R)$ and $f=0$ on Δ. For a regular exhaustion $\{R_n\}_1^\infty$ of R we construct continuous functions u_n on R with $u_n|R_n\in H(R_n)$ and $u_n|(R-R_n)=f$. Then $u_n\in \mathrm{M}(R)$ and $|u_n|\le \sup_R|f|$ on R. By Green's formula

$$D_R(u_{n+p}-u_n, u_{n+p})=\int_{\partial R_{n+p}}(u_{n+p}-u_n)*du_{n+p}=0$$

and therefore $D_R(u_{n+p}-u_n)=D_R(u_n)-D_R(u_{n+p})$. Hence $\{u_n\}_1^\infty$ is D-Cauchy. By choosing a suitable subsequence we may assume that $u=BD\text{-}\lim_n u_n$ exists. Clearly $u\in HD(R)$. Since $u_n-f\in \mathrm{M}_0(R)$ we have $u-f\in \mathrm{M}_\Delta(R)$ or $u=f=0$ on Δ. By maximum principle III, $u\equiv 0$ on R and $f\in \mathrm{M}_\Delta(R)$.

3. Orthogonal Projection

3 A. Quasi-Dirichlet Finiteness. A real-valued function f on R will be called *quasi-Dirichlet finite* if $D_R((f\cap c)\cup(-d))$ exists and is finite for all nonnegative numbers c and d. A complex-valued function f on R will be called quasi-Dirichlet finite if $\mathrm{Re}\, f$ and $\mathrm{Im}\, f$ have this property.

A Tonelli function f on R is quasi-Dirichlet finite if and only if $(f\cap c)\cup(-d)\in \mathrm{M}(R)$ for every pair of nonnegative numbers c and d.

If f is quasi-Dirichlet finite then clearly so is αf for any real number α. If f_1 and f_2 are nonnegative and quasi-Dirichlet finite then the same is true of f_1+f_2. In fact we must show that $D_R((f_1+f_2)\cap c)<\infty$. To do this let $A_j=\{p \mid p\in R,\, f_j(p)>c\}$ $(j=1,2)$ and $B=\{p \mid p\in R,\, f_1(p)+f_2(p)>c\}$. Obviously $B\supset A_j$ $(j=1,2)$ and therefore

$$\begin{aligned}\sqrt{D_R((f_1+f_2)\cap c)}&=\sqrt{D_{R-B}(f_1+f_2)}\le\sqrt{D_{R-B}(f_1)}+\sqrt{D_{R-B}(f_2)}\\&\le\sqrt{D_{R-A_1}(f_1)}+\sqrt{D_{R-A_2}(f_2)}=\sqrt{D_R(f_1\cap c)}+\sqrt{D_R(f_2\cap c)}<\infty.\end{aligned}$$

THEOREM. *If f is continuous on R, a Tonelli function on $R-\{|f|=\infty\}$, and quasi-Dirichlet finite on R then f has a continuous extension to R^*.*

First assume that $f\ge 0$ on R. For each $n=1,2,\ldots$ we have $f\cap n\in \mathrm{M}(R)$ and thus $f\cap n$ can be continuously extended to R^* in a unique manner. We denote by f_n this continuous extension. Since $\{f_n\}_1^\infty$ is nondecreasing on R^* we can define $h(p)=\lim_n f_n(p)$ on R^*. For $m>n$ we have $(f\cap m)\cap n=f\cap n$ on R and thus $f_m\cap n=f_n$. Assume that $h(p_0)<\infty$ for some $p_0\in R^*$ and take an integer n such that $h(p_0)<n$. Then $f_n(p_0)<n$. Choose a

neighborhood U of p_0 such that $f_n(p)<n$ on U. Then $f_m(p)\cap n=f_n(p)<n$ on U implies that $f_m(p)=f_n(p)$ on U for all $m>n$. Therefore $h|U=f_n|U$ is continuous on U. Next assume $h(p_0)=\infty$. Then for any $c>0$ there exists an f_n such that $f_n(p_0)>c$. Let U be a neighborhood of p_0 with $f_n(p)>c$ on U. Then $h(p)>c$ on U, i.e. h is continuous at p_0. Obviously $f_n=f\cap n$ on R and thus $h|R=f$.

In the general case of a real-valued f, $g=f\cup 0$ and $h=-(f\cap 0)$ satisfy the conditions of the theorem if f does, and g and h can be continuously extended to R^*. We use the same notation g and h for these extensions. It is easily seen that $g-h$ has a definite meaning on R^* and is continuous on R^*. Since $(g-h)|R=f$ the assertion is true for f.

If f is complex-valued then we only have to consider $\operatorname{Re} f$ and $\operatorname{Im} f$.

COROLLARY. *Dirichlet finite Tonelli functions on R have continuous extensions to R^*.*

3 B. Orthogonal Decomposition. Let K be a compact set in R^* such that $\overline{K\cap R}=K$ and for any $z\in\partial(K\cap R)$ there is a disk U such that $U\cap\partial(K\cap R)$ is a piecewise analytic arc joining two different points in ∂U. We call such a compact set K in R^* a *distinguished compact set.*

We make the following convention: *if $R\in O_G$, then $HD(R)=\{0\}$.* This convention which may seem somewhat artificial will turn out to be in a sense quite natural (cf. end of 3 C). At this point we remark that

$$O_G\subset O_{HD}. \tag{27}$$

Thus $HD(R)=\{0\}$ for $R\in O_G$ means that we shall not consider nonzero constants as harmonic functions.

The relation (27) can be proved as follows. By Theorem 2 F, $R\in O_G$ implies that $1\in \mathrm{M}_\Delta(R)$, i.e. there exists a sequence $\{\varphi_n\}\subset \mathrm{M}_0(R)$ such that $1=BD\text{-}\lim_n\varphi_n$. Let $u\in HD(R)$ and $u_m=(u\cap m)\cup(-m)$ for $m=1,2,\ldots$. By Green's formula $D_R(\varphi_n u_m, u)=\int_{\partial G}\varphi_n u_m * du=0$ for $G\supset\operatorname{supp}\varphi_n$. Clearly $u_m=BD\text{-}\lim_n\varphi_n u_m$ and hence $D_R(u_m,u)=0$. It is also easy to see that $u=CD\text{-}\lim_m u_m$. Thus $D_R(u)=\lim_m D(u_m,u)=0$, and u is constant, i.e. $R\in O_{HD}$.

We now establish the following decomposition theorem, one of the fundamental theorems in the study of HD-functions.

THEOREM. *Let f be a Dirichlet finite Tonelli function on R and $K\subset R^*$ a distinguished compact set which may be empty. Then*

(a) *f extended to R^* can be uniquely decomposed into the form $f=u+g$ where u and g are Dirichlet finite Tonelli functions on R with $u\in HD(R-K)$ and $g=0$ on $K\cup\Delta$,*

(b) $D_R(f)=D_R(u)+D_R(g)$,

(c) $|u|\le\sup_{(\partial(K\cap R))\cup\Delta}|f|$ *on $R-K$,*

(d) $D_R(u,\varphi)=0$ *for any Dirichlet finite Tonelli function* φ *on* R *with* $\varphi=0$ *on* $K\cup\Delta$,

(e) *if* $R\notin O_G$ *or* $K\neq\emptyset$ *and* v *is superharmonic (resp. subharmonic) on* $R-K$ *with* $v\geq f$ *(resp.* $v\leq f$*) on* $R-K$ *then* $v\geq u$ *(resp.* $v\leq u$*) on* $R-K$.

If $K=\emptyset$ and $R\in O_G$ then our assertion is trivial. The same is true if R is closed and $K\neq\emptyset$. We therefore exclude these cases. Moreover we may assume that f is real-valued.

Let $\{R_n\}_0^\infty$ be a regular exhaustion of R with $\bar{R}_0\subset R-K$, and for $n\geq 1$ let u_n, u_n', and u_n'' be harmonic on R_n-K and continuous on R with

$$u_n\ (u_n' \text{ or } u_n'')=f \quad (f\cup 0 \text{ or } -(f\cap 0)) \qquad \text{on } R-(R_n-K).$$

Since $u_n', u_{n+p}'\in \mathbb{M}(R_{n+p+1})$ we may use Green's formula:

$$D_R(u_{n+p}'-u_n', u_{n+p}')=\int_{\partial(R_{n+p}-K)}(u_{n+p}'-u_n')*du_{n+p}'=0.$$

Thus $D_R(u_{n+p}'-u_n')=D_R(u_n')-D_R(u_{n+p}')$ and $\{u_n'\}_1^\infty$ is D-Cauchy. For $g_n'=f\cup 0-u_n'$ we have similarly

$$D_R(u_n')+D_R(g_n')=D_R(f\cup 0)\leq D_R(f). \tag{28}$$

Let w_n be continuous on R with $w_n|\bar{R}_0=1$, $w_n|(R-(R_n-K))=0$, and $w_n\in H(R_n-K-\bar{R}_0)$. By Green's formula

$$\begin{aligned} D_R(g_n', w_n) &= \int_{\partial(R_n-R_0)} g_n' * dw_n = \int_{-\partial R_0} g_n' * dw_n \\ &= \int_{-\partial R_0} (f\cup 0) * dw_n - \int_{-\partial R_0} u_n' * dw_n. \end{aligned}$$

On setting $a_n=\inf_{\partial R_0} u_n'$ and $b=\sup_{\partial R_0}(f\cup 0)$ we see that

$$\begin{aligned} a_n D_R(w_n) &= a_n \int_{-\partial R_0} * dw_n \leq \int_{-\partial R_0} u_n' * dw_n \\ &\leq b\int_{-\partial R_0} * dw_n - D_R(g_n', w_n) \leq b\, D_R(w_n)+\sqrt{D_R(f)\cdot D_R(w_n)}. \end{aligned}$$

Hence we obtain $a_n\leq b+\sqrt{D_R(f)/D_R(w_n)}$. By our assumption R is not in O_G if $K=\emptyset$, and thus $\limsup_n a_n<\infty$. In view of Harnack's inequality $\{u_n'\}$ is bounded on R_0. On choosing a suitable subsequence we may therefore assume that the sequence $\{u_n'\}_1^\infty$ is C-Cauchy on R. If $K\neq\emptyset$ and $R\in O_G$, then again $\{u_n'\}$ is C-Cauchy because of (28) and the fact that $u_{n+p}'-u_n'$ vanishes along some subarc of $\partial(K\cap R)$. It is thus legitimate to suppose that $\{u_n'\}$ is CD-Cauchy on R.

Similarly $\{u_n''\}$ may be assumed to be CD-Cauchy on R and the same is true of $\{u_n\}$. Let

$$u=CD\text{-}\lim_n u_n$$

on R. Then u is a Dirichlet finite Tonelli function on R and harmonic on $R-K$, with $u=f$ on $R\cap K$ and hence on $K=\overline{R\cap K}$.

Let $g_n=f-u_n$ and $g=f-u$. Then $g_n\in \mathrm{M}_0(R)$, $g_n=0$ on K, and $g=CD\text{-}\lim_n g_n$ on R. Thus $g=0$ on K. It is also easy to see that

$$\frac{g}{1+|g|}=BD\text{-}\lim_n \frac{g_n}{1+|g_n|}$$

on R, and since $g_n/(1+|g_n|)\in \mathrm{M}_0(R)$, $g/(1+|g|)\in \mathrm{M}_\Delta(R)$. Hence $g=0$ on Δ, and $f=u+g$ is a required decomposition.

Let $f=u'+g'$ be another decomposition. Then $u-u'=0$ on $\Delta\cup K$ and therefore $u\equiv u'$ on $R-K$ by maximum principle III. Hence $u\equiv u'$ on R and the decomposition is unique. This completes the proof of (a).

By the same maximum principle, (c) is trivially valid. Assertion (e) is clear if we observe that $v\geq u_n$ (resp. $v\leq u_n$) on R_n-K.

Finally we prove (d), from which (b) will follow. Let $\varphi=0$ on $K\cup\Delta$. By the above proof φ is the CD-limit of a sequence $\{\varphi_n\}_1^\infty\subset \mathrm{M}_0(R)$ such that $\varphi_n=0$ on K. By Green's formula $D_R(\varphi_n, u)=0$ and thus $D_R(\varphi, u)=0$ in the limit.

3 C. Reformulation. We denote by $\tilde{\mathrm{M}}(R)$ the class of Dirichlet finite Tonelli functions, i.e. functions f on R satisfying (M. 2) and (M. 3) of 1 B. Thus $\tilde{\mathrm{M}}(R)\supset \mathrm{M}(R)$ and also $\tilde{\mathrm{M}}(R)\subset C(R^*)$. Let $\tilde{\mathrm{M}}_{\Delta\cup K}(R)$ be the set of functions $f\in\tilde{\mathrm{M}}(R)$ such that $f=0$ on $K\cup\Delta$, where K is a distinguished compact set in R^*. If $K=\emptyset$, then we write $\tilde{\mathrm{M}}_\Delta(R)$ instead of $\tilde{\mathrm{M}}_{\Delta\cup\emptyset}(R)$. From Theorem 3 B and its proof we deduce:

Corollary 1. *The CD-closure of $\{f\,|\,f\in \mathrm{M}_0(R), f=0$ on $K\}$ is $\tilde{\mathrm{M}}_{\Delta\cup K}(R)$ where K is a distinguished compact set in R^*.*

Corollary 2. *The orthogonal decomposition $\tilde{\mathrm{M}}(R)=HD(R-K)+\tilde{\mathrm{M}}_{\Delta\cup K}(R)$ holds for any distinguished compact set K in R^* including $K=\emptyset$.*

Corollary 3. $\mathrm{M}(R)=HBD(R)+\mathrm{M}_\Delta(R)$.

At this point we make the following observation. For any $f\in\tilde{\mathrm{M}}(R)$ $\big($resp. $\tilde{\mathrm{M}}_\Delta(R)\big)$ it is easy to see that $(f\cap n)\cup(-n)\in \mathrm{M}(R)$ $\big($resp. $\mathrm{M}_\Delta(R)\big)$ and $f=CD\text{-}\lim_{n\to\infty}(f\cap n)\cup(-n)$. Thus $\mathrm{M}(R)=\mathrm{M}_\Delta(R)$ is equivalent to $\tilde{\mathrm{M}}(R)=\tilde{\mathrm{M}}_\Delta(R)$ and we conclude that

$$R\in O_G \text{ if and only if } \tilde{\mathrm{M}}(R)=\tilde{\mathrm{M}}_\Delta(R).$$

If $R\notin O_G$, we have $\tilde{\mathrm{M}}(R)=HD(R)+\tilde{\mathrm{M}}_\Delta(R)$ in the proper sense. In order that this formula be valid even for $R\in O_G$ it is necessary and sufficient to assume $HD(R)=\{0\}$ for $R\in O_G$. This is the meaning of our convention in 3 B.

3 D. Orthogonal Projection. Let K be a distinguished compact set in R^*. For $f \in \widetilde{\mathrm{M}}(R)$ we denote by $\pi_K f$ the function u of Theorem 3 B: $\pi_K f$ is in $HD(R-K)$, coincides with f on $K \cup \Delta$, and is continuous on R^*. For $K=\emptyset$ we simply write $\pi = \pi_\emptyset$. We call $\pi_K f$ the *harmonic projection* of f on $R-K$, and the operator π_K the *harmonizer* on $R-K$.

3 E. *HD*-Minimal Functions. A positive HD-function u on R which is not identically zero is called *HD-minimal* if for any $v \in HD(R)$, $u \geq v \geq 0$ on R implies the existence of a constant c_v such that $v = c_v u$ on R.

If $R \in O_G$ then $HD(R) = \{0\}$ by our convention and therefore there is no HD-minimal function on R. If $R \in O_{HD} - O_G$ then each $f \in HD(R)$ is constant, and HD-minimal unless $f \leq 0$.

The HD-minimality is characterized in terms of Δ as follows:

THEOREM. *An HD-function u on R is HD-minimal if and only if there exists an isolated point $p \in \Delta$ such that $0 < u(p) < \infty$ and $u = 0$ on $\Delta - p$. In particular an HD-minimal function is automatically strictly positive and bounded.*

Assume that $u \in HD(R)$, $u(p) > 0$ and $u = 0$ on $\Delta - p$ for an isolated point p in Δ. By maximum principle III, $u > 0$ on R. By Theorem 2 C there exists an $f \in \mathrm{M}(R)$ such that $f(p) = 1$ and $f = 0$ on $\Delta - p$. The function $u_0 = \pi f$ is in $HBD(R)$ with $u_0(p) = 1$ and $u_0 = 0$ on $\Delta - p$. For any c in $(0, u(p))$ the difference $u - c u_0$ is in $HD(R)$ and is nonnegative on Δ. Thus by maximum principle III, $u \geq c u_0$ on R. Fix a point $z_0 \in R$. Then $c \leq u(z_0)/u_0(z_0)$. This means that $0 < u(p) < \infty$, i.e. $u \in HBD(R)$. For any $v \in HD(R)$ with $u \geq v \geq 0$ on R we have $v = 0$ on $\Delta - p$ and $0 \leq v(p) < \infty$. As a consequence $c_v u - v$ with $c_v = v(p)/u(p)$ vanishes identically on Δ, i.e. $v = c_v u$ and u is an HD-minimal function.

Conversely assume that u is HD-minimal on R. Of course $u \geq 0$ and $u \not\equiv 0$ on Δ. Thus we can find a point $p \in \Delta$ with $u(p) > 0$. Suppose there exists a point $q \in \Delta$, $q \neq p$, such that $u(q) > 0$. Take $f \in \mathrm{M}(R)$ such that $f(p) = 1$, $f(q) = 0$, and $0 \leq f \leq 1$ on R^*. Consider $v = \pi(fu)$. Clearly $0 \leq v = fu \leq u$ on Δ. Therefore $0 \leq v \leq u$ on R and hence we can find a constant c_v such that $v = c_v u$. We now encounter the contradiction $0 = v(q) = c_v u(q) > 0$. It follows that u vanishes on Δ except at p. Continuity of u on Δ implies that p is isolated in Δ, $u(p) > 0$, and $u = 0$ on $\Delta - p$. From the first part of the proof we have $u \in HBD(R)$.

COROLLARY. *Let $p \in \Delta$ be isolated in Δ. There always exists a $u \in HBD(R)$ such that $u(p) = 1$ and $u = 0$ on $\Delta - p$. Moreover any HD-function v on R has a finite value at p.*

Only the second part requires proof. The functions $v_1 = \pi(v \cup 0)$ and $v_2 = -\pi(v \cap 0)$ are in $HD(R)$ and are nonnegative. Since $v_1 - v_2 = v \cup 0 + v \cap 0 = v$ on Δ, $v = v_1 - v_2$ on R. Thus we may assume that $v > 0$

on R. Suppose $v(p)=\infty$. Then $v-nu\geq 0$ on Δ and therefore $v\geq nu$ on R for any $n=1,2,\ldots$. This implies that $v\equiv\infty$, a contradiction.

3 F. A Characterization of O_{HD}. We have seen that $R\in O_G$ if and only if $\Delta=\emptyset$. Since $O_G\subset O_{HD}$, $\Delta=\emptyset$ is a sufficient condition for $R\in O_{HD}$. What can be said about $R\in O_{HD}-O_G$? The answer is given by the following result of ROYDEN [3]:

THEOREM. *$R\in O_{HD}$ if and only if Δ consists of at most one point. More precisely $R\in O_G$ if and only if $\Delta=\emptyset$, and $R\in O_{HD}-O_G$ if and only if Δ consists of one point.*

Assume that $R\in O_{HD}-O_G$. Suppose Δ contains two points p and q, $p\neq q$. Take $f\in \mathrm{M}(R)$ with $f(p)=1$, $f(q)=0$, and $0\leq f\leq 1$ on R^*. Then by maximum principle III, $u=\pi f\in HD(R)$ takes its maximum 1 and minimum 0 on Δ. Thus u cannot be constant, in violation of $R\in O_{HD}$, and we conclude that Δ consists of only one point. Conversely if Δ is a point then first of all $R\notin O_G$. Since $u\in HD(R)$ takes its maximum and minimum on Δ, u is constant, i.e., $R\in O_{HD}$.

3 G. Space *HD* of Finite Dimension. We consider real-valued harmonic functions only. Clearly $HD(R)$ is a real vector space. We are interested in the case $0\leq \dim HD(R)<\infty$. For the class O_{HD}^n of surfaces R with $\dim HD(R)\leq n<\infty$ we have $O_{HD}^n\subset O_{HD}^{n+1}$. By the convention $HD(R)=\{0\}$ for $R\in O_G$, $O_G=O_{HD}^0$. It is also clear that $O_{HD}=O_{HD}^1$. As a generalization of Theorem 3 F we state (KUSUNOKI-MORI [4], CONSTANTINESCU [13]):

THEOREM. *A surface $R\in O_{HD}^n-O_{HD}^{n-1}$ $(1\leq n<\infty)$ if and only if Δ consists of n points. In this case $HD(R)=HBD(R)$.*

Assume that $R\in O_{HD}^n-O_{HD}^{n-1}$, i.e. $\dim HD(R)=n$. First we show that Δ cannot contain more than n points. Contrary to the assertion assume that there exist $n+1$ distinct points $p_1,\ldots,p_{n+1}$ in Δ. Take neighborhoods V_j of p_j such that $\overline{V}_j\cap\overline{V}_{j'}=\emptyset$ $(j\neq j')$ in R^*. We can find $f_j\in\mathrm{M}(R)$ such that $f_j(p_j)=1$, $f_j=0$ on R^*-V_j and $0\leq f_j\leq 1$. Let $u_j=\pi f_j\in HD(R)$. Consider a linear combination

$$c_1u_1+c_2u_2+\cdots+c_{n+1}u_{n+1}=0$$

on R. This is also true on Δ and since $u_{j'}=0$ on V_j for $j'\neq j$ we must have $c_ju_j(p_j)=0$, i.e. $c_j=0$. This contradicts $\dim HD(R)=n$.

Thus Δ consists of m points, $0<m\leq n$. By Corollary 3 E, $HD(R)=HBD(R)$. From $\mathrm{M}(R)=HBD(R)+\mathrm{M}_\Delta(R)$ we conclude that $HD(R)$ is isomorphic to $\mathrm{M}(R)|\Delta$, which coincides with the class $B(\Delta)$ of bounded

continuous functions on Δ since Δ consists of m points and Urysohn's property is valid for $\mathrm{MI}(R)$. Now it is easy to see that $\dim B(\Delta)=m$ implies $m=n$.

From the above argument it can also be deduced that $HD(R)=HBD(R)$ has dimension n if Δ consists of n points.

The existence of surfaces R in $O_{HD}^{n}-O_{HD}^{n-1}$ $(1\leq n<\infty)$ will be established in 5. The essential part is the proof of $O_{HD}-O_{G}\neq\emptyset$ from which the remainder can be seen easily by forming a suitable double.

3 H. Evans' Superharmonic Function. We have seen thus far that the set $\Gamma-\Delta$ is in a sense harmonically negligible. This is best illustrated by the following property of $\Gamma-\Delta$ (CONSTANTINESCU [13]):

THEOREM. *Let F be an arbitrary nonempty compact set in $\Gamma-\Delta$. There exists a continuous positive superharmonic function $v\in\tilde{\mathrm{MI}}(R)$ with $v=\infty$ on F and $v=0$ on Δ.*

Let K be a distinguished compact set in R^* such that $K\cap\Delta=\emptyset$ and the interior $\mathring{K}$ of K contains F. For a regular exhaustion $\{R_n\}_1^\infty$ of R set $K_n=K-R_n$. Take an $f\in\mathrm{MI}(R)$ such that $f|K=1$, $f|\Delta=0$, and $0\leq f\leq 1$ on R^*. Let $v_n=\pi_{K_n}f\in\mathrm{MI}(R)$. Since $\pi_{K_{n+p}}(\pi_{K_n}f)=\pi_{K_{n+p}}f$, by Theorem 3 B and maximum principle III we obtain $D_R(v_n-v_{n+p})=D_R(v_n)-D_R(v_{n+p})$ and $0\leq v_{n+p}\leq v_n\leq 1$ on R^*. Hence $\{v_n\}_1^\infty$ is BD-Cauchy. Let $v_0=BD\text{-}\lim_n v_n$ on R. Then $v_0\in HBD(R)$ and since $0\leq v_0\leq v_n$ on R, $v_0=0$ on Δ, i.e.

$$BD\text{-}\lim_{n\to\infty} v_n\equiv 0$$

on R.

Let z_0 be a fixed point in R_1. We can choose a subsequence $\{n_k\}_1^\infty$ of positive integers such that $v_{n_k}(z_0)<2^{-k}$ and $\sqrt{D_R(v_{n_k})}<2^{-k}$. Let $v'_m=\sum_1^m v_{n_k}$ and $v=\sum_1^\infty v_{n_k}$ on R. Clearly

$$v=CD\text{-}\lim_{m\to\infty} v'_m$$

on R and thus $v\in\tilde{\mathrm{MI}}(R)$. It is also easy to see that v is positive and superharmonic on R. Observe that

$$\frac{v}{1+v}=BD\text{-}\lim_{m}\frac{v'_m}{1+v'_m}$$

on R. Since v'_m is in $\mathrm{MI}_\Delta(R)$ so is $v'_m/(1+v'_m)$ and hence by Theorem 1 G $v/(1+v)\in\mathrm{MI}_\Delta(R)$, i.e. $v=0$ on Δ. On the other hand $v>v'_m=m$ on $K_{n_m}\supset F$. Therefore $v=\infty$ on F.

One naturally asks whether v can be chosen to be harmonic. That this is actually the case will be seen in V. § 3.

3 I. Maximum Principle IV. The foregoing theorem will be useful for many purposes. We now generalize Theorem 2 H as follows:

THEOREM. *Let G be a subregion of R (G may be R) and u a superharmonic function on G bounded from below. Suppose that*

$$\liminf_{z\in G, z\to p} u(z) \geq m \tag{29}$$

for any $p\in(\bar{G}\cap\Delta)\cup\partial G$. Then $u\geq m$ on G.

The analogue, mutatis mutandis, is true for subharmonic functions.

We have only to consider the case where u is superharmonic. We define a function u' on $\gamma=\bar{G}-G$ by

$$u'(p)=\liminf_{z\in G, z\to p} u(z);$$

it is lower semicontinuous on γ. Let c be an arbitrary real number with $m>c$ and consider the open set

$$U=\{p\,|\,p\in\gamma, u'(p)>c\}$$

in γ. Since $u'(p)\geq m>c$ for every $p\in(\bar{G}\cap\Delta)\cup\partial G\subset\gamma$, $U\supset(\bar{G}\cap\Delta)\cup\partial G$. Hence $F=\gamma-U$ is a compact subset of $\Gamma-\Delta$.

Let v be as in Theorem 3 H for F. For $n=1, 2, \ldots$ set $w=u+v/n$ on G. Then w is superharmonic, bounded from below on G, and

$$\liminf_{z\in G, z\to p} w(z)>c$$

for every $p\in\gamma$. Thus $w(z)>c$ on G. On letting $n\to\infty$ we obtain the desired conclusion.

3 J. Dirichlet Integral of the Harmonic Measure. Let K be a distinguished compact set in R, and w_K the lower envelope of the family of nonnegative superharmonic functions on R which dominate 1 on K. Then w_K is called the *harmonic measure* of K relative to R. It is easy to see that $w_K\in\mathrm{MI}_\Delta(R)$, $w_K|K=1$, $w_K|\Delta=0$, and

$$D_R(w_K)=\int_{-\partial K} *dw_K=\int_{-\partial R_0} *dw_K$$

where R_0 is a regular region containing K.

This simple fact is very useful in the study of HD-functions, and we have already employed it in the proof of Theorem 3 B. However the limitation $K\subset R$ makes the applicable area small and thus it is desirable to remove this restriction. The only requirement turns out to be that $K\cap\Delta=\emptyset$:

THEOREM. *Let K be a nonempty distinguished compact set in R^* with $K \cap \Delta = \emptyset$ and set $u(z) = \inf\{v(z) \mid v \in \mathscr{F}_K\}$ on R where $\mathscr{F}_K$ is the family of nonnegative superharmonic functions v on R such that $v \mid K \cap R \geq 1$. The function u has the following properties:*

(a) *$u \in \mathrm{M}(R)$, $u \mid K = 1$, $u \mid \Delta = 0$, and $u \in HD(R-K)$;*

(b) *$D_R(u) = \int_{-\partial K} * du$;*

(c) *if K' is a distinguished compact set in R^* such that the interior $\mathring{K}'$ of K' contains K, $K' \cap \Delta = \emptyset$ and $\int_{\partial K'} |* du| < \infty$ then*

$$\int_{\partial K'} * du = \int_{\partial K} * du.$$

Since $\mathscr{F}_K$ is a Perron family (cf. 5 A) on $R-K$ and each point of $\partial(R-K)$ is regular for the Dirichlet problem u is nonnegative and harmonic on $R-K$, $u=1$ on K, and u is continuous on R. Let $f \in \mathrm{M}(R)$ such that $f \mid K = 1$, $f \mid \Delta = 0$, and $0 \leq f \leq 1$ on R^*. Then

$$\liminf_{z \in R, z \to p} \big(u(z) - (\pi_K f)(z)\big) \geq 0$$

for every $p \in \Delta \cup \partial K$. Thus by maximum principle IV, $u \geq \pi_K f$ on R and since $\pi_K f \in \mathscr{F}_K$, $u = \pi_K f$ on R. This proves (a).

Let $\{R_n\}_1^\infty$ be a regular exhaustion of R and set $K_n = \bar{R}_n \cap K$ ($n = 1, 2, \ldots$), $K_\infty = K$. Then $\pi_{K_n}(\pi_{K_{n+p}} f) = \pi_{K_n} f$ ($p = 1, 2, \ldots, \infty$). Hence by Theorem 3 B we obtain $D(\pi_{K_{n+p}} f - \pi_{K_n} f) = D(\pi_{K_{n+p}} f) - D(\pi_{K_n} f)$ and $D(\pi_{K_n} f) \leq D(\pi_K f) = D(u)$. Thus $\{\pi_{K_n} f\}_1^\infty$ is D-Cauchy. Since $\{\pi_{K_n} f\}$ is nondecreasing and $\pi_{K_n} f = u$ on K_n it is easy to see that

$$u = \pi_K f = BD\text{-}\lim_{n \to \infty} \pi_{K_n} f$$

on R.

For a fixed n and for $m > n$ let $v_m \in \mathrm{M}_0(R)$ such that $v_m \mid K_n = 1$, $v_m \mid (R - R_m) = 0$, and $v_m \in H(R_m - K_n)$. As in the proof of Theorem 3 B we can readily deduce that $\pi_{K_n} f = BD\text{-}\lim_m v_m$. Since $u \mid K = 1$ and $v_m \mid (R - R_m) = 0$ we obtain by Green's formula

$$D_R(v_m, u) = D_{R_m - K}(v_m, u) = \int_{\partial(R_m - K)} v_m * du = \int_{-\partial K} v_m * du.$$

From $* du \geq 0$ along $-\partial K$ and the fact that $\{v_m\}_m^\infty$ is nondecreasing on ∂K we conclude by Lebesgue's convergence theorem:

$$D_R(\pi_{K_n} f, u) = \lim_m D_R(v_m, u) = \lim_m \int_{-\partial K} v_m * du = \int_{-\partial K} \pi_{K_n} f * du.$$

Again since $\pi_{K_n} f$ increases to 1 on ∂K it follows that

$$D_R(u) = \lim_n D_R(\pi_{K_n} f, u) = \lim_n \int_{-\partial K} \pi_{K_n} f * du = \int_{-\partial K} * du.$$

Thus we have proved (b).

We turn to (c). Let $(\mathring{K}'-K)\cap R=\bigcup_1^\infty S_n$ be the decomposition into components. If we can prove that $\int_{(\partial S_n)\cap(\partial(\mathring{K}'-K))} *du=0$ for each S_n then we shall have

$$\int_{\partial(\mathring{K}'-K)} *du=\sum_1^\infty \int_{(\partial S_n)\cap(\partial(\mathring{K}'-K))} *du=0$$

since $\int_{\partial(\mathring{K}'-K)}|*du|<\infty$. Therefore we may assume without loss of generality that $S=(\mathring{K}'-K)\cap R$ is connected. In view of $\bar{S}\cap\Delta=\emptyset$ we infer by Theorem 2 G that $\hat{S}\in O_G$. Let $\{R_n\}_0^\infty$ be an exhaustion of R such that $\bar{R}_0\subset S$. Take a function w_n continuous on $S\cup\partial S$ and such that $w_n|\bar{R}_0=1$, $w_n|(S-R_n)=0$, $w_n\in H(R_n\cap S-\bar{R}_0)$, and $*dw_n=0$ on ∂S.

The symmetric extension $\hat{w}_n$ of w_n to $\hat{S}$ is the harmonic measure of $\bar{R}_0\cup\bar{G}_0$ with respect to $(S\cap R_n)\cup H_n$, where G_0 and H_n are the reflections about ∂S of R_0 and $S\cap R_n\cup((\partial S)\cap R_n)$ in $\hat{S}$. From $\hat{S}\in O_G$ it follows that BD-$\lim_n w_n=1$ on S. By Green's formula $D_S(w_n, u)=\int_{\partial S} w_n*du-\int_{\partial R_0}*du$. Since u is harmonic on $\bar{R}_0$

$$D_S(w_n, u)=\int_{\partial S} w_n*du.$$

From $|w_n*du|\le|*du|$ on ∂S, $\int_{\partial S}|*du|<\infty$, and $\lim_n w_n=1$ on ∂S we conclude by Lebesgue's convergence theorem that

$$\int_{\partial S} *du=\lim_n \int_{\partial S} w_n*du=\lim_n D_S(w_n, u)=D_S(1, u)=0.$$

Remark. The concepts which we call Royden's compactification and Royden's algebra were introduced in ROYDEN [3]. The essence of the theory, i.e. the notion of harmonic boundary and potential subalgebra appeared initially in his dissertation [2]. The most important first steps toward the present form of the theory were taken in MORI-ÔTA [1] and then in KUSUNOKI-MORI [4].

In the history of the theory of Riemann surfaces the compactifications due to ALEXANDROFF and STOÏLOW (cf. IV.5 D) have been traditionally significant. The first precise boundary used in the theory was the 2-dimensional version of the MARTIN [1] compactification employed by HEINS [1] and PARREAU [4]. The KURAMOCHI [13] compactification is another mode of modifying that of MARTIN.

These compactifications are "naive" in the sense that the second countability axiom is satisfied. In contrast Royden's compactification as well as that of WIENER to be discussed in Chapter IV does not satisfy this axiom. But this is compensated for by the fact that the construction of harmonic functions in the classes HD or HB can be carried out as on regular plane regions. Since the classification problem for H-classes is nothing more than determining whether specified constructions of harmonic functions are possible Royden's and Wiener's compactifications are the most powerful tools in this part of the classification theory.

For a general theory of ideal boundaries of Riemann surfaces we refer the reader to the comprehensive monograph of CONSTANTINESCU-CORNEA [17].

§ 2. Dirichlet's Problem

The benefit of the existence of a boundary of a Riemann surface lies in the possibility of integral representation. In the case of the unit disk the Poisson formula gives explicitly the behavior of harmonic functions. Similar representations which are appropriate for HD-functions will be furnished in 4. In this connection we shall show that HBD is dense in HD in the CD-topology, which again implies $O_{HB} \subset O_{HBD} = O_{HD}$. The strictness of this inclusion will be shown by an example.

For the integral representation we shall need the existence of a standard measure which we call the harmonic measure on Royden's boundary. In contrast with topological degeneracy, viz. the existence of an isolated point, we are naturally led to measure-theoretic degeneracy, the existence of a point with positive measure. To study it we consider a monotonically completed class $\widetilde{HD}$ of HD. This class produces a new kind of degenerate class $U_{\widetilde{HD}}$ of Riemann surfaces R characterized by the existence of $\widetilde{HD}$-minimal functions on R.

The connection between Dirichlet's principle and Perron's method is of considerable interest. We shall discuss Dirichlet's problem from the Perron viewpoint in 5. This is used to study ideal boundary points of subregions. As an application we give a test for a given surface to have what we shall call almost finite genus. As a special case we see that a surface $R \in O_{HD}$ of finite genus can always be obtained from a closed surface by removing a compact set of capacity zero.

An important role in the study of harmonic functions on a disk is played by the limits of functions along the radii of the disk. A generalization of radius is the concept of the Green's line. In 6 we give a characterization of O_{HD} in terms of the indivisibility of the space of Green's lines.

We close the section by discussing in 7 and 8 the invariance of certain null classes under homeomorphisms of Royden's compactifications.

4. Harmonic Measure and Kernel

4 A. Harmonic Measure on Γ. We shall assume throughout this no. that R is an open Riemann surface. The *harmonic measure* μ on Royden's boundary Γ of R with respect to a fixed point $z_0 \in R$ is the measure on Γ such that

(μ.1) *μ is a positive regular Borel measure on Γ,*
(μ.2) *every superharmonic function $v \in \mathbf{M}(R)$ has the Gauss property $v(z_0) \geq \int_\Gamma v(p)\, d\mu(p)$.*

The point z_0 will be referred to as the *center* of the measure μ. We denote by S_μ the support of μ in Γ. As an immediate consequence of (μ.2) we have

$$u(z_0) = \int_\Gamma u(p)\, d\mu(p) \tag{30}$$

for every $u \in HBD(R)$. From Theorem 3 B and property (μ.3) proved below we see that

$$v(z_0) \geq (\pi v)(z_0) = \int_\Gamma (\pi v)(p)\, d\mu(p) = \int_\Gamma v(p)\, d\mu(p),$$

and thus (μ.2) and (30) are equivalent.

We are ready to state:

THEOREM. *Suppose that $R \notin O_G$. There exists a unique harmonic measure μ on Γ with respect to an arbitrarily fixed center $z_0 \in R$. It satisfies the following conditions:*

(μ.3) $S_\mu = \Delta$,
(μ.4) $\mu(\Delta) = 1$,
(μ.5) *for any open set U in Γ with $U \cap \Delta \neq \emptyset$, $\mu(U) > 0$.*

First assume the existence of a μ with (μ.1) and (μ.2). Take an arbitrary nonempty compact set K in $\Gamma - \Delta$ and let $v_n \in \mathbf{M}(R)$ be the function $v \cap n$ obtained from v of Theorem 3 H for K. By (μ.2)

$$\frac{1}{n} v(z_0) \geq \frac{1}{n} v_n(z_0) \geq \int_\Gamma \frac{v(p) \cap n}{n}\, d\mu(p) \geq \mu(K).$$

Thus on letting $n \to \infty$ we obtain $\mu(K) = 0$. By the regularity of μ we see that $S_\mu \subset \Delta$. Let U be open in Γ with $U \cap \Delta \neq \emptyset$, and F an arbitrary compact set in $U \cap \Delta$. Take an $f \in \mathbf{M}(R)$ with $f|F = 1$, $f|(\Delta - U \cap \Delta) = 0$, and $0 \leq f \leq 1$ on R^*. Set $u = \pi f \in HBD(R)$. By (30) and $S_\mu \subset \Delta$

$$u(z_0) = \int_\Gamma u(p)\, d\mu(p) = \int_\Delta u(p)\, d\mu(p) \leq \int_U d\mu(p) = \mu(U).$$

Since $u(z_0) > 0$ we have (μ.5) and further (μ.3). By setting $u \equiv 1$ in (30) we obtain (μ.4).

If μ' is another measure with (μ.1) and (μ.2) then as we have seen above, μ' satisfies (μ.3), (μ.4), and (μ.5). Let $f \in \mathbf{M}(R)$. By (30) and (μ.3)

$$(\pi f)(z_0) = \int_\Delta f(p)\, d\mu(p) = \int_\Delta f(p)\, d\mu'(p).$$

Let $B(\Delta)$ be the space of bounded continuous functions on Δ. Since $\mathrm{M\!I}(R)|\Delta$ is dense in $B(\Delta)$ with respect to the U-topology we conclude that $\mu=\mu'$.

Finally we shall prove the existence of a μ with $(\mu.1)$ and $(\mu.2)$ and a fortiori $(\mu.3)$. Observe that

$$\sup_{z\in R}|(\pi f)(z)|=\sup_{p\in\Delta}|f(p)| \tag{31}$$

and that $\mathrm{M\!I}(R)$ is dense in $B(R^*)$ with respect to the U-topology, i.e. the topology induced by $\sup_{z\in R}|f(z)|$. Thus π can be extended to $B(R^*)$, (31) is also satisfied, and $\pi f\in HB(R)\cap B(R^*)$. Let $f\in B(\Delta)$ and $f'\in B(R^*)$ with $f=f'|\Delta$. We set $\rho f=\pi f'$. By (31), ρf is independent of the choice of f' and $\rho f\in HB(R)\cap B(R^*)$ with

$$\sup_{z\in R}|(\rho f)(z)|=\sup_{p\in\Delta}|f(p)|.$$

Thus $f\to(\rho f)(z_0)$ is a bounded linear functional on $B(\Delta)$ and hence there exists a measure μ on Δ with $(\mu.1)$ and $(\mu.3)$ such that

$$(\rho f)(z_0)=\int_\Delta f(p)\,d\mu(p).$$

If $u\in HBD(R)$ then $\rho u=u$ on R^*. On setting $\mu(\Gamma-\Delta)=0$ we obtain (30) and conclude that μ satisfies $(\mu.2)$.

4 B. Harmonic Kernel. Suppose that $R\notin O_G$ and the center of the harmonic measure μ is $z_0\in R$. The harmonic kernel $P(z,p)$ associated with (R,μ,z_0) is the real-valued function on $R\times\Gamma$ such that

(P.1) $P(z_0,p)=1$ *on* Δ,

(P.2) $P(z,p)=0$ *on* $R\times(\Gamma-\Delta)$,

(P.3) $P(z,p)\in HP(R)$ *for every fixed* $p\in\Gamma$,

(P.4) *for every fixed* $z\in R$, $P(z,p)$ *is a nonnegative bounded Borel function on* Γ,

(P.5) *for every superharmonic function* $v\in\mathrm{M\!I}(R)$ *and every point* $z\in R$ *the inequality* $v(z)\geq\int_\Gamma P(z,p)\,v(p)\,d\mu(p)$ *is valid.*

In particular (P.5) implies that

$$u(z)=\int_\Gamma P(z,p)\,u(p)\,d\mu(p) \tag{32}$$

for every $z\in R$ and $u\in HBD(R)$. By considering πv we can deduce (P.5) from (32), i.e. (P.5) and (32) are equivalent.

We claim:

THEOREM. *There exists a unique harmonic kernel $P(z,p)$ on $R\times\Gamma$ with respect to an arbitrary fixed point $z_0\in R$.*

Here the uniqueness means that for two harmonic kernels P and P' there exists a set E of μ-measure zero in Δ with $P \equiv P'$ on $R \times (\Gamma - E)$.

In fact, let $\{z_n\}_1^\infty$ be dense in R and $Q(z,p) = P(z,p) - P'(z,p)$. Since $Q(z,p) \equiv 0$ on $R \times (\Gamma - \Delta)$ it follows from (32) and the denseness of $\mathbf{M}(R)|\Delta$ in $B(\Delta)$ that

$$\int_\Delta Q(z,p)\, f(p)\, d\mu(p) = 0 \tag{33}$$

for any $f \in B(\Delta)$. Thus for fixed $z \in R$, $Q(z,p) = 0$ on Δ except for a set $E(z)$ of μ-measure zero. Let $E = \bigcup_1^\infty E(z_n)$. Then $\mu(E) = 0$ and $Q(z_n, p) = 0$ for $p \in \Gamma - E$ and for every $n = 1, 2, \ldots$. For a fixed $p \in \Gamma - E$, $Q(z,p) \in H(R)$. Thus the denseness of $\{z_n\}_1^\infty$ in R implies $Q(z,p) \equiv 0$ for every $z \in R$ and a fixed $p \in \Gamma - E$. Since p is arbitrary we have established the assertion.

The existence proof will be given in 4 C – 4 D.

4 C. Harnack's Function. Let k be *Harnack's function* on $R \times R$, i.e. the function k defined by

$$k(z', z) = \inf\{c \mid c^{-1} u(z) \le u(z') \le c\, u(z) \text{ for every } u \in HP(R)\} \tag{34}$$

for $(z', z) \in R \times R$. Then $\log k(z', z)$ *is finitely continuous on* $R \times R$. It is also a *semimetric* on R, i.e. a metric in which zero distance does not necessarily imply the identity of points.

In fact set $z' - z = re^{i\theta}$ in a closed parametric disk $|z' - z| \le 1$. For any $u \in HP(R)$ we have

$$u(z') = \frac{1}{2\pi} \int_0^{2\pi} \frac{1 - r^2}{1 - 2r\cos(\theta - t) + r^2}\, u(e^{it} + z)\, dt$$

and thus

$$\frac{1-r}{1+r}\, u(z) \le u(z') \le \frac{1+r}{1-r}\, u(z).$$

Therefore $1 \le k(z', z) \le (1 + |z' - z|)/(1 - |z' - z|) \to 1$ as $z' \to z$ and consequently $\log k(z', z) \to 0$ as $z' \to z$. It follows easily from the definition that $\log k(z', z)$ is a semimetric on R. By the triangle inequality $\log k(z', z)$ is finitely continuous on $R \times R$ since $\log k(z', z) \to 0$ as $z' \to z$.

Let μ_z be the harmonic measure on Γ with center z; $\mu_{z_0} = \mu$. For any nonnegative function $f \in B(\Delta)$, $(\pi f)(z) = \int_\Delta f(p)\, d\mu_z(p) \in HP(R)$. Thus

$$k(z, z_0)^{-1} \int_\Delta f(p)\, d\mu(p) \le \int_\Delta f(p)\, d\mu_z(p) \le k(z, z_0) \int_\Delta f(p)\, d\mu(p).$$

From this it follows at once that

$$k(z, z_0)^{-1}\, \mu(X) \le \mu_z(X) \le k(z, z_0)\, \mu(X) \tag{35}$$

for every Borel set X in Δ and hence in Γ. Let $\tilde{P}(z,p)$ be a function on $R \times \Gamma$ which can be used as the Radon-Nikodym density of μ_z with respect to μ. It is nonnegative and Borel measurable on Γ. Moreover

$$k(z, z_0)^{-1} \leq \tilde{P}(z, p) \leq k(z, z_0) \tag{36}$$

μ-a.e. on Γ for fixed $z \in R$. We may assume that $\tilde{P}(z_0, p) \equiv 1$ on Δ and $\tilde{P}(z, p) \equiv 0$ on $\Gamma - \Delta$ for every $z \in R$. From (35) and the definition of k we easily see that

$$k(z, z')^{-1} \tilde{P}(z, p) \leq \tilde{P}(z', p) \leq k(z, z') \tilde{P}(z, p) \tag{37}$$

μ-a.e. on Δ for fixed $z, z' \in R$.

Let T be a countable dense subset of R. Take a Borel subset $E(z, z')$ of Δ such that (37) is true for $p \in \Gamma - E(z, z')$ and $\mu(E(z, z')) = 0$. Set $E = \bigcup_{z, z' \in T} E(z, z')$. Again $\mu(E) = 0$ and (37) is valid on $\Gamma - E$ for every $z, z' \in T$. In particular (36) holds on $\Gamma - E$ for all $z \in T$. Thus for $z, z' \in T$ and $p \in \Gamma - E$

$$|\tilde{P}(z, p) - \tilde{P}(z', p)| \leq k(z, z_0) \max\left(k(z, z') - 1, 1 - k(z, z')^{-1}\right). \tag{38}$$

Since $\log k(z, z')$ is a continuous semimetric on R, (38) implies

$$\lim_{z, z' \in T;\, z, z' \to z''} |\tilde{P}(z, p) - \tilde{P}(z', p)| = 0 \tag{39}$$

for every $p \in \Gamma - E$. This means that $\lim_{z \in T, z \to z'} \tilde{P}(z, p)$ exists for every $z' \in R$ and $p \in \Gamma - E$, and if $z' \in T$ then the limit is $\tilde{P}(z', p)$.

Therefore it is possible to define

$$P(z, p) = \lim_{z' \in T, z' \to z} \tilde{P}(z, p) \tag{40}$$

for $z \in R$ and $p \in \Gamma - E$. By (39), $P(z, p)$ is finitely continuous on R for each fixed $p \in \Gamma - E$.

4 D. Harmonicity of $P(z, p)$. For $z \in R$ take a sequence $\{z_n\}_1^\infty \subset T$ converging to z. For any $u \in HBD(R)$ we see by (36), Lebesgue's convergence theorem, and the definition of $P(z, p)$ that

$$\begin{aligned} u(z) &= \lim_n u(z_n) = \lim_n \int_{\Delta - E} \tilde{P}(z_n, p)\, u(p)\, d\mu(p) \\ &= \int_{\Delta - E} \lim_n \tilde{P}(z_n, p)\, u(p)\, d\mu(p) = \int_\Gamma P(z, p)\, u(p)\, d\mu(p). \end{aligned} \tag{41}$$

Let z be a local parameter in $|z| < 1$ and $\{z_n\}_1^\infty$ a dense sequence in it. Then for any $f \in B(\Delta)$, since $\int_\Delta P(z, p) f(p)\, d\mu(p) \in H(|z| < 1)$, we have by Fubini's theorem

$$\begin{aligned} \int_\Delta P(z_n, p) f(p)\, d\mu(p) &= \frac{1}{2\pi} \int_0^{2\pi} \left[\int_\Delta P(z_n + r_m e^{i\theta}, p) f(p)\, d\mu(p)\right] d\theta \\ &= \int_\Delta \left[\frac{1}{2\pi} \int_0^{2\pi} P(z_n + r_m e^{i\theta}, p)\, d\theta\right] f(p)\, d\mu(p) \end{aligned}$$

where $\{r_m\}_1^\infty$ is a dense sequence in $(0, 1-|z_n|)$. Hence there exists a set $F_{n,m}$ in Δ with $\mu(F_{n,m})=0$ such that for any $p\in\Delta-F_{n,m}$

$$P(z_n, p)=\frac{1}{2\pi}\int_0^{2\pi} P(z_n+r\,e^{i\theta}, p)\,d\theta \tag{42}$$

with $r=r_m$. Let $F_n=(\bigcup_{m=1}^\infty F_{n,m})\cup E$. Then $\mu(F_n)=0$ and (42) holds for every $r=r_m$ $(m=1, 2, \ldots)$ and p in $\Delta-F_n$. By the continuity of $P(z, p)$ in z for a fixed $p\in\Delta-E$ we obtain (42) for every $r\in(0, 1-|z_n|)$ and $p\in\Gamma-F_n$. Finally let $F=\bigcup_1^\infty F_n$. Again $\mu(F)=0$ and (42) holds for all $n=1, 2, \ldots$ and for all $r\in(0, 1-|z_n|)$ if z_n is fixed. By the continuity of $P(z, p)$ in z we conclude that

$$P(z, p)=\frac{1}{2\pi}\int_0^{2\pi} P(z+r\,e^{i\theta}, p)\,d\theta \tag{43}$$

for every $z\in\{|z|<1\}$ and $r\in(0, 1-|z|)$ with an arbitrarily fixed $p\in\Gamma-F$.

We set $P(z, p)\equiv 1$ on R for $p\in F$ and leave P unchanged for $p\in\Gamma-F$. Then (P. 1) and (P. 2) are satisfied in view of the definition of P, and (P. 3) follows from (43). By (36) and the definition of P, (P. 4) is seen to be valid. Finally (41) gives (32) and thus (P. 5).

4 E. Integral Representation. By using the harmonic kernel we can prove a Schwarz-type theorem:

THEOREM. *If f is μ-integrable on Γ then*

$$u(z)=\int_\Gamma P(z, p)\,f(p)\,d\mu(p) \tag{44}$$

is on R a harmonic function which is a C-limit of HBD-functions on R.

If in addition f is bounded on Δ and continuous at $q\in\Delta$ as a function on Δ then

$$\lim_{z\in R,\, z\to q} u(z)=f(q).$$

Since the restrictions of HBD-functions to Δ are dense in $B(\Delta)$ with respect to the U-topology and $B(\Delta)$ is dense in $L^1(\Delta, \mu)$ with respect to the L^1-norm there exists a sequence $\{u_n\}_1^\infty$ in $HBD(R)$ such that

$$\lim_{n\to\infty}\int_\Delta |u_n(p)-f(p)|\,d\mu(p)=0.$$

Since $u_n(z)=\int_\Delta P(z, p)\,u_n(p)\,d\mu(p)$ is in $HBD(R)$ and $u=C\text{-}\lim_n u_n$ on R we have $u\in H(R)$.

Let $\varepsilon>0$. If f is continuous at $q\in\Delta$ as a function on Δ then we can find a neighborhood U of q in Δ such that $|f(p)-f(q)|<\varepsilon$ for every $p\in U$. If f is bounded then there exists a $w\in HBD(R)$ with $w(q)=0$ such that $|f(p)-f(q)| \leq w(p)+\varepsilon$ on Δ. It follows that

$$\left|\int_\Delta P(z,p)\, f(p)\, d\mu(p) - f(q)\right| \leq \int_\Delta P(z,p)\, |f(p)-f(q)|\, d\mu(p)$$
$$\leq \int_\Delta P(z,p)\big(w(p)+\varepsilon\big)\, d\mu(p) = w(z)+\varepsilon.$$

Hence $\limsup_{z\in R,\, z\to q} |u(z)-f(q)| \leq w(q)+\varepsilon = \varepsilon$.

4 F. Vector Lattice *HD*. For two real-valued harmonic functions u_1 and u_2 on R denote by $u_1 \vee u_2$ (resp. $u_1 \wedge u_2$) the least harmonic majorant (resp. the greatest harmonic minorant) of u_1 and u_2, i.e.

$$(u_1 \vee u_2)(z) = \inf\{v(z) \mid v \in H(R),\ v \geq u_1, u_2 \text{ on } R\}$$

and

$$(u_1 \wedge u_2)(z) = \sup\{v(z) \mid v \in H(R),\ v \leq u_1, u_2 \text{ on } R\}$$

provided these functions are harmonic. Functions with these properties do not always exist; if we write $u_1 \vee u_2$ or $u_1 \wedge u_2$ we presuppose the existence.

For real-valued functions in HD this existence is always assured:

THEOREM. *Every HD-function u on R has μ-integrable boundary values $u(p)$ on Γ and*

$$u(z) = \int_\Gamma P(z,p)\, u(p)\, d\mu(p) \tag{45}$$

on R. If u_1 and u_2 are HD-functions on R then

$$\begin{aligned}(u_1 \vee u_2)(z) &= \int_\Gamma P(z,p)\big(u_1(p) \cup u_2(p)\big)\, d\mu(p),\\ (u_1 \wedge u_2)(z) &= \int_\Gamma P(z,p)\big(u_1(p) \cap u_2(p)\big)\, d\mu(p)\end{aligned} \tag{46}$$

are also HD-functions on R.

Therefore $HD(R)$ constitutes a vector lattice with respect to the lattice operations $\vee$ and $\wedge$.

First assume that $u \geq 0$ on R. Then

$$u(z_0) \geq \pi(u \cap n)(z_0) = \int_\Gamma \big(u(p) \cap n\big)\, d\mu(p)$$

and thus $u(z_0) \geq \int_\Gamma u(p)\, d\mu(p)$, i.e. $u(p)$ is μ-integrable. Similarly

$$u(z) \geq \int_\Gamma P(z,p)\, u(p)\, d\mu(p) \geq \lim_{n\to\infty} \pi(u \cap n)(z). \tag{47}$$

By Theorem 3 B, $u \wedge n = \pi(u \cap n) \in HBD(R)$ and clearly the boundary values of $\lim_n (u \wedge n)$ on Δ dominate $u(p) \cap n$ for every $n = 1, 2, \ldots$. Thus $u(z) = \lim_{n\to\infty}(u \wedge n)(z)$ and (47) implies (45). For a general u we have only to observe that $u=\pi(u\cup 0)+\pi(u\cap 0)$ and $(\pi(u\cup 0))(p)=u(p)\cup 0$ and $(\pi(u\cap 0))(p)=u(p)\cap 0$ on Δ.

To prove (46) we may obviously assume that $u_1=u$, $u_2=0$, and have only to consider $u\vee 0$. But for this function (46) is obvious from the above observation.

4 G. The Identity $O_{HBD}=O_{HD}$. Once more we return to the Virtanen identity $O_{HBD}=O_{HD}$. This immediately yields

$$O_{HB}\subset O_{HD}=O_{HBD}. \tag{48}$$

We now give an alternate proof of these relations as a corollary of the following more general assertion:

THEOREM. *The space $HBD(R)$ is dense in $HD(R)$ in the CD-topology.*

Take $u\in HD(R)$. We have to show the existence of a sequence $\{u_n\}_1^\infty\subset HBD(R)$ such that $u=CD\text{-}\lim_n u_n$ on R. By Theorem 4 F we may suppose that $u>0$. The required sequence $\{u_n\}_1^\infty$ is then obtained by $u_n=u\wedge n$. Again by Theorem 4 F, $u=C\text{-}\lim_n u_n$. By Theorem 3 B, $D_R(u-u_n)\leq D_R(u-u\cap n)=D_{(u>n)}(u)\to 0$ as $n\to\infty$, i.e. $u=D\text{-}\lim_n u_n$.

4 H. Strict Inclusion $O_{HB}<O_{HD}$. Here we digress to prove that the inclusion (48) is strict (TÔKI [2]):

$$O_{HB}<O_{HD}. \tag{49}$$

By (48) we must prove the existence of a Riemann surface R which admits a nonconstant HB-function but does not carry any nonconstant HBD-functions. The surface which will be constructed is a complete covering surface R of the unit disk. In fact not only is $R\notin O_{HB}$ but also $R\notin O_{AB}$ since the composite function of $f(z)\equiv z$ with the projection map belongs to $AB(R)$. Hence the surface to be constructed also shows that

$$O_{HD}\not\subset O_{AB}. \tag{50}$$

We shall see in 5 that $O_{AB}\not\subset O_{HD}$ even for planar surfaces and thus there is no inclusion relation between O_{HD} and O_{AB}.

The following ingenious example due to TÔKI [2] is one of the most important achievements in classification theory.

First observe that $(m,n)\to\mu=2^{m-1}(2n-1)$ gives a one-to-one mapping of the set of pairs of positive integers onto the set of positive integers. For each pair (m,n) with $m,n=1,2,\ldots$ consider the $2^{2\mu}$ line segments

$$S_{mn}^\nu=\{z=r\,e^{i\theta}\,|\,-2^{-2\mu}\leq\log r\leq -2^{-(2\mu+1)},\ \theta=\nu\cdot 2\pi\cdot 2^{-2\mu}\} \tag{51}$$

for $\nu=1,\ldots,2^{2\mu}$ with $\mu=2^{m-1}(2n-1)$. Denote by R_0 the radial slit disk obtained from $|z|<1$ by deleting all slits S_{mn}^ν (see Fig. 6).

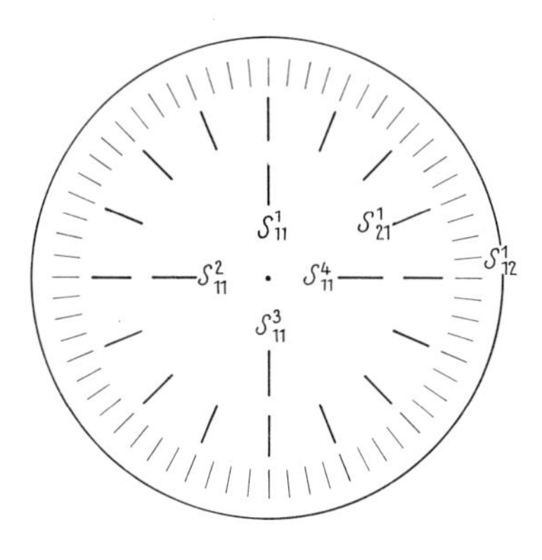
S^1_{11}
S^1_{21}
S^1_{12}
S^2_{11}
S^4_{11}
S^3_{11}

Fig. 6

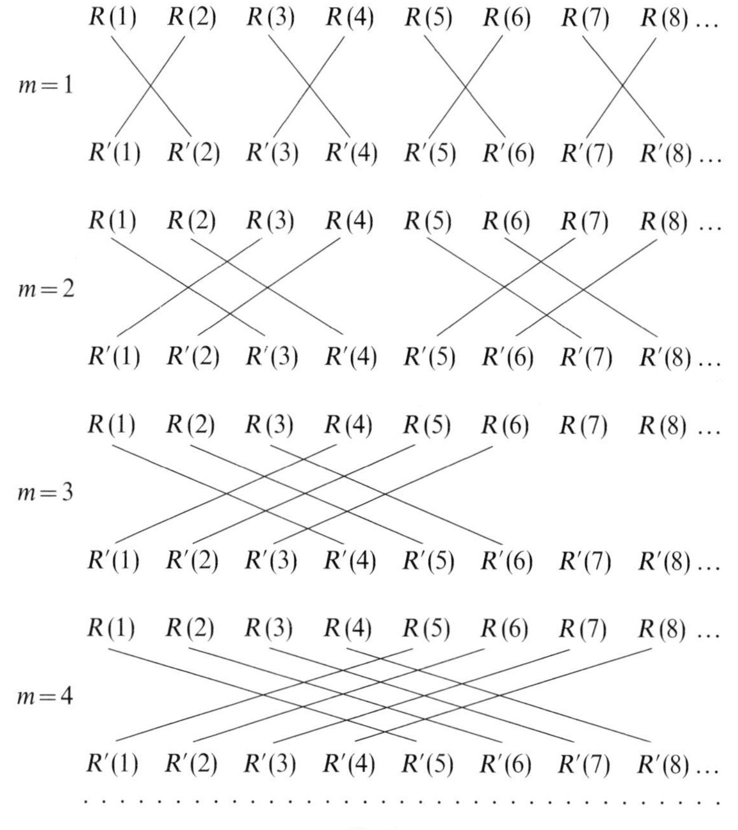
$R(1)$ $R(2)$ $R(3)$ $R(4)$ $R(5)$ $R(6)$ $R(7)$ $R(8)\ldots$
$m=1$
$R'(1)$ $R'(2)$ $R'(3)$ $R'(4)$ $R'(5)$ $R'(6)$ $R'(7)$ $R'(8)\ldots$
$R(1)$ $R(2)$ $R(3)$ $R(4)$ $R(5)$ $R(6)$ $R(7)$ $R(8)\ldots$
$m=2$
$R'(1)$ $R'(2)$ $R'(3)$ $R'(4)$ $R'(5)$ $R'(6)$ $R'(7)$ $R'(8)\ldots$
$R(1)$ $R(2)$ $R(3)$ $R(4)$ $R(5)$ $R(6)$ $R(7)$ $R(8)\ldots$
$m=3$
$R'(1)$ $R'(2)$ $R'(3)$ $R'(4)$ $R'(5)$ $R'(6)$ $R'(7)$ $R'(8)\ldots$
$R(1)$ $R(2)$ $R(3)$ $R(4)$ $R(5)$ $R(6)$ $R(7)$ $R(8)\ldots$
$m=4$
$R'(1)$ $R'(2)$ $R'(3)$ $R'(4)$ $R'(5)$ $R'(6)$ $R'(7)$ $R'(8)\ldots$

Fig. 7

Let $\{R(h)\}_1^\infty$ and $\{R'(h)\}_1^\infty$ be two sequences of duplicates of R_0. Successively for each fixed $m=1,2,\ldots$ and subsequently fixed $j=0,1,\ldots$ and $i=1,\ldots,m$ join $R(i+mj)$ with $R'(i+m+mj)$ for even j and $R(i+mj)$ with $R'(i-m+mj)$ for odd j, crosswise along every slit S_{mn}^ν ($n=1,2,\ldots$; $\nu=1,\ldots,2^{2\mu}$). This rather complicated procedure is clarified by the scheme of Fig. 7.

The covering surface R over $|z|<1$ thus constructed is the required example. As already remarked $R\notin O_{AB}$ and a fortiori $R\notin O_{HB}$.

We next show that any HB-function on R takes the same value at all points on R which lie over the same point of the unit disk. Then we can conclude that there is no nonconstant HBD-function on R, i.e. $R\in O_{HD}$. To this end consider the open set R_{mn} in R which lies over the annulus

$$-2^{-(2\mu-1)}<\log|z|<-2^{-(2\mu+2)}, \qquad \mu=2^{m-1}(2n-1) \tag{52}$$

and the set $L_{mn}\subset R_{mn}$ which lies over the circle

$$\log|z|=-\tfrac{3}{4}\cdot 2^{-2\mu}. \tag{53}$$

From (51) we see that R_{mn} contains all copies of S_{mn}^ν ($\nu=1,\ldots,2^{2\mu}$), and L_{mn} passes through all S_{mn}^ν. Clearly each component $R_{mn;j}$ of R_{mn} is a 2-sheeted covering surface of (52).

By a v-function on R_{mn} we shall understand an HB-function v on R_{mn} such that $|v|\leq M$ on R_{mn} for some constant M and $v=0$ at every branch point, i.e. end point of some S_{mn}^ν. We assert that

$$|v|\leq aM \tag{54}$$

on L_{mn} where a is a constant $0<a<1$ independent of m, n, v, and M.

It is clear that we have only to show this for $R_{mn;j}$ and $L_{mn;j}=L_{mn}\cap R_{mn;j}$ rather than for R_{mn} and L_{mn}. We may also replace $R_{mn;j}$ by the part $R'_{mn;j}$ which lies above

$$-2\frac{2\pi}{2^{2\mu}}<\theta<2\frac{2\pi}{2^{2\mu}},$$

and $L_{mn;j}$ by the part $L'_{mn;j}$ which lies above

$$-\frac{2\pi}{2^{2\mu}}<\theta<\frac{2\pi}{2^{2\mu}}.$$

Note that the configurations $(R'_{mn;j}, L'_{mn;j})$ are conformally equivalent for every m, n, and j. If (54) were not true for $R'_{mn;j}$ and $L'_{mn;j}$ then there would exist a sequence $\{v_k\}_1^\infty$ of v-functions on $R'_{mn;j}$ such that

$$\lim_{n\to\infty}\max_{L'_{mn;j}}|v_k|=M.$$

We may assume by choosing a subsequence that $\{v_k\}$ converges to a v-function v_0 on $R'_{mn;j}$. Clearly $\max_{L'_{mn;j}}|v_0|=M$ and by $|v_0|\le M$ on $R'_{mn;j}$, v_0 must be constant. But this contradicts the vanishing of v_0 at the branch points of $R'_{mn;j}$.

Let T_m be the cover transformation obtained by interchanging the sheets as indicated in Fig. 7: points in $R(i+mj)$ or $R'(i+m+mj)$ are sent to points, with the same projections, in $R'(i+m+mj)$ or $R(i+mj)$ for even j, and points in $R(i+mj)$ or $R'(i-m+mj)$ are sent to points in $R'(i-m+mj)$ or $R(i+mj)$ for odd j, where $j=0, 1, \ldots$ and $i=1, \ldots, m$.

Let $u\in HB(R)$. Consider the function

$$u_m(p)=\tfrac{1}{2}\big(u(p)-u(T_m(p))\big)$$

which is again in $HB(R)$ and qualifies as v in (54). From $|u_m|\le aM$ on L_{mn} with $M=\sup_R|u|$ we obtain $|u_m|\le aM$ on $R_{m,n-1}$, and again by (54), $|u_m|\le a^2M$ on $L_{m,n-1}$. On continuing this process we infer that $|u_m|=a^nM$ on L_{m1}. Since n is arbitrary we conclude that $u_m=0$ on L_{m1} and thus $u_m=0$ on R. Therefore

$$u(p)=u(T_m(p))$$

for any $p\in R, m=1, 2, \ldots$. This means that $u\notin HBD(R)$ unless u is constant. The proof is herewith complete.

4 I. The Class $\widetilde{HD}$. We denote by $\widetilde{HD}(R)$ the class of harmonic functions $u\ge 0$ on R such that $u=\inf F$ where F is a decreasingly directed subfamily of $HD(R)$, i.e. $u_1, u_2\in F$ implies $u_1\wedge u_2\in F$. It is readily seen that $u\in\widetilde{HD}(R)$ if and only if there exists a nonincreasing sequence $\{u_n\}_1^\infty$ of nonnegative functions in $HD(R)$ such that $u=C\text{-}\lim_n u_n$ on R.

An $\widetilde{HD}$-minimal function u is a positive function in $\widetilde{HD}(R)$ such that for any $v\in\widetilde{HD}(R)$ with $0\le v\le u$ there exists a constant c_v with $v=c_v u$ on R. From the definition it follows that HD-minimal functions are $\widetilde{HD}$-minimal (see also 4 M). We denote by $U_{\widetilde{HD}}$ (resp. U_{HD}) the class of Riemann surfaces R on which an $\widetilde{HD}$-minimal (resp. HD-minimal) function exists. These classes were introduced in CONSTANTINESCU-CORNEA [3]. Clearly

$$O_G\cap U_{\widetilde{HD}}=\emptyset, \qquad O_G\cap U_{HD}=\emptyset, \tag{55}$$

since $\widetilde{HD}(R)=HD(R)=\{0\}$ for $R\in O_G$. If $R\in O^n_{HD}-O^{n-1}_{HD}$ (see 3 G) then $HD(R)=\widetilde{HD}(R)=\mathbb{E}^n$ (n-dimensional Euclidean space) and thus

$$U_{\widetilde{HD}}\supset U_{HD}\supset O^{n+1}_{HD}-O_G\supset O^n_{HD}-O_G \tag{56}$$

for every $n=0, 1, \ldots$.

We are going to investigate the behavior of $u\in\widetilde{HD}(R)$ at Royden's boundary, in particular at the harmonic boundary Δ.

4 J. Upper Semicontinuous Functions on Δ. We denote by $U(\Delta)$ the set of nonnegative functions f on Δ such that

$$f(p)=\inf_{v\in F_f} v(p) \tag{57}$$

for every $p\in\Delta$, where

$$F_f=\{v\,|\,v\in HD(R),\, v\geq f \text{ on } \Delta\}. \tag{58}$$

Clearly F_f is a decreasingly directed subfamily of $HD(R)$. Thus we see that

$f\in U(\Delta)$ is upper semicontinuous and μ-integrable on Δ.

Moreover $U(\Delta)$ is a half vector space and forms a lattice:

$f_1\cup f_2,\ f_1\cap f_2\in U(\Delta)$ along with $f_1,\ f_2\in U(\Delta)$.

In fact let $q\in\Delta$ and $\varepsilon>0$. We can find $v_i\in F_{f_i}$ such that $v_i(q)<f_i(q)+\varepsilon$. Since $w=v_1\wedge v_2\in HD(R)$ and $w(p)=v_1(p)\cap v_2(p)\geq f_1(p)\cap f_2(p)$ on Δ, $w\in F_{f_1\cap f_2}$. In view of $w(q)<(f_1(q)+\varepsilon)\cap(f_2(q)+\varepsilon)=f_1(q)\cap f_2(q)+\varepsilon$ we have $(f_1\cap f_2)(p)=\inf_{v\in F_{f_1\cap f_2}} v(p)$ on Δ, i.e. $f_1\cap f_2\in U(\Delta)$. Similarly $f_1\cup f_2\in U(\Delta)$.

Whereas $f\in U(\Delta)$ is upper semicontinuous on Δ the converse is not necessarily true. However in a manner of speaking this is almost the case:

If $f\geq 0$ is bounded and upper semicontinuous on Δ then $f\in U(\Delta)$.

To see this let $g\geq f$ be in the family $B(\Delta)$ of bounded continuous functions on Δ. Since $\mathrm{M}(R)|\Delta$ is dense in $B(\Delta)$ we can find $g_n\in\mathrm{M}(R)$ such that $f<g_n<g+1/n$ on Δ. Let $u_n=\pi g_n$. Then $f<u_n<g+1/n$ on Δ. Thus

$$f(p)\leq \inf_{u\in F_f} u(p)\leq g(p).$$

By the upper semicontinuity of f, $f(p)=\inf\{g(p)\,|\,g\in B(\Delta),\, g\geq f\}$, and we have (57) and $f\in U(\Delta)$.

The role of $U(\Delta)$ becomes significant by virtue of the following result:

THEOREM. *A function u belongs to $\widetilde{HD}(R)$ if and only if there exists an $f\in U(\Delta)$ such that the integral representation*

$$u(z)=\int_\Delta P(z,p)\,f(p)\,d\mu(p) \tag{59}$$

with the boundary function f is valid on R.

First assume that $u\in\widetilde{HD}(R)$ and set $F=\{v\,|\,v\in HD(R),\, v\geq u \text{ on } R\}$. Then F is decreasingly directed and $u(z)=\inf_{v\in F} v(z)$ on R. Set $f(p)=\inf_{v\in F} v(p)$ on Δ. Clearly $F_f=F$ and $f\in U(\Delta)$. On interchanging the directed infimum and the integration we obtain

$$u(z)=\inf_{v\in F} v(z)=\inf_{v\in F} \int_{\Delta} P(z,p)\,v(p)\,d\mu(p)$$
$$=\int_{\Delta} P(z,p)\big(\inf_{v\in F} v(p)\big)\,d\mu(p)=\int_{\Delta} P(z,p)\,f(p)\,d\mu(p).$$

Conversely assume that u is given by (59). Again we interchange the directed infimum and the integration:

$$u(z)=\int_{\Delta} P(z,p)\big(\inf_{v\in F_f} v(p)\big)\,d\mu(p)$$
$$=\inf_{v\in F_f} \int_{\Delta} P(z,p)\,v(p)\,d\mu(p)=\inf_{v\in F_f} v(z).$$

Let $F=\{v \mid v\in HD(R), v\geq u \text{ on } R\}$. Then $F\supset F_f$ and $u(z)=\inf_{v\in F} v(z)$, i.e. $u\in \widetilde{HD}(R)$.

4 K. Boundary Function. For a real-valued function f on R we write

$$\bar{f}(p)=\limsup_{z\in R,\, z\to p} f(z) \tag{60}$$

with $p\in\Delta$. Clearly $\bar{f}$ is upper semicontinuous on Δ. In the representation (59) the function f is not uniquely determined since we may change an f on a set in Δ of measure zero so that the resulting function is still upper semicontinuous. However we can find the least possible f as follows:

THEOREM. *Assume that $u\in\widetilde{HD}(R)$ is expressed as (59) with an $f\in U(\Delta)$. Then*

$$\bar{u}(p)\leq f(p) \tag{61}$$

everywhere on Δ and

$$\bar{u}(p)=f(p) \tag{62}$$

μ-almost everywhere on Δ. In particular

$$u(z)=\int_{\Delta} P(z,p)\,\bar{u}(p)\,d\mu(p). \tag{63}$$

We observe as before that

$$u(z)=\int_{\Delta} P(z,p)\,f(p)\,d\mu(p)=\inf_{v\in F_f} \int_{\Delta} P(z,p)\,v(p)\,d\mu(p)=\inf_{v\in F_f} v(z)$$

on R. For any $v\in F_f$, $u\leq v$ on R implies $\bar{u}(p)\leq v(p)$ on Δ. Thus by $f(p)=\inf_{v\in F_f} v(p)$ we obtain (61).

To prove (62) we first assume that f and consequently u is bounded. Suppose $\bar{u}(p)<f(p)-\varepsilon$ $(\varepsilon>0)$ on a compact set $K\subset\Delta$. We wish to conclude that $\mu(K)=0$. Suppose $\mu(K)>0$. Then the function

$$w(z)=\varepsilon\int_K P(z,p)\,d\mu(p)$$

is in $HP(R)$ and $0<w(z)\le\varepsilon$ on R. By Theorem 4 E, $\lim_{z\in R,z\to q} w(z)=0$ for every $q\in\Delta-K$. Thus

$$\limsup_{z\in R,z\to q}\big(u(z)+w(z)\big)=\bar{u}(q)\le f(q)$$

for $q\in\Delta-K$, and

$$\limsup_{z\in R,z\to q}\big(u(z)+w(z)\big)\le\bar{u}(q)+\varepsilon<f(q)$$

for $q\in K$. Therefore for every $v\in F_f$ and all $q\in\Delta$

$$\liminf_{z\in R,z\to q}\big(v(z)-(u(z)+w(z))\big)\ge 0$$

which by maximum principle III (2 I) gives $v\ge u+w$ on R. We conclude that $u(z)\ge u(z)+w(z)$ and in particular $u(z_0)\ge u(z_0)+w(z_0)=u(z_0)+\varepsilon\,\mu(K)$; this contradicts $\mu(K)>0$.

We now drop the assumption that f is bounded and set

$$u_n(z)=\int_\Delta P(z,p)\big(f(p)\cap n\big)\,d\mu(p)$$

for $n=1,2,\ldots$. Since $u_n\in\widetilde{HD}(R)$ by virtue of $f\cap n\in U(\Delta)$, we have $\bar{u}(p)\ge\bar{u}_n(p)=f(p)\cap n$ on Δ μ-almost everywhere. It follows that $\bar{u}(p)\ge f(p)$ on Δ μ-almost everywhere. By (61) we obtain (62), and assertion (63) is trivial.

4 L. Semivector Lattice $\widetilde{HD}$. Corresponding to Theorem 4 F we can show that $\widetilde{HD}(R)$ forms a semivector lattice:

THEOREM. *If u_1 and u_2 belong to $\widetilde{HD}(R)$ then $u_1\vee u_2$ and $u_1\wedge u_2$ belong to $\widetilde{HD}(R)$. Moreover if*

$$u_i(z)=\int_\Delta P(z,p)\,f_i(p)\,d\mu(p) \tag{64}$$

for $f_i\in U(\Delta)$ then

$$\begin{aligned}(u_1\vee u_2)(z)&=\int_\Delta P(z,p)\big(f_1(p)\cup f_2(p)\big)\,d\mu(p),\\ (u_1\wedge u_2)(z)&=\int_\Delta P(z,p)\big(f_1(p)\cap f_2(p)\big)\,d\mu(p).\end{aligned} \tag{65}$$

In view of $f_1\cap f_2\in U(\Delta)$ we infer that

$$a(z)=\int_\Delta P(z,p)\big(f_1(p)\cap f_2(p)\big)\,d\mu(p)$$

belongs to $\widetilde{HD}(R)$ and $a \leq u_1, u_2$ on R. Let $w \in H(R)$ satisfy $w \leq u_1, u_2$ on R. Then $w \leq v_i \in F_{f_i}$ and thus $w \leq v_1 \wedge v_2$ on R. In the proof that $U(\Delta)$ forms a lattice we saw that $(f_1 \cap f_2)(p) = \inf(v_1 \wedge v_2)(p)$ for $v_i \in F_{f_i}$. As in the proof of Theorem 4 K we conclude that $a = \inf(v_1 \wedge v_2)$ on R with $v_i \in F_{f_i}$, and therefore $w \leq a$ on R. This means that $a = u_1 \wedge u_2$. Thus $u_1 \wedge u_2 \in \widetilde{HD}(R)$ and the latter part of (65) is valid.

We turn to $u_1 \wedge u_2$. First let us assume that f_i and consequently u_i is bounded on R. Set

$$b(z) = \int_\Delta P(z,p)\big(f_1(p) \cup f_2(p)\big)\, d\mu(p)$$

on R. Clearly $b \geq u_1, u_2$ and $f_1 \cup f_2 \in U(\Delta)$ implies $b \in \widetilde{HD}(R)$. Let $w \in H(R)$ with $w \geq u_1, u_2$. We are going to show that $w \geq b$, i.e. $b = u_1 \vee u_2$.

By Lusin's theorem we can find a compact set K_n in Δ with the following properties:

$$\mu(\Delta - K_n) < \frac{1}{n}$$

and there exist $x_n, y_n \in B(\Delta)$ such that

$$x_n = \bar{u}_1, \qquad y_n = \bar{u}_2 \tag{66}$$

on K_n and

$$0 \leq x_n, y_n \leq k \tag{67}$$

on Δ with a finite constant k independent of n. Let

$$w_n(z) = w(z) + \int_{\Delta - K_n} P(z,p)\big(x_n(p) + y_n(p)\big)\, d\mu(p).$$

It is easy to see that

$$w_n(z) \geq \int_\Delta P(z,p)\, x_n(p)\, d\mu(p), \ \int_\Delta P(z,p)\, y_n(p)\, d\mu(p)$$

on R. Therefore for any $q \in \Delta$

$$\liminf_{z \in R, z \to q} w_n(z) \geq x_n(q), y_n(q)$$

and thus

$$\liminf_{z \in R, z \to q} w_n(z) \geq x_n(q) \cup y_n(q)$$

which in turn gives

$$\liminf_{z \in R, z \to q} \Big(w_n(z) - \int_\Delta P(z,p)(x_n \cup y_n)(p)\, d\mu(p)\Big) \geq 0.$$

By maximum principle II (in 2 H)

$$w_n(z) \geq \int_\Delta P(z,p)(x_n \cup y_n)(p)\, d\mu(p)$$

and consequently

$$w(z)+k\int_{\Delta-K_n} P(z,p)\,d\mu(p)\geq \int_{K_n} P(z,p)(\bar{u}_1\cup\bar{u}_2)(p)\,d\mu(p)$$

on R. On letting $n\to\infty$ and noting that $f_i=\bar{u}_i$ μ-a.e. on Δ we obtain

$$w(z)\geq\int_\Delta P(z,p)(\bar{u}_1\cup\bar{u}_2)(p)\,d\mu(p)=\int_\Delta P(z,p)\big(f_1(p)\cup f_2(p)\big)\,d\mu(p)=b(z).$$

Here we observe the following. Let $u\in\widetilde{HD}(R)$ and

$$u(z)=\int_\Delta P(z,p)\,f(p)\,d\mu(p)$$

with $f\in U(\Delta)$. From

$$(u\wedge n)(z)=\int_\Delta P(z,p)\big(\bar{u}(p)\cap n\big)\,d\mu(p)$$

we obtain $u(z)=C\text{-}\lim(u\wedge n)(z)$.

Finally we shall remove the boundedness assumption concerning the u_i. Let $v=\lim_n(u_1\wedge n)\vee(u_2\wedge n)$ on R. Since $u_i\wedge n\in\widetilde{HD}(R)$ is bounded $(u_1\wedge n)\vee(u_2\wedge n)$ exists and belongs to $\widetilde{HD}(R)$ as seen above. Hence v is well-defined on R. In view of $(u_1\wedge n)\vee(u_2\wedge n)\in\widetilde{HD}(R)$ we also have

$$\begin{aligned}\big((u_1\wedge n)\vee(u_2\wedge n)\big)(z)&=\int_\Delta P(z,p)\big((f_1(p)\cap n)\cup(f_2(p)\cap n)\big)\,d\mu(p)\\&=\int_\Delta P(z,p)\big((f_1(p)\cup f_2(p))\cap n\big)\,d\mu(p).\end{aligned}$$

On letting $n\to\infty$ we obtain

$$v(z)=\int_\Delta P(z,p)\big(f_1(p)\cup f_2(p)\big)\,d\mu(p)$$

which belongs to $\widetilde{HD}(R)$ by Theorem 4 J. From this it is clear that $v\geq u_1,u_2$ on R. Let $w\in H(R)$ with $w\geq u_1,u_2$ on R. Then $w\geq(u_1\wedge n)\vee(u_2\wedge n)$. Again for $n\to\infty$ we have $w\geq v$, i.e. $v=u_1\vee u_2$.

4 M. Characterization of $\widetilde{HD}$-Minimality. We have seen in 3 E that isolated points in Δ and HD-minimal functions correspond in a one-to-one manner. As a counterpart of this we shall prove:

THEOREM. *If u is $\widetilde{HD}$-minimal on R with $u(z_0)=1$ then there exists a point $q\in\Delta$ such that $\mu(q)>0$ and $u(z)=P(z,q)$ on R. Conversely if $q\in\Delta$ satisfies $\mu(q)>0$ then $P(z,q)$ is $\widetilde{HD}$-minimal.*

An isolated point q in Δ clearly has positive μ-measure and thus HD-minimal functions are $\widetilde{HD}$-minimal. Consequently $U_{\widetilde{HD}}\supset U_{HD}$.

First assume that u is $\widetilde{HD}$-minimal on R with $u(z_0)=1$. Recall that $u(z)=\int_\Delta P(z,p)\,\bar{u}(p)\,d\mu(p)$. Set $c=1/\sup_R u$ and $K_n=\{p\,|\,p\in\Delta,\bar{u}(p)\geq 1/n\}$, a compact set. Since the characteristic function of K_n on Δ belongs to $U(\Delta)$ we have

$$u(z) \geq \frac{1}{n} \int_{K_n} P(z, p)\, d\mu(p) \in \widetilde{HD}(R).$$

Therefore we can find a constant c_n such that $\int_{K_n} P(z, p)\, d\mu(p) = c_n\, u(z)$ on R. Since $\mu\left(\bigcup_1^\infty K_n\right) > 0$ there exists an n with $\mu(K_n) > 0$ and a fortiori $c_n > 0$. For the sake of simplicity we set $K_n = K$ and $w(z) = \int_K P(z, p)\, d\mu(p)$. Note that $\bar{w}(p) = 1$ μ-almost everywhere on K. Hence $\sup_R w = 1$ and by $1 = \sup_R w = c_n \sup_R u = c_n/c$ we see that $w = c\,u$.

Take an arbitrary compact set $A \subset K$ with $\mu(K - A) > 0$. As above, $u(z) \geq (1/n) \int_A P(z, p)\, d\mu(p) \in \widetilde{HD}(R)$, and if $\mu(A) > 0$ then $\int_A P(z, p)\, d\mu(p) = c\,u(z)$. Thus $c\,\bar{u}(p) = 0$ μ-a.e. on $\Delta - A$. This is a contradiction since $K - A \subset \Delta - A$, $\mu(K - A) > 0$, and $c\,\bar{u}(p) = 1$ μ-a.e. on $K - A$. Hence $\mu(A) = 0$. It follows that there exists a point $q \in K$ with $\mu(q) > 0$ and therefore

$$0 < P(z, q) = \frac{1}{\mu(q)} \int_q P(z, p)\, d\mu(p) \in \widetilde{HD}(R)$$

and $P(z, q) \leq w(z)/\mu(q)$. From this we deduce as before that $P(z, q) = k\,u(z)$ for some constant k. On setting $z = z_0$ we obtain $k = 1$, i.e. $P(z, q) = u(z)$.

Conversely let $q \in \Delta$ with $\mu(q) > 0$. Then $P(z, q)$ is positive on R and belongs to $\widetilde{HD}(R)$. If $v \in \widetilde{HD}(R)$ satisfies $P(z, q) \geq v(z)$ on R then $\bar{P}(p, q) \geq \bar{v}(p)$ on Δ implies $\bar{v}(p) = 0$ μ-a.e. on $\Delta - q$. Hence on setting $\bar{v}(q) = c$ we obtain

$$v(z) = \int_\Delta P(z, p)\, \bar{v}(p)\, d\mu(p) = \int_q P(z, p)\, c\, d\mu(p) = c\,\mu(q)\, P(z, q),$$

i.e. $P(z, q)$ is $\widetilde{HD}$-minimal.

4 N. Characterization of $U_{\widetilde{HD}}$. We are now able to obtain at once the following characterization of the class $U_{\widetilde{HD}}$:

THEOREM. *A hyperbolic Riemann surface R belongs to $U_{\widetilde{HD}}$ (resp. U_{HD}) if and only if Royden's boundary of R contains a point with positive harmonic measure (resp. a point which is isolated on the harmonic boundary).*

5. Perron's Method

5 A. Perron's Family. Let $R \notin O_G$ and consider a bounded real-valued function f on Royden's boundary Γ of R. Denote by $\mathfrak{A}(R^*, f)$ the set of superharmonic functions v on R such that

$$\liminf_{z \in R,\, z \to p} v(z) \geq f(p) \tag{68}$$

for any point $p\in\Gamma$. The class $\mathfrak{A}(R^*, f)$ shall be referred to as *Perron's family* with respect to R^* and f. Set

$$\overline{H}(z; R^*, f)=\overline{H}(z; f)=\inf\{v(z)\,|\,v\in\mathfrak{A}(R^*, f)\} \tag{69}$$

and

$$\underline{H}(z; R^*, f)=\underline{H}(z; f)=-\overline{H}(z; -f) \tag{70}$$

for $z\in R$. It is easy to see that $\overline{H}(z; f)$ and $\underline{H}(z; f)$ are harmonic on R and $\overline{H}(z; f)\geq\underline{H}(z; f)$ on R. If $\overline{H}(z; f)=\underline{H}(z; f)$ then we denote this common function by $H(z; R^*, f)=H(z; f)$. In this case we say that f is *resolutive* with respect to R^*.

A point $p\in\Gamma$ is said to be a *regular point* for the Dirichlet problem with respect to R^* if

$$\lim_{z\in R, z\to p} H(z; f)=f(p) \tag{71}$$

for any resolutive function f on Γ which is continuous at p. Which points in Γ are regular points? We answer this by the following:

THEOREM. *A bounded real Borel function f on Γ is resolutive and*

$$H(z; f)=\int_\Gamma P(z, p)\, f(p)\, d\mu(p) \tag{72}$$

on R. The set of regular points in Γ coincides with Δ.

First assume that f is continuous on Γ. Then

$$v(z)=\int_\Gamma P(z, p)\, f(p)\, d\mu(p)\in HB(R)$$

is continuous on R^* and $v(p)=f(p)$ on Δ (Theorem 4 E). For any positive constant $\varepsilon>0$ there exists an open set W in Γ such that $W\supset\Delta$ and

$$-\varepsilon<v(p)-f(p)<\varepsilon$$

for every $p\in W$. Let $K=\Gamma-W$. Then K is compact with $K\subset\Gamma-\Delta$ and there exists on R a positive superharmonic function u which is continuous on R^* with $u=\infty$ on K and $u=0$ on Δ (Theorem 3 H). Clearly $v+\varepsilon+\varepsilon u\in\mathfrak{A}(R^*, f)$ and thus $\overline{H}(z; f)\leq v(z)+\varepsilon+\varepsilon u(z)$ on R. On letting $\varepsilon\to 0$ we obtain

$$\overline{H}(z; f)\leq v(z) \tag{73}$$

on R. Since $-f$ is continuous and $-v(z)=\int_\Gamma P(z, p)(-f(p))\, d\mu(p)$ we obtain as above $\overline{H}(z; -f)\leq -v(z)$. By definition $\underline{H}(z; f)=-\overline{H}(z; -f)$, hence $\underline{H}(z; f)\geq v(z)$. This with (73) gives $\overline{H}(z; f)=\underline{H}(z; f)=v(z)$, and (72) has been proved for continuous f on Γ.

It is easy to see that $H(z; f)$ is for each fixed z a continuous linear functional on the family of resolutive functions f. Hence (72) remains valid for Borel functions f.

Next assume that $q \in \Gamma$ is regular. We wish to show that $q \in \Delta$. If $q \in \Gamma - \Delta$ then we can find a positive superharmonic function s on R such that s is continuous on R^* and $s(q)=1$, $s|\Delta=0$ (Theorem 3 H). Clearly $f=s|\Gamma$ is resolutive and

$$H(z;f)=\int_\Gamma P(z,p)\,s(p)\,d\mu(p)=\int_\Delta P(z,p)\,s(p)\,d\mu(p)=0.$$

Thus $\lim_{z\in R, z\to q} H(z;f)=0 \neq f(q)=1$, which contradicts the regularity of q.

Conversely assume that $q \in \Delta$. Let f be an arbitrary resolutive function on Γ, continuous at q on Γ and a fortiori on Δ. We may suppose that $f(q)=0$. For an arbitrary $\varepsilon>0$ there exist functions $g_1, g_2 \in B(\Gamma)$ such that

$$g_1-\varepsilon<f<g_2+\varepsilon$$

on Γ and $g_1(q)=g_2(q)=0$. Clearly

$$\int_\Gamma P(z,p)\,g_1(p)\,d\mu(p)-\varepsilon \le H(z;f) \le \int_\Gamma P(z,p)\,g_2(p)\,d\mu(p)+\varepsilon$$

and we conclude that

$$-\varepsilon \le \liminf H(z;f) \le \limsup H(z;f) \le \varepsilon$$

as $z \in R$ tends to q. Thus $\lim_{z\in R, z\to q} H(z;f)=0=f(q)$, i.e. q is regular.

5 B. Compactification of Subregions. Let R be an arbitrary Riemann surface and G a subregion of R. Here the case $G=R$ is admitted. We denote by $\bar{G}$ the closure of G in R^* and by G^* Royden's compactification of the Riemann surface G. We give a relation between G^* and $\bar{G}$:

THEOREM. *There exists a unique continuous mapping $j=j(G^*, \bar{G})$ of G^* onto $\bar{G}$ fixing G elementwise.*

The mapping $j(G^*, \bar{G})$ shall be referred to as the *projection* of G^* onto $\bar{G}$. A set $\bar{E} \subset \bar{G}$ such that $\bar{E}=j(E^*)$ with $E^* \subset G^*$ is called the projection of E^*. The set $j^{-1}(\bar{p})$, with $\bar{p}$ in $\bar{G}$, is the *fiber* over $\bar{p}$.

The uniqueness of such a j is clear since $j(p)=p$ for every $p \in G$ and G is dense in both G^* and $\bar{G}$. Hence we have only to show the existence. Let $\mathrm{M}(G;R)=\mathrm{M}(R)|G$. Observe that $f \in \mathrm{M}(G;R)$ can be extended to $\bar{G}$ and to G^* as continuous functions $\bar{f}$ and f^*. Let $p^* \in G^*$. Then $f^* \to f^*(p^*)$ is a character X^* on $\mathrm{M}(G)$ (cf. 2 B), and $\bar{X}=X^*|\mathrm{M}(G;R)$ is a character on $\mathrm{M}(G;R)$. We prove the existence of a unique $\bar{p} \in \bar{G}$ such that

$$\bar{X}(\bar{f})=\bar{f}(\bar{p}) \tag{74}$$

for all $\bar{f} \in \mathrm{M}(G;R)$. Consider $\mathscr{I}=\{\bar{f} \mid \bar{f} \in \mathrm{M}(G;R), \bar{X}(\bar{f})=0\}$. We shall deduce a contradiction from the assumption that there exists an $\bar{f}_q \in \mathscr{I}$

for every $q \in \bar{G}$ such that $\bar{f}_q(q) \neq 0$. Since $\mathscr{I}$ is an ideal of $\mathrm{MI}(G; R)$ we may assume that $\bar{f}_q \geq 0$ on $\bar{G}$ by taking $\bar{f}_q^2$ instead of $\bar{f}_q$. Moreover we may suppose that $\bar{f}_q(q) > 1$. By the compactness of $\bar{G}$ we can find points $q_1, \ldots, q_n$ such that $\bar{f} = \sum_{j=1}^n \bar{f}_{q_j} \geq 1$ on $\bar{G}$. Then $1 = \bar{f} \cdot (1/\bar{f}) \in \mathscr{I}$ and $X(1) = 0$, a contradiction. Thus there exists a $\bar{p} \in \bar{G}$ with $\bar{f}(\bar{p}) = 0$ for all $\bar{f} \in \mathscr{I} \subset \mathrm{MI}(G; R)$. Since $\bar{f} - \bar{f}(\bar{p}) \in \mathscr{I}$, (74) has been proved. The uniqueness follows from the fact that $\mathrm{MI}(G; R)$ separates points in $\bar{G}$.

We denote by $j(p^*)$ the point $\bar{p}$ in (74). We have shown that

$$f^*(p^*) = \bar{f}(j(p^*)) \tag{75}$$

for all $f \in \mathrm{MI}(G; R)$. From this the continuity of j follows.

Finally we have to prove that j is onto. Let $\bar{p} \in \bar{G}$ and set

$$\mathscr{J} = \{f \mid f \in \mathrm{MI}(G),\ \lim_{z \in G, z \to \bar{p}} f(z) = 0\}.$$

$\mathscr{J}$ contains $\{f \mid f \in \mathrm{MI}(G; R), \bar{f}(\bar{p}) = 0\}$ and $1 \notin \mathscr{J}$. Thus $\mathscr{J}$ is a nontrivial ideal of the Banach algebra $\mathrm{MI}(G)$. By virtue of Gelfand's theorem (see e.g. Yosida [1]) there exists a character X^* on $\mathrm{MI}(G)$ such that $X^* \mid \mathscr{J} = 0$. Corollary 2 in 2 B gives the existence of a point $p^* \in G^*$ such that $X^*(f) = f^*(p^*)$. Take an arbitrary $f \in \mathrm{MI}(G; R)$. Then $f - \bar{f}(\bar{p}) \in \mathscr{J}$ and thus $\bar{f}(\bar{p}) = f^*(p^*)$. We conclude by (75) that $\bar{p} = j(p^*)$.

Again by (75) it is clear that $j(z) = j^{-1}(z) = z$ for $z \in G$.

5 C. Coincidence of Boundary Points. The boundary $\bar{G} - G$ of G consists of two parts: the "relative" boundary $\overline{\partial G}$ and the "ideal" boundary

$$b_G = (\bar{G} - \overline{\partial G}) \cap \Gamma. \tag{76}$$

Correspondingly the boundary $G^* - G$ of G can be divided into two parts: $j^{-1}(\overline{\partial G})$ and $j^{-1}(b_G)$. The structures $\overline{\partial G}$ and $j^{-1}(\overline{\partial G})$ are quite different since every f in $\mathrm{MI}(G)$ is continuous at each point of $j^{-1}(\overline{\partial G})$ but not at $\overline{\partial G}$. In contrast we can prove:

Theorem. *The projection $j = j(G^*, \bar{G})$ is a homeomorphism of $G \cup j^{-1}(b_G)$ onto $G \cup b_G$.*

Since $G \cup j^{-1}(b_G)$ and $G \cup b_G$ are locally compact and since j is a continuous map of $G \cup j^{-1}(b_G)$ onto $G \cup b_G$, we must show that the fiber $j^{-1}(\bar{p})$ consists of one point for every $\bar{p} \in b_G$. Assume that $j^{-1}(\bar{p})$ contained two distinct points p_1^* and p_2^*. Then there would exist an $f \in \mathrm{MI}(G)$ such that $f(p_i^*) = i$ $(i = 1, 2)$. Let U be an open neighborhood of $\bar{p}$ in R^* such that $U \subset G \cup b_G$. Let $\varphi \in \mathrm{MI}(R)$ such that $\varphi \mid U = 1$ and $\varphi \mid V = 0$ where V is an open set containing $(R - \bar{G}) \cup \overline{\partial G}$, with $\bar{U} \cap \bar{V} = \emptyset$ (see 2 C). Then clearly $f \circ \varphi \in \mathrm{MI}(R)$, it being understood that $f \circ \varphi = 0$

on $R-G$. Thus we may consider that $f\circ\varphi$ is in $\mathrm{M}(G;R)$. By (75)

$$i=f^*(p_i^*)=f^*(p_i^*)\,\varphi^*(p_i^*)=(f\circ\varphi)^*(p_i^*)$$
$$=\overline{(f\circ\varphi)}\big(j(p_i^*)\big)=\overline{(f\circ\varphi)}(\bar{p})$$

for $i=1,2$. This is a contradiction.

5 D. Correspondence of Harmonic Measures I. We denote by μ_R (resp. μ_G) the harmonic measure on Royden's boundary of R (resp. G) and fix the point $z_0\in G\subset R$ as the center of μ_R and μ_G (see 4 A).

The following result shows that $G\cup j^{-1}(b_G)$ and $G\cup b_G$ are the same not only topologically but also harmonically.

THEOREM. *Let E be a Borel set in b_G. Then $\mu_R(E)>0$ if and only if $\mu_G(j^{-1}(E))>0$.*

Since μ_R and μ_G are regular we may suppose that E is a compact set. Note that $E\cap\overline{\partial G}=\emptyset$ in $\bar{G}-G$. Thus $\big(j^{-1}(E)\big)\cap\big(j^{-1}(\overline{\partial G})\big)=\emptyset$ in Γ_G. From the regularity of μ_R and μ_G and the fact that j is a homeomorphism between $G\cup j^{-1}(b_G)$ and $G\cup b_G$ it follows that there exists a sequence $\{V_n\}_1^\infty$ of open sets V_n in R^* such that $G\cup b_G\supset V_n\supset\bar{V}_{n+1}\supset E$ and

$$\begin{aligned}\mu_R(E)&=\lim_n \mu_R(V_n\cap\Gamma_R),\\ \mu_G\big(j^{-1}(E)\big)&=\lim_n \mu_G\big(j^{-1}(V_n)\cap\Gamma_G\big).\end{aligned}\tag{77}$$

Let $\{f_n\}_1^\infty\subset\mathrm{M}(R)$ satisfy $f_n|\bar{V}_{n+1}=1$, $f_n|(R^*-V_n)=0$, $0\le f_n\le 1$, and $f_n\ge f_{n+1}$ (see 2 C). Clearly f_n may be viewed as an element of $\mathrm{M}(G)$, and $f_n|j^{-1}(\bar{V}_{n+1})=1$, $f_n|\big(G^*-j^{-1}(V_n)\big)=0$. Let

$$u_n(z)=\int_{\Gamma_R} P_R(z,p)\,f_n(p)\,d\mu_R(p)\tag{78}$$

for $z\in R$ and

$$v_n(z)=\int_{\Gamma_G} P_G(z,p)\,f_n(p)\,d\mu_G(p)\tag{79}$$

for $z\in G$. Take a regular exhaustion $\{R_m\}_1^\infty$ of R and set

$$v_{nm}=\pi_{R^*-R_m\cap G}\,f_n$$

on R. By Theorem 3 B we see that $\{v_{nm}\}_{m=1}^\infty$ is BD-Cauchy on R and a fortiori on G. By the property of j, $v_{nm}-f_n$ vanishes on Γ_G and hence $v_{nm}-f_n\in\mathrm{M}_0(G)$. Thus $BD\text{-}\lim_m(v_{nm}-f_n)\in\mathrm{M}_\Delta(G)$ and it follows that $BD\text{-}\lim_m v_{nm}=f_n$ on Δ_G. In view of this and (79) we obtain

$$v_n=BD\text{-}\lim_m v_{nm}\qquad\text{on } G.\tag{80}$$

Since $v_{nm}-f_n=0$ on $R^*-R_m\cap G$ we have $v_n\in\mathrm{M}(R)$ if we define $v_n=0$ on R outside of G.

On the other hand $v_{nm}-f_n\in\mathbf{M}_0(R)$ and, since

$$v_n=BD\text{-}\lim_m v_{nm} \qquad \text{on } R, \tag{81}$$

$v_n-f_n\in\mathbf{M}_\Delta(R)$. Therefore

$$v_n=f_n \tag{82}$$

on Δ_R.

As above we can show that

$$u_n=BD\text{-}\lim_m u_{nm} \tag{83}$$

on R where

$$u_{nm}=\pi_{R^*-R_m}f_n=\pi_{R^*-R_m}v_n.$$

By the construction of $\{f_n\}_1^\infty$ and by the maximum principle we conclude that $u_n\geq u_{n+1}$ on R, $v_n\geq v_{n+1}$ on G, and $u_n\geq v_n$ on G. In particular $u_n(z_0)\geq v_n(z_0)$. This means that $\mu_R(V_n\cap\Gamma_R)\geq\mu_G\big(j^{-1}(V_{n+1})\cap\Gamma_G\big)$ and by the regularity of the measures μ_R and μ_G

$$\mu_R(E)\geq\mu_G\big(j^{-1}(E)\big). \tag{84}$$

In particular $\mu_G\big(j^{-1}(E)\big)>0$ implies $\mu_R(E)>0$.

Next assume that $\mu_G\big(j^{-1}(E)\big)=0$. Then $\lim_n v_n(z_0)=\mu_G\big(j^{-1}(E)\big)=0$. Since the continuous function on R^* which is harmonic on R_m and equals v_1-v_n on R^*-R_m must coincide with $u_{1m}-u_{nm}$ and since $u_{1m}-u_{nm}\geq v_1-v_n$ we conclude by the maximum principle that $u_{1m}-u_{nm}\leq u_{1,m+1}-u_{n,m+1}$ and $\lim_m(u_{1m}-u_{nm})=u_1-u_n$. Hence $u_1-u_n\geq u_{1m}-u_{nm}$. From $U\text{-}\lim_n v_n=0$ on ∂R_m, and $u_{nm}=v_n$ on ∂R_m it follows that $U\text{-}\lim_n u_{nm}=0$ on R_m. Therefore $\lim_n(u_1-u_n)\geq u_{1m}$ on R_m. On letting $m\to\infty$ we infer that $\lim_n(u_1-u_n)\geq u_1$ on R. Since $u_n\geq 0$ we finally conclude that $\lim_n u_n(z_0)=0$, that is $\mu_R(E)=0$.

5 E. Correspondence of Harmonic Measures II. Let $\omega_{\partial G}$ be the *harmonic measure* on ∂G with respect to G for the center $z_0\in G$, i.e. for each Borel set X in ∂G

$$\omega_{\partial G}(X)=\inf\{s(z_0)\,|\,s\in\mathfrak{A}(\bar{G},f_X)\} \tag{85}$$

where f_X is the characteristic function of X on $\bar{G}-G$ and $\mathfrak{A}(\bar{G},f_X)$ is the set of superharmonic functions s on G with

$$\liminf_{z\in G,\,z\to p} s(z)\geq f_X(p) \tag{86}$$

for every $p\in\bar{G}-G$. It is easy to see that $\omega_{\partial G}$ is a regular Borel measure for Borel subsets in ∂G.

THEOREM. *For every Borel set $E\subset\partial G$, $\mu_G\big(j^{-1}(E)\big)=\omega_{\partial G}(E)$.*

It is readily verified that $\mathfrak{A}(G^*,f_E\circ j)=\mathfrak{A}(\bar{G},f_E)$ and the assertion follows from Theorem 5 A.

5 F. Surfaces of Almost Finite Genus. We shall give a sufficient condition for an open Riemann surface not to belong to $U_{\widetilde{HD}} \supset U_{HD}$.

A Riemann surface R is said to be of *almost finite genus* if there exists a finite or countably infinite sequence $\{A_n\}$ of relatively compact annuli in R such that

(α) $\bar{A}_n \cap \bar{A}_m = \emptyset$ $(n \neq m)$,

(β) $R - \bigcup_n \bar{A}_n$ *is a planar subregion of* R,

(γ) $\sum_n (1/\log \operatorname{mod} A_n) < \infty$.

By an annulus A on a Riemann surface R we mean a region which is conformally equivalent to a doubly connected plane region.

Surfaces of finite genus are, of course, of almost finite genus. There are surfaces of almost finite genus which are not of finite genus. As an example let $\{a_n\}_1^\infty$ and $\{b_n\}_1^\infty$ be defined by

$$a_n = 3n - e^{-n^2}, \qquad b_n = 3n + e^{-n^2}.$$

Consider the region obtained from $|z| < \infty$ by deleting the intervals $[a_n, b_n]$, $n = 1, 2, \ldots$. Join two such copies crosswise along $[a_n, b_n]$, $n = 1, 2, \ldots$, so as to obtain a 2-sheeted covering surface R of $|z| < \infty$. On the upper sheet, say, let A_n be the annulus which is the 1-sheeted covering surface of the annulus

$$B_n = \{z \mid e^{-(n+1)^2} < |z - 3(n+1)| < 1\}.$$

Then $\{A_n\}_1^\infty$ satisfies (α) and (β). Moreover $\log \operatorname{mod} A_n = \log \operatorname{mod} B_n = (n+1)^2$, and ($\gamma$) is satisfied. Thus R is of almost finite genus.

From each sheet of this R remove a disk $|z| < 1$ and let R' be the resulting surface. Again R' is of almost finite genus. Obviously $R \in O_G$ and $R' \notin O_{HD}$.

We assert:

THEOREM. *If R is a Riemann surface of almost finite genus then $R \notin U_{\widetilde{HD}}$.*

We may suppose that R is open. Let A_{nk} $(k = 1, 2)$ be a subdivision of A_n into two disjoint annuli such that $\bar{A}_n = \bar{A}_{n1} \cup \bar{A}_{n2}$ and $\log \operatorname{mod} A_{nk} = \frac{1}{2} \log \operatorname{mod} A_n$. Let $f_n \in \mathrm{M}_0(R)$ such that $f_n | (R - A_n) = 0$, $f_n | (A_{n1} \cup A_{n2}) \in H(A_{n1} \cup A_{n2})$, and $f_n | (\bar{A}_{n1} \cap \bar{A}_{n2}) = 1$. Then

$$D_R(f_n) = D_{A_{n1}}(f_n) + D_{A_{n2}}(f_n) = \frac{2\pi}{\log \operatorname{mod} A_{n1}} + \frac{2\pi}{\log \operatorname{mod} A_{n2}}$$

$$= \frac{8\pi}{\log \operatorname{mod} A_n}.$$

Set $f=\sum_n f_n$. If $\{A_n\}$ is finite then $f\in \mathbf{M}_0(R)\subset \mathbf{M}_\Delta(R)$. If $\{A_n\}$ is infinite then $f=BD\text{-}\lim_m \sum_1^m f_n$, and by $\sum_1^m f_n\in \mathbf{M}_0(R)$, $f\in \mathbf{M}_\Delta(R)$.

From the assumption $R\in U_{H\tilde{D}}$ we shall deduce a contradiction. By Theorem 4 M there exists a point $q\in\Delta_R$ such that $\mu_R(q)>0$. Let $G=\{p\,|\,p\in R, f(p)<\frac{1}{2}\}$. It is easy to see that $G\supset R-\bigcup_n \bar{A}_n$, G is planar, and $\bar{G}-\partial G$ is a neighborhood of q since $f(q)=0$. Take $j=j(G^*, \bar{G})$. In view of $q\in b_G=(\bar{G}-\partial\bar{G})\cap\Gamma_R$, $\mu_R(q)>0$ implies $\mu_G(q^*)>0$ where $q^*=j^{-1}(q)$, a point in Γ_G. Thus again by Theorem 4 M, $G\in U_{H\tilde{D}}$. Since G is planar it is conformally equivalent to a plane region F. Let $\bar{F}$ be the closure in the Riemann sphere S. Observe that $S^*=S$ and thus $\partial F=\bar{F}-F$. Let $j'=j(F^*, \bar{F})$. Since $F\in U_{H\tilde{D}}$, Γ_F contains a point p^* with $\mu_F(p^*)>0$. Hence $p=j'(p^*)$ is a point on ∂F with $\omega_{\partial F}(p)>0$ (Theorem 5 E). However no single point on the boundary of a plane region can have positive harmonic measure, a contradiction.

We shall make use of Theorem 5 F in VI.5 C.

5 G. $O_G=O_{HD}$ for Almost Finite Genus. In 2 H.(22) and Theorem 4 G we obtained the inclusion relations

$$O_G\subset O_{HB}\subset O_{HD}=O_{HBD} \tag{87}$$

and actually in 4 H, $O_{HB}<O_{HD}$. In contrast for surfaces of almost finite genus and consequently for surfaces of finite genus

$$O_G=O_{HB}=O_{HD}=O_{HBD}. \tag{88}$$

This is a consequence of the following result:

THEOREM. *For almost finite genus and consequently for finite genus*

$$O_G=O_{HD}. \tag{89}$$

We have only to show that if $R\in O_{HD}$ has almost finite genus then $R\in O_G$. But if $R\notin O_G$ then $R\in(O_{HD}-O_G)\subset U_{HD}$, in violation of Theorem 5 F.

We remark in passing that (87) and, for finite genus, (88) can also be obtained by means of principal functions in analogy with the reasoning in II.9 A.

5 H. Boundary Theorem of Riesz-Lusin-Privaloff Type. In order to study the relationship between degeneracy classes concerning HD-functions and those connected with analytic functions we shall first prove:

THEOREM. *Let E be a Borel subset of Γ with $\mu_R(E)>0$, and G a subregion of R such that $\bar{G}$ is a neighborhood of E. Every meromorphic function f on G with boundary value zero at each point of E vanishes identically on G.*

Contrary to the assertion assume that $f \not\equiv 0$ on G. We may suppose that E is a compact subset of Δ. By Theorem 5 D we may also take $G=R$. Consider

$$F=\{z \mid z\in R,\ |f(z)|<1\}$$

and let $F=\bigcup_1^N F_k$ $(N\le\infty)$ be the decomposition into components. Set $E_k=E\cap\bar{F}_k$. Since $\bar{F}-\overline{\partial F}$ is a neighborhood of E and $\partial F_k\subset\partial F$, $E_k\subset \bar{F}_k\cap\Gamma-\overline{\partial F}_k$. Let $j_k=j(F_k^*,\bar{F}_k)$. By Theorem 5 C, j_k is a homeomorphism on $j_k^{-1}(\bar{F}_k\cap\Gamma-\overline{\partial F}_k)$. We write $j_k^{-1}(E_k)=E_k^*$.

We now wish to show that

$$\mu_{F_k}(E_k^*)>0 \tag{90}$$

for at least one k.

If this were not so then by Theorem 5 D we would have $\mu_R(E_k)=0$ for every $k=1,\dots,N$. In this case $N=\infty$ since if $N<\infty$ then $E=\bigcup_1^N E_k$ and therefore $\mu_R(E)=0$, a contradiction.

Let U be an open set in R^* such that $E\subset U\subset\bar{U}\subset\bar{F}-\overline{\partial F}$ and

$$\mu_R(U\cap\Gamma-E)<\tfrac{1}{2}\mu_R(E). \tag{91}$$

Take $f_\infty\in \mathrm{M\!I}(R)$ such that $0\le f_\infty\le 1$ on R^*, $f_\infty|E=1$, and $f_\infty|(R^*-U)=0$. Let f_n $(1\le n<\infty)$ be defined by $f_n|\bigcup_1^n F_k=f_\infty$ and $f_n|(R-\bigcup_1^n \bar{F}_k)=0$. Clearly $f_n\in \mathrm{M\!I}(R)$, $f_n|(\bigcup_1^n E_k)=1$, and $f_n|((R^*-U)\cup(U-\bigcup_1^n \bar{F}_k))=0$. Observe that $\{f_n\}_1^\infty$ is increasing and dominated by f_∞. Thus the same is true of $\{\pi f_n\}_1^\infty$ and πf_∞. Since

$$D_R(\pi f_\infty-\pi f_n)\le D_R(f_\infty-f_n)=\sum_{k=n+1}^{\infty} D_{F_k}(f_\infty)\to 0$$

as $n\to\infty$ we conclude that $BD\text{-}\lim_n(\pi f_\infty-\pi f_n)=c$ exists and is a non-negative constant. Again $\pi f_\infty-\pi f_n=f_\infty=f_n=0$ on $E_{k_0}\subset\Delta$ for some E_{k_0}. Thus $c=0$, i.e.

$$\pi f_\infty=BD\text{-}\lim_{n\to\infty}\pi f_n$$

on R. Therefore

$$\begin{aligned}\mu_R(E)&=\int_E d\mu_R(p)\le\int_\Gamma(\pi f_\infty)(p)\,d\mu_R(p)=(\pi f_\infty)(z_0)\\&=\lim_n(\pi f_n)(z_0)=\lim_n\int_\Gamma(\pi f_n)(p)\,d\mu_R(p)\le\lim_n\int_{H_n}d\mu_R(p)\end{aligned}$$

where $H_n=(\bigcup_{k=1}^n \bar{F}_k)\cap\Gamma\cap U$. Hence

$$\begin{aligned}\mu_R(E)&\le\lim_n\mu_R(H_n)=\lim_n\big(\mu_R(H_n\cap E)+\mu_R(H_n-E)\big)\\&\le\lim_n\left(\sum_1^n\mu_R(E_k)+\mu_R(U\cap\Gamma-E)\right)\le\tfrac{1}{2}\mu_R(E)\end{aligned}$$

which is a contradiction. We have proved (90).

The function $w(z) = -\log|f(z)|$ is positive superharmonic on F_k with (90), and $w = \infty$ at each point of E_k^*. Clearly $w/n \in \mathfrak{A}(F_k^*, f_{E_k^*})$ for every $n = 1, 2, \ldots$ where $f_{E_k^*}$ is the characteristic function of E_k^* on Γ_{F_k}. Thus

$$\mu_{F_k}(E_k^*) = H(z_0; f_{E_k^*}) \leq \frac{w(z_0)}{n}$$

for every $n = 1, 2, \ldots$, which contradicts (90).

5 I. The Inclusion $U_{\widetilde{HD}} < O_{AD}$. As an application of Theorem 5 H we obtain (CONSTANTINESCU-CORNEA [3]):

THEOREM. *The following inclusion relation holds:*

$$U_{\widetilde{HD}} < O_{AD}. \tag{92}$$

If $R \in U_{\widetilde{HD}}$ then Γ contains a point q with $\mu(q) > 0$. Every function $f \in AD(R)$ has a continuous extension f to R^* and $f - f(q)$ vanishes at q. Thus by Theorem 5 H, $f - f(q)$ vanishes identically on R, i.e. $R \in O_{AD}$.

The strictness of the inclusion (92) is furnished by the finite plane $R: |z| < \infty$. By Theorem 5 F, $R \notin U_{\widetilde{HD}}$, whereas clearly $R \in O_{AD}$.

As a consequence of (92) we obtain the following interesting result of KURAMOCHI [10]:

COROLLARY. *The ideal boundary of $R \in O_{HD} - O_G$ is so intricate that the surface obtained from R by deleting any compact set still belongs to O_{AD}.*

Let $R' = R - K$ be connected, with K a compact set in R. By Theorems 5 D and 4 N and the fact $R \in U_{\widetilde{HD}}$ we see that $R' \in U_{\widetilde{HD}} \subset O_{AD}$.

Remark. The above corollary remains valid if we replace D by B; this will be discussed in IV. 7 G. Historically these two results of KURAMOCHI were an incentive to the second evolution of classification theory around 1960, and especially to the recent development of the theory of ideal boundaries of Riemann surfaces.

Many alternate proofs of the Kuramochi theorem have been given from various view points. CORNEA [1] proved it using the normal operator method. Constantinescu-Cornea's proof in their joint work [3] is especially important because of the introduction of the notion of $\widetilde{HD}$-minimality. Approaches similar to the one in this book were used by KUSUNOKI-MORI [4], S. MORI [2], KUSUNOKI [7], and HAYASHI [2]. All these proofs, including Kuramochi's original one, are more or less based on the $\widetilde{HD}$-minimality. MATSUMOTO [4] pointed out that Kuramochi's theorem can actually be derived from more general properties: the existence of a boundary point of STOÏLOW of positive harmonic measure and the Iversen property for analytic functions. This topic will be discussed in Chapter VI.

5 J. Examples of O_{HD}^n-Surfaces. The Tôki surface R of 4 H belongs to $O_{HD}-O_{HB}$ and a fortiori to $O_{HD}-O_G$. Therefore by Theorem 3 F the harmonic boundary $\Delta=\Delta(R)$ consists of exactly one point.

Take two disks α and β with disjoint closures on R. Let R_1 (resp. R_n) be the surface obtained from R by removing $\bar{\alpha}$ (resp. $\bar{\alpha}$ if n is even, $\bar{\beta}$ if n is odd). Similarly $R_2, \ldots, R_{n-1}$ are surfaces obtained from R by removing $\bar{\alpha}\cup\bar{\beta}$. Identify $\partial\alpha$ in R_1 with $\partial\alpha$ in R_2. Next identify $\partial\beta$ in R_2 with $\partial\beta$ in R_3. On repeating this process up to R_n we obtain a Riemann surface R^n.

By Theorem 5 D we see that the harmonic boundary $\Delta(R^n)$ of R^n consists of exactly n points. Therefore by Theorem 3 G

$$R^n \in O_{HD}^n - O_{HD}^{n-1}$$

for $n=1, 2, \ldots$. Thus in particular

$$O_G < O_{HD} = O_{HD}^1 < \cdots < O_{HD}^n < \cdots < U_{HD} \cup O_G \subset U_{\widetilde{HD}} \cup O_G.$$

Remark. Using Theorem 3 G we may define a limit class O_{HD}^∞ of the O_{HD}^n as the class of Riemann surfaces R whose harmonic boundaries Δ are closures of countable sets of isolated points in Δ. The limit surface $\lim R^n$ will give an example of a surface in $O_{HD}^\infty - \bigcup_1^\infty O_{HD}^n$, the details being left to the reader.

6. Green's Lines

6 A. Polar Coordinates. Let $R \notin O_G$, $z_0 \in R$, and let Ω_0 be a parametric disk $|z-z_0|<1$ in R. Take a regular region Ω such that $\bar{\Omega}_0 \subset \Omega \subset R$ and let $\omega_\Omega = \omega(z; \Omega-\bar{\Omega}_0)$ be the harmonic measure of $\partial\Omega$ with respect to $\Omega-\bar{\Omega}_0$ (cf. 2 F), with $\omega_\Omega=0$ on $\bar{\Omega}_0$ and $\omega_\Omega=1$ on $R-\Omega$. By $R \notin O_G$ the directed sequence $\{\omega_\Omega\}$ converges to the harmonic measure $\omega = \omega(z; R-\bar{\Omega}_0)$ of the ideal boundary of R such that $0<\omega<1$ on $R-\bar{\Omega}_0$.

Take two regular regions Ω and Ω' with $\bar{\Omega}_0 \subset \Omega \subset \bar{\Omega} \subset \Omega'$. Let s be $\log(1/|z-z_0|)$ on $\bar{\Omega}_0$ and $a\,\omega_{\Omega'}$ on $\Omega'-\Omega$ with $a=2\pi/\int_{\partial\Omega} * d\omega_{\Omega'}$. By Theorem I.7 A there exists an $h \in H(\Omega'-z_0)$ such that $h-s$ is bounded on $(\bar{\Omega}_0-z_0) \cup (\Omega'-\Omega)$. Let $g_\Omega = g_\Omega(z, z_0)$ be the function obtained from h by subtracting the harmonic function on Ω with boundary values h on $\partial\Omega$. Then $g_\Omega|\bar{\Omega}_0 = \log(1/|z-z_0|) + \text{harmonic function}$, and $g_\Omega|\partial\Omega=0$; thus g_Ω is the *Green's function* on Ω with pole at z_0.

Set $g_\Omega=0$ on $R-\bar{\Omega}$. Clearly $\{g_\Omega\}$ constitutes an increasingly directed set with $\int_\alpha * dg_\Omega = -2\pi$ where $\alpha=\partial\Omega_0$. By Green's formula

$$\int_\alpha (1-\omega_\Omega) * dg_\Omega + \int_\alpha g_\Omega * d\omega_\Omega = 0$$

and Harnack's inequality provides us with a constant $0<k<\infty$ such that

$$k^{-1}(\max_{\alpha} g_\Omega)\int_{\alpha} *d\omega_\Omega \leq \int_{\alpha} g_\Omega * d\omega_\Omega = 2\pi.$$

Since $\int_\alpha * d\omega_\Omega = \int_{\partial\Omega} \omega_\Omega * d\omega_\Omega = \int_{\partial(\Omega-\bar{\Omega}_0)} \omega_\Omega * d\omega_\Omega = D_R(\omega_\Omega)$ converges to $D_R(\omega)>0$ as $\Omega \to R$ we conclude that

$$\lim_{\Omega\to R}(\max_{\alpha} g_\Omega) \leq \frac{2\pi k}{D_R(\omega)} < \infty.$$

Hence $\{g_\Omega\}$ converges to a harmonic function $g=g(z, z_0)$ on $R-z_0$ with

$$g(z; z_0)|\Omega_0 = \log\frac{1}{|z-z_0|} + \text{harmonic function}. \tag{93}$$

Note that

$$0<g_\Omega \leq \frac{2\pi k}{D_R(\omega_\Omega)}(1-\omega_\Omega)$$

on $\Omega - \bar{\Omega}_0$ and thus

$$0<g(z; z_0) \leq \frac{2\pi k}{D_R(\omega)}(1-\omega(z; R-\bar{\Omega}_0)) \tag{94}$$

on $R-\bar{\Omega}_0$. Since $1-\omega_\Omega \in \mathrm{M}_0(R)$ and $1-\omega = BD\text{-}\lim_{\Omega\to R}(1-\omega_\Omega)$ we infer that $1-\omega=0$ on Δ and therefore

$$g(p, z_0)=0 \tag{95}$$

on Δ. Obviously g has a finite Dirichlet integral over the complement of a neighborhood of z_0.

By maximum principle II in 2 H the harmonic function on $R-z_0$ with (93) and (95) is unique. Such a function $g(z, z_0)$ is called the *Green's function* on R with pole at z_0.

Suppose $g_\Omega(\cdot, z_0)$ converges to $g(\cdot, z_0)$ and let $m=\min_{\partial\Omega_0} g(\cdot, z_0)$. Clearly $g_\Omega(\cdot, z_0) \geq m(1-\omega_\Omega)$ on $\Omega-\Omega_0$ and consequently $g(\cdot, z_0) \geq m(1-\omega)$ on $R-\Omega_0$. Since $\inf_{R-\bar{\Omega}_0} g < m$ we infer that ω is not constant. We conclude:

$R \in O_G$ if and only if there exists no Green's function.

We shall discuss Green's functions more extensively in V. 8.

On a surface $R \notin O_G$ consider the pair $(r(z), \theta(z))$ of local functions of local parameters z defined by

$$\frac{dr(z)}{r(z)} = -dg(z, z_0), \qquad d\theta(z) = -*dg(z, z_0). \tag{96}$$

The function $r(z)$ is uniquely determined as a global function on R if we choose $r(z)=e^{-g(z,z_0)}$. Obviously $0\le r(z)<1$ on R. We set

$$G_\rho=\{z|z\in R, r(z)<\rho\}, \qquad C_\rho=\partial G_\rho$$

for every ρ with $0<\rho<1$ and call C_ρ (and G_ρ) *regular* if $dg(z,z_0)\neq 0$ on C_ρ. Except for a countable number of ρ in $0<\rho<1$, C_ρ and G_ρ have this property. Note that C_ρ and $G_\rho\cup C_\rho$ need not be compact.

THEOREM. *For any regular* C_ρ, $(2\pi)^{-1}\int_{C_\rho} d\theta(z)=1$.

Let $\{R_n\}_0^\infty$ be a regular exhaustion of R with $z_0\in R_0\subset \bar{R}_0\subset G_\rho$ and let w_n be the continuous function on $\bar{G}_\rho$ such that $w_n|\bar{R}_0=1$, $w_n|(\bar{G}_\rho-R_n)=0$, $w_n\in H(\bar{G}_\rho\cap(R_n-\bar{R}_0))$, and $*dw_n=0$ on $C_\rho\cap R_n$. Since by (95), $\bar{G}_\rho\cap\Delta=\emptyset$, Theorem 2 G shows that the double $\hat{G}_\rho$ of G_ρ about C_ρ is parabolic. Therefore $w_n\to 1$ on $G_\rho\cup C_\rho$ and $D_{G_\rho}(w_n)\to 0$ as $n\to\infty$. Observe that

$$D_{G_\rho}(w_n,g)=\int_{C_\rho\cap R_n} w_n(z)*dg(z,z_0)-\int_{\partial R_0} *dg(z,z_0).$$

On using $\lim_n D_{G_\rho}(w_n,g)=0$, $\int_{\partial R_0} *dg(z,z_0)=-2\pi$, and Lebesgue's convergence theorem we obtain the required equality.

6 B. Space of Green's Lines. Although $\theta(z)$ is not single-valued it is harmonic on $R-z_0$. A *level arc* of $\theta(z)$ is an open arc on which $d\theta(z)\neq 0$ and $\theta(z)$ is constant. We call a level arc of $\theta(z)$ a *Green's arc*. The totality of Green's arcs forms a set partially ordered by inclusion. In this sense a maximal Green's arc is called a *Green's line*.

Denote by

$$\mathbb{G}=\mathbb{G}(R,z_0)$$

the totality of Green's lines L issuing from z_0; then $z_0\in\bar{L}$. We also denote by

$$\mathbb{G}'=\mathbb{G}'(R,z_0)$$

the set consisting of the point z_0 and all points in R which lie on Green's lines belonging to $\mathbb{G}(R,z_0)$. Clearly $\mathbb{G}'(R,z_0)$ is a simply connected subregion of R. It is called a *Green's star region* in R with respect to z_0.

For a sufficiently small $\varepsilon>0$ the set G_ε is regular and relatively compact in R, and $w=f(z)=(1/\varepsilon)\,r(z)\,e^{i\theta(z)}$ is a conformal mapping of $\bar{G}_\varepsilon$ onto the unit disk $\{w||w|\le 1\}$. Hereafter we fix such an ε and use the notation

$$\mathscr{E}=G_\varepsilon=\{z|z\in R, r(z)<\varepsilon\}, \qquad \mathbb{J}=\partial\mathscr{E}.$$

The inverse function $z=f^{-1}(w)$ maps $\{w||w|\le 1\}$ onto $\mathscr{E}\cup\mathbb{J}$. We represent each point $z\in\mathbb{J}$ by the coordinate $\theta\in[0,2\pi)$ where $z=f^{-1}(e^{i\theta})$.

Using this we can designate each $L \in \mathbb{G}(R, z_0)$ by

$$L = L_\theta$$

with θ the coordinate of the point $L \cap \mathbb{J}$. We may identify $\mathbb{J}$ with $[0, 2\pi)$ and write

$$\mathbb{G} = \mathbb{G}(R, z_0) = \{L_\theta \mid \theta \in \mathbb{J}\}.$$

Since the totality of points in R at which $d\theta(z) = 0$ is a discrete set in R,

$$\mathbb{E}_0 = \mathbb{E}_0(R, z_0) = \{\theta \mid L_\theta \in \mathbb{G}(R, z_0),\ \bar{L}_\theta \text{ is compact in } R\}$$

is a countable subset of $\mathbb{J}$. For each $L_\theta \in \mathbb{G}$ we write

$$d_\theta = \sup\{r(z) \mid z \in L_\theta\}.$$

Clearly $\varepsilon < d_\theta \leq 1$. If $d_\theta < 1$ then we call L_θ a *singular Green's line*. It is easy to see that

$$\mathbb{E} = \mathbb{E}(R, z_0) = \{\theta \in \mathbb{J} \mid d_\theta < 1\}$$

is an F_σ-set in $\mathbb{J}$, i.e. a countable union of closed sets in $\mathbb{J}$, and

$$\mathbb{E}_0 \subset \mathbb{E} \subset \mathbb{J}.$$

We may use $r(z)\, e^{i\theta(z)}$ as a local parameter at each point of R except at a countable number of isolated points at which $d\theta(z) = 0$. If we take the branch of $\theta(z)$ at $z \in L_\theta$ with $\theta(z) = \theta$ then we can use the single-valued function

$$r(z)\, e^{i\theta(z)} = r\, e^{i\theta}$$

on $\mathbb{G}'(R, z_0)$ as a global coordinate which may be interpreted as a *polar coordinate* on R.

We call the normalized Lebesgue measure

$$dm(\theta) = \frac{1}{2\pi}\, d\theta$$

on $\mathbb{J}$ a *Green's measure*. We may consider m as a measure on $\mathbb{G}(R, z_0)$. Using these concepts we first give the mean value theorem of Gauss:

THEOREM. *Let $\rho \in (0, 1)$ be a number with a regular C_ρ and let u be a function in $HB(G_\rho)$ continuous on $G_\rho \cup C_\rho$. Then*

$$u(z_0) = \int_0^{2\pi} u(\rho\, e^{i\theta})\, dm(\theta). \tag{97}$$

Take a regular exhaustion $\{R_n\}_1^\infty$ of R with $\bar{\mathscr{E}} \subset R_1$ and denote by $g_n(z, z_0)$ the Green's function on $G_\rho \cap R_n$ with pole at z_0. Set $g_n = 0$ on $G_\rho - R_n$. Clearly $g(z, z_0) + \log \rho$ is the Green's function on G_ρ. Hence

$g_n(z, z_0) \to g(z, z_0) + \log \rho$ on $G_\rho \cup C_\rho$ and $0 \geq *dg_n(z, z_0) \geq *dg(z, z_0)$ on $C_\rho \cap R_n$. Consider the harmonic function w_n on $G_\rho \cap R_n - \mathscr{E}$ with the continuous boundary values 0 on $\mathbb{J}$, 1 on $\partial R_n \cap G_\rho$, and $*dw_n = 0$ on $C_\rho \cap R_n$. Set $w_n = 0$ on $\mathscr{E}$ and $w_n = 1$ on $G_\rho - R_n$. Since $\bar{G}_\rho \cap \Delta = \emptyset$, we conclude by Theorem 2 F that $w_n \to 0$ on $G_\rho \cup C_\rho$ and $D_{G_\rho}(w_n) \to 0$ as $n \to \infty$.

By Green's formula

$$\Big| \int_{\partial R_n \cap G_\rho} *dg_n(z, z_0) + \int_{C_\rho \cap R_n} w_n(z) * dg_n(z, z_0) \Big|^2$$
$$= \Big| \int_{\partial(G_\rho \cap R_n - \mathscr{E})} w_n(z) * dg_n(z, z_0) \Big|^2$$
$$\leq D_{G_\rho - \bar{\mathscr{E}}}(g_n)\, D_{G_\rho}(w_n) \leq D_{G_\rho - \bar{\mathscr{E}}}(g)\, D_{G_\rho}(w_n) \to 0$$

as $n \to \infty$.

On the other hand $0 \geq w_n * dg_n \geq *dg$ on $C_\rho \cap R_n$, $w_n * dg_n \to 0$ on C_ρ, and $*dg(z, z_0) = -d\theta(z)$ is integrable on C_ρ. Hence by Lebesgue's convergence theorem $\lim_n \int_{C_\rho \cap R_n} w_n(z) * dg_n(z, z_0) = 0$ and we obtain

$$\lim_n \int_{\partial R_n \cap G_\rho} *dg_n(z, z_0) = 0.$$

Again by Green's formula

$$\begin{aligned} -u(z_0) &= \frac{1}{2\pi} \int_{\partial(G_\rho \cap R_n)} u(z) * dg_n(z, z_0) \\ &= \frac{1}{2\pi} \int_{C_\rho \cap R_n} u(z) * dg_n(z, z_0) + \frac{1}{2\pi} \int_{\partial R_n \cap G_\rho} u(z) * dg_n(z, z_0). \end{aligned} \tag{98}$$

Here

$$\Big| \int_{\partial R_n \cap G_\rho} u(z) * dg_n(z, z_0) \Big| \leq -(\sup_{G_\rho} |u|) \int_{\partial R_n \cap G_\rho} *dg_n(z, z_0) \to 0$$

as $n \to \infty$. Observe that $|u(z) * dg_n(z, z_0)| \leq -(\sup_{G_\rho} |u|) * dg(z, z_0)$ along $C_\rho \cap R_n$, the right hand term is integrable on C_ρ, and $\lim_n u * dg_n = u * dg$ on C_ρ. Thus by Lebesgue's convergence theorem $\int_{C_\rho \cap R_n} u * dg_n \to \int_{C_\rho} u * dg$ as $n \to \infty$ in (98) and it follows that $-u(z_0) = (2\pi)^{-1} \int_{C_\rho} u(z) * dg(z, z_0)$. On setting $*dg(z, z_0) = -d\theta(z)$, $z = \rho e^{i\theta}$, and $dm(\theta) = (2\pi)^{-1} d\theta$ we finally obtain (97).

6 C. Ends of Green's Lines. For $L_\theta \in \mathbb{G}(R, z_0)$ the set

$$e_\theta = \bar{L}_\theta - L_\theta - \{z_0\}$$

is compact in R^*. We call it the *end* of L_θ in R^*. Clearly $e_\theta \in R$ if and only if $\theta \in \mathbb{E}_0$. In this case e_θ consists of one point in R at which $d\theta = 0$. If $\theta \in \mathbb{J} - \mathbb{E}_0$ then $e_\theta \subset \Gamma$ and e_θ is a proper continuum as is seen by the proof of Theorem 2 D. For any subset $S \subset \Gamma$ we write

$$\tilde{S} = \{\theta \in \mathbb{J} \mid e_\theta \cap S \neq \emptyset\}, \qquad \check{S} = \{\theta \in \mathbb{J} \mid e_\theta \subset S\}.$$

Obviously $\check{S} \subset \tilde{S} \subset \mathbb{J}$. We assert:

THEOREM. *Let K be a compact subset in $\Gamma - \Delta$. Then $m(\tilde{K}) = 0$.*

Take an open neighborhood U of K in R^* such that $\bar{U} \cap \Delta = \emptyset$ and $\bar{U}$ is a distinguished compact subset of R^*. Let $U_n = U - \bar{R}_n$ where $\{R_n\}_1^\infty$ is a regular exhaustion of R. Then there exists a unique $w_n \in \mathbb{M}(R)$ harmonic on $R - \bar{U}_n$ such that $w_n \mid \bar{U}_n = 1$, $w_n \mid \Delta = 0$. As in 3 H, $w_n \to 0$ and $D_R(w_n) \to 0$ for $n \to \infty$. Here we set

$$U_n' = \{\theta \mid L_\theta \in \mathbb{G}(R, z_0), L_\theta \cap (U_n \cap R) \neq \emptyset\} \subset \mathbb{J}.$$

Clearly U_n' is open in $\mathbb{J}$ and $U_n' \supset U_{n+1}' \supset \tilde{K}$. For each $\theta \in U_n'$ we choose a point $z_\theta \in L_\theta \cap (U_n \cap R)$. Then $w_n(z_\theta) = 1$ and

$$1 - w_n(\varepsilon e^{i\theta}) = \int_\varepsilon^{r(z_\theta)} \frac{\partial}{\partial r} w_n(r e^{i\theta})\, dr.$$

There exists an $a > 0$ such that $1 - w_n > a$ on $\mathbb{J}$ for all sufficiently large n, that is $a < \int_\varepsilon^{r(z_\theta)} (\partial w_n(r e^{i\theta})/\partial r)\, dr$. By Schwarz's inequality

$$\begin{aligned} a^2 &< \int_\varepsilon^{r(z_\theta)} \left| \frac{\partial}{\partial r} w_n(r e^{i\theta}) \right|^2 r dr \int_\varepsilon^{r(z_\theta)} \frac{dr}{r} \\ &\le (-\log \varepsilon) \int_\varepsilon^{d_\theta} \left| \frac{\partial}{\partial r} w_n(r e^{i\theta}) \right|^2 r dr. \end{aligned}$$

Hence

$$\int_{U_n'} a^2\, d\theta \le (-\log \varepsilon) \int_{U_n'} \int_\varepsilon^{d_\theta} \left| \frac{\partial}{\partial r} w_n(r e^{i\theta}) \right|^2 r dr d\theta \le (-\log \varepsilon) D_R(w_n)$$

and therefore $m(U_n') \le c\, D_R(w_n)$, $c = (-\log \varepsilon)/2\pi a^2$. We have proved that $m(\tilde{K}) = 0$.

The following result is due to BRELOT-CHOQUET [1].

COROLLARY. *The set of singular Green's lines has Green's measure zero, i.e. $m(\mathbb{E}) = 0$ or equivalently $d_\theta = 1$ a.e. on $\mathbb{J}$.*

The set $K_n = \{p \mid p \in \Gamma, g(p, z_0) \ge 1/n\}$ is compact in $\Gamma - \Delta$ and hence $m(\tilde{K}_n) = 0$. Clearly $\widetilde{\bigcup_1^\infty K_n} = \bigcup_1^\infty \tilde{K}_n$ and $\mathbb{E} \subset \mathbb{E}_0 \cup (\widetilde{\bigcup_1^\infty K_n})$. Since $\mathbb{E}_0$ is countable we see that $m(\mathbb{E}) \le m(\mathbb{E}_0) + \sum_1^\infty m(\tilde{K}_n) = 0$.

Remark. MAEDA [1] proved that e_θ reduces to one point for almost all θ if we replace R^* by its factor space which is metrizable. In particular this is the case on Kuramochi's compactification.

6 D. Radial Limits. We say that a real- or complex-valued function f on R possesses a radial limit almost everywhere on $\mathbb{J}$ if

$$f(\theta)=\lim_{r\to 1} f(r\,e^{i\theta})=\lim_{z\in L_\theta,\, r(z)\to 1} f(z) \tag{99}$$

exists a.e. on $\mathbb{J}-\mathbb{E}$. We maintain (GODEFROID [1]):

THEOREM. *Every Dirichlet finite Tonelli function f on R possesses a radial limit a.e. on $\mathbb{J}$. In particular this is true of AD- and HD-functions.*

Let $L^2(\mathbb{J}, m)$ be the Hilbert space of square integrable functions on $\mathbb{J}$ with respect to the norm $\|\varphi\|_2=(\int_0^{2\pi}|\varphi(\theta)|^2\,dm(\theta))^{\frac{1}{2}}$. Set $f_r(\theta)=f(r\,e^{i\theta})$ $(0<r<1)$. In view of the definition of a Tonelli function

$$f_b(\theta)-f_a(\theta)=\int_a^b \frac{\partial}{\partial r} f(r\,e^{i\theta})\,dr, \qquad 0<a<b<1,$$

for almost every fixed $\theta\in\mathbb{J}-\mathbb{E}$. By Schwarz's inequality

$$|f_b(\theta)-f_a(\theta)|^2\le\left(\log\frac{b}{a}\right)\int_a^b\left|\frac{\partial}{\partial r} f(r\,e^{i\theta})\right|^2 r\,dr.$$

On integrating both sides with respect to $dm(\theta)$ over $\mathbb{J}$ we obtain

$$\|f_b-f_a\|_2\le\left(\log\frac{b}{a}\right)^{\frac{1}{2}}(D_{G_{ab}}(f))^{\frac{1}{2}}$$

where $G_{ab}=G_b-\bar{G}_a$ $(0<a<b<1)$.

Let $\{b_k\}_{k=1}^\infty$ be an arbitrary strictly increasing sequence of positive numbers with $\lim_k b_k=1$. Set

$$h(\theta)=\sum_{k=1}^\infty |f_{b_{k+1}}(\theta)-f_{b_k}(\theta)|$$

for $\theta\in\mathbb{J}-\mathbb{E}$. By the triangle inequality

$$\begin{aligned}\left(\int_0^{2\pi}|h(\theta)|^2\,d\theta\right)^{\frac{1}{2}}&\le\sum_{k=1}^\infty\|f_{b_{k+1}}-f_{b_k}\|_2\le\sum_{k=1}^\infty\left(\log\frac{b_{k+1}}{b_k}\right)^{\frac{1}{2}}(D_{G_{b_k b_{k+1}}}(f))^{\frac{1}{2}}\\&\le\sum_{k=1}^\infty\left(\log\frac{b_{k+1}}{b_k}+D_{G_{b_k b_{k+1}}}(f)\right)\le\log\frac{1}{b_1}+D_R(f),\end{aligned}$$

i.e. $h(\theta)<\infty$ a.e. on $\mathbb{J}$. Hence $\sum_{k=1}^{\infty}\left(f_{b_{k+1}}(\theta)-f_{b_k}(\theta)\right)$ converges a.e. on $\mathbb{J}$. Since $\sum_{k=1}^{\infty}\left(f_{b_{k+1}}(\theta)-f_{b_k}(\theta)\right)=\lim_k f_{b_k}(\theta)-f_{b_1}(\theta)$, $\lim_k f_{b_k}(\theta)$ exists a.e. on $\mathbb{J}$. From this and the fact that $f_r(\theta)$ is continuous in r for each fixed θ in $\mathbb{J}-\mathbb{E}$ it follows that (99) exists for almost all $\theta\in\mathbb{J}-\mathbb{E}$.

6 E. Lattice of Radial Limits. As a counterpart of Theorem 4 F we shall prove:

THEOREM. *For every u and v in $HD(R)$, $u\vee v$ and $u\wedge v$ also belong to $HD(R)$, and the identities*

$$(u\vee v)(\theta)=u(\theta)\cup v(\theta),\qquad (u\wedge v)(\theta)=u(\theta)\cap v(\theta) \tag{100}$$

are valid for $\theta\in\mathbb{J}$ a.e.

The first half was already obtained in 4 F.

Since $(u\vee v)(\theta)=-((-u)\wedge(-v))(\theta)$ and

$$u(\theta)\cup v(\theta)=-((-u(\theta))\cap(-v(\theta)))$$

we have only to prove the second identity (100). We may assume that $u, v\geq 0$ on R. First we consider the case in which v is bounded. Then $f(z)=u(z)\cap v(z)$ is a nonnegative bounded superharmonic function on R belonging to $\mathbb{M}(R)$. Let $\{R_n\}_1^{\infty}$ be a regular exhaustion of R with $z_0\in R_1$ and take a $0<\rho<1$ with a regular C_ρ. Set $w_{\rho n}=\pi_{R^*-G_\rho\cap R_n} f$. Clearly $f\geq w_{\rho n}\geq w_{\rho, n+1}\geq u\wedge v$ on R and $w_{\rho n}\geq w_{\rho' n}$ $(\rho'>\rho)$. Thus $w_\rho=\lim_n w_{\rho n}$ exists on R, is harmonic on G_ρ, and continuous on R; moreover $w_\rho(z)=u(z)\cap v(z)$ on C_ρ, and $f\geq w_\rho\geq w_{\rho'}\geq u\wedge v$ $(\rho'>\rho)$. It follows that

$$\lim_{\rho\to 1} w_\rho=u\wedge v$$

on R. By (97)

$$w_\rho(z_0)=\int_0^{2\pi} w_\rho(\rho e^{i\theta})\,dm(\theta)=\int_0^{2\pi} u(\rho e^{i\theta})\cap v(\rho e^{i\theta})\,dm(\theta).$$

Since $w_\rho(z_0)\to(u\wedge v)(z_0)$ as $\rho\to 1$ Lebesgue's convergence theorem gives

$$(u\wedge v)(z_0)=\int_0^{2\pi} u(\theta)\cap v(\theta)\,dm(\theta).$$

Again by (97), $(u\wedge v)(z_0)=\int_0^{2\pi}(u\wedge v)(\rho e^{i\theta})\,dm(\theta)$ and on letting $\rho\to 1$ we obtain

$$(u\wedge v)(z_0)=\int_0^{2\pi}(u\wedge v)(\theta)\,dm(\theta).$$

In view of $(u\wedge v)(\rho e^{i\theta})\leq u(\rho e^{i\theta})\cap v(\rho e^{i\theta})$ we have $(u\wedge v)(\theta)\leq u(\theta)\cap v(\theta)$ a.e. on $\mathbb{J}-\mathbb{E}$. Hence $\int_0^{2\pi}\left(u(\theta)\cap v(\theta)-(u\wedge v)(\theta)\right)dm(\theta)=0$ implies (100) a.e. on $\mathbb{J}-\mathbb{E}$.

Next we remove the boundedness assumption on v. From the above reasoning it follows that $(u \wedge v)(\theta) \cap n = ((u \wedge v) \wedge n)(\theta) = (u \wedge (v \wedge n))(\theta) = u(\theta) \cap (v \wedge n)(\theta) = u(\theta) \cap (v(\theta) \cap n) = (u(\theta) \cap v(\theta)) \cap n$, i.e. $(u(\theta) \cap v(\theta) - (u \wedge v)(\theta)) \cap n = 0$ a.e. on $\mathbb{J} - \mathbb{E}$ for every $n = 1, 2, \ldots$. Thus on letting $n \to \infty$ we obtain (100).

6 F. Gauss' Property of Radial Limits. As an application of the preceding theorem we shall prove the following Gauss property:

$$u(z_0) = \int_0^{2\pi} u(\theta)\, dm(\theta) \tag{101}$$

for every $u \in HD(R)$.

In fact by (100) we may assume that $u \geq 0$ on R. By Theorem 6 B, $(u \wedge n)(z_0) = \int_0^{2\pi} (u \wedge n)(\rho e^{i\theta})\, dm(\theta)$ for every $n = 1, 2, \ldots$. For $\rho \to 1$ we obtain by (100)

$$(u \wedge n)(z_0) = \int_0^{2\pi} (u \wedge n)(\theta)\, dm(\theta) = \int_0^{2\pi} u(\theta) \cap n\, dm(\theta).$$

Assertion (101) follows on letting $n \to \infty$.

6 G. Functions with Radial Limits Zero. For Dirichlet finite Tonelli functions f we can consider two kinds of boundary values: $f(p)$ on Royden's boundary, and the radial limit $f(\theta)$. They are related as follows:

THEOREM. *Let f be a Dirichlet finite Tonelli function on R. It vanishes on Δ if and only if $f(\theta) = 0$ a.e. on $\mathbb{J}$.*

We may assume that $f \geq 0$. First suppose $f | \Delta = 0$. Then the set $K_n = \{p \mid p \in \Gamma, f(p) \geq 1/n\}$ is compact in $\Gamma - \Delta$ and hence by Theorem 6 C, $m(\tilde{K}_n) = 0$. Let $E_n = \{\theta \mid \theta \in \mathbb{J}, f(\theta) \geq 1/n\}$. Evidently the set $\tilde{K}_n \cup E_n - \tilde{K}_n \cap E_n$ has m-measure zero. Therefore $m(E_n) = 0$. Since $\{\theta \mid \theta \in \mathbb{J} - \mathbb{E}, f(\theta) > 0\} = \bigcup_1^\infty E_n$ has m-measure zero $f(\theta) = 0$ a.e. on $\mathbb{J}$.

Conversely assume that $f(\theta) = 0$ a.e. on $\mathbb{J}$. By Theorem 3 B, $f_1 = f - \pi f$ vanishes on Δ. Thus $f_1(\theta) = 0$ a.e. on $\mathbb{J}$ and a fortiori $(\pi f)(\theta) = 0$ a.e. on $\mathbb{J}$. In view of (101) we obtain $(\pi f)(z_0) = 0$. From the assumption $f \geq 0$ on R it follows that $\pi f = f \geq 0$ on Δ and therefore $\pi f \geq 0$. Hence we have $\pi f \equiv 0$ on R, i.e. $f = f_1$ vanishes on Δ.

6 H. Harmonic and Green's Measures. The outer and inner measures on $\mathbb{J}$ induced by m will be denoted by $\overline{m}$ and $\underline{m}$ and called the *outer* and *inner Green's measures* on $\mathbb{G}(R, z_0)$. We state:

THEOREM. *For every F_σ-set K in Γ, $\overline{m}(\tilde{K}) \leq \mu(K)$. For every G_δ-set U in Γ, $\underline{m}(\check{U}) \geq \mu(U)$.*

We first prove the latter assertion. By the decreasing monotone continuity of $\underline{m}$ and μ we may assume that U is open in Γ. Let $F=\Gamma-U$. For an arbitrary $\eta>0$ there exists a compact set $H\subset\Delta\cap U$ such that $\mu(H)<\mu(U)<\mu(H)+\eta$. Let $u\in HBD(R)$ such that $0\leq u\leq 1$ on R^*, $u|H=1$, and $u|(\Delta-U)=0$. The set $F_n=\{p|p\in F, u(p)\geq 1/n\}$ is compact in $\Gamma-\Delta$, and by Theorem 6 C, $m(\tilde{F}_n)=0$.

Let $F_0=F-\bigcup_1^\infty F_n=\{p|p\in F, u(p)=0\}$. Assume $u(\theta)$ exists for a $\theta\in\tilde{F}_0$. Since $\bar{L}_\theta\cap F_0\neq\emptyset$, $u(\theta)=0$. Thus $u(\theta)=0$ a.e. on $\tilde{F}_0$. By $m(\widetilde{\bigcup_1^\infty F_n})=m(\bigcup_1^\infty \tilde{F}_n)=0$ we have $m(\tilde{F}-\tilde{F}_0)=0$ and therefore $u(\theta)=0$ a.e. on $\tilde{F}$. Since $\Gamma=F\cup U$ and a fortiori $\mathbb{J}-\mathbb{E}_0=\tilde{F}\cup\check{U}$ we conclude that the set $V=\{\theta|u(\theta)>0\}$ satisfies $m(V-\check{U})=0$ and is measurable. Hence $\underline{m}(\check{U})\geq m(V)$.

On the other hand

$$m(V)=\int_V dm(\theta)\geq\int_0^{2\pi} u(\theta)\,dm(\theta)=u(z_0)$$

$$=\int_\Gamma u(p)\,d\mu(p)\geq\int_H d\mu(p)=\mu(H).$$

Therefore $\underline{m}(\check{U})\geq\mu(U)-\eta$ and on letting $\eta\to 0$ we conclude that $\underline{m}(\check{U})\geq\mu(U)$.

Next assume that K is an F_σ-set in Γ. Then $U=\Gamma-K$ is a G_δ-set in Γ and $\check{U}=(\mathbb{J}-\mathbb{E}_0)-\tilde{K}$. Thus

$$\bar{m}(\tilde{K})=1-\underline{m}(\check{U})\leq 1-\mu(U)=\mu(K).$$

6 I. Boundary Theorem of Riesz Type. As a counterpart of Theorem 5 H we shall prove:

THEOREM. *Let $f\in AD(R)$ and let $\mathbb{Z}$ be a subset of $\mathbb{J}-\mathbb{E}$ with positive measure. Assume that $f(\theta)=0$ for $\theta\in\mathbb{Z}$. Then $f\equiv 0$ on R.*

For $S=\bigcup_{\theta\in\mathbb{Z}} e_\theta$ the set $K=\bar{S}$ is compact in Γ with $\tilde{K}\supset\check{K}\supset\check{S}\supset\mathbb{Z}$. Hence by Theorem 6 H, $\mu(K)\geq\bar{m}(\tilde{K})\geq m(\mathbb{Z})>0$. Since $f=0$ on e_θ with $\theta\in\mathbb{Z}$, $f=0$ on S. By continuity $f=0$ on K. Therefore by Theorem 5 H, $f\equiv 0$ on R.

6 J. Blocks. We assume that $R\notin O_{HD}$. For each point $p\in\Delta$ we set

$$\Lambda_p=\{q|q\in\Gamma, u(q)=u(p) \text{ for all } u\in HBD(R)\}$$

and call Λ_p the *block* at p. Since $HBD(R)$ separates points in Δ we deduce that

$$\Lambda_p\cap\Delta=p,\quad \Lambda_p\cap\Lambda_q=\emptyset\qquad (p\neq q).$$

Take $u \in HBD(R)$ such that $u > 0$ on R and $u(p) = 0$ for a $p \in \Delta$; then $u = 0$ on Λ_p. On multiplying u by a suitable constant a we obtain $au(z) > g(z, z_0)$ on $\partial\mathscr{E}$. Since $au \geq g$ on Δ we conclude that $au \geq g$ on $R^* - \mathscr{E}$. Thus

$$g(q, z_0) | \Lambda_p = 0.$$

In particular if $e_\theta \cap \Lambda_p \neq \emptyset$ then $d_\theta = 1$ or equivalently $\theta \in \mathbb{J} - \mathbb{E}$.

For the magnitude of blocks we have the following evaluation:

THEOREM. *The block Λ_p at $p \in \Delta$ has measurable $\tilde{\Lambda}_p$ and*

$$m(\tilde{\Lambda}_p) = \mu(p). \tag{102}$$

First assume that $\mu(p) = 0$. Since $\overline{m}(\tilde{\Lambda}_p) \leq \mu(p) = 0$, $\tilde{\Lambda}_p$ is measurable and (102) is valid.

Next suppose $\mu(p) > 0$. Set $u(z) = (1/\mu(p)) \int_p P(z, q)\, d\mu(q) = P(z, p)$. By Theorems 4 K and 4 E, $u \in \widetilde{HD}(R)$, $0 < u < 1$ on R, $\limsup_{z \in R, z \to p} u(z) = 1$, and $\lim_{z \in R, z \to q} u(z) = 0$ for $q \neq p$ in Δ. Let $\{u_n\}_1^\infty \subset HD(R)$ such that $0 < u \leq u_{n+1} \leq u_n < 1$ and $\lim_n u_n = u$ on R. Clearly

$$\bar{u}(\theta) = \limsup_{\rho \to 1} u(\rho e^{i\theta}) \leq u_n(\theta).$$

Since $u(\rho e^{i\theta})$ is continuous in ρ for $0 < \rho < 1$ and measurable in θ on $\mathbb{J} - \mathbb{E}$, $\bar{u}(\theta)$ is measurable on $\mathbb{J}$. Set $a_n = u_n(z_0) - u(z_0)$. Then

$$\int_0^{2\pi} (u_n(\rho e^{i\theta}) - u(\rho e^{i\theta}))\, dm(\theta) = a_n$$

and by Fatou's lemma

$$\int_0^{2\pi} (u_n(\theta) - \bar{u}(\theta))\, dm(\theta) \leq a_n.$$

For $v(\theta) = \lim_n u_n(\theta)$ we have $v(\theta) \geq \bar{u}(\theta)$ and

$$\int_0^{2\pi} (v(\theta) - \bar{u}(\theta))\, dm(\theta) = 0.$$

Hence $v(\theta) = \bar{u}(\theta)$ a.e. on $\mathbb{J}$. Let

$$U_{nk} = \{q \mid q \in \Gamma, u_n(q) > 1 - 1/k\},$$

$$F_n = \{q \mid q \in \Gamma, u_n(q) = 1\}, \qquad F = \bigcap_{n=1}^\infty F_n.$$

Then $F_n = \bigcap_{k=1}^\infty U_{nk} \supset \Lambda_p$ and hence $F \supset \Lambda_p$. By Theorem 6 D there is a set $\mathbb{J}' \subset \mathbb{J}$ such that $m(\mathbb{J} - \mathbb{J}') = 0$ and $u_n(\theta)$ exists on $\mathbb{J}'$ for $n = 1, 2, \ldots$. For all n, $u_n(\theta) = 1$ on $\tilde{F}_n \cap \mathbb{J}'$ and a fortiori $u_n(\theta) = 1$ for any $\theta \in \tilde{F} \cap \mathbb{J}'$. Thus

$v(\theta)=1$ on $\tilde{F}\cap\mathbb{J}'$. Therefore there exists a set $\mathbb{J}''\subset\mathbb{J}$ with $m(\mathbb{J}-\mathbb{J}'')=0$ and $\bar{u}(\theta)=1$ on $\tilde{F}\cap\mathbb{J}''$.

Let $w\in HBD(R)$ and $c=(\sup_{z\in R}|w(z)-w(p)|)^{-1}$. In view of the boundary behavior we have

$$\liminf_{z\in R,\, z\to q}((1-u(z))\pm c(w(z)-w(p)))\geq 0$$

for any $q\in\Delta$. By maximum principle II (see 2 H)

$$c|w(z)-w(p)|\leq 1-u(z)$$

on R. Hence for $\theta\in\tilde{F}\cap\mathbb{J}''$ there exists a sequence $r_n\to 1$ such that

$$\lim_n u(r_n e^{i\theta})=\limsup_{r\to 1} u(r e^{i\theta})=\bar{u}(\theta)=1.$$

Let $z_n=r_n e^{i\theta}$ and let q be an accumulation point of $\{z_n\}_1^\infty$. Then $q\in e_\theta$ and since $c|w(z_n)-w(p)|\leq 1-u(z_n)$ we obtain $w(q)=w(p)$. This is true for all $w\in HBD(R)$ and thus $q\in\Lambda_p$, or $e_\theta\cap\Lambda_p\neq\emptyset$. Hence $\theta\in\tilde{\Lambda}_p$ which implies $\tilde{F}\cap\mathbb{J}''\subset\tilde{\Lambda}_p\subset\tilde{F}$. This shows that $m(\tilde{F}-\tilde{\Lambda}_p)=0$. Since $\tilde{F}_n\to\tilde{F}$, $\underline{m}(\tilde{\Lambda}_p)=\underline{m}(\tilde{F})=\lim_n\underline{m}(\tilde{F}_n)$. In view of $\tilde{U}_{nk}\to\tilde{F}_n$ as $k\to\infty$, $\underline{m}(\tilde{F}_n)=\lim_k\underline{m}(\tilde{U}_{nk})$ $\geq\lim_k\underline{m}(\check{U}_{nk})\geq\lim_k\mu(U_{nk})=\mu(F_n)\geq\mu(\Lambda_p)=\mu(p)$. It follows that $\underline{m}(\tilde{\Lambda}_p)$ $\geq\mu(p)$.

On the other hand since Λ_p is compact Theorem 6 H gives $\overline{m}(\tilde{\Lambda}_p)\leq$ $\mu(\Lambda_p)=\mu(p)$. Thus $\mu(p)\leq\underline{m}(\tilde{\Lambda}_p)\leq\overline{m}(\tilde{\Lambda}_p)\leq\mu(p)$ which shows that $\tilde{\Lambda}_p$ is measurable and $m(\tilde{\Lambda}_p)=\mu(p)$.

6 K. Another Characterization of $U_{\widetilde{HD}}$. In addition to 4 M we give a characterization of the class $U_{\widetilde{HD}}$ using Green's lines:

THEOREM. *In order that $R\notin O_G$ belong to $U_{\widetilde{HD}}$ it is necessary and sufficient that there exist a measurable set $\mathbb{P}\subset\mathbb{J}$ with $m(\mathbb{P})>0$ such that for every $u\in HD(R)$, $u(\theta)=\lim_{r\to 1}u(r e^{i\theta})$ is constant a.e. on $\mathbb{P}$.*

Let $R\in U_{\widetilde{HD}}$. If $R\in O_{HD}$ then our assertion is clear. Thus we may assume that $R\notin O_{HD}$. By Theorem 4 M there exists a $p\in\Gamma$ with $\mu(p)>0$. Since $p\in\Delta$ we can consider the block Λ_p at p. We shall show that $\mathbb{P}=\tilde{\Lambda}_p$ is the required set. Note that $m(\mathbb{P})=m(\tilde{\Lambda}_p)=\mu(p)>0$. Let $u\in HD(R)$. We must show that $u(\theta)$ is constant a.e. on $\mathbb{P}$. We may assume that $u>0$ on R. Let $u_c=u\wedge c\in HD(R)$ for $c>0$, a constant. By Theorem 6 E, $u_c(\theta)=u(\theta)\cap c$ a.e. on $\mathbb{J}$. We also have $u_c(q)=u(q)\cap c$ on Δ (Theorem 4 F). Since $\mu(p)>0$, $u(p)<\infty$. Hence for all $c>u(p)$, $u_c(q)=u(p)$ for $q\in\Lambda_p$.

Let $\{c_n\}_1^\infty$ be a sequence with $u(p)<c_n\to\infty$. Take $\mathbb{J}'\subset\mathbb{J}$ with $m(\mathbb{J}-\mathbb{J}')=0$ such that $u_{c_n}(\theta)$ exists for every $\theta\in\mathbb{J}'$ and $n=1,2,\ldots$. If $\theta\in\mathbb{P}\cap\mathbb{J}'=\tilde{\Lambda}_p\cap\mathbb{J}'$ then u_{c_n} is constant on e_θ and $e_\theta\cap\Lambda_p\neq\emptyset$. Thus $u_{c_n}=u(p)$ on Λ_p and hence $u_{c_n}(\theta)=u(\theta)\cap c_n=u(p)$. We conclude that $u(\theta)=u(p)$, a constant, a.e. on $\mathbb{P}$.

Conversely assume the existence of $\mathbb{P}$ with the property stated in the theorem. Let $T(\mathbb{P})=\{u \mid u \in HD(R), 0<u\leq 1 \text{ on } R, u(\theta)=1 \text{ a.e. on } \mathbb{P}\}$. Take $u, v \in T(\mathbb{P})$. Clearly $0 \leq u \wedge v \leq 1$ on R, and $(u \wedge v)(\theta)=u(\theta) \cap v(\theta)=1$ a.e. on $\mathbb{P}$. Hence $u \wedge v>0$ and $u \wedge v \in T(\mathbb{P})$. It follows that

$$s(z)=\inf\{u(z) \mid u \in T(\mathbb{P})\}$$

belongs to $H(R)$ and there exists a decreasing sequence $\{u_n\}_1^\infty \subset T(\mathbb{P})$ such that $s=\lim_n u_n$ on R. For this reason $s \in \widetilde{HD}(R)$. Since

$$s(z_0)=\lim_n u_n(z_0)=\lim_n \int_0^{2\pi} u_n(\theta)\, dm(\theta) \geq \int_{\mathbb{P}} dm(\theta)=m(\mathbb{P})>0,$$

$s>0$ on R.

We wish to show that s is $\widetilde{HD}$-minimal on R. To this end take an arbitrary $t \in \widetilde{HD}(R)$ with $s \geq t \geq 0$ on R. Let $\{v_n'\}_1^\infty$ be a decreasing sequence of functions in $HD(R)$ such that $\lim_n v_n'=t$. Note that $v_n=u_n \wedge v_n' \in HD(R)$, $v_n \to s \wedge t=t$, and $1 \geq u_n \geq v_n>0$ on R. By the property of $\mathbb{P}$, $v_n(\theta)=c_n$, a constant, a.e. on $\mathbb{P}$. Clearly $0 \leq c_n \leq 1$ and $c=\lim_n c_n$ exists with $0 \leq c \leq 1$. If $c=1$ then $0<v_n \leq 1$ on R and $v_n(\theta)=1$ a.e. on $\mathbb{P}$. Hence $v_n \in T(\mathbb{P})$ and a fortiori $v_n \geq s$. Therefore $t \geq s$ on R, which implies $t=s$.

If $c<1$ then we may assume that $c_n<1$. Observe that $(u_n-v_n)/(1-c_n)$, $((u_n+v_n)/(1+c_n)) \wedge 1 \in T(\mathbb{P})$. Hence $(u_n-v_n)/(1-c_n) \geq s$ and $(u_n+v_n)/(1+c_n) \geq s$ on R. On letting $n \to \infty$ we obtain $(s-t)/(1-c) \geq s$ and $(s+t)/(1+c) \geq s$ on R. These in turn imply $t=cs$ on R. Thus s is $\widetilde{HD}$-minimal, i.e. $R \in U_{\widetilde{HD}}$.

Remark. Using uniformization onto the unit disk U one can embed $HD(R)$ as a subset X of $H(U)$. Constantinescu-Cornea [3] proved that each function in X has radial limits, and that $R \in U_{\widetilde{HD}}$ if and only if all functions in X have constant radial limits on a set of positive measure.

6 L. Second Proof of $U_{\widetilde{HD}} \subset O_{AD}$. Here we give an alternate proof of $U_{\widetilde{HD}} \subset O_{AD}$ (Theorem 5 I) as an application of Green's lines.

Let $R \in U_{\widetilde{HD}}$. By Theorem 6 K there exists a $\mathbb{P} \subset \mathbb{J}$ with positive m-measure such that $u(\theta)=$const. a.e. on $\mathbb{P}$ for every $u \in HD(R)$. Take any $f \in AD(R)$. Then $f(\theta)=$const. a.e. on $\mathbb{P}$. It follows by Theorem 6 I that $f \equiv$const. on R, i.e. $R \in O_{AD}$.

§ 3. Invariance under Deformation

Thus far we have studied the class O_{HD} and related classes using Royden's algebra $\mathbb{M}(R)$ and Royden's compactification R^*. It is interesting to determine which properties of $\mathbb{M}(R)$ and R^* are sufficient for a given R to belong or not belong to O_{HD}.

In 7 we shall see that the algebraic structure of $\mathbf{M}(R)$ characterizes the quasiconformal structure of R, which determines the topology of R^*. In 8 we discuss a topological equivalence of Riemann surfaces which characterizes the topology of R^*. It will be shown that this equivalence actually implies the invariance of O_{HD}.

Quasiconformal invariances or noninvariances are then compiled in 8, the only new classes here being O_G, O_{HD}, and U_{HD}.

7. Algebraic Structure

7 A. Quasiconformal Mappings. Let R_j $(j=1,2)$ be Riemann surfaces. A topological mapping T of R_1 onto R_2 is called *quasiconformal* if the quantity

$$K(T)=\inf\{c \mid c^{-1} \bmod Q \leq \bmod TQ \leq c \bmod Q,\ Q \in \{Q\}\} \tag{103}$$

is finite. Here $\{Q\}$ is the family of quadrilaterals Q in R_1, each Q consisting of a Jordan region Q' and four distinguished points $z_1, \ldots, z_4$ on $\partial Q'$. Map the region Q' conformally onto a rectangle Q'' such that $z_1, \ldots, z_4$ correspond to the four vertices of Q''. Let a and b be the lengths of the sides of Q'' which correspond to $z_1 z_2$ and $z_2 z_3$. The ratio a/b is determined uniquely by Q and is denoted by $\bmod Q$.

The quantity $K(T)$, called the *maximal dilatation* of T, is equal to

$$K(T)=\inf\{c \mid c^{-1} \log \bmod A \leq \log \bmod TA \leq c \log \bmod A,\ A \in \{A\}\} \tag{104}$$

where $\{A\}$ is the family of annuli A on R_1 such that A is contained in a simply connected region D_A in R_1 (Gehring-Väisälä [2], Reich [2]). Thus we may define T to be quasiconformal if the quantity $K(T)$ in (104) is finite. Clearly $K(T^{-1})=K(T)$, and T^{-1} is quasiconformal along with T. Moreover $K(S_1 \circ T \circ S_2)=K(T)$ if the S_j are conformal either directly or indirectly.

A topological mapping T of R_1 onto R_2 is (directly or indirectly) conformal if and only if the maximal dilatation $K(T)=1$ (Ahlfors [13]).

We say that R_1 and R_2 are *quasiconformally equivalent* or that R_1 and R_2 have the same *quasiconformal structure* if there exists a quasiconformal mapping of R_1 onto R_2.

That the above definition of quasiconformality is equivalent to the one given in II.14 in terms of Dirichlet mappings will be seen in 7 D – 7 G.

For a systematic account of the theory of quasiconformal mappings we refer the reader to the monographs of Ahlfors [13] and Künzi [1].

7 B. Annular Functions. Let $A \in \{A\}$ be an annulus contained in a simply connected region D_A of a Riemann surface R. We define a continuous function f_A on R by the conditions $f_A|A'=1$, $f_A|A \in H(A)$, and

$f_A|(R-A\cup A')=0$ where A' is the component of $R-A$ in D_A. We call f_A the *annular function* with *base* A. First we observe:

The annular function f_A *belongs to* $\mathrm{M}(R)$ *and*

$$\log \operatorname{mod} A = \frac{2\pi}{D_R(f_A)}. \tag{105}$$

Moreover if $f\in \mathrm{M}(R)$ *such that* $f=f_A$ *on* $R-A$ *then* $D_R(f_A)\le D_R(f)$.

Let $A_n=\{z|z\in R,\ 1/n<f_A(z)<1-1/n\}$ $(n=3,4,\ldots)$. Then $\bar{A}_n$ is a distinguished compact set in R, and clearly $f_{A_n}\in \mathrm{M}(R)$, $f_A=B\text{-}\lim_n f_{A_n}$ on R. Furthermore $\{f_{A_n}\}$ is D-Cauchy on R since

$$D_R(f_{A_n}-f_{A_{n+p}}, f_{A_{n+p}})=\int_{\partial A_{n+p}} (f_{A_n}-f_{A_{n+p}})*df_{A_{n+p}}=0$$

and thus $D_R(f_{A_n}-f_{A_{n+p}})=D_R(f_{A_n})-D_R(f_{A_{n+p}})$. Therefore $f_A=BD\text{-}\lim_n f_{A_n}$, and the BD-completeness of $\mathrm{M}(R)$ implies $f_A\in \mathrm{M}(R)$. The identity (105) follows from I.1 A.

It is easy to see that $f_A=BD\text{-}\lim_n \pi_{R^*-A_n} f$ on R. Hence $D_R(\pi_{R^*-A_n} f)\le D_R(f)$ implies $D_R(f_A)\le D_R(f)$.

7 C. Algebraic Characterization. We are now in a position to state the main result of No. 7:

THEOREM. *The algebraic structure of Royden's algebra* $\mathrm{M}(R)$ *characterizes the quasiconformal structure of an arbitrary Riemann surface* R.

Explicitly a quasiconformal mapping T *of a Riemann surface* R_1 *onto another* R_2 *induces an algebraic isomorphism* $\sigma: f\to f^\sigma$ *of* $\mathrm{M}(R_1)$ *onto* $\mathrm{M}(R_2)$ *such that*

$$f^\sigma=f\circ T^{-1}. \tag{106}$$

Conversely an algebraic isomorphism $\sigma: f\to f^\sigma$ *of* $\mathrm{M}(R_1)$ *onto* $\mathrm{M}(R_2)$ *induces a quasiconformal mapping* T *of* R_1 *onto* R_2 *for which* (106) *is valid.*

In either case T *can be extended to a homeomorphism* T^* *of Royden's compactification* R_1^* *of* R_1 *onto Royden's compactification* R_2^* *of* R_2 *such that* T^* *preserves the harmonic boundaries* Δ_j *of the* R_j $(j=1,2)$, *i.e.*

$$T^*\Delta_1=\Delta_2. \tag{107}$$

Moreover every $f\in \mathrm{M}(R_1)$ *satisfies*

$$K(T)^{-1}D_{R_1}(f)\le D_{R_2}(f\circ T^{-1})\le K(T)\,D_{R_1}(f) \tag{108}$$

where $K(T)$ *is the maximal dilatation of* T.

The proof will be given in 7 D – 7 G.

7 D. Analytic Properties. Assume that there exists a quasiconformal mapping T of R_1 onto R_2 with maximal dilatation $K(T)$. Then T^{-1} is also a quasiconformal mapping of R_2 onto R_1 with maximal dilatation $K=K(T^{-1})=K(T)$. Let $w=T^{-1}(z)$ be the local representation of T^{-1}, i.e.

$$u=u(z)=u(x,y), \qquad v=v(z)=v(x,y)$$

where $z=x+iy$ and $w=u+iv$ are local parameters in $D_z\subset R_2$ and $D_w\subset R_1$ respectively.

It is readily seen that u and v are Tonelli functions on D_z and totally differentiable a.e. on D_z with

$$\sup_{dz}\frac{|dw|^2}{|dz|^2}\leq KJ(z) \tag{109}$$

at z. Here $w=T^{-1}(z)=u(z)+iv(z)$ is totally differentiable and

$$J(z)=|u_x(z)\,v_y(z)-u_y(z)\,v_x(z)|,$$

the Jacobian of $T^{-1}(z)$ at z, is positive a.e. on D_z (A. Mori [6], Bers [2], Pfluger [7]). Moreover $T^{-1}(z)$ is a measurable mapping and

$$\int_{T^{-1}X} du\,dv=\int_X J(z)\,dx\,dy$$

for every measurable set $X\subset D_z$ (Bers [1], Morrey [1], Jenkins [1]).

7 E. Existence of the Isomorphism. For $f\in \mathrm{M}(R_1)$ define the function f^σ by (106). If $f^\sigma\in \mathrm{M}(R_2)$ then clearly $f\to f^\sigma$ is an algebraic isomorphism of $\mathrm{M}(R_1)$ into $\mathrm{M}(R_2)$, and by reversing the process we see that it is actually onto. Therefore we have only to show that $f^\sigma\in \mathrm{M}(R_2)$.

To achieve this, first take $f\in \mathrm{M}(R_1)\cap C^\infty(R_1)$. Since $f\in C^\infty(R_1)$ it is easy to see that $f^\sigma(z)=f\big(u(x,y),v(x,y)\big)$ satisfies (T.1) and (T.2) in 1 A. Since u and v are totally differentiable a.e. the same is true of $f^\sigma(z)$. Hence

$$f_x^\sigma dx+f_y^\sigma dy=f_u du+f_v dv \tag{110}$$

a.e. Suppose $|\mathrm{grad}\,f^\sigma|\neq 0$ at z where (109) is valid. Choose $dz=dx+idy$ so as to satisfy

$$f_x^\sigma : f_y^\sigma = dx : dy$$

at z. Then by Schwarz's inequality and (110)

$$\begin{aligned}\big((f_x^\sigma)^2+(f_y^\sigma)^2\big)\big((dx)^2+(dy)^2\big)&=(f_x^\sigma dx+f_y^\sigma dy)^2\\&=(f_u du+f_v dv)^2\leq (f_u^2+f_v^2)\big((du)^2+(dv)^2\big).\end{aligned}$$

It follows that

$$|\text{grad}_z f^\sigma|^2 \le |\text{grad}_w f|^2 \frac{|dw|^2}{|dz|^2}.$$

This with (109) gives

$$|\text{grad}_z f^\sigma|^2 \le K\,|\text{grad}_w f|^2 J(z).$$

Therefore

$$D_{R_2}(f^\sigma) \le K D_{R_1}(f) \tag{111}$$

for every $f \in \mathrm{M}(R_1) \cap C^\infty(R_1)$. In particular we conclude that $f^\sigma \in \mathrm{M}(R_2)$.

Next let f be an arbitrary function in $\mathrm{M}(R_1)$. By Theorem 1 D there exists a sequence $\{f_n\}_1^\infty \subset \mathrm{M}(R_1) \cap C^\infty(R_1)$ such that $f = UD\text{-}\lim_n f_n$ on R_1. Clearly $f^\sigma = U\text{-}\lim_n f_n^\sigma$ on R_2. By (111) we infer that $\{f_n^\sigma\}$ is D-Cauchy on R_2 along with $\{f_n\}$. Hence $f^\sigma = UD\text{-}\lim_n f_n^\sigma$. Since $f_n^\sigma \in \mathrm{M}(R_2)$ and $\mathrm{M}(R_2)$ is UD-complete we infer that $f^\sigma \in \mathrm{M}(R_2)$, and (111) is also valid for this f. By reversing the process we obtain (108).

7 F. Existence of a Topological Map. Conversely assume that an algebraic isomorphism $\sigma: f \to f^\sigma$ of $\mathrm{M}(R_1)$ onto $\mathrm{M}(R_2)$ is given. We recall that $\mathrm{M}(R_j)$ is a Banach algebra with the norm

$$\|f\| = \|f\|_{R_j} = \sup_{R_j} |f| + \sqrt{D_{R_j}(f)}$$

for $j = 1, 2$. Therefore the isomorphism σ is automatically bicontinuous (GELFAND [1]; see also RICKART [1], LOOMIS [1]), i.e.

$$1 \le k < \infty \tag{112}$$

where

$$k = k(\sigma) = \inf\{c \mid c^{-1}\|f\| \le \|f^\sigma\| \le c\|f\|,\ f \in \mathrm{M}(R_1)\}. \tag{113}$$

Let $p \in R_2^*$. Since $f \to X_p(f) = f^\sigma(p)$ is clearly a character on $\mathrm{M}(R_1)$, there exists by 2 B a unique point $S^*(p) \in R_1^*$ such that $X_p(f) = f(S^*(p))$ for every $f \in \mathrm{M}(R_1)$. Thus S^* is a mapping of R_2^* into R_1^* with

$$f^\sigma = f \circ S^* \tag{114}$$

for every $f \in \mathrm{M}(R_1)$. From this we deduce that S^* is continuous.

On repeating the same argument for $\tau = \sigma^{-1}$ we obtain a continuous mapping T^* of R_1^* onto R_2^* such that

$$f^\tau = f \circ T^* \tag{115}$$

for every $f \in \mathrm{M}(R_2)$. From (114) and (115) it follows that S^* and T^* are onto mappings and $S^* = (T^*)^{-1}$. Hence in particular S^* and T^* are homeomorphisms between R_1^* and R_2^*.

By Theorem 2 D, $S^*(p) \in R_1$ if and only if $p \in R_2$. Thus $S = S^*|R_2$ is a homeomorphism of R_2 onto R_1. Similarly $T = T^*|R_1$ is a homeomorphism of R_1 onto R_2, and $S = T^{-1}$. In view of (114) we have proved the existence of a topological mapping T of R_1 onto R_2 with property (106).

7 G. Quasiconformality. The proof of Theorem 7 C will be complete if we show that T is quasiconformal. In fact by 7 E, T will then satisfy (108) and the function $f \to f \circ T^{-1}$ will be bicontinuous in the BD-topology. The fact that $f \in \mathrm{M}_0(R_1)$ if and only if $f \circ T^{-1} \in \mathrm{M}_0(R_2)$ implies therefore that $f \in \mathrm{M}_\Delta(R_1)$ if and only if $f \circ T^{-1} \in \mathrm{M}_\Delta(R_2)$. This means that (107) holds.

To prove the quasiconformality of T take an $A \in \{A\}$, i.e. an annulus contained in a simply connected region $D_A \subset R_1$. Let f_A be the annular function with base A. Then $f_A \in \mathrm{M}(R_1)$ and by (113)

$$k^{-1} \| f_A \| \le \| f_A \circ T^{-1} \| \le k \| f_A \| \tag{116}$$

for every $A \in \{A\}$. By 7 B

$$D_{R_2}(f_{TA}) \le D_{R_2}(f_A \circ T^{-1}). \tag{117}$$

For convenience set

$$a = \sup_{\{A\}} \frac{\log \operatorname{mod} A}{\log \operatorname{mod} TA}$$

and

$$a_n = \sup_{\{A\}_n} \frac{\log \operatorname{mod} A}{\log \operatorname{mod} TA}$$

where $\{A\}_n$ is the subset of $\{A\}$ such that $\log \operatorname{mod} A \le 2\pi/n^2$. Clearly

$$a \ge a_1 \ge \cdots \ge a_n \ge \cdots. \tag{118}$$

We shall show that for any fixed number $\varepsilon > 0$ there exists a positive integer m such that

$$a_m < k^2 + \varepsilon. \tag{119}$$

Contrary to the assertion assume that $a_n \ge k^2 + \varepsilon$ $(n = 1, 2, \ldots)$. Then there exist positive numbers b and c such that

$$a_n > b^2 > c^2 > k^2$$

for every n. Let $n > c/(b-c)$. By the definition of a_n there exists an $A_n \in \{A\}_n$ such that

$$b^2 \log \operatorname{mod} TA_n < \log \operatorname{mod} A_n.$$

Since $\log \operatorname{mod} A_n = 2\pi/D_{R_1}(f_{A_n})$, $\log \operatorname{mod} TA_n = 2\pi/D_{R_2}(f_{TA_n})$, and (117) is valid, we infer that

$$\sqrt{D_{R_2}(f_{A_n}\circ T^{-1})} \geq \sqrt{D_{R_2}(f_{TA_n})} > b\sqrt{D_{R_1}(f_{A_n})}.$$

Note that $\sup_{R_1}|f_{A_n}| = \sup_{R_2}|f_{TA_n}| = \sup_{R_2}|f_{A_n}\circ T^{-1}| = 1$. From this and $D_{R_1}(f_{A_n}) \geq n^2$ we see on setting $f^\sigma_{A_n} = (f_{A_n})^\sigma$ that

$$\begin{aligned}\|f^\sigma_{A_n}\|_{R_2} \geq \|f_{TA_n}\|_{R_2} > \sqrt{D_{R_2}(f_{TA_n})} > b\sqrt{D_{R_1}(f_{A_n})} \\ = (b-c)\sqrt{D_{R_1}(f_{A_n})} + c\sqrt{D_{R_1}(f_{A_n})} \\ \geq (b-c)\,n + c\sqrt{D_{R_1}(f_{A_n})} \\ > c + c\sqrt{D_{R_1}(f_{A_n})} = c\,\|f_{A_n}\|_{R_1},\end{aligned}$$

i.e. $\|f^\sigma_{A_n}\|_{R_2} > c\,\|f_{A_n}\|_{R_1}$. On the other hand $\|f^\sigma_{A_n}\|_{R_2} \leq k\,\|f_{A_n}\|_{R_1}$. Hence $k > c$, which contradicts the choice of $c \geq k$. We have established (119).

Next we shall prove that

$$\log \operatorname{mod} A \leq a_m \log \operatorname{mod} TA \tag{120}$$

for every $A \in \{A\}$ where m is the integer of (119).

Let n satisfy

$$2^{-n} \log \operatorname{mod} A < 2\pi/m^2 \tag{121}$$

for an arbitrarily fixed $A \in \{A\}$ and consider a subdivision of A into disjoint annuli $A_1, A_2, \ldots, A_{2^n}$ such that

$$\log \operatorname{mod} A_k = 2^{-n} \log \operatorname{mod} A < 2\pi/m^2$$

for $k = 1, \ldots, 2^n$. Then

$$\log \operatorname{mod} A_k \leq a_m \log \operatorname{mod} TA_k.$$

On observing that $TA_1, \ldots, TA_{2^n}$ is a subdivision of TA we infer that

$$\log \operatorname{mod} A = \sum_{k=1}^{2^n} \log \operatorname{mod} A_k \leq a_m \sum_{k=1}^{2^n} \log \operatorname{mod} TA_k \leq a_m \log \operatorname{mod} TA$$

(cf. I.1 C), and have proved (120).

By (119) and (120) we see that $\log \operatorname{mod} A < (k^2+\varepsilon) \log \operatorname{mod} TA$ for any fixed $A \in \{A\}$. Since $\varepsilon > 0$ is arbitrary we have shown that $a \leq k^2$, i.e. $\log \operatorname{mod} A \leq k^2 \log \operatorname{mod} TA$ for every $A \in \{A\}$.

On reversing the process we also obtain $\log \operatorname{mod} TA \leq k^2 \log \operatorname{mod} A$ for every $A \in \{A\}$. Hence we have proved that

$$K(T) \leq k^2, \tag{122}$$

i.e. T is quasiconformal. The proof of Theorem 7 C is herewith complete.

7 H. Conformal Equivalence. From (108) and (122) it follows that

$$K(T)=\inf\{c^2 | c^{-1}\|f\| \leq \|f\circ T^{-1}\| \leq c\|f\|,\ f\in \mathrm{M}(R_1)\}$$
$$=\inf\{c | c^{-1} D_{R_1}(f) \leq D_{R_2}(f\circ T^{-1}) \leq c\, D_{R_1}(f),\ f\in \mathrm{M}(R_1)\}.$$

As a corollary of Theorem 7 C we obtain for $K(T)=1$:

THEOREM. *The Banach algebraic structure of Royden's algebra* $\mathrm{M}(R)$ *characterizes the (direct or indirect) conformal structure of an arbitrary Riemann surface R.*

More precisely, every isometric isomorphism σ *of* $\mathrm{M}(R_1)$ *onto* $\mathrm{M}(R_2)$ *induces (and is induced by) a conformal mapping* T *of* R_1 *onto* R_2 *satisfying* (106).

8. Topological Structure

8 A. *A*-Sets. Consider two nonempty open sets G_1 and G_2 in R such that for any $z\in\partial G_j$ there exists a parametric disk U about z with the property that $(\partial G_j)\cap\bar{U}$ is a Jordan arc joining two different boundary points of ∂U $(j=1,2)$ and that $G_1\supset\bar{G}_2\cap U$. We call the pair $A=(G_1, G_2)$ an *A-set* in R. We often use the same notation A for the set $G_1-\bar{G}_2$. An A-set generalizes the notion of annulus.

We define the *logarithmic modulus* of an A-set as

$$\log \operatorname{mod} A=\frac{2\pi}{\inf D_A(\varphi)} \tag{123}$$

where φ runs over all continuous functions φ on $\bar{A}\cap R=\bar{G}_1\cap R-G_2$ such that $\varphi|\partial G_j=j$ $(j=1,2)$ and φ is a Tonelli function on A.

By I.1 A and the Dirichlet principle, (123) gives the original logarithmic modulus for an annulus A. Clearly

$$0\leq \log \operatorname{mod} A<\infty$$

where $\log \operatorname{mod} A=0$ can occur only for a noncompact A in R (see Fig. 8).

$$R=\{z|\operatorname{Re} z>0\}$$

8 B. Royden's Mapping. Consider a topological mapping T of a Riemann surface R_1 onto another R_2. For an A-set $A=(G_1, G_2)$ in R_1, $TA\subset R_2$ can be considered as an A-set in R_2 in a natural manner: $TA=(TG_1, TG_2)$. We call T a *Royden mapping* if there exists a constant $K(T, A)\geq 1$ such that

$$K(T, A)^{-1}\log \operatorname{mod} A\leq \log \operatorname{mod} TA\leq K(T, A)\log \operatorname{mod} A \tag{124}$$

for every A-set $A\subset R_1$. Here $K(T, A)$ may vary with A. Thus (124) is equivalent to the condition: $\log \operatorname{mod} TA>0$ if and only if $\log \operatorname{mod} A>0$.

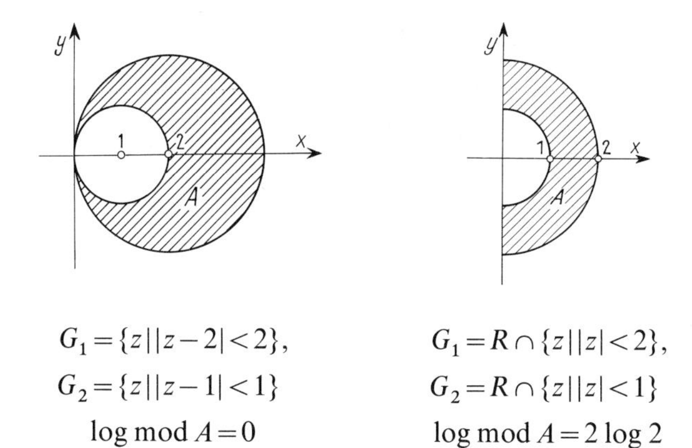

$G_1=\{z \mid |z-2|<2\}$,
$G_2=\{z \mid |z-1|<1\}$
$\log \operatorname{mod} A=0$

$G_1=R\cap\{z \mid |z|<2\}$,
$G_2=R\cap\{z \mid |z|<1\}$
$\log \operatorname{mod} A=2\log 2$

Fig. 8

First observe:

A quasiconformal mapping is a Royden mapping but not conversely.

Suppose that T is a quasiconformal mapping of R_1 onto R_2. Let $\{\varphi\}$ be the totality of competing functions in (123) for TA. Then by 7 D and 7 E, $\{\varphi\circ T\}$ is that for A and $K(T)^{-1}D_A(\varphi\circ T)\le D_{TA}(\varphi)\le K(T)D_A(\varphi\circ T)$ for every $\varphi\in\{\varphi\}$. Thus (124) is valid for T with $K(T,A)=K(T)$.

Next modify T on a relatively compact open set $G\subset R_1$ such that the modified T is not totally differentiable between G and TG at any point. It is easy to see that the modification can be so chosen that the modified T, while not quasiconformal, is still a Royden mapping.

8 C. Topological Characterization. As a counterpart of Theorem 7 C in terms of Royden's compactification instead of Royden's algebra we obtain the following:

THEOREM. *There exists a Royden mapping between Riemann surfaces if and only if Royden's compactifications of these two surfaces are homeomorphic.*

Explicitly a Royden mapping T of a Riemann surface R_1 onto another R_2 can be extended to a homeomorphism T^* of R_1^* onto R_2^*.

Conversely if there exists a homeomorphism T^* of R_1^* onto R_2^* then $T=T^*|R_1$ is a Royden mapping of R_1 onto R_2.

In either case T^* preserves the harmonic boundaries Δ_j of the R_j $(j=1,2)$, i.e.

$$T^*\Delta_1=\Delta_2. \tag{125}$$

The proof will be given in 8 D – 8 G.

8 D. Topological Extension. Suppose that there exists a Royden mapping T of R_1 onto R_2. Let $p \in \Gamma_1 = R_1^* - R_1$. We shall show that the cluster set $C_{R_1}(T, p)$ of T at p is a one-point set, where

$$C_{R_1}(T, p) = \bigcap_{\{U\}} \overline{T(U \cap R_1)} \tag{126}$$

with $\{U\}$ the set of open neighborhoods of p in R_1^*.

Contrary to the assertion assume that (126) contains two distinct points q_1 and q_2 in R_2^*. Clearly these points belong to $\Gamma_2 = R_2^* - R_2$. Take open neighborhoods V_j of the q_j in R_2^* $(j = 1, 2)$ such that $\overline{V}_1 \cap \overline{V}_2 = \emptyset$ and the $\overline{V}_j$ are distinguished compact sets in R_2^*. Let $G_1 = R_2 - \overline{V}_1$ and $G_2 = V_2 \cap R_2$. Then $A = (G_1, G_2)$ is an A-set in R_2. There exists an $f \in \mathrm{MI}(R_2)$ such that $f | \overline{V}_1 = 1$ and $f | \overline{V}_2 = 2$ (cf. 2 C). Since $f | \overline{A}$ belongs to the family of competing functions in (123) and $0 < D_A(f | A) < \infty$ we have

$$\log \operatorname{mod} A > 0. \tag{127}$$

Let $U_j = T^{-1}(V_j \cap R_2) \subset R_1$. Then $T^{-1} G_1 = R_1 - \overline{U}_1$ and $T^{-1} G_2 = U_2$. Since T^{-1} is also a Royden mapping along with T, (124) and (127) imply that

$$\log \operatorname{mod} T^{-1} A > 0. \tag{128}$$

Therefore by (123) there exists a Tonelli function φ on $\overline{T^{-1} A} = \overline{T^{-1} G_1 - T^{-1} G_2} = R_1 - \overline{U}_1 - \overline{U}_2$ continuous on $\overline{T^{-1} A} \cap R_1$ such that $\varphi = 1$ on $\partial(T^{-1} G_1) = \partial U_1$, $\varphi = 2$ on $\partial(T^{-1} G_2) = \partial U_2$, and $D_{T^{-1}A}(\varphi) < \infty$. Thus if we define the function g on R_1 by $g | \overline{T^{-1} A} \cap R_1 = \varphi$, $g | U_1 = 1$, and $g | U_2 = 2$ then clearly $g \in \mathrm{MI}(R_1)$. This implies that g is continuous on R_1^* and $g = j$ on $\overline{U}_j$. Therefore

$$\overline{U}_1 \cap \overline{U}_2 = \emptyset. \tag{129}$$

On the other hand since $q_j \in C_{R_1}(T, p)$ there exists a directed net $\{z_\lambda^j\} \subset R_1$ converging to p such that $\{T z_\lambda^j\} \subset R_2$ converges to q_j. Since V_j is a neighborhood of q_j in R_2^* we may assume that $\{T z_\lambda^j\} \subset V_j \cap R_2$ and consequently $\{z_\lambda^j\} \subset U_j$. Then the limit p of $\{z_\lambda^j\} \subset U_j$ belongs to $\overline{U}_j$, i.e. $p \in \overline{U}_1 \cap \overline{U}_2$, which contradicts (129).

We conclude that the cluster set (126) consists of a single point, which we denote by $T^*(p)$. We set $T^*(z) = T(z)$ for $z \in R_1$.

The mapping T^* of R_1^* into R_2^* thus obtained is continuous. In fact take an open neighborhood V of $T^*(p)$ in R_2^* for $p \in \Gamma_1$. The set (126) consists of the point $T^*(p)$ and hence there exists an open neighborhood U of p such that

$$\overline{T(U \cap R)} \subset V. \tag{130}$$

By the definition of T^*, $T^*(\bar{U}) = \overline{T(U \cap R)}$ and (130) means that $T^*(U) \subset V$, i.e. T^* is continuous.

On repeating the same process for $S = T^{-1}$ we deduce that S^* is continuous. Since $S^* \circ T^*$ and $T^* \circ S^*$ are identities on R_1 and R_2 respectively the same is true on R_1^* and R_2^*. Thus T^* is onto and bicontinuous.

8 E. Restriction. Suppose there exists a homeomorphism T^* of R_1^* onto R_2^*. By Theorem 2 D, $T^* p \in \Gamma_2$ if and only if $p \in \Gamma_1$. Thus $T^* R_1 = R_2$ and $T = T^*|R_1$ is a homeomorphism of R_1 onto R_2.

We have to prove that T is a Royden mapping. For this purpose take an A-set $A = (G_1, G_2)$ in R_1 such that

$$\log \operatorname{mod} A > 0. \tag{131}$$

It suffices to show that

$$\log \operatorname{mod} TA > 0 \tag{132}$$

since the implication of (131) by (132) can be deduced similarly.

The sets $U_1 = R_1^* - \bar{G}_1$ and $U_2 = \bar{G}_2 - \overline{\partial G_2}$ are both open in R_1^*. By the same argument as in 8 D we see that (131) is equivalent to $\bar{U}_1 \cap \bar{U}_2 = \emptyset$. Thus $\overline{(T^* U_1)} \cap \overline{(T^* U_2)} = \emptyset$ and since $T^* U_1 = R_2^* - \overline{T^* G_1}$ and $T^* U_2 = \overline{T^* G_2} - \overline{\partial (T^* G_2)}$, (132) follows for the same reason.

8 F. Boundary Behavior. Let T^* be a homeomorphism of R_1^* onto R_2^*. Then $T = T^*|R_1$ is a Royden mapping of R_1 onto R_2. We are going to prove (125). To this end it suffices to show that $q = T^*(p) \in \Gamma_2 - \Delta_2$ if $p \in \Gamma_1 - \Delta_1$ since the converse can be demonstrated similarly.

Contrary to the assertion suppose that $q = T^*(p) \in \Delta_2$.

Let L be an open set in Γ_1 with $p \in L \subset \bar{L} \subset \Gamma_1 - \Delta_1$. Choose a sequence $\{W_n\}_1^\infty$ of open sets in R_1^* such that

$$W_n \supset \overline{W}_{n+1} \supset W_{n+1} \supset L$$

for every n, $\overline{W}_1 \cap \Delta_1 = \emptyset$, and $\overline{T^* W_n}$ is a distinguished compact set in R_2^*.

Let $f_{np} \in \mathrm{M}(R_1)$ such that $f_{np}|(R_1^* - W_n) = 0$ and $f_{np}|\overline{W}_{n+p} = 1$. It is easy to see that there exists a unique harmonic function u_{np} on $(W_n - \overline{W}_{n+p}) \cap R_1$ such that u_{np} is continuous on $(\overline{W}_n - W_{n+p}) \cap R$, $u_{np}|\partial(R_1 - W_n) = 0$, $u_{np}|\partial W_{n+p} = 1$, and $D_{np} = D_{(W_n - \overline{W}_{n+p}) \cap R_1}(u_{np}) \le D_{(W_n - \overline{W}_{n+p}) \cap R_1}(\varphi)$ for every $\varphi \in \mathrm{M}(R_1)$ with $\varphi = f_{np}$ on R_1^* outside of $W_n - \overline{W}_{n+p}$. By maximum principle II in 2 H, $\lim_p u_{np} = 0$ on $W_n \cap R_1$ and $D_{np} \ge D_{n,\, p+1}$. Thus we can make D_{np} arbitrarily small by taking p sufficiently large.

On choosing a subsequence of $\{W_n\}$ and renumbering we may thus also assume that $\{W_n\}_1^\infty$ satisfies

$$D_{n1} = D_{(W_n - \overline{W}_{n+1}) \cap R_1}(u_{n1}) = \inf_\varphi D_{(W_n - \overline{W}_{n+1}) \cap R_1}(\varphi) < 2^{-n} \tag{133}$$

for every n, where the φ are continuous functions on $(\overline{W}_n - W_{n+1}) \cap R_1$ and Tonelli functions on $(W_n - \overline{W}_{n+1}) \cap R_1$ such that $\varphi|\partial W_n$ and $\varphi|\partial W_{n+1}$ are constants and the difference of these constants is 1.

Set

$$G_1 = \bigcup_{n=0}^{\infty} (W_{4n+1} - \overline{W}_{4n+4}) \cap R_1,$$

$$G_2 = \bigcup_{n=0}^{\infty} (W_{4n+2} - \overline{W}_{4n+3}) \cap R_1,$$

and consider the A-set $A = (G_1, G_2)$. From (133) we conclude that

$$\log \bmod A > 0 \tag{134}$$

and therefore

$$\log \bmod TA > 0. \tag{135}$$

8 G. Boundary Behavior (continued). Let $W_n' = T^* W_n$ and let $\{R_2^m\}_1^\infty$ be a regular exhaustion of R_2. Denote by w_{nm} the harmonic function on $(W_1' - \overline{W}_n') \cap R_2^m$ with boundary values $w_{nm} = 0$ at $\partial W_1' \cap \overline{R_2^m}$, $w_{nm} = 1$ at $\partial W_n' \cap \overline{R_2^m}$, and $*dw_{nm} = 0$ at $\partial R_2^m \cap (W_1' - \overline{W}_n')$. Let $1 \le s < t \le n$ and $\varphi \in \mathrm{M}(R_2)$ such that $\varphi|(R_2 - W_s') = 0$ and $\varphi|\overline{W}_t' = 1$. By Green's formula

$$D_{(W_1' - \overline{W}_n') \cap R_2^m}(\varphi - w_{nm}, w_{nm}) = 0. \tag{136}$$

From this it follows that $\{w_{nm}\}_{m=1}^{\infty}$ is D-Cauchy and we obtain the existence of

$$w_n = \lim_m w_{nm}$$

on $(W_1' - \overline{W}_n') \cap R_2$. Clearly $D_{(W_1' - \overline{W}_n') \cap R_2}(w_n) < \infty$, $w_n \ge 0$, $w_n|\partial W_1' = 0$, $w_n|\partial W_n' = 1$, and $w_n \in H((W_1' - \overline{W}_n') \cap R_2)$. Set $w_n = 0$ on $R_2 - W_1'$ and $w_n = 1$ on $\overline{W}_n' \cap R_2$. Then $w_n \in \mathrm{M}(R_2)$. By (136) we see that

$$D_{(W_1' - \overline{W}_n') \cap R_2}(\varphi - w_n, w_n) = 0. \tag{137}$$

In view of this we infer that $\{w_n\}$ is D-Cauchy and obtain the existence of

$$w = BD\text{-}\lim_n w_n \ge 0$$

on R_2. Since $w_n = 1$ on $L' \cap \Delta_2 \neq \emptyset$ with $L' = T^* L$ we see that $w = 1$ on $L' \cap \Delta_2$. Moreover $w \in H(W_1 \cap R_2)$, $w|(R_2 - W_1) = 0$, $w \ge 0$, and $w \in \mathrm{M}(R_2)$. From (137) it follows that

$$D_{R_2}(\varphi - w, w) = 0. \tag{138}$$

In particular we obtain

$$0 < D_{R_2}(w) \le D_{R_2}(\varphi). \tag{139}$$

Let φ be an arbitrary competing function in (123) for TA. We extend φ by setting $\varphi=1$ on $R_2-\overline{G_1'}$ with $G_1'=T^*G_1$, and $\varphi=2$ on $G_2'=T^*G_2$. Let $\varphi_n=\varphi$ on $(W_{2n-1}'-\overline{W}_{2n}')\cap R_2$, $\varphi_n=1$ on R_2-W_{2n-1}', and $\varphi_n=2$ on $\overline{W}_{2n}'\cap R_2$, $n=1,2,\ldots$. Then

$$D_{TA}(\varphi)=\sum_1^\infty D(\varphi_n). \tag{140}$$

For odd n, φ_n-1 qualifies as φ in (139), and for even n the same is true of $2-\varphi_n$. Hence we obtain $D(\varphi_n)\geq D(w)>0$ for every n and (140) implies that $D_{TA}(\varphi)=\infty$. Therefore log mod $TA=0$, which contradicts (135).

The proof of Theorem 8 C is herewith complete.

8 H. Invariance of O_G, O_{HD}, O_{HD}^n, and U_{HD}. By 2 and 3 the properties $R\in O_G, O_{HD}, O_{HD}^n$ $(1\leq n\leq\infty)$, and U_{HD} are characterized by the set-theoretic and topological structure of Δ. Thus we obtain by Theorem 8 C:

THEOREM. *Classes O_G, O_{HD}, O_{HD}^n, and U_{HD} are invariant under Royden's mappings.*

8 I. Boundary Property. Assume that two Riemann surfaces R_1 and R_2 have conformally equivalent ideal boundary neighborhoods R_1' and R_2'. By 5 we see that the harmonic boundaries Δ_i of the R_i are homeomorphic in such a way that the positiveness of the harmonic measures of subsets of the Δ_i is preserved. We conclude by 2, 3, and 4:

THEOREM. *Belonging to any one of the classes O_G, O_{HD}, O_{HD}^n, U_{HD}, and $U_{\widetilde{HD}}$ is a property of the ideal boundary.*

Chapter IV

Other Classes of Harmonic Functions

From the class of Dirichlet finite harmonic functions we proceed to other classes of harmonic functions related to certain boundedness properties. Two central ones are boundedness in absolute value and positiveness. Two derived boundedness properties, quasiboundedness and essential positiveness, will also be considered. These fall into the general category of Φ-boundedness.

The class HD was studied on the basis of Dirichlet's principle and its generalized version, the theory of Royden's compactification. In contrast the study of the classes mentioned above depends more or less on the Perron-Wiener-Brelot method. Accordingly the chapter opens with an introduction to an adaption of the method for general Riemann surfaces, i.e. the theory of Wiener's compactification. This is developed in § 1 largely as in the case of Royden's compactification. The new classes O_{HP} and O_{HB} are also introduced and the strict inclusion $O_{HP} < O_{HB}$ is stated; its proof will be based on the relation $O_G < O_{HP}$ to be established in V.7 C.

In § 2 we discuss the behavior at Wiener's boundary of harmonic functions "bounded" in the above sense. In particular the class $H\Phi$ of Φ-bounded harmonic functions and the corresponding null class $O_{H\Phi}$ are studied in detail.

We then take up in § 3 a relation between "bounded" harmonic functions and "bounded" meromorphic functions.

The chapter closes in § 4 with the invariance of null classes under a deformation of surfaces which is closely related to the structure of Wiener's compactification.

§ 1. Wiener's Compactification

In the Perron-Wiener-Brelot method essential use is made of the family of superharmonic functions. The importance of these generating functions lies in that they are "harmonizable." As a counterpart of Royden's algebra Wiener's algebra consisting of bounded continuous harmonizable functions will be considered in 1. Based on this algebra we construct in 2 Wiener's compactification in analogy with Royden's compactification. The fact that Tonelli functions are harmonizable implies that Wiener's compactification is a fiber space of Royden's compactification.

The classes HB of bounded harmonic functions, HP of nonnegative harmonic functions, HB' of quasibounded harmonic functions, and HP' of essentially positive harmonic functions are then introduced in 3. Among the corresponding null classes the important proper inclusion $O_{HP}<O_{HB}$ is given. A decomposition theorem, a counterpart of the orthogonal decomposition of Tonelli functions, is then proved for continuous functions on Wiener's compactification. As applications we discuss HB-minimal functions, in particular tests for U_{HB} and O_{HB}.

1. Wiener's Algebra

1 A. Harmonizable Functions. A subregion G of a Riemann surface R is, by definition, *normal* if ∂G is regular for the Dirichlet problem. Let f be a real-valued function on R. Take a normal subregion $G \notin O_G$ and denote by $\bar{U}(G,f)$ (resp. $\underline{U}(G,f)$) the class of superharmonic (resp. subharmonic) functions s on G for which there exists a compact set $K_s \subset G$ with $s \geq f$ (resp. $s \leq f$) on $G-K_s$. If neither $\bar{U}(G,f)$ nor $\underline{U}(G,f)$ is empty then these classes are Perron families and consequently

$$\begin{aligned} \overline{W}_f^G(z) &= \inf_{s \in \bar{U}(G,f)} s(z), \\ \underline{W}_f^G(z) &= \sup_{s \in \underline{U}(G,f)} s(z) \end{aligned} \tag{1}$$

are harmonic on G and satisfy $\overline{W}_f^G \geq \underline{W}_f^G$ there. If these two functions coincide then we denote the common function by W_f^G.

The function f on R is called *harmonizable* on R if W_f^G exists for every normal subregion $G \notin O_G$. A complex-valued function f is called harmonizable if $\operatorname{Re} f$ and $\operatorname{Im} f$ are harmonizable.

Every function in $H(R)$ is harmonizable. The same is true of $f \in A(R)$. It is also easily seen that nonnegative superharmonic functions are harmonizable and hence so are bounded superharmonic functions and bounded subharmonic functions.

1 B. Definition of Wiener's Algebra. Consider the family $\mathbb{N}(R)$ of real- (or complex-) valued functions f on R satisfying the following conditions:

($\mathbb{N}$.1) *f is bounded on R,*

($\mathbb{N}$.2) *f is continuous on R,*

($\mathbb{N}$.3) *f is harmonizable on R.*

We call $\mathbb{N}(R)$ the real (or complex) *Wiener algebra* associated with R. That $\mathbb{N}(R)$ actually forms an algebra will be seen in 1 D.

When does $\mathbb{N}(R)$ coincide with the family $B(R)$ of bounded continuous functions on R? We shall show:

THEOREM. *The identity $\mathbb{N}(R)=B(R)$ is valid if and only if $R\in O_G$.*

Assume that $R\notin O_G$. Let $\{R_n\}_1^\infty$ be a regular exhaustion of R and let $f\in B(R)$ be such that $0\le f\le 1$, $f|\bigcup_1^\infty(\bar{R}_{4n-2}-R_{4n-3})=1$, and $f|\bigcup_1^\infty(\bar{R}_{4n}-R_{4n-1})=0$. For an arbitrary $s\in\bar{U}(R,f)$, $s\ge f$ on $\bar{R}_{4(n+p)-2}-R_{4n-2}$ for some n and all $p>0$. Since $s\ge f=1$ on $\partial(R_{4(n+p)-2}-\bar{R}_{4n-2})$ we conclude that $s\ge 1$ on R. For an $s\in\underline{U}(R,f)$ we similarly deduce by considering $\bar{R}_{4(n+p)}-R_{4n}$ that $s\le 0$ on R. Thus $\underline{W}_f^R\le 0<1\le\bar{W}_f^R$, i.e. $f\notin\mathbb{N}(R)$ and therefore $\mathbb{N}(R)\neq B(R)$.

Next suppose that $R\in O_G$. Take an $f\in B(R)$. We have only to show that $f\in\mathbb{N}(R)$ since $\mathbb{N}(R)\subset B(R)$. To this end we may clearly assume that R is open and $f\ge 0$. Consider a normal subregion $G\notin O_G$. To show the existence of W_f^G we may suppose that $G\cup\partial G$ is noncompact. Let v_n be a continuous function on $G\cup\partial G-(\partial G)\cap\partial R_n$ with $v_n|R_n\cap\partial G=0$, $v_n|G-\bar{R}_n=1$, and $v_n|G\cap R_n\in H(G\cap R_n)$. Clearly $0<v_{n+p}\le v_n\le 1$ on G and $v_0=B\text{-}\lim_n v_n$ exists on G with $v_0=0$ at ∂G. Since $R\in O_G$, $v_0\equiv 0$ on G. Therefore by choosing a suitable subsequence of $\{R_n\}_1^\infty$ we may assume that

$$v(z)=\sum_1^\infty v_n(z)$$

converges on G. Evidently v is superharmonic and $v>0$ on G. Moreover

$$v(z)\ge n \tag{2}$$

on $G-\bar{R}_n$.

Let u_n be the continuous function on $G\cup\partial G-(\partial G)\cap\partial R_n$ with $u_n|R_n\cap\partial G=f$, $u_n|G-\bar{R}_n=0$, and $u_n|G\cap R_n\in H(G\cap R_n)$. Clearly $0\le u_n\le u_{n+p}\le\sup_G f$ on G and $u=B\text{-}\lim_n u_n$ exists on G with $u=f$ at ∂G. For every $\varepsilon>0$ it follows from (2) that $u+\varepsilon v+\varepsilon\in\bar{U}(G,f)$ and $u-\varepsilon v-\varepsilon\in\underline{U}(G,f)$. Therefore

$$u+\varepsilon v+\varepsilon\ge\bar{W}_f^G\ge\underline{W}_f^G\ge u-\varepsilon v-\varepsilon.$$

On letting $\varepsilon\to 0$ we conclude that W_f^G exists.

1 C. Potential Subalgebra. For $f\in\mathbb{N}(R)$, W_f^R has meaning only if $R\notin O_G$. *If $R\in O_G$ then we define* $W_f^R=0$. For an arbitrary R we set

$$\mathbb{N}_\Delta(R)=\{f\,|\,f\in\mathbb{N}(R),\ W_f^R=0\}$$

and call it the *potential subalgebra* of $\mathbb{N}(R)$. It will be seen in 1 D that $\mathbb{N}_\Delta(R)$ is an ideal of $\mathbb{N}(R)$. As the first and actually the essential step to this end we prove:

The potential subalgebra $\mathbb{N}_\Delta(R)$ *is a vector space such that* $\mathbb{N}(R)\cdot\mathbb{N}_\Delta(R)=\mathbb{N}_\Delta(R)$. *Moreover for every* $f\in\mathbb{N}(R)$, $f-W_f^R\in\mathbb{N}_\Delta(R)$.

If $R\in O_G$ then by Theorem 1 B and the definition $W_f^R=0$ for every $f\in\mathbb{N}(R)$, $\mathbb{N}(R)=\mathbb{N}_\Delta(R)=B(R)$. Therefore the above assertion is trivial and we have only to deal with the case $R\notin O_G$.

Let f and g be in $\mathbb{N}(R)$, and $s\in\overline{U}(R,f)$, $s'\in\overline{U}(R,g)$. Clearly

$$s+s'\in\overline{U}(R,f+g)\quad\text{and}\quad \overline{W}^R_{f+g}\le W_f^R+W_g^R.$$

Similarly we see that $\underline{W}^R_{f+g}\ge W_f^R+W_g^R$. Thus W^R_{f+g} exists and

$$W^R_{f+g}=W_f^R+W_g^R. \tag{3}$$

This is also true for every normal subregion $G\notin O_G$, and therefore $f+g\in\mathbb{N}(R)$. In particular if $f,g\in\mathbb{N}_\Delta(R)$ then (3) implies that $W^R_{f+g}=0$, i.e. $f+g\in\mathbb{N}_\Delta(R)$.

Again let $f\in\mathbb{N}(R)$ and take a real number a. If $a>0$ then $as\in\overline{U}(G,af)$ for every $s\in\overline{U}(G,f)$ and vice versa. Therefore $af\in\mathbb{N}(R)$ and

$$W^R_{af}=a\,W_f^R. \tag{4}$$

It is clear that $W^R_{-f}=-W_f^R$. Thus (4) remains valid for $a<0$ and trivially for $a=0$. In particular $f\in\mathbb{N}_\Delta(R)$ implies $af\in\mathbb{N}_\Delta(R)$. We have proved that $\mathbb{N}(R)$ and $\mathbb{N}_\Delta(R)$ are vector spaces.

It is obvious that $\mathbb{N}(R)\cdot\mathbb{N}_\Delta(R)\supset\mathbb{N}_\Delta(R)$ since $1\in\mathbb{N}(R)$. Let $g\in\mathbb{N}(R)$ and $f\in\mathbb{N}_\Delta(R)$. We have to show that $gf\in\mathbb{N}_\Delta(R)$. Set $c=\inf_R g$. Since $(g-c)f=gf-cf$ and $cf\in\mathbb{N}_\Delta(R)$, $(g-c)f\in\mathbb{N}_\Delta(R)$ implies $gf\in\mathbb{N}_\Delta(R)$. Hence we can assume that $g\ge 0$ on R. Set $k=\sup_R g$. For arbitrary $s\in\overline{U}(R,f)$ and $s'\in\underline{U}(R,f)$

$$s'\le\underline{W}_f^R=W_f^R=\overline{W}_f^R\le s$$

and $W_f^R=0$. Thus $s'\le 0\le s$ on R. Therefore

$$k\,s'\le gf\le k\,s$$

on $R-K_s\cup K_{s'}$ and thus $k\,s\in\overline{U}(R,gf)$ and $k\,s'\in\underline{U}(R,gf)$. As a consequence

$$k\,\underline{W}_f^R\le\underline{W}^R_{gf}\le\overline{W}^R_{gf}\le k\,\overline{W}_f^R,$$

and $\underline{W}_f^R=\overline{W}_f^R=W_f^R=0$ implies that W^R_{gf} exists and equals zero, i.e. $gf\in\mathbb{N}_\Delta(R)$.

Finally take $f\in\mathbb{N}(R)$ and set $g=f-W_f^R$. Let $s\in\overline{U}(R,g)$ and $s'\in\underline{U}(R,g)$. Clearly $s+W_f^R\in\overline{U}(R,f)$ and $s'+W_f^R\in\underline{U}(R,f)$, and the converse is also true. Thus

$$\overline{W}_f^R = \inf_{s \in \bar{U}(R,g)} (s + W_f^R) = \inf_{s \in \bar{U}(R,g)} s + W_f^R = \overline{W}_g^R + W_f^R$$

and similarly

$$\underline{W}_f^R = \underline{W}_g^R + W_f^R.$$

Since $\overline{W}_f^R = \underline{W}_f^R = W_f^R$ we conclude that $\overline{W}_g^R = \underline{W}_g^R = W_g^R = 0$, i.e. $g \in \mathbb{N}_\Delta(R)$.

1 D. Properties of $\mathbb{N}(R)$. We are now able to show that Wiener's algebra $\mathbb{N}(R)$ is indeed an algebra, as anticipated in 1 B.

THEOREM. *Wiener's algebra $\mathbb{N}(R)$ forms an algebra with respect to the usual addition, multiplication, and scalar multiplication of functions. The potential subalgebra $\mathbb{N}_\Delta(R)$ is an ideal of $\mathbb{N}(R)$.*

Both $\mathbb{N}(R)$ and $\mathbb{N}_\Delta(R)$ are closed under the operations $\cup$ and $\cap$. Moreover $\mathbb{N}(R)$ is a Banach algebra under the supnorm

$$\|f\|_\infty = \sup_{z \in R} |f(z)|,$$

and $\mathbb{N}_\Delta(R)$ is a closed subset of $\mathbb{N}(R)$.

In 1 C we saw that $\mathbb{N}(R)$ is a vector space. Therefore it remains to prove the closedness of $\mathbb{N}(R)$ under multiplication. As the first step we shall show that $f^2 \in \mathbb{N}(R)$ if $f \in \mathbb{N}(R)$. For simplicity we set $u = W_f^R$ and $g = f - u$. By 1 C, $g \in \mathbb{N}_\Delta(R)$ and $(2u+g)g \in \mathbb{N}_\Delta(R)$. Moreover $f^2 = u^2 + (2u+g)g$, where $u^2 \in \mathbb{N}(R)$ because of the boundedness and subharmonicity of u^2.

Next let $f, g \in \mathbb{N}(R)$. Since f^2, g^2, and $(f+g)^2 \in \mathbb{N}(R)$, $fg = \frac{1}{2}(f+g)^2 - \frac{1}{2}(f^2 + g^2)$ belongs to $\mathbb{N}(R)$. Thus $\mathbb{N}(R)$ is an algebra. It follows from 1 C that $\mathbb{N}_\Delta(R)$ is an ideal of $\mathbb{N}(R)$.

The remainder of the proof will be given in 1 E – 1 F.

1 E. Completeness. In addition to (3) and (4) we can easily see from the definition that

$$W_f^G \leq W_g^G \tag{5}$$

on G for every normal $G \subset R$ with $G \notin O_G$ and for all $f, g \in \mathbb{N}(R)$ with $f \leq g$ on R.

Suppose that $\{f_n\} \subset \mathbb{N}(R)$ is U-Cauchy. Then $f = U\text{-}\lim_n f_n$ belongs to $B(R)$. Clearly

$$W_{f_n}^G - \|f - f_n\|_\infty \leq \underline{W}_f^G \leq \overline{W}_f^G \leq W_{f_n}^G + \|f - f_n\|_\infty.$$

Since $\|f - f_n\|_\infty \to 0$ we deduce that $\underline{W}_f^G = \overline{W}_f^G$ for every normal $G \notin O_G$. Thus $f \in \mathbb{N}(R)$, i.e. $\mathbb{N}(R)$ is U-complete. We also have

$$\|W_f^G - W_g^G\|_\infty \leq \|f - g\|_\infty \tag{6}$$

and in particular $\|W_f^R - W_{f_n}^R\|_\infty \leq \|f - f_n\|_\infty$. If $W_{f_n}^R = 0$ then $W_f^R = 0$ follows from this inequality on letting $n \to \infty$, i.e. $\mathbb{N}_\Delta(R)$ is U-closed in $\mathbb{N}(R)$.

From the proof we also conclude:

COROLLARY. *If* $\{f_n\} \subset \mathbb{N}(R)$ *and* $f = U\text{-}\lim_n f_n$ *on* R *then* $f \in \mathbb{N}(R)$ *and* $W_f^R = U\text{-}\lim_n W_{f_n}^R$ *on* R.

1 F. Lattice. The lattice property of $\mathbb{N}$ can be deduced directly (cf. 2 H). Here we present a proof using a lesser property of $\mathbb{N}$.

Recall that the Taylor expansion

$$\sqrt{1-x} = 1 - \tfrac{1}{2}x - \tfrac{1}{2}\sum_{k=1}^{\infty} \frac{1 \cdot 3 \cdot 5 \cdot \cdots \cdot (2k-1)}{2^k (k+1)!} x^{k+1}$$

is uniformly convergent for $0 \leq x \leq 1$. Hence if we set

$$p_n(x) = 1 - \tfrac{1}{2}x - \tfrac{1}{2}\sum_{k=1}^{n} \frac{1 \cdot 3 \cdot 5 \cdot \cdots \cdot (2k-1)}{2^k (k+1)!} x^{k+1}$$

then $\sqrt{1-x} = U\text{-}\lim_n p_n(x)$ on $[0, 1]$.

Let $f \in \mathbb{N}(R)$ (resp. $\mathbb{N}_\Delta(R)$) with $\|f\|_\infty < 1$. We have $1 - f^2(z) \in [0, 1]$ on R and $|f(z)| = \sqrt{1 - (1 - f^2(z))} = U\text{-}\lim_n p_n(1 - f^2(z))$ on R. Since $p_n(1 - f^2(z))$ is a polynomial in $f(z)$, $p_n(1 - f^2(z)) \in \mathbb{N}(R)$ (resp. $\mathbb{N}_\Delta(R)$). By closedness under the U-limit, $|f| \in \mathbb{N}(R)$ (resp. $\mathbb{N}_\Delta(R)$).

Note that

$$f \cup g = \max(f, g) = \frac{(f+g) + |f-g|}{2},$$

$$f \cap g = \min(f, g) = \frac{(f+g) - |f-g|}{2}.$$

Therefore $f \cup g$, $f \cap g \in \mathbb{N}(R)$ (resp. $\mathbb{N}_\Delta(R)$) along with $f, g \in \mathbb{N}(R)$ (resp. $\mathbb{N}_\Delta(R)$).

The proof of Theorem 1 D is herewith complete.

1 G. The Inclusion $\mathbb{M}(R) \subset \mathbb{N}(R)$. At this point we state the relation of Wiener's algebra to Royden's algebra:

THEOREM. *Royden's algebra* $\mathbb{M}(R)$ *is a subalgebra of Wiener's algebra* $\mathbb{N}(R)$, *i.e.* $\mathbb{M}(R) \subset \mathbb{N}(R)$. *For potential subalgebras the relation* $\mathbb{M}_\Delta(R) = \mathbb{M}(R) \cap \mathbb{N}_\Delta(R)$ *is valid.*

Take an $f \in \mathbb{M}(G)$ with $G \notin O_G$ and let $\overline{\mathfrak{A}}(G^*, f)$ and $\underline{\mathfrak{A}}(G^*, f)$ be the corresponding Perron families of superharmonic and subharmonic functions (cf. III.5 A). Clearly $\overline{U}(G,f)\subset\overline{\mathfrak{A}}(G^*,f)$, $\underline{U}(G,f)\subset\underline{\mathfrak{A}}(G^*,f)$, and $s+\varepsilon\in\overline{U}(G,f)$ (resp. $s-\varepsilon\in\underline{U}(G,f)$) for every $s\in\overline{\mathfrak{A}}(G^*,f)$ (resp. $\underline{\mathfrak{A}}(G^*,f)$) and $\varepsilon>0$. Therefore $\overline{H}(z;G^*,f)=\overline{W}_f^G(z)$ and $\underline{H}(z;G^*,f)=\underline{W}_f^G(z)$. By III.5 A, $\overline{H}=\underline{H}$ and thus $\overline{W}_f^G=\underline{W}_f^G$.

If $f\in\mathbb{M}(R)$ then f can be considered to belong to $\mathbb{M}(G)$ for every subregion $G\notin O_G$ and $\overline{W}_f^G=\underline{W}_f^G$. Hence $f\in\mathbb{N}(R)$, i.e. $\mathbb{M}(R)\subset\mathbb{N}(R)$.

If $R\in O_G$ then $\mathbb{M}_\Delta(R)=\mathbb{M}(R)=\mathbb{M}(R)\cap\mathbb{N}(R)=\mathbb{M}(R)\cap\mathbb{N}_\Delta(R)$. Next suppose $R\notin O_G$. For $f\in\mathbb{M}(R)$ we have seen above that

$$H(z;R^*,f)=W_f^R(z). \tag{7}$$

If $f\in\mathbb{M}_\Delta(R)$ then $H(z;R^*,f)=\int_\Delta P(z,p)\,f(p)\,d\mu(p)=0$ and we infer by (7) that $f\in\mathbb{N}_\Delta(R)$. Conversely $f\in\mathbb{M}(R)\cap\mathbb{N}_\Delta(R)$ and (7) imply $H(z;R^*,f)=0$. Since $H(q;R^*,f)=f(q)$ on Δ we have $f=0$ on Δ, i.e. $f\in\mathbb{M}_\Delta(R)$ by III.2 J.

Remark. Wiener's algebra $\mathbb{N}(R)$ and its potential subalgebra $\mathbb{N}_\Delta(R)$ were introduced by S. Mori [3, 4]. The concept of harmonizability is due to Constantinescu-Cornea [17].

2. Wiener's Compactification

2 A. Definition of Wiener's Compactification. Let R be a Riemann surface. Consider a topological space $R_{\mathbb{N}}^*$ with the following properties:

($R_{\mathbb{N}}^*$.1) *$R_{\mathbb{N}}^*$ is a compact Hausdorff space,*
($R_{\mathbb{N}}^*$.2) *R is an open dense subspace of $R_{\mathbb{N}}^*$,*
($R_{\mathbb{N}}^*$.3) *every function in $\mathbb{N}(R)$ can be continuously extended to $R_{\mathbb{N}}^*$, i.e. $\mathbb{N}(R)\subset B(R_{\mathbb{N}}^*)$,*
($R_{\mathbb{N}}^*$.4) *$\mathbb{N}(R)$ separates points in $R_{\mathbb{N}}^*$.*

If there is no ambiguity we shall often simply write R^* instead of $R_{\mathbb{N}}^*$.

2 B. Characters. Recall that a character X on an algebra with unit 1 is a multiplicative linear functional X such that $X(1)=1$ (cf. III.2 B). By considering the space of characters on $\mathbb{N}(R)$ with the weak* topology we can prove in the same manner as in III.2 B:

Theorem. *There exists a unique Wiener compactification $R_{\mathbb{N}}^*$ for every Riemann surface R.*

Again as in III.2 B we conclude:

Corollary. *A point $p\in R_{\mathbb{N}}^*$ gives a character X_p by $X_p(f)=f(p)$. Conversely for every character X on $\mathbb{N}(R)$ there exists a unique point $p\in R_{\mathbb{N}}^*$ such that $X=X_p$.*

Wiener's compactification was introduced by S. MORI [3], HAYASHI [2], KUSUNOKI [7], and CONSTANTINESCU-CORNEA [17].

2 C. The Identity $\mathbb{N}(R) = B(R_{\mathbb{N}}^*)$. Royden's algebra $\mathbb{M}(R)$ almost exhausts $B(R_{\mathbb{M}}^*)$ in the sense that Urysohn's property is valid for $\mathbb{M}(R)$ as we saw in III.2 C. However $\mathbb{M}(R) < B(R_{\mathbb{M}}^*)$. The first essential distinction between $\mathbb{M}$ and $\mathbb{N}$ lies in the following convenient relation:

THEOREM. *Wiener's algebra has the property* $\mathbb{N}(R) = B(R_{\mathbb{N}}^*)$.

Since $\mathbb{N}(R)$ is a subalgebra of $B(R_{\mathbb{N}}^*)$ which is U-closed, separates points in $R_{\mathbb{N}}^*$, and contains 1, we obtain the asserted identity as an immediate consequence of the Stone-Weierstrass theorem.

2 D. Čech Compactification. The Čech compactification $\check{R}$ of R is defined by replacing $R_{\mathbb{N}}^*$ and $\mathbb{N}(R)$ by $\check{R}$ and $B(R)$ respectively in $(R_{\mathbb{N}}^*.1)-(R_{\mathbb{N}}^*.4)$. Its unique existence is shown exactly as in III.2 B. As a counterpart of Theorem 1 B we see at once:

THEOREM. *Wiener's compactification of R coincides with the Čech compactification if and only if $R \in O_G$.*

2 E. Wiener's Boundary. The compact set $\Gamma_{\mathbb{N}}(R) = R_{\mathbb{N}}^* - R$ is called *Wiener's boundary* of R. We sometimes write $\Gamma(R)$ or Γ if no confusion is to be feared. To be precise we also use $\Gamma_{\mathbb{M}}(R) = \Gamma_{\mathbb{M}}$ for Royden's boundary.

Consider the set

$$\Delta_{\mathbb{N}}(R) = \Delta_{\mathbb{N}} = \{p \in R_{\mathbb{N}}^* | f(p) = 0 \text{ for every } f \in \mathbb{N}_{\Delta}(R)\}. \tag{8}$$

Clearly the subalgebra $\mathbb{N}_0(R)$ of $\mathbb{N}(R)$ consisting of functions with compact supports in R is contained in $\mathbb{N}_{\Delta}(R)$. Hence $\Delta = \Delta_{\mathbb{N}}$ is a compact subset of $\Gamma_{\mathbb{N}}$. We call $\Delta_{\mathbb{N}}$ *Wiener's harmonic boundary* of R. We also use the notation $\Delta_{\mathbb{M}}$ for Royden's harmonic boundary.

THEOREM. *Wiener's potential subalgebra has the duality property*

$$\mathbb{N}_{\Delta}(R) = \{f \in \mathbb{N}(R) | f(\Delta_{\mathbb{N}}) = 0\}. \tag{9}$$

Here we understand that $f(\emptyset) = 0$ for every function f.

We have only to show that if $f \in \mathbb{N}(R)$ and $f(\Delta_{\mathbb{N}}) = 0$ then $f \in \mathbb{N}_{\Delta}(R)$. We may assume that $f \geq 0$. Take the neighborhood $U_n = \{p \in R_{\mathbb{N}}^* | f(p) < 1/n\}$ of $\Delta_{\mathbb{N}}$. Let $f_n \in \mathbb{N}(R) = B(R_{\mathbb{N}}^*)$ such that $f_n | R_{\mathbb{N}}^* - U_n = f - 1/n$ and $f_n | U_n = 0$, i.e. $f_n = (f - 1/n) \cup 0$. Clearly $f = U\text{-}\lim_n f_n$ on $R_{\mathbb{N}}^*$. If $f_n \in \mathbb{N}_{\Delta}(R)$ then by the U-closedness of $\mathbb{N}_{\Delta}(R)$, $f \in \mathbb{N}_{\Delta}(R)$.

Thus we may assume that $f=0$ in a neighborhood U of Δ_{N}. Let V be an open neighborhood of Δ_{N} such that $\bar{V}\subset U$. For each $p\in R^*_{\mathrm{N}}-V$ we can find by (8) a function $f_p\in \mathrm{N}_\Delta(R)$ such that $f_p(p)>1$. Since $f_p^2\in \mathrm{N}_\Delta(R)$ we may assume that $f_p\geq 0$ on R^*_{N}. Since $\bigcup_{p\in R^*_{\mathrm{N}}-V}\{q|f_p(q)>1\}$ covers $R^*_{\mathrm{N}}-V$, a compact set, there exists a finite subcovering $\bigcup_1^n\{q|f_{p_k}(q)>1\}$. Thus $g=\sum_1^n f_{p_k}\in \mathrm{N}_\Delta(R)$ and $g>1$ on $R^*_{\mathrm{N}}-V$. Since $\mathrm{N}(R)$ is a vector lattice $\varphi=g\cap 1\in \mathrm{N}_\Delta(R)$. Because of $\varphi|R^*_{\mathrm{N}}-V=1$, $f\varphi=f$ on R^*_{N} and we infer that $\varphi\in \mathrm{N}_\Delta(R)$ implies $f=f\varphi\in \mathrm{N}_\Delta(R)$.

COROLLARY. *A surface $R\in O_G$ if and only if $\Delta_{\mathrm{N}}=\emptyset$.*

Assume that $R\in O_G$. By 1 C, $\mathrm{N}_\Delta(R)=\mathrm{N}(R)$. Thus the existence of the function 1 in $\mathrm{N}_\Delta(R)$ assures in (8) that $\Delta_{\mathrm{N}}=\emptyset$. Conversely if $\Delta_{\mathrm{N}}=\emptyset$ then it follows from (9) that $\mathrm{N}_\Delta(R)=\mathrm{N}(R)$. We deduce from Theorem 1 G that $\mathrm{M}_\Delta(R)=\mathrm{M}(R)$, i.e. $\Delta_{\mathrm{M}}=\emptyset$. Thus by III.2 F, $R\in O_G$.

2 F. The Fiber Space (R^*_{N}, R^*_{M}, ρ). Since $\mathrm{M}\subset\mathrm{N}$, $\mathrm{M}(R)\subset B(R^*_{\mathrm{N}})$. Therefore if we identify points in R^*_{N} which are not separated by $\mathrm{M}(R)$ and introduce a suitable topology in this quotient space it gives rise to Royden's compactification R^*_{M}, and the natural map gives a continuous mapping of R^*_{N} onto R^*_{M}. We state this in a more comprehensive form:

THEOREM. *There exists a unique continuous mapping ρ of Wiener's compactification R^*_{N} of R onto Royden's compactification R^*_{M} of R such that ρ is the identity on R. Moreover $\rho(\Delta_{\mathrm{N}})=\Delta_{\mathrm{M}}$.*

We call ρ the *projection* of R^*_{N} onto R^*_{M}. The set $\rho^{-1}(E)$ for $E\subset R^*_{\mathrm{M}}$ is called the *fiber* over E.

To construct the mapping ρ take a $p\in R^*_{\mathrm{N}}$ and let X_p be the character on $\mathrm{N}(R)$ defined by $X_p(f)=f(p)$. Then $X'_p=X_p|\mathrm{M}(R)$ is a character on $\mathrm{M}(R)$ and thus by III.2 B there exists a unique point, say $\rho(p)$, in R^*_{M} with $X'_p(f)=f(\rho(p))$ for every $f\in\mathrm{M}(R)$. We thus obtain a mapping ρ of R^*_{N} into R^*_{M} such that

$$f(p)=f(\rho(p)) \tag{10}$$

for every $f\in\mathrm{M}(R)\subset\mathrm{N}(R)$. From this we conclude that ρ is continuous, and $\rho(z)=\rho^{-1}(z)=z$ for $z\in R$ since $\mathrm{N}_0(R)|\Gamma_{\mathrm{N}}=\mathrm{M}_0(R)|\Gamma_{\mathrm{M}}=0$.

Let q be an arbitrary point in Γ_{M} and set

$$F_q=\{f\in\mathrm{N}(R)|\limsup_{z\in R,\, z\to q} f(z)=0\}.$$

Since $F_q\supset\mathrm{N}_0(R)$ and $1\notin F_q$, F_q is a proper ideal of $\mathrm{N}(R)$. Suppose there do not exist points in R^*_{N} at which every function in F_q vanishes. For each $p\in R^*_{\mathrm{N}}$ we can find an $f_p\in F_q$ such that $f_p(p)=1$ and $f_p\geq 0$ on R^*_{N}. From the open covering $\bigcup_{p\in R^*_{\mathrm{N}}}U_p=R^*_{\mathrm{N}}$ we can choose a finite subcovering $\bigcup_1^m U_{p_n}=R^*_{\mathrm{N}}$ where $U_p=\{r\in R^*_{\mathrm{N}}|f_p(r)>\frac{1}{2}\}$. Let $f=\sum_1^m f_{p_n}$. Then

$f>\frac{1}{2}$ on $R^*_{\mathbb{N}}$ and $1/f\in\mathbb{N}(R)$. Hence $1=(1/f)f\in F_q$ along with $f\in F_q$, a contradiction. Thus we must have a point $p\in R^*_{\mathbb{N}}$ such that $F_q|p=0$. For $f\in\mathbb{M}(R)$ evidently $f-f(q)\in F_q$ and therefore $f(p)=f(q)$, i.e. $q=\rho(p)$. It follows that ρ is an onto mapping.

Finally let $p\in\Delta_{\mathbb{N}}$. By (10) and 1 G, $f(\rho(p))=0$ for every $f\in\mathbb{M}_\Delta(R)$, i.e. $\rho(\Delta_{\mathbb{N}})\subset\Delta_{\mathbb{M}}$. Assume that $\Delta_{\mathbb{M}}-\rho(\Delta_{\mathbb{N}})$ is not empty. Since $\Delta_{\mathbb{M}}-\rho(\Delta_{\mathbb{N}})$ is open there is a function $u\in HBD(R)$ such that $u>0$ on R and $u=0$ on $\rho(\Delta_{\mathbb{N}})$ (see e.g. III.4 E). Note that $u\in\mathbb{N}(R)$. On considering $u\circ\rho$ we find that $u=0$ on $\Delta_{\mathbb{N}}$. Thus by Theorem 2 E, $u\in\mathbb{N}_\Delta(R)$, i.e. $W_u^R=0$. On the other hand $W_u^R=u>0$, a contradiction. Hence $\rho(\Delta_{\mathbb{N}})=\Delta_{\mathbb{M}}$.

This completes the proof.

2 G. Remarks on $\Gamma_{\mathbb{N}}$ and $\Delta_{\mathbb{N}}$. We append here some topological properties of $\Gamma_{\mathbb{N}}$ and $\Delta_{\mathbb{N}}$. As a counterpart of Theorem III.2 D we first state:

A point $p\in R^*_{\mathbb{N}}$ belongs to $\Gamma_{\mathbb{N}}$ if and only if p is not a G_δ-set.

To see this let $p\in\Gamma_{\mathbb{N}}$ and $\rho(p)=q$. If p is G_δ then there exists a sequence $\{z_n\}_1^\infty\subset R$ such that $z_n\neq z_m$ $(n\neq m)$ and $p=\lim_n z_n$. Consequently $\lim_n z_n=\lim_n\rho(z_n)=\rho(p)=q$ in $R^*_{\mathbb{M}}$. Let D_{2n} with $\bar{D}_{2n}\subset R$ be a disk about z_{2n} and D'_{2n} a concentric smaller disk such that $D(w_{2n})<2^{-n}$ where $w_{2n}\in\mathbb{M}_0(R)$ with $w_{2n}|R-D_{2n}=0$, $w_{2n}|D'_{2n}=1$, and $w_{2n}|D_{2n}-D'_{2n}\in H(D_{2n}-D'_{2n})$. We can choose the D_{2n} pairwise disjoint, with $\bigcup_1^\infty D_{2n}\cap\{z_{2n-1}\}_1^\infty=\emptyset$. Then $w=\sum_1^\infty w_{2n}\in\mathbb{M}_\Delta(R)$. Clearly $\lim_n w(z_{2n})=1$ and $\lim_n w(z_{2n-1})=0$. This contradicts $\lim_n w(z_n)=w(q)$.

From the proof we also have an analogue of Theorem III.2 E:

The set $\Delta_{\mathbb{N}}$ is topologically so small that $\overline{\Gamma_{\mathbb{N}}-\Delta_{\mathbb{N}}}=\Gamma_{\mathbb{N}}$.

In fact suppose there is an open neighborhood U of $p\in\Delta_{\mathbb{N}}$ such that $U\cap(\Gamma_{\mathbb{N}}-\Delta_{\mathbb{N}})=\emptyset$. Let V be an open neighborhood of p with $\bar{V}\subset U$. Take a sequence $\{z_n\}_1^\infty\subset R\cap V$ with $z_n\neq z_m$ $(n\neq m)$ and $p=\lim_n z_n$. Construct w as above by taking $\bar{D}_{2n}\subset V\cap R$. Since $w\in\mathbb{M}_\Delta(R)\subset\mathbb{N}_\Delta(R)$ and $\bar{V}\cap\Gamma_{\mathbb{N}}\subset U\cap\Gamma_{\mathbb{N}}\subset\Delta_{\mathbb{N}}$ we have $w|\bar{V}\cap\Gamma_{\mathbb{N}}=0$. Observe that $w=0$ on $R^*_{\mathbb{N}}-\bar{V}$. Thus $w=0$ on $\Gamma_{\mathbb{N}}$ and $\{p\in R^*_{\mathbb{N}}|w(p)\geq\frac{1}{2}\}$ is a compact subset of R. This is a contradiction since $w(z_{2n})=1$ for all n.

2 H. The Class $\mathscr{W}$ for $R\notin O_G$. Given $R\notin O_G$ denote by $\mathscr{W}(R)$ the class of continuous harmonizable functions f on R for which there exists a continuous superharmonic function s_f with a discrete $\{s_f=\infty\}$ and with

$$s_f\geq|f| \tag{11}$$

on R. Clearly $\mathbb{N}(R)\subset\mathscr{W}(R)$. The class $\mathscr{W}$ is localizable: if $f\in\mathscr{W}(R)$ and G is a subregion of R then $f|G\in\mathscr{W}(G)$.

THEOREM. *The class $\mathscr{W}(R)$ is a vector lattice such that for every $f, g \in \mathscr{W}(R)$ and every real number a the following relations are valid:*

$$W_{af}^R = a\, W_f^R, \qquad W_{f+g}^R = W_f^R + W_g^R, \tag{12}$$

$$W_{f\cup g}^R = W_f^R \vee W_g^R, \qquad W_{f\cap g}^R = W_f^R \wedge W_g^R, \tag{13}$$

$$\| W_f^R \|_\infty \le \| f \|_\infty . \tag{14}$$

Moreover every function in $\mathscr{W}(R)$ is continuously extendable to $R_{\mathbb{N}}^*$, i.e.

$$\mathscr{W}(R) \subset C(R_{\mathbb{N}}^*). \tag{15}$$

Identities (12) are easily proved in the same manner as 1 C. (3) and (4).

Clearly (11) is satisfied by $f \cup g$ and $f \cap g$ since $|f \cup g|$ and $|f \cap g|$ are dominated by $s_f + s_g$. It follows from the definition that

$$\underline{W}_{f\cup g}^R \ge \underline{W}_f^R \vee \underline{W}_g^R = W_f^R \vee W_g^R.$$

On the other hand for every $\varepsilon > 0$ and $z_0 \in R$ we can find an $s \in \bar{U}(R, f)$ and an $s' \in \bar{U}(R, g)$ such that $0 < s(z_0) - W_f^R(z_0)$, $s'(z_0) - W_g^R(z_0) < \varepsilon$. By virtue of $(s - W_f^R) + (s' - W_g^R) + W_f^R \vee W_g^R \ge s, s'$ on R we have $(s - W_f^R) + (s' - W_g^R) + W_f^R \vee W_g^R \in \bar{U}(R, f \cup g)$. It follows that $2\varepsilon + (W_f^R \vee W_g^R)(z_0) \ge \overline{W}_{f\cup g}^R(z_0)$. Since $z_0 \in R$ and $\varepsilon > 0$ are arbitrary we conclude that

$$W_f^R \vee W_g^R \ge \overline{W}_{f\cup g}^R.$$

We have obtained the first identity (13) and also the fact that $f \cup g \in \mathscr{W}(R)$. The second identity (13) and thus the relation $f \cap g \in \mathscr{W}(R)$ can be deduced similarly.

By (13) we infer that if $f \le g$ then

$$W_f^R \le W_g^R. \tag{16}$$

Therefore the relations $-\|f\|_\infty \le f \le \|f\|_\infty$, (12), and $W_1^R = 1$ imply (14).

As in III. 3 A we see that (15) is valid.

2 I. The Class $\mathscr{W}$ for $R \in O_G$. In the case of $R \in O_G$ the class $\mathscr{W}$ can be defined without reference to harmonizability. In fact denote by $\mathscr{W}(R)$ the totality of continuous functions f on R such that for any normal subregion $G \subset R$ with $G \notin O_G$ there exists a continuous superharmonic function $s_{G,f}$ on G with a discrete $\{s_{G,f} = \infty\}$ such that

$$s_{G,f} \ge |f| \,|\, G. \tag{17}$$

In view of the definition $W_f^R \equiv 0$, Theorem 2 H remains valid. However the localizability is true only for bounded functions in $\mathscr{W}(R)$. In fact $\mathscr{W}(R) \supset B(R) = \mathbb{N}(R)$, and we now give an example of an $f \in \mathscr{W}(R)$ such that $f|G \notin \mathscr{W}(G)$ for some $G \subset R$ with $G \notin O_G$. Such an f is necessarily unbounded.

Let $R = \{z|0<|z|\leq\infty\}$ and $G=\{z|0<|z|<1\}$. Consider the continuous function f on R such that $f \in H(R - \bigcup_{n=0}^{\infty} l_n)$, $f|l_{2k}=0$, and $f|l_{2k+1}=2k+1$ for $k=0, 1, \ldots$ where $l_n = \{z|\log(1/|z|)=n\}$, $n=0,1,\ldots$.

Take an arbitrary normal subregion $F \subset R$ with $F \notin O_G$. Let α be a nonempty subarc of ∂F and g the Green's function of $\{z||z|\leq\infty\}-\alpha$ with pole at 0. Since $0 \leq f \leq \log(1/|z|)$ on G, $0 \leq f \leq g + a$ on $\{z||z|\leq\infty\}-\alpha$ and a fortiori on F, with $a = (\max_\alpha \log(1/|z|)) \cup 0$. Thus (17) is satisfied for f with respect to F, i.e. $f \in \mathscr{W}(R)$.

Let $s \in \bar{U}(G, f)$. Since $s \geq f$ in a neighborhood of $|z|=1$ and $z=0$ on G, $s>0$ on G and $s \geq 2k+1$ on every l_{2k+1} for sufficiently large k. Thus $s \geq \log 1/|z|$ on G and

$$\overline{W}_f^G = \log \frac{1}{|z|}.$$

On the other hand let $s \in \underline{U}(G, f)$. In view of $s \leq f$ on G outside of $\{z|1/m \leq |z| \leq 1-1/m\}$ for sufficiently large m, $s \leq 0$ between l_0 and l_{2k} for sufficiently large k. From this it follows that $s \leq 0$ on G, whence

$$\underline{W}_f^G \equiv 0.$$

Thus we have seen that $\overline{W}_f^G \neq \underline{W}_f^G$, i.e. $f|G \notin \mathscr{W}(G)$.

3. Harmonic Projection

3 A. Positive Harmonic Functions. We introduce the classes

$$HP(R) = \{u \in H(R) | u \geq 0 \text{ on } R\},$$

$$HP'(R) = \{u_1 - u_2 | u_1, u_2 \in HP(R)\}.$$

We call $u \in HP'$ an *essentially positive* harmonic function. Clearly $u \in H(R)$ belongs to $HP'(R)$ if and only if $|u|$ admits a harmonic majorant. We denote by $\langle u \rangle$ the least harmonic majorant of $|u|$. The space $HP'(R)$ forms a conditionally complete vector lattice with lattice operations $\vee$ and $\wedge$, and $\langle u \rangle = u \vee 0 - u \wedge 0$. By Theorem 2 H

$$HP(R) \subset HP'(R) \subset \mathscr{W}(R) \tag{18}$$

and thus every function in $HP'(R)$ is continuous on $R^*_{\mathbb{N}}$.

Let O_{HP} be the class of Riemann surfaces R on which the HP-functions reduce to constants. The class $O_{HP'}$ is defined similarly and we have

$$O_{HP} = O_{HP'}. \tag{19}$$

3 B. Bounded Harmonic Functions. We designate by $HB(R)$ the family of bounded harmonic functions on R, and by O_{HB} the corresponding null class of Riemann surfaces.

We now define the operator B' on $HP'(R)$ into itself as follows. First for $u \in HP(R)$ set

$$(B'u)(z) = \sup\{v(z) \mid v \in HB(R),\ v \leq u \text{ on } R\}. \tag{20}$$

It exists since $v_1 \vee v_2 \leq u$ for $v_1, v_2 \leq u$. Next for $u \in HP'(R)$ set

$$B'u = B'u_1 - B'u_2 \tag{21}$$

if $u = u_1 - u_2$; $u_1, u_2 \in HP(R)$. We have to show that (21) is independent of the decomposition of u. To this end let $u = u_1' - u_2'$. Then $u_1 + u_2' = u_1' + u_2$. Clearly B' is additive on HP and hence $B'u_1 + B'u_2' = B'u_1' + B'u_2$. Thus $B'u_1 - B'u_2 = B'u_1' - B'u_2'$.

The following identities are easily verified:

$$B'^2 u = B'(B'u) = B'u, \tag{22}$$

$$B'(au) = a B'u, \tag{23}$$

$$B'(u_1 + u_2) = B'u_1 + B'u_2, \tag{24}$$

$$B'(u_1 \vee u_2) = (B'u_1) \vee (B'u_2), \qquad B'(u_1 \wedge u_2) = (B'u_1) \wedge (B'u_2), \tag{25}$$

$$\|B'u\|_\infty \leq \|u\|_\infty, \tag{26}$$

$$B'u = \lim_n B'u_n, \tag{27}$$

for $\{u_n\} \subset HP'(R)$ converging increasingly to $u \in HP'(R)$.

We set

$$HB'(R) = \{u \in HP'(R) \mid B'u = u\}$$

and call $u \in HB'(R)$ a *quasibounded* harmonic function. The corresponding null class is

$$O_{HB'} = O_{HB}. \tag{28}$$

Clearly $HP'(R) \supset HB'(R) \supset HB(R)$. In contrast the elements of the class

$$HP''(R) = \{u \in HP'(R) \mid B'u = 0\}$$

are called *singular*.

By virtue of $I = B' + (I - B')$ with (22), where I is the identity operator on $HP'(R)$, we obtain the direct sum decomposition:

$$HP'(R) = HB'(R) + HP''(R). \tag{29}$$

Remark. Quasibounded and singular harmonic functions together with the decomposition (29) were introduced by PARREAU [4]. Recently DOOB [1] extended quasiboundedness to functions in $\mathscr{W}(R)$ which he characterized as "uniformly integrable" functions.

3 C. Strict Inclusion $O_{HP} < O_{HB}$. Since $HB' \subset HP'$ we have $O_{HP'} \subset O_{HB'}$, and trivially $O_{HP} \subset O_{HB}$. Moreover we claim (TÔKI [2] and others) that

$$O_{HP} < O_{HB}. \tag{30}$$

In V.7 C we shall construct a surface $R \in O_{HP}$ on which the Green's function $g(z, z_0)$ with pole $z_0 \in R$ exists. Take the surface $R' = R - z_0$. Then $g|R' \in HP(R')$ and $R' \notin O_{HP}$. Next take a $u \in HB(R')$. We may assume that $u > 1$ on R'. Let D be a disk about z_0 in R, and v a harmonic conjugate of u on $\bar{D} - z_0$. Then $f = e^{(u+iv)/\alpha} \in AB(D - z_0)$ where $\alpha = 1$ if $\int_{\partial D} * du = \int_{\partial D} dv = 0$; $\alpha = (1/2\pi) \int_{\partial D} * du$ if $\int_{\partial D} * du \neq 0$. By Cauchy's formula f can be continued so as to belong to $AB(D)$. Since $f \neq 0$ on $\bar{D}$, a branch of $\log f$ is in $A(D)$. Thus $u = \mathrm{Re}(\alpha \log f) \in H(D)$, i.e. $u \in HP(R)$. Hence u must be constant: $R' \in O_{HB}$.

3 D. Maximum Principle V. As a counterpart of III.2 I we now prove the maximum principle for HB'-functions with respect to Wiener's harmonic boundary (S. MORI [3], HAYASHI [3], KUSUNOKI [7]):

THEOREM. *Let G be a subregion of a Riemann surface R, and $u \in HB'(G)$. If*

$$m \leq \liminf_{z \in G, z \to p} u(z) \leq \limsup_{z \in G, z \to p} u(z) \leq M \tag{31}$$

for every point $p \in (\partial G) \cup (\Delta_{\mathbb{N}} \cap \bar{G})$ then $m \leq u \leq M$ on G.

In particular every HB'-function on R takes on its maximum and minimum on $\Delta_{\mathbb{N}}$.

We have only to show that the left-hand inequality in (31) implies $u \geq m$ on G. If $m = -\infty$ there is nothing to prove. Hence we assume the existence of a number c such that $-\infty < c < m$. Suppose there exists a nonempty component F of $\{z \in G | u(z) < c\}$. By (31), F is a normal subregion of R with $\bar{F} \cap \Delta_{\mathbb{N}} = \emptyset$. The function $v = c - u$ is again an HB'-function on F and $v = 0$ on ∂F. Since $v \in HP(F)$

$$v = B' v = \sup \{ w \in HB | w \leq v \text{ on } F \}.$$

If $w \in HB(F)$, and $w \leq v$ on F then $w \vee 0 \leq v$ on F and $w \vee 0 \in HB(F)$. Thus we can find a $w \in HB(F)$ with $0 < w < v$ on F and boundary values 0 at ∂F. Extend w to all of R by setting $w = 0$ on $R - F$. The resulting function w is subharmonic and belongs to $\mathbb{N}(R)$. Since $\bar{F} \cap \Delta_{\mathbb{N}} = \emptyset$, $w = 0$ on $\Delta_{\mathbb{N}}$, i.e. $w \in \mathbb{N}_{\Delta}(R)$. Thus $W_w^R = 0$. However if $\Delta_{\mathbb{N}} \neq \emptyset$, i.e. if $R \notin O_G$ then $w \in \underline{U}(R, w)$ and $w \leq W_w^R$. In particular $W_w^R \geq w > 0$ on F, a contradiction. Therefore $R \in O_G$.

Take a disk $R_0 \subset R$ such that $\bar{R}_0 \subset R - F$. Then

$$k\, w(z) \leq \omega(z;\, R_n - \bar{R}_0)$$

where $\{R_n\}_0^\infty$ is an exhaustion of R, $\omega(z;\, R_n - \bar{R}_0)$ is the harmonic measure of ∂R_n with respect to $R_n - \bar{R}_0$, and $k = 1/\sup_{z\in F} w(z)$. It follows that $\lim_n \omega(z;\, R_n - \bar{R}_0) > 0$ on $R - \bar{R}_0$ and we have $R \notin O_G$. This is again a contradiction. Consequently $\{z \in G \mid u(z) < c\}$ is empty, i.e. $u \geq c$. On letting $c \to m$ we conclude that $u \geq m$.

3 E. Harmonic Decomposition. In analogy with the convention $HD(R) = \{0\}$ for $R \in O_G$ we now make the agreement that $HP'(R) = \{0\}$ and consequently $HB'(R) = \{0\}$ for $R \in O_G$. This will also turn out to be natural.

In the same manner as in III. 3 B we define distinguished compact sets $K \subset R_{\mathbb{N}}^*$. For an open set $\mathcal{O}$, $f \in HB'(\mathcal{O})$ means that f is an HB'-function on each component of $\mathcal{O}$.

THEOREM. *Let $f \in \mathscr{W}(R)$ and let K be a distinguished compact set in $R_{\mathbb{N}}^*$ which may be empty. Then*

(a) *f can be decomposed uniquely into the form $f = u + g$ where $u \in HB'(R - K) \cap \mathscr{W}(R)$ and $g \in \mathscr{W}(R)$ with $g = 0$ on $K \cup \Delta_{\mathbb{N}}$,*

(b) $\|u\|_{\infty, R} \leq \|f\|_{\infty, K \cup \Delta_{\mathbb{N}}}$.

The proof will be given in 3 F – 3 G.

3 F. The Space $\mathscr{W}_\Delta(R)$. First we treat the case $K = \emptyset$. If $R \in O_G$ then $HB'(R) = \{0\}$ and $\Delta_{\mathbb{N}} = \emptyset$ and there is nothing to prove. Thus we suppose $R \notin O_G$.

The uniqueness in (a) and the inequality (b) follow from Theorem 3 D. We have only to show the possibility of the decomposition (a). We may assume that $f \geq 0$ on R. Let $g'' = f - W_f^R$. The mapping $s \to s - W_f^R$ is one-to-one and onto between $\bar{U}(R, f)$ and $\bar{U}(R, g'')$, and thus $W_{g''}^R = 0$. By virtue of $W_{((-n)\cup g'')\cap n}^R = ((-n) \vee W_{g''}^R) \wedge n = 0$ we have $((-n) \cup g'') \cap n \in \mathbb{N}_\Delta(R)$ and $((-n) \cup g'') \cap n = 0$ on $\Delta_{\mathbb{N}}$ for $n = 1, 2, \ldots$. Thus $g'' = 0$ on $\Delta_{\mathbb{N}}$.

Since $f \geq 0$, $W_f^R \geq 0$ on R. Let $u = B' W_f^R$, $g' = (I - B') W_f^R = W_f^R - u$, and $g = g' + g''$. To conclude that $f = u + g$ is the desired decomposition we have to show that $g' | \Delta_{\mathbb{N}} = 0$. Note that $g' \in HP''(R)$ and $g' \geq 0$. In view of $g' \wedge n \in HB(R)$, $g' \geq g' \wedge n \geq 0$ implies that $g' \wedge n = 0$. Here $g' \wedge n = W_{g'}^R \wedge W_n^R = W_{g'\cap n}^R = 0$ gives $g' \cap n \in \mathbb{N}_\Delta$ or equivalently $g' \cap n = 0$ on $\Delta_{\mathbb{N}}$ for $n = 1, 2, \ldots$. Thus $g' | \Delta_{\mathbb{N}} = 0$.

This completes the proof of Theorem 3 E for the case $K = \emptyset$.

Here we introduce the subspaces

$$\mathscr{W}_\Delta(R) = \{f \in \mathscr{W}(R) \mid f | \Delta_{\mathbb{N}} = 0\}, \qquad \mathscr{W}_\Delta'(R) = \{f \in \mathscr{W}(R) \mid W_f^R = 0\}.$$

COROLLARY. *The direct sum decompositions*

$$\mathscr{W}(R)=HB'(R)+\mathscr{W}_{\Delta}(R), \qquad \mathscr{W}_{\Delta}(R)=HP''(R)+\mathscr{W}'_{\Delta}(R), \tag{32}$$

$$\mathbb{N}(R)=HB(R)+\mathbb{N}_{\Delta}(R) \tag{33}$$

are valid on arbitrary Riemann surfaces.

3 G. The Space $\mathscr{W}_{\Delta\cup K}(R)$. We turn to the case $K\neq\emptyset$. Let $F=R-K$ and denote by F_i the components of F. The uniqueness in (a) and the inequality (b) are again direct consequences of Theorem 3 D. Hence we have only to give the decomposition in (a). It is obvious for $f\in HB'(R)$.

For $f\in HP''(R)$ we may assume that $f>0$ on R. The function $B'_{F_i}f$ is clearly in $HB'(F_i)$. Let $\{R_m\}_1^\infty$ be an exhaustion of R, and u_{nm} the continuous function on $\bar{F}_i\cap R$ such that $u_{nm}|F_i\cap R_m\in H(F_i\cap R_m)$ and $u_{nm}|\bar{F}_i\cap R-F_i\cap R_m=f\cap n$. Then $\{u_{nm}\}_m$ is a decreasing sequence and $u_n=\lim_m u_{nm}$ exists and is harmonic on F_i. Clearly $u_n=f\cap n$ on ∂F_i. Thus $f\geq B'_{F_i}f\geq u_n$ on $\bar{F}_i\cap R$ and $f\geq B'_{F_i}f\geq f\cap n$ on ∂F_i for every n. Hence $B'_{F_i}f=f$ on ∂F_i.

Let $u=f$ on $R-\bigcup F_i$ and $u=B'_{F_i}f$ on each F_i. Then u is nonnegative superharmonic on R and thus $u\in\mathscr{W}(R)$ and $u|F\in HB'(F)$. Set $g=f-u$. Since $f\geq u\geq 0$ and $f=0$ on $\Delta_{\mathbb{N}}$, $u|\Delta_{\mathbb{N}}=0$. Hence $g|\Delta_{\mathbb{N}}=0$. Clearly $g=0$ on $K\cap R$ and by continuity $g=0$ on $K=\overline{K\cap R}$.

By (32) it remains to prove (a) for $f\in\mathscr{W}'_{\Delta}(R)$. It may again be assumed that $f\geq 0$ on R.

First consider the case $R\notin O_G$.

Let u'_n be the continuous function on R with $u'_n|R_n-K\in H(R_n-K)$ and $u'_n|R-(R_n-K)=f$. Take $s\in\bar{U}(F_i,f)$ and $s'\in\underline{U}(F_i,f)$. Then $s'\leq u'_n\leq s$ on F_i for sufficiently large n. Hence

$$s'\leq\liminf_n u'_n\leq\limsup_n u'_n\leq s$$

on F_i and consequently $\underline{W}_f^{F_i}\leq\liminf_n u'_n\leq\limsup_n u'_n\leq\overline{W}_f^{F_i}$ on F_i. We conclude that $u'=\lim_n u'_n\geq 0$ exists and is continuous on R, $u'|F_i=W_f^{F_i}\in HP(F_i)$, and $u'|K\cap R=f$. By the same argument as in the case of $f\in HP''(R)$, $B'_{F_i}W_f^{F_i}$ has the boundary values f at ∂F_i. On each F_i replace $u'|F_i$ by $B'_{F_i}W_f^{F_i}$ and denote by u the resulting function. Then u is continuous on R, $u|K\cap R=f$, $u'\geq u\geq 0$ on R, and $u|F\in HB'(F)$.

Set $g=f-u$. We claim that $f=u+g$ is the required decomposition. All we have to show is that $u\in\mathscr{W}'_{\Delta}(R)$, since $f\in\mathscr{W}'_{\Delta}(R)$.

There exists a superharmonic function w' such that $0\leq f\leq w'$ on R. Let w_{np} be the solution of the generalized Dirichlet problem for the open set $R_{n+p}-\bar{R}_n$ with boundary values w' on ∂R_n and 0 on ∂R_{n+p}. Clearly $0\leq w_{np}\leq w_{n,p+1}\leq w'$ and thus $w_n=\lim_p w_{np}$ exists on $R-\bar{R}_n$. Set $w_n=w'$ on $\bar{R}_n$. Then w_n is superharmonic on R. Consider $\omega_{np}\in\mathbb{M}_0(R)$ with

$\omega_{np}|\bar{R}_{n+1}=1$, $\omega_{np}|R-R_{n+p}=0$, and $\omega_{np}|R_{n+p}-\bar{R}_{n+1}\in H(R_{n+p}-\bar{R}_{n+1})$. The function $\omega_n=BD\text{-}\lim_p \omega_{np}$ exists, $\omega_n\in \mathbb{M}_\Delta(R)\subset\mathbb{N}_\Delta(R)$, and $W^R_{\omega_n}=0$. On setting $c=\max_{\partial R_{n+1}} w_n$ we obtain $w_{np}\le c\,\omega_{np}$ on $R_{n+p}-\bar{R}_{n+1}$ and hence $0\le w_n\le c\,\omega_n$ on $R-\bar{R}_{n+1}$. From this we see that $0\le W^R_{w_n}\le c\,W^R_{\omega_n}=0$, i.e.

$$W^R_{w_n}=0. \tag{34}$$

Since $0\le u\le w'$ on R, (11) is satisfied by u. Suppose $s_n\in\bar{U}(R,f)$ with $s_n\ge f$ on $R-\bar{R}_n$. Then $f\le s_n+w_n$ on R and consequently $u\le s_n+w_n$. By (34), $0\le \bar{W}^R_u\le \bar{W}^R_{s_n+w_n}=W^R_{s_n}+W^R_{w_n}=W^R_{s_n}\le s_n$ and we obtain

$$0\le \underline{W}^R_u\le \bar{W}^R_u\le s_n.$$

Since s_n is arbitrary and $\inf_{\bar{U}(R,f)} s_n=0$ we infer that

$$W^R_u=0. \tag{35}$$

Next let G be a normal subregion of R with $G\notin O_G$. We shall show the existence of W^G_u. Let h_m be continuous on $\bar{G}\cap R$ such that $h_m|G\cap R_m\in H(G\cap R_m)$ and $h_m|\bar{G}\cap R-G\cap R_m=u$. Since $0\le u\le s_n+w_n$, $0\le h_m\le s_n+w_n$. On choosing a suitable subsequence of $\{R_m\}$ we may assume that $h=\lim_m h_m$ exists. Clearly $h\in H(G)$ and $h|\partial G=u$. In view of $0\le h\le s_n+w_n$ we have $h-s_n-w_n-\varepsilon\in\underline{U}(G,u)$ and $h+s_n+w_n+\varepsilon\in\bar{U}(G,u)$ for any $\varepsilon>0$. Thus

$$h-s_n-w_n-\varepsilon\le\underline{W}^G_u\le\bar{W}^G_u\le h+s_n+w_n+\varepsilon.$$

On letting $\varepsilon\to 0$ we obtain

$$-s_n-w_n\le\underline{W}^G_u-h\le\bar{W}^G_u-h\le s_n+w_n.$$

The function $\bar{W}^G_u-\underline{W}^G_u$ has boundary values 0 at ∂G. On setting $\bar{W}^G_u-\underline{W}^G_u=0$ outside of G on R we have

$$0\le\bar{W}^G_u-\underline{W}^G_u\le 2s_n+2w_n$$

on all of R, with $v=\bar{W}^G_u-\underline{W}^G_u$ subharmonic on R. Consequently

$$0\le v\le W^R_v\le 2W^R_{s_n}+2W^R_{w_n}.$$

In view of (34) and $W^R_{s_n}\le s_n$ we obtain

$$0\le\bar{W}^G_u-\underline{W}^G_u\le 2s_n.$$

Since $\inf_{\bar{U}(R,f)} s_n=0$ we infer that $\bar{W}^G_u=\underline{W}^G_u$, i.e. W^G_u exists.

We turn to the case $R\in O_G$.

Let u_{nm} be the continuous function on R with $u_{nm}|R_m-K\in H(R_m-K)$ and $u_{nm}|R-(R_m-K)=f\cap n$. Note that $f\cap n\in\mathbb{N}_\Delta(R)=\mathbb{N}(R)$. As in the case $R\notin O_G$ we obtain the existence of $u_n=\lim_m u_{nm}$. Clearly $u_n|F\in HB(F)$, $u_n|K\cap R=f\cap n$, and u_n is continuous on R. There exists a superharmonic function w on F such that $0 \le f \le w$ and hence $0 < u_n \le u_{n+1} \le w$ on F. Here $u_n \le u_{n+1}$ follows from Theorem 3 D, and $u = \lim_n u_n \ge 0$ exists on R. Again u is in $HB'(F)$ and continuous on R.

To conclude that $f = u + g$ is the required decomposition we must prove that $u \in \mathscr{W}(R)$. Take an arbitrary normal subregion $G \notin O_G$. We have only to show that u has a continuous superharmonic majorant on G.

If $R \cap K - G \neq \emptyset$ then take an arc α in this set so that $R - \alpha \notin O_G$. There exists a continuous superharmonic function w with $w \ge f \ge 0$ on $R - \alpha \supset G$. From the construction of u_{nm}, u_n and u we see that $u_{nm} \le w$, $u_n \le w$, and $u \le w$ on $R - \alpha$ and consequently on G.

If $R \cap K - G = \emptyset$, i.e. $G \supset K \cap R$ then take an arc α in $R - G \subset R - K$. There exists again a superharmonic function w on $R - \alpha \supset G$. For $c = \max_\alpha u$ we have $u_{nm} \le w + c$, $u_n \le w + c$ and $u \le w + c$ on $R - \alpha \supset G$.

This completes the proof of Theorem 3 E.

We introduce the subspaces

$$\mathscr{W}_{\Delta \cup K}(R) = \{f \in \mathscr{W}_\Delta(R) \mid f \mid K = 0\}, \qquad HB'(R, K) = \mathscr{W}(R) \cap HB'(R - K),$$

$$\mathbb{N}_{\Delta \cup K}(R) = \{f \in \mathbb{N}_\Delta(R) \mid f \mid K = 0\}, \qquad HB(R, K) = \mathbb{N}(R) \cap HB(R - K).$$

Corollary. *The spaces $\mathscr{W}(R)$ and $\mathbb{N}(R)$ have the direct sum decompositions*

$$\mathscr{W}(R) = HB'(R, K) + \mathscr{W}_{\Delta \cup K}(R), \tag{36}$$

$$\mathbb{N}(R) = HB(R, K) + \mathbb{N}_{\Delta \cup K}(R). \tag{37}$$

3 H. Harmonic Projection. Take $f \in \mathscr{W}(R)$ and let K be a distinguished compact set in $R^*_{\mathbb{N}}$. Denote by $\pi_K f$ the function in $\mathscr{W}(R)$ such that $\pi_K f \mid K \cup \Delta_{\mathbb{N}} = f \mid K \cup \Delta_{\mathbb{N}}$ and $\pi_K f \in HB'(R - K)$ (see Theorem 3 E). If it is necessary to distinguish this π_K from the orthogonal projection in III.3 D then we write $\pi_K^{\mathbb{N}}$. We shall call $\pi_K^{\mathbb{N}}$ the *harmonic projection*. In particular $\pi_\emptyset^{\mathbb{N}}$ will be simply denoted by $\pi = \pi^{\mathbb{N}}$.

3 I. Evans' Superharmonic Function. As an analogue of Theorem III.3 H we shall establish the following result which shows the function-theoretic smallness of the set $\Gamma_{\mathbb{N}} - \Delta_{\mathbb{N}}$. Recall that it is topologically large: $\overline{\Gamma_{\mathbb{N}} - \Delta_{\mathbb{N}}} = \Gamma_{\mathbb{N}}$.

Theorem. *Let $R \notin O_G$. For every compact set $F \subset \Gamma_{\mathbb{N}} - \Delta_{\mathbb{N}}$ there exists a finitely continuous positive superharmonic function s_F on R such that $s_F \mid \Delta_{\mathbb{N}} = 0$ and $s_F \mid F = \infty$.*

Take an open neighborhood $V \supset F$ in $R^*_{\mathbb{N}}$ such that $\overline{V} \cap \Delta_{\mathbb{N}} = \emptyset$ and $\overline{V}$ is a distinguished compact set in $R^*_{\mathbb{N}}$. For a regular exhaustion $\{R_n\}_1^\infty$ of R the set $K_n = \overline{V} - R_n$ is compact in $R^*_{\mathbb{N}}$. Choose $f \in \mathbb{N}_\Delta(R)$ such that $f \mid \overline{V} = 1$. Set $u_n = \pi_{K_n}^{\mathbb{N}} f$. By 3 D, $\{u_n\}_{n=1}^\infty$ is a decreasing sequence and thus

$u=\lim_n u_n \in HB(R)$. Since $0 \le u \le u_n$ and $u_n|\Delta_{\mathbb{N}} = f|\Delta_{\mathbb{N}} = 0$ the same is true of u. Again by 3 D, $u \equiv 0$. Let $z_0 \in R_1$. On taking a suitable subsequence of $\{R_n\}$ we may assume that $u_n(z_0) < 2^{-n}$ $(n=1,2,\ldots)$. Then

$$s_F = \sum_{n=1}^{\infty} u_n$$

is a finitely continuous positive superharmonic function on R, and clearly $s_F|F = \infty$.

Here we make the following remark. Let s be a continuous superharmonic function on R bounded from below and set $m = \min_{\Delta_{\mathbb{N}}} s$. Then $s \ge m$ on R. To see this let $m > c > -\infty$ and $K = \{p \in \Gamma_{\mathbb{N}} | s(p) \le c\}$. Then K is compact in $\Gamma_{\mathbb{N}} - \Delta_{\mathbb{N}}$. Let s_K be as above for $F = K$. For any positive number $\varepsilon > 0$, $s + \varepsilon s_K$ is superharmonic on R and

$$\lim_{z \in R,\, z \to p} \big(s(z) + \varepsilon s_K(z)\big) \ge c$$

for every $p \in \Gamma_{\mathbb{N}}$. Thus $s + \varepsilon s_K \ge c$ on R. On first letting $\varepsilon \to 0$ and then $c \to m$ we draw the desired conclusion.

We return to s_F. Since $\sum_{n=1}^{N} u_n | \Delta_{\mathbb{N}} = 0$ the inequality $s_F \ge v$ for an HB-function v on R implies $\sum_{n=N+1}^{\infty} u_n \ge v$. On letting $N \to \infty$ we obtain $0 \ge v$ on R.

Now suppose contrary to the assertion that $s_F | \Delta_{\mathbb{N}} \not\equiv 0$. Then $v = \pi^{\mathbb{N}}(s_F \cap 1) \in HB(R)$ and $v > 0$. Since $(s_F - v)|\Delta_{\mathbb{N}} \ge 0$ we infer that $s_F \ge v > 0$ on R. This is a contradiction since $s_F \ge v$ implies $v \le 0$.

3 J. Maximum Principle VI. In the course of the proof of Theorem 3 I we anticipated the following statement:

THEOREM. *Let G be a subregion of R and let s be a superharmonic function on G bounded from below. If*

$$\liminf_{z \in G,\, z \to p} s(z) \ge m \tag{38}$$

for every $p \in (\Delta_{\mathbb{N}} \cap \bar{G}) \cup \partial G$ then $s \ge m$ on G.

The proof is similar to that of Theorem III. 3 I.

3 K. The Class U_{HB}. Let $u \in HB(R)$ (resp. $HB'(R)$) and $u > 0$. Suppose that for every $v \in HB(R)$ (resp. $HB'(R)$), $u \ge v \ge 0$ implies the existence of a constant c_v with $c_v u = v$. Then the function u is called *HB-minimal* (resp. *HB'-minimal*) on R. Clearly HB-minimality implies HB'-minimality, and the converse is true for bounded functions.

We denote by U_{HB} the class of Riemann surfaces R on which there exists at least one HB-minimal function. By the convention $HB(R) = HB'(R) = \{0\}$ for $R \in O_G$ we see that $U_{HB} \cap O_G = \emptyset$. Clearly $U_{HB} \supset O_{HB} - O_G$.

We also denote by O_{HB}^n the class of Riemann surfaces such that $HB(R)$ has at most dimension n. Observe that $O_{HB}^0 = O_G$ and $O_{HB}^1 = O_{HB}$.

The following result is due to S. MORI [3] and HAYASHI [3]:

THEOREM. *A function $u \in HB(R)$ is HB-minimal if and only if there exists an isolated point $p \in \Delta_{\mathrm{N}}$ such that $u(p) > 0$ and $u|(\Delta_{\mathrm{N}} - p) = 0$.*

The proof is analogous to that of Theorem III. 3 E.

COROLLARY 1. *A surface $R \in U_{HB}$ if and only if Δ_{N} contains an isolated point.*

COROLLARY 2. *A surface $R \in O_{HB}^n - O_{HB}^{n-1}$ $(1 \leq n < \infty)$ if and only if Δ_{N} consists of n points.*

3 L. Relative Classes SO_{HB} and SO_{HD}. Let SO_{HB} be the class of normal subregions $(G, \partial G)$ of Riemann surfaces such that every HB-function on G with vanishing continuous boundary values on ∂G reduces to the constant zero. The class SO_{HD} is defined similarly. Note that we here drop the requirement of analyticity of ∂G used in I. 1 F; it is easily verified that our reasoning will remain valid in this more general setting. We also consider the class SO_{HBD} defined analogously. In view of the orthogonal decomposition III. 3 B applied to $n \cap (f \cup -n)$ we conclude at once (A. MORI [2], BADER-PARREAU [1]) that

$$SO_{HD} = SO_{HBD} \tag{39}$$

and hence $SO_{HB} \subset SO_{HD}$. For convenience we also say that a normal open set $\bigcup_j (G_j, \partial G_j) \in SO_{HB}$ (resp. SO_{HD}) if every $(G_j, \partial G_j) \in SO_{HB}$ (resp. SO_{HD}) where the $G_j \cup \partial G_j$ are assumed to be disjoint.

We claim (KUSUNOKI-MORI [4, 5], HAYASHI [3], S. MORI [3]):

THEOREM. *Let G be a normal subregion of R. Then $(G, \partial G) \in SO_{HB}$ (resp. SO_{HD}) if and only if $(\bar{G} - \overline{\partial G}) \cap \Delta_{\mathrm{N}} = \emptyset$ (resp. $(\bar{G} - \overline{\partial G}) \cap \Delta_{\mathrm{M}} = \emptyset$).*

Assume that $(\bar{G} - \overline{\partial G}) \cap \Delta_{\mathrm{N}} = \emptyset$. Let $u \in HB(G \cup \partial G)$ with $u|\partial G = 0$. Set $\gamma = \bar{G} - G$ and $u'(p) = \liminf_{z \in G, z \to p} u(z)$ for $p \in \gamma$. Let $K = \{p \in \Gamma_{\mathrm{N}} \cap \bar{G} | u'(p) \leq c < 0\}$. If $K = \emptyset$ then $u > c$. If $K \neq \emptyset$ take s_K of 3 I for the surface $R - \alpha$ where α is a closed subarc of ∂G. Then $u' + \varepsilon s_K > c$ on γ and $u > c - \varepsilon s_K$ for every $\varepsilon > 0$. It follows that $u \geq c$ and therefore $u \geq 0$. On considering $-u$ instead of u we conclude that $u \leq 0$ and a fortiori $u \equiv 0$.

The necessity is a direct consequence of the harmonic decomposition 3 E.

3 M. Two Region Test. The property of a Riemann surface R belonging or not belonging to O_{HB} (or O_{HD}) can be determined by the structure of $\Delta_{\mathbb{N}}$ (or $\Delta_{\mathbb{M}}$). However it is not easy in general to visualize $\Delta_{\mathbb{N}}$ (or $\Delta_{\mathbb{M}}$). In this regard the following criterion (A. MORI [2], S. MORI [3], MATSUMOTO [2]) is sometimes useful:

THEOREM. *A surface $R \notin O_{HB}^n$ (resp. O_{HD}^n) if and only if there exist at least $n+1$ disjoint normal open subsets G_j of R such that $(G_j, \partial G_j) \notin SO_{HB}$ (resp. SO_{HD}).*

If there exist $n+1$ such sets G_j then by Theorem 3 L, $\Delta_{\mathbb{N}}$ contains at least $n+1$ points, and by Corollary 2 in 3 K, $R \notin O_{HB}^n$. Similarly if $R \notin O_{HB}^n$ then $\Delta_{\mathbb{N}}$ contains at least $n+1$ points p_j. Let $f \in \mathbb{N}(R)$ such that $f(p_j)=j$. Set $G_j^* = \{p \in R_{\mathbb{N}}^* | j-\frac{1}{2} < f(p) < j+\frac{1}{2}\}$ and $G_j = G_j^* \cap R$. Then $(G_j, \partial G_j) \notin SO_{HB}$, and the G_j are disjoint.

The assertion for O_{HD}^n can be proved similarly.

COROLLARY (two region test). *If there exist two disjoint normal subregions G_1 and G_2 of R such that $(G_j, \partial G_j) \notin SO_{HB}$ (resp. SO_{HD}) $(j=1,2)$ then $R \notin O_{HB}$ (resp. O_{HD}).*

At this point we make the following observation (OHTSUKA). The disk R: $|z|<1$ is clearly in neither O_{HB} nor O_{HD}. Let G_1 be a subregion of R whose boundary ∂G_1 consists of two spirals starting from 0 and approaching the unit circumference of R as indicated in Fig. 9. Set

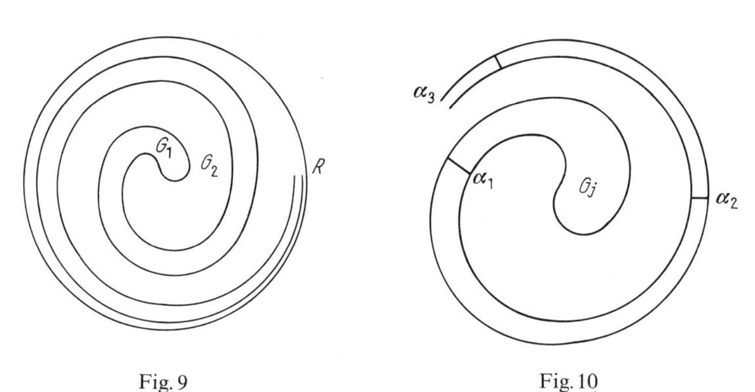

Fig. 9 Fig. 10

$G_2 = R - \bar{G}_1$. Then $R = G_1 \cup G_2 \cup \partial G_1$. In view of the two region test one might suspect that at least one of G_1 and G_2 does not belong to SO_{HB} (or SO_{HD}). However this is not the case. In fact since G_j is simply connected there is a conformal mapping φ_j of G_j onto $|z|<1$, with the Carathéodory boundary of G_j consisting of two kinds of points: each point in ∂G_j and one Carathéodory boundary component α determined by the sequence of cross cuts α_n indicated in Fig. 10. Hence by Carathéodory's mapping theorem φ_j gives a topological mapping of $G_j \cup (\partial G_j) \cup \{\alpha\}$ onto $|z| \leq 1$. Let $\varphi_j(\alpha)=1$. Then $(G_j, \partial G_j)=(\{|z|<1\}, \{|z|=1, z \neq 1\})$, and clearly $(G_j, \partial G_j) \in SO_{HB} \subset SO_{HD}$.

We shall return to SO_{HB} and SO_{HD} in V.7 I.

3 N. Strict Inclusions. We wish to show that $U_{HB} < U_{\widetilde{HD}}$. To this end we consider once more Tôki's example R discussed in III.4 H. It is a complete infinitely many sheeted covering surface of the unit disk U: $|z|<1$ with the property $HB(R)=HB(U)\circ\pi$, where π is the natural projection of R onto U.

Since $R \in O_{HD} - O_{HB} \subset O_{HD} - O_G$, the harmonic boundary Δ_{MI} of R consists of a point of positive harmonic measure and therefore $R \in U_{\widetilde{HD}}$. To show that $R \notin U_{HB}$ suppose this were not true. By the relation $HB(R)=HB(U)\circ\pi$, $HB(U)$ would contain an HB-minimal function u. Here u can be expressed as

$$u(z)=\frac{1}{2\pi}\int_{|\zeta|=1} \mathrm{Re}\left(\frac{\zeta+z}{\zeta-z}\right)\varphi(\zeta)\,|d\zeta|$$

with φ a nonnegative integrable function on ∂U. We claim that $E=\{\zeta \in \partial U \mid \varphi(\zeta)>0\}$ is of linear measure zero, i.e. $\varphi=0$ a.e. on ∂U and thus $u \equiv 0$, a contradiction.

If this were not the case then there would exist a decomposition $E=E_1 \cup E_2$, $E_1 \cap E_2=\emptyset$, with the measures of E_1 and E_2 strictly positive. Let u_1 be the harmonic function on U given by the boundary function φ_1 such that $\varphi_1|E_1=\varphi$ and $\varphi_1|\partial U - E_1=0$. Since $u>u_1>0$ there must exist a positive constant c with $u_1 \equiv c\,u$ on U and consequently $\varphi_1=c\,\varphi$ a.e. on ∂U. However this is not possible since $\varphi_1|E_2=0$ and $\varphi|E_2>0$.

Thus we have proved the strict inclusion

$$U_{HB} < U_{\widetilde{HD}}. \tag{40}$$

The proof of $U_{HB} \subset U_{\widetilde{HD}}$ will be given in 5 B.

Actually the above R belongs to O_{HD} and a fortiori to O_{HD}^n $(n \geq 1)$ but $HB(R)$ has infinite dimension. Therefore we also conclude that

$$O_{HB}^n < O_{HD}^n \tag{41}$$

for all $n \geq 1$.

Finally take a surface $R \in O_{HB} - O_G$, the existence of which will be shown in V.7. By the same method as in III.5 J we can prove the strict inclusion relations

$$O_{HB}^{n-1} < O_{HB}^n \tag{42}$$

for every $n \geq 1$.

Let us summarize here what we have established (cf. III.5 I, 5 J) and also include a number of relations we are going to prove:

$$\begin{array}{ccccccccccccccc} O_G < O_{HP} < & O_{HB} & = & O_{HB}^1 & < & O_{HB}^2 & < \cdots < & O_{HB}^n & < \cdots < & U_{HB} \cup O_G & < & O_{AB} \\ & \wedge & & \wedge & & \wedge & & \wedge & & \wedge & & \wedge \\ & O_{HD} & = & O_{HD}^1 & < & O_{HD}^2 & < \cdots < & O_{HD}^n & < \cdots < & U_{H\widetilde{D}} \cup O_G & < & O_{AD}. \end{array} \tag{43}$$

The relation $O_G < O_{HP}$ will be demonstrated in V.7, and $U_{HB} < O_{AB}$ in 7.

Remark. As a limiting class of O_{HB}^n we may define O_{HB}^∞ as the class of Riemann surfaces R whose harmonic boundaries $\Delta_{\mathbb{N}}$ are closures of countable sets of isolated points in $\Delta_{\mathbb{N}}$. As in Remark III.5 J we see that $O_{HB}^\infty - \bigcup_1^\infty O_{HB}^n$ is not empty.

§ 2. Dirichlet's Problem

Integral representations of HB'-functions in terms of their boundary values on Wiener's boundary will be discussed in 4. To this end the harmonic measure on Wiener's boundary and the harmonic kernel are introduced. The entire technique is quite similar to that in III.4. However an essential distinction lies in the fact that Wiener's harmonic boundary is Stonean, i.e. the closure of an open set is again open.

Wiener's harmonic boundary is seen to be the set of all regular points with respect to the Dirichlet problem. This gives the important measure correspondence by pairs between Royden's, Wiener's, and Stoïlow's compactifications. These topics with an explicit test for quasiboundedness are given in 5.

The most general boundedness property, which we call Φ-boundedness, is introduced in 6. Despite its generality the corresponding null class $O_{H\Phi}$ turns out to be O_{HB} or O_{HP} depending on the choice of Φ.

4. Harmonic Measure and Kernel

4 A. Harmonic Measure on $\Gamma_{\mathbb{N}}$. We assume that R is an open Riemann surface. The *harmonic measure* $\mu = \mu_{\mathbb{N}}$ on Wiener's boundary $\Gamma_{\mathbb{N}}$ of R with respect to a fixed point $z_0 \in R$ is the measure on $\Gamma_{\mathbb{N}}$ satisfying these conditions:

(μ.1) *μ is a positive regular Borel measure on $\Gamma_{\mathbb{N}}$,*

(μ.2) *for every superharmonic function $v \in \mathbb{N}(R)$ the Gauss property $v(z_0) \geq \int_{\Gamma_{\mathbb{N}}} v(p)\, d\mu(p)$ is valid.*

We call z_0 the *center* of the measure and denote by S_μ the support of μ in $R_{\mathbb{N}}^*$. From (μ.2) we conclude that

$$u(z_0) = \int_{\Gamma_{\mathbb{N}}} u(p)\, d\mu(p) \tag{44}$$

for every $u \in HB(R)$. In the same way as in III. 4 A we see that (44) in turn implies (μ.2), and thus (μ.2) and (44) are equivalent.

Also as in III. 4 A we can show:

THEOREM. *If $R \notin O_G$ then there exists a unique harmonic measure μ on $\Gamma_{\mathbb{N}}$ with respect to an arbitrarily fixed center $z_0 \in R$. It satisfies the conditions*

(μ.3) $S_\mu = \Delta_{\mathbb{N}}$,

(μ.4) $\mu(\Delta_{\mathbb{N}}) = 1$,

(μ.5) *for any open set U in $\Gamma_{\mathbb{N}}$ with $U \cap \Delta_{\mathbb{N}} \neq \emptyset$, $\mu(U) > 0$.*

4 B. Harmonic Kernel. For an $R \notin O_G$ let z_0 be the center of the harmonic measure μ. We associate with (R, μ, z_0) the *harmonic kernel* $P(z, p) = P_{\mathbb{N}}(z, p)$, i.e. the real-valued function on $R \times \Gamma_{\mathbb{N}}$ with the properties

(P.1) $P(z_0, p) = 1$ *on* $\Delta_{\mathbb{N}}$,

(P.2) $P(z, p) = 0$ *on* $R \times (\Gamma_{\mathbb{N}} - \Delta_{\mathbb{N}})$,

(P.3) $P(z, p) \in HP(R)$ *for every fixed* $p \in \Gamma_{\mathbb{N}}$,

(P.4) *for every fixed $z \in R$, $P(z, p)$ is a nonnegative bounded Borel function on $\Gamma_{\mathbb{N}}$,*

(P.5) *for every superharmonic function $v \in \mathbb{N}(R)$ and every point $z \in R$, $v(z) \geq \int_{\Gamma_{\mathbb{N}}} P(z, p)\, v(p)\, d\mu(p)$.*

It is easily seen that (P.5) is equivalent to

$$u(z) = \int_{\Gamma_{\mathbb{N}}} P(z, p)\, u(p)\, d\mu(p) \tag{45}$$

for every $z \in R$ and $u \in HB(R)$.

As in III. 4 B – 4 D we conclude:

THEOREM. *There exists a unique harmonic kernel $P(z, p)$ on $R \times \Gamma_{\mathbb{N}}$ with respect to an arbitrary center $z_0 \in R$.*

4 C. Stonean Space $\Delta_{\mathbb{N}}$. Since $B(R^*_{\mathbb{N}}) = \mathbb{N}(R)$, $B(\Delta_{\mathbb{N}}) = \mathbb{N}(R)|\Delta_{\mathbb{N}}$. By 3 F.(33), b: $u \to u|\Delta_{\mathbb{N}}$ is a one-to-one mapping of $HB(R)$ onto $B(\Delta_{\mathbb{N}})$. Hence $B(\Delta_{\mathbb{N}})$ is *conditionally complete*, i.e. for any subfamily $\mathscr{F} \subset B(\Delta_{\mathbb{N}})$ which is bounded from above there exists an $f_{\mathscr{F}} \in B(\Delta_{\mathbb{N}})$ such that $f_{\mathscr{F}} \geq f$ for every $f \in \mathscr{F}$, and $f' \geq f$ for every $f \in \mathscr{F}$ with $f' \in B(\Delta_{\mathbb{N}})$ implies $f_{\mathscr{F}} \leq f'$. Such a space $\Delta_{\mathbb{N}}$ is called a *Stonean space*. We shall prove (S. MORI [3], HAYASHI [3]):

THEOREM. *The Wiener harmonic boundary $\Delta_{\mathbb{N}}$ is Stonean. The closure of every open set in $\Delta_{\mathbb{N}}$ is again open in $\Delta_{\mathbb{N}}$.*

The second assertion is known to be equivalent to the Stonean property (cf., e.g. RICKART [1]). Here we prove only that $\bar{U}$ is open in $\Delta_{\mathbb{N}}$

if U is open in Δ_{N}. Let $\mathscr{F}$ be the class of functions f in $B(\Delta_{\mathrm{N}})$ with $0 \leq f \leq 1$ and $f|U=0$. Since Δ_{N} is Stonean $f_{\mathscr{F}}$ exists. Let U_1 and U_2 be arbitrary open sets in Δ_{N} such that $\bar{U}_1 \subset U \subset \bar{U} \subset U_2$. Take $f_i \in B(\Delta_{\mathrm{N}})$, $i=1,2$, such that $0 \leq f_i \leq 1$, $f_1|U_1=0$, $f_1|(\Delta_{\mathrm{N}}-U)=1$, $f_2|U=0$, and $f_2|(\Delta_{\mathrm{N}}-U_2)=1$. Clearly $f_1 \geq f_{\mathscr{F}} \geq f_2$ on Δ_{N} and thus $f_{\mathscr{F}}=0$ on U_1 and $f_{\mathscr{F}}=1$ on $\Delta_{\mathrm{N}}-U_2$. Since U_1 and U_2 are arbitrary $f_{\mathscr{F}}|U=0$, $f_{\mathscr{F}}|(\Delta_{\mathrm{N}}-U)=1$. Therefore $f_{\mathscr{F}}$ takes on only the two values 0 and 1. By continuity the set

$$\bar{U}=\{p \in \Delta_{\mathrm{N}} | f_{\mathscr{F}}(p)=0\}=\{p \in \Delta_{\mathrm{N}} | f_{\mathscr{F}}(p)<\tfrac{1}{2}\}$$

must be open.

4 D. Integral Representation. Observe that more precise information about the boundary behavior is obtainable in the case of Wiener's boundary than of Royden's boundary. As a counterpart of III.4 E we shall deduce the following result:

THEOREM. *If $f(p)$ is μ-integrable on Γ_{N} then*

$$u(z)=\int_{\Gamma_{\mathrm{N}}} P(z,p)\, f(p)\, d\mu(p) \tag{46}$$

belongs to $HB'(R)$ and

$$u(p)=f(p) \tag{47}$$

μ-a.e. on Δ_{N} (and hence on Γ_{N}). If in addition $f|\Delta_{\mathrm{N}}$ is continuous at $p \in \Delta_{\mathrm{N}}$ then (47) *is valid at this point.*

The last assertion follows easily from (47) and (μ.5). To prove (46) and (47) we may assume that $f \geq 0$. Clearly $B(\Delta_{\mathrm{N}})=HB(R)|\Delta_{\mathrm{N}}$ is dense in $L^1(\Delta_{\mathrm{N}}, \mu)$ with respect to the L^1-norm. Therefore (46) is harmonic (cf. III.4 E). Since

$$u_n(z)=\int_{\Gamma_{\mathrm{N}}} P(z,p)\big(f(p) \cap n\big)\, d\mu(p) \tag{48}$$

belongs to $HB(R)$ and $\{u_n\}$ converges to u increasingly we obtain $u \in HB'(R)$.

For the proof of (47) set $\mathscr{F}=\{v \in HB(R)\,|\, v|\Delta_{\mathrm{N}} \leq f \text{ on } \Delta_{\mathrm{N}}\}$. By Perron's theorem there exists an increasing sequence $\{v_n\}_1^\infty \subset \mathscr{F}$ such that $v_0(z)=\sup_{v \in \mathscr{F}} v(z)=\lim_n v_n(z)$ for every $z \in R$. Set $v_0^*(p)=\lim_n v_n(p)$ for $p \in \Delta_{\mathrm{N}}$. Clearly $v_0^*(p) \leq f(p)$ on Δ_{N} and since

$$v_n(z)=\int_{\Delta_{\mathrm{N}}} P(z,p)\, v_n(p)\, d\mu(p) \leq \int_{\Delta_{\mathrm{N}}} P(z,p)\, f(p)\, d\mu(p)=u(z)$$

we conclude that

$$v_0(z)=\int_{\Delta_{\mathrm{N}}} P(z,p)\, v_0^*(p)\, d\mu(p) \leq u(z). \tag{49}$$

Suppose we can show that $v_0=u$. By Theorem 3 E we see that $(u\wedge n)|\Delta_{\mathbb{N}}=(u\cap n)|\Delta_{\mathbb{N}}$ and since $(u\wedge n)(z)=\int_{\Delta_{\mathbb{N}}}P(z,p)(u\wedge n)(p)\,d\mu(p)=\int_{\Delta_{\mathbb{N}}}P(z,p)\big(u(p)\cap n\big)\,d\mu(p)$ we have

$$u(z)=\int_{\Delta_{\mathbb{N}}}P(z,p)\,u(p)\,d\mu(p)=\int_{\Delta_{\mathbb{N}}}P(z,p)\,v_0^*(p)\,d\mu(p)$$
$$=\int_{\Delta_{\mathbb{N}}}P(z,p)\,f(p)\,d\mu(p).$$

On observing that $u(p)\geq v_n(p)$ we obtain $u(p)\geq v_0^*(p)$. In view of this and $f(p)-v_0^*(p)\geq 0$ on $\Delta_{\mathbb{N}}$ the above equalities give

$$\int_{\Delta_{\mathbb{N}}}\big(u(p)-v_0^*(p)\big)\,d\mu(p)=\int_{\Delta_{\mathbb{N}}}\big(f(p)-v_0^*(p)\big)\,d\mu(p)=0.$$

It follows that $u(p)=v_0^*(p)=f(p)$ μ-a.e. on $\Delta_{\mathbb{N}}$.

Hence we have only to show that $v_0=u$. Since $v_0\leq u$ it suffices to prove $v_0(z_0)=u(z_0)$. By the definition of integration there exists a finite decomposition $\Delta_{\mathbb{N}}=\bigcup_{i=1}^n E_i$ of $\Delta_{\mathbb{N}}$ into disjoint Borel sets E_i such that

$$0<\int_{\Delta_{\mathbb{N}}}f(p)\,d\mu(p)-\sum_{i=1}^n a_i\,\mu(E_i)<\frac{\varepsilon}{2} \tag{50}$$

where $\varepsilon>0$ is arbitrarily given and $a_i=\inf_{p\in E_i}f(p)$. By the regularity of μ there are compact sets K_i such that $K_i\subset E_i$ and $\mu(E_i-K_i)<\varepsilon/2a$, where $a=1+\sum_{i=1}^n a_i$.

Let $\{h\}$ be the totality of functions h in $B(\Delta_{\mathbb{N}})$ with $0\leq h\leq 1$ and $h=0$ on K_i. Then $\{h\}$ is increasingly directed and if we consider $\{h\}$ as a subset of $HB(R)$ in a natural manner, we can choose an increasing $\{h_n\}_1^\infty\subset\{h\}$ such that $h_0(z)=\lim_n h_n(z)$ exists on R and $h_0\geq h$ for every $h\in\{h\}$. Clearly $\int_{\Delta_{\mathbb{N}}}h_n(p)\,d\mu(p)\leq\int_{\Delta_{\mathbb{N}}-K_i}d\mu(p)$ and thus

$$\int_{\Delta_{\mathbb{N}}}h_0(p)\,d\mu(p)\leq\mu(\Delta_{\mathbb{N}}-K_i). \tag{51}$$

Again by the regularity of μ we can find for any $\eta>0$ a compact set K in $\Delta_{\mathbb{N}}-K_i$ such that $0<\mu(\Delta_{\mathbb{N}}-K_i)-\mu(K)<\eta$. Clearly there exists an $h_K\in\{h\}$ such that $h_K|K=1$. Then $\int_{\Delta_{\mathbb{N}}}h_0(p)\,d\mu(p)\geq\int_{\Delta_{\mathbb{N}}}h_k(p)\,d\mu(p)\geq\mu(K)>\mu(\Delta_{\mathbb{N}}-K_i)-\eta$. From this with (51) and the arbitrary choice of $\eta>0$ we obtain

$$\int_{\Delta_{\mathbb{N}}}h_0(p)\,d\mu(p)=\mu(\Delta_{\mathbb{N}}-K_i). \tag{52}$$

Since $h_K|K=1$ and $K\subset\Delta_{\mathbb{N}}-K_i$ is arbitrary, $h_0|(\Delta_{\mathbb{N}}-K_i)=1$. Thus by continuity, $h_0|(\overline{\Delta_{\mathbb{N}}-K_i})=1$. In view of $\mu(\overline{\Delta_{\mathbb{N}}-K_i})=\int_{\overline{\Delta_{\mathbb{N}}-K_i}}d\mu(p)\leq\int_{\Delta_{\mathbb{N}}}h_0(p)\,d\mu(p)$, (52) implies that $\mu(\overline{\Delta_{\mathbb{N}}-K_i})=\mu(\Delta_{\mathbb{N}}-K_i)$. Therefore $K_i'=\Delta_{\mathbb{N}}-(\overline{\Delta_{\mathbb{N}}-K_i})$ satisfies $\mu(K_i')=\mu(K_i)$ and $K_i'\subset K_i\subset E_i$. Since $\overline{\Delta_{\mathbb{N}}-K_i}$ is both open and closed in Δ_N, so is K_i'. The function v defined by $v|K_i'=a_i$ and $v|(\Delta_N-\bigcup_{i=1}^n K_i')=0$ is continuous on Δ_N and $v\leq f$ on Δ_N. By virtue of $\mu(E_i-K_i')=\mu(E_i-K_i)<\varepsilon/2a$, (50) implies

$$0<\int_{\Delta_N} f(p)\,d\mu(p)-\int_{\Delta_N} v(p)\,d\mu(p)<\varepsilon. \tag{53}$$

Clearly $v_0^*\geq v$ since $v\in B(\Delta_N)$ may be considered as belonging to $HB(R)$. Hence $u(z_0)-v_0(z_0)\leq u(z_0)-v(z_0)<\varepsilon$ and therefore $u(z_0)=v_0(z_0)$.

COROLLARY 1. *A function $f\in C(\Delta_N)$ is μ-integrable if and only if there exists a $u\in HB'(R)$ with $u|\Delta_N=f$. In this case*

$$u(z)=\int_{\Delta_N} P(z,p)\,u(p)\,d\mu(p). \tag{54}$$

For $u, v\in HB'(R)$, $u\vee v$ and $u\wedge v$ belong to $HB'(R)$ and

$$\begin{aligned}(u\vee v)(z)&=\int_{\Delta_N} P(z,p)\big(u(p)\cup v(p)\big)\,d\mu(p),\\ (u\wedge v)(z)&=\int_{\Delta_N} P(z,p)\big(u(p)\cap v(p)\big)\,d\mu(p).\end{aligned} \tag{55}$$

COROLLARY 2. *For an open set U, $\mu(\bar{U})=\mu(U)$.*

4 E. Operator B'. Recall (3 B) that $B'^2=B'$ and $I=B'+(I-B')$ give the direct sum decomposition $HP'(R)=HB'(R)+HP''(R)$. In view of Theorem 3 E, the space $HP''(R)$ is characterized by $HP''(R)=\{u\in HP'(R)\,|u|\,\Delta_N=0\}$. Therefore for $u\in HP'(R)$, $u|\Delta_N=B'u|\Delta_N+(I-B')\,u|\Delta_N=B'u|\Delta_N$.

THEOREM. *For every $u\in HP'(R)$*

$$(B'u)(z)=\int_{\Delta_N} P(z,p)\,u(p)\,d\mu(p). \tag{56}$$

From this and (55) we obtain:

COROLLARY. *For $u\in HP'(R)$*

$$B'u=\lim_{n,m\to\infty}(u\wedge n)\vee(-m). \tag{57}$$

4 F. Evans' Harmonic Function for a Set in Δ_N. A set of μ-measure zero in Δ_N can be characterized as follows:

THEOREM. *A Borel set $E\subset\Delta_N$ is of μ-measure zero if and only if there exists a positive function $u\in HB'(R)$ such that $u|E=\infty$.*

Let $\{U_n\}$ be a sequence of open sets $U_n \supset E$ with $\mu(U_n) \to \mu(E) = 0$, $U_n \supset U_{n+1}$, and $\mu(U_n) < 2^{-n}$. The function

$$u_n(z) = \int_{U_n} P(z, p)\, d\mu(p) \tag{58}$$

belongs to $HB(R)$ and $u_n|E = 1$. Set

$$u(z) = \sum_{n=1}^{\infty} u_n(z). \tag{59}$$

Since $u_n > 0$ and $u(z_0) < 1$ we infer that $u \in HB'(R)$ and $u|E = \infty$.

Conversely assume the existence of such a function u. Then

$$n\mu(E) \leq \int_{\Delta_{\mathrm{N}}} (u(p) \cap n)\, d\mu(p) \leq \int_{\Delta_{\mathrm{N}}} u(p)\, d\mu(p) = u(z_0)$$

for every $n = 1, 2, \ldots$. Thus $\mu(E) = 0$.

Remark. The theorem is also true, mutatis mutandis, for Royden's boundary if we replace $u|E = \infty$ by $\lim_{z \in R, z \to p} u(z) = \infty$ for every $p \in E$. The proof is the same.

5. Perron's Method

5 A. Perron's Family. Consider a real-valued bounded function f on Wiener's boundary Γ_{N} of a surface $R \notin O_G$. The family $\mathfrak{A}(R^*_{\mathrm{N}}, f)$ of superharmonic functions v on R with

$$\liminf_{z \in R, z \to p} v(z) \geq f(p) \tag{60}$$

for every point $p \in \Gamma_{\mathrm{N}}$ shall be referred to as *Perron's family* with respect to R^*_{N} and f. The functions

$$\overline{H}(z; R^*_{\mathrm{N}}, f) = \overline{H}(z; f) = \inf\{v(z) | v \in \mathfrak{A}(R^*_{\mathrm{N}}, f)\} \tag{61}$$

and

$$\underline{H}(z; R^*_{\mathrm{N}}, f) = \underline{H}(z; f) = -\overline{H}(z, -f) \tag{62}$$

for $z \in R$ are harmonic on R and $\overline{H} \geq \underline{H}$. If $\overline{H} = \underline{H}$ then we denote by $H(z; R^*_{\mathrm{N}}, f) = H(z; f)$ the common function. In this case f is said to be *resolutive* with respect to R^*_{N}. A point $p \in \Gamma_{\mathrm{N}}$ is by definition *regular* for the Dirichlet problem with respect to R^*_{N} if

$$\lim_{z \in R, z \to p} H(z; f) = f(p) \tag{63}$$

for every resolutive function f on Γ_{N} which is continuous at p.

Using Evans' harmonic function of 3 I we can prove the following analogue of Theorem III. 5 A:

THEOREM. *Every bounded real Borel function f on Γ_{N} is resolutive and*

$$H(z; R_{\mathrm{N}}^*, f) = \int_{\Gamma_{\mathrm{N}}} P(z,p)\, f(p)\, d\mu(p) \tag{64}$$

on R. The class of regular points in Γ_{N} coincides with Δ_{N}.

5 B. Measure Correspondence between Γ_{N} and Γ_{MI}. Let $(R_{\mathrm{N}}^*, R_{\mathrm{MI}}^*, \rho)$ be the fiber space defined in 2 F. The projection ρ maps R_{N}^* onto R_{MI}^* and $\rho(\Delta_{\mathrm{N}}) = \Delta_{\mathrm{MI}}$.

THEOREM. *For any Borel set $E \subset \Gamma_{\mathrm{MI}}$, $\rho^{-1}(E)$ is also a Borel set in Γ_{N} and*

$$\mu_{\mathrm{N}}(\rho^{-1}(E)) = \mu_{\mathrm{MI}}(E). \tag{65}$$

Let f be the characteristic function of E on R_{MI}^*. Then $f \circ \rho$ is the characteristic function of $\rho^{-1}(E)$ on R_{N}^*. Clearly $\mathfrak{A}(R_{\mathrm{N}}^*, f \circ \rho) = \mathfrak{A}(R_{\mathrm{MI}}^*, f)$. Thus $H(z; R_{\mathrm{N}}^*, f \circ \rho) = H(z; R_{\mathrm{MI}}^*, f)$, and (64) at $z = z_0$ implies (65).

By (65) we conclude that $U_{HB} \subset U_{\widetilde{HD}}$. In fact let $R \in U_{HB}$. Then by Theorem 3 K, Γ_{N} contains a point $p \in \Delta_{\mathrm{N}}$ isolated in Δ_{N} such that $\mu_{\mathrm{N}}(p) > 0$ and therefore $\mu_{\mathrm{N}}(\rho^{-1}(\rho(p))) > 0$. Clearly by (65), $q = \rho(p)$ is a point in Γ_{MI} with $\mu_{\mathrm{MI}}(q) > 0$. Hence $R \in U_{\widetilde{HD}}$.

At this point we remark:

If $p \in \Gamma_{\mathrm{N}}$ has positive μ_{N}-measure then p is isolated in Δ_{N}.

Obviously $p \in \Delta_{\mathrm{N}}$. Since $\mu_{\mathrm{N}}(\overline{\Delta_{\mathrm{N}} - p}) = \mu_{\mathrm{N}}(\Delta_{\mathrm{N}} - p)$ and $\overline{\Delta_{\mathrm{N}} - p}$ is both open and closed in Δ_{N}, $p = \Delta_{\mathrm{N}} - (\overline{\Delta_{\mathrm{N}} - p})$ is again open and closed in Δ_{N}, i.e. p is isolated in Δ_{N}.

5 C. Compactifications of Subsurfaces. Let $R \notin O_G$ be an open Riemann surface and G a normal subregion. Denote by $\bar{G} = \bar{G}^{\mathrm{N}}$ the closure of G in R_{N}^* and set $b_G = b_G^{\mathrm{N}} = (\bar{G} - \partial G) \cap \Gamma_{\mathrm{N}}$. The harmonic measure $\omega_{\partial G}$ is defined as in III. 5 E.

THEOREM. *There exists a unique continuous mapping $j = j(G_{\mathrm{N}}^*, \bar{G}^{\mathrm{N}})$ of G_{N}^* onto $\bar{G}$ which fixes G elementwise. The projection j is a homeomorphism of $G \cup j^{-1}(b_G^{\mathrm{N}})$ onto $G \cup b_G^{\mathrm{N}}$. If E is a Borel set in b_G^{N} then $\mu_R(E) > 0$ if and only if $\mu_G(j^{-1}(E)) > 0$. If F is a Borel set in ∂G then $\omega_{\partial G}(F) = \mu_G(j^{-1}(F))$.*

The proof is similar to that of Theorems III. 5 B – 5 E.

5 D. Stoïlow's Compactification R_S^*. Let R be an open Riemann surface and $\{R_n\}_1^\infty$ an exhaustion of R such that each open set $R - \bar{R}_n$ is decomposed into a finite number of noncompact components

$K_1^{(n)}, \ldots, K_{N(n)}^{(n)}$. A *determining sequence* is a sequence $\{K_{i_n}^{(n)}\}_1^\infty$ such that

$$K_{i_1}^{(1)} \supset K_{i_2}^{(2)} \supset \cdots \supset K_{i_n}^{(n)} \supset \cdots. \tag{66}$$

Another exhaustion $\{R_n'\}_1^\infty$ gives another determining sequence $\{K_{i_n}'^{(n)}\}_1^\infty$. Two determining sequences $\{K_{i_n}^{(n)}\}_1^\infty$ and $\{K_{i_n}'^{(n)}\}_1^\infty$ are said to be equivalent if for any n there exists an m such that $K_{i_n}^{(n)} \supset K_{i_m}'^{(m)}$ and conversely. An equivalence class of determining sequences is called a *Stoïlow end.*

For a fixed exhaustion $\{R_n\}_1^\infty$ the totality of determining sequences corresponds in a one-to-one and onto manner to the totality of ends of R. Let a determining sequence $\{K_{i_n}^{(n)}\}_1^\infty$ give Stoïlow's end $e\{K_{i_n}^{(n)}\}$ and let R_S^* be the union of R and the family of ends $e\{K_{i_n}^{(n)}\}$. Denote by $\tilde{e}(K_{i_n}^{(n)})$ the totality of ends which correspond to determining sequences containing $K_{i_n}^{(n)}$. For the base of the neighborhood system at $e\{K_{i_n}^{(n)}\}$ choose $K_{i_n}^{(n)} \cup \tilde{e}(K_{i_n}^{(n)})$. Then R_S^* is a Hausdorff space. From the theorem we are going to prove below it will follow that R_S^* is compact. The space R_S^* is called the *Stoïlow compactification* of R. A point in $R_S^* - R$ is often referred to as a Stoïlow ideal boundary component or as a Stoïlow ideal boundary point.

Let R^* be Royden's (resp. Wiener's) compactification of R and let Γ be Royden's (resp. Wiener's) boundary of R. Take the decomposition $\Gamma = \bigcup_k \Gamma_k$ of Γ into components Γ_k. Consider each Γ_k as one point and choose an open set in R^* containing Γ_k as a neighborhood of the point Γ_k. The resulting quotient space, which we denote by R^*/S, is again a compact Hausdorff space. The natural mapping φ of R^* onto R^*/S defined by $\varphi(p) = p$ for $p \in R$ and $\varphi(p) = \Gamma_k$ for $p \in \Gamma_k$ is continuous. Therefore as the image of the compact set R^*, R^*/S is compact.

Let $\alpha\{K_{i_n}^{(n)}\} = \cap \bar{K}_{i_n}^{(n)}$ in R^*.

THEOREM. *The set α $\{K_{i_n}^{(n)}\}$ coincides with a Γ_k, and every Γ_k is obtained in this manner.*

Since $\bigcap \bar{K}_{i_n}^{(n)}$ is a connected compact set in R^*, $\alpha\{K_{i_n}^{(n)}\}$ is a continuum in R^* and actually in Γ. There exists a component Γ_k with $\alpha\{K_{i_n}^{(n)}\} \cap \Gamma_k \neq \emptyset$. We wish to show that $\alpha = \alpha\{K_{i_n}^{(n)}\} = \Gamma_k$. Clearly $\alpha \subset \Gamma_k$ since Γ_k is a (connected) component. Contrary to the assertion assume that there exists a point $p_0 \in \Gamma_k - \alpha$. Take an n with $p \notin \bar{K}_{i_n}^{(n)}$ and an analytic Jordan curve $C \subset K_{i_n}^{(n)}$ homologous to $\partial K_{i_n}^{(n)}$. Denote by D the part of $K_{i_n}^{(n)}$ bounded by $C \cup \partial K_{i_n}^{(n)}$. Let $w \in \mathbb{M}(R) \subset \mathbb{N}(R)$ such that $w|(R - K_{i_n}^{(n)}) = 1$, $w|(K_{i_n}^{(n)} - \bar{D}) = 0$, and $w|D \in H(D)$. Clearly w takes on only the two values 0 and 1 on Γ and $w = 0$ on $\Gamma \cap \bar{K}_{i_n}^{(n)} \supset \alpha$. Since $w|\alpha = 0$, $w(p_0) = 1$, and $\alpha \cup p_0 \subset \Gamma_k$, Γ_k is not connected, a contradiction.

Conversely take Γ_k and $p_0 \in \Gamma_k$. It follows from the behavior of the above function w that $\bar{K}_1^{(n)}, \bar{K}_2^{(n)}, \ldots, \bar{K}_{N(n)}^{(n)}$ are pairwise disjoint. Hence there exists only one $\bar{K}_{i_n}^{(n)}$ containing p_0. By this selection we obtain a determining sequence $\{K_{i_n}^{(n)}\}_1^\infty$ and in the same manner as above conclude that $\Gamma_k = \alpha\{K_{i_n}^{(n)}\}$.

The proof is herewith complete.

Identification of $e\{K_{i_n}^{(n)}\}$ with $\alpha\{K_{i_n}^{(n)}\}$ clearly gives the identification of R_S^* with R^*/S including topologies. Thus R_S^* is a compact Hausdorff space. Let I be the map of R^*/S onto R_S^* given by this identification. Then $\rho = \rho(R^*, R_S^*) = I \circ \varphi$ defines the fiber space (R^*, R_S^*, ρ). Since R is dense in both R^* and R_S^* such a map ρ is unique.

COROLLARY. *There exists a unique continuous mapping $\rho = \rho(R_{\mathfrak{N}}^*, R_S^*)$ (resp. $\rho(R_{\mathfrak{M}}^*, R_S^*)$) of $R_{\mathfrak{N}}^*$ (resp. $R_{\mathfrak{M}}^*$) onto R_S^* such that ρ is the identity between R and $\rho^{-1}(R) = R$.*

It is easy to see that

$$\rho(R_{\mathfrak{N}}^*, R_S^*) = \rho(R_{\mathfrak{M}}^*, R_S^*) \circ \rho(R_{\mathfrak{N}}^*, R_{\mathfrak{M}}^*). \tag{67}$$

5 E. Harmonic Measure on Γ_S. We call $\Gamma_S = R_S^* - R$ the *Stoïlow boundary* of R. Exactly as in III.5 A and IV.5 A we define $\mathfrak{A}(R_S^*, f)$ and $\overline{H}(z; R_S^*, f)$. For a Borel set $E \subset \Gamma_S$ let f_E be its characteristic function. The *harmonic measure* $\mu_S(E)$ of E on Γ_S taken at $z_0 \in R$ is given by

$$\mu_S(E) = \overline{H}(z_0; R_S^*, f_E). \tag{68}$$

In analogy with 5 B we can easily see:

THEOREM. *For a Borel set E in Γ_S*

$$\mu_S(E) = \mu_{\mathfrak{M}}\big(\rho(R_{\mathfrak{M}}^*, R_S^*)^{-1}(E)\big) = \mu_{\mathfrak{N}}\big(\rho(R_{\mathfrak{N}}^*, R_S^*)^{-1}(E)\big). \tag{69}$$

5 F. Test for Quasiboundedness. Once more we return to the property of quasiboundedness. Let $u \in HP(R)$, $R \notin O_G$, and take a positive number a. Consider the *a-level line* of u:

$$\mathscr{L}(u; a) = \{z \in R \mid u(z) = a\}. \tag{70}$$

The *harmonic measure* of an arbitrary closed subset F of R relative to R taken at $z \in R$ is

$$\omega(F; z) = \inf s(z) \tag{71}$$

where s runs over all positive superharmonic functions on R such that $s \geq 1$ on F. Clearly $a\omega(\mathscr{L}(u; a); z) \leq u(z)$ on $\{z \in R \mid u(z) \leq a\}$, i.e.

$$\omega(\mathscr{L}(u; a); z) = O\left(\frac{1}{a}\right)$$

as $a \to \infty$ for fixed $z \in R$.

If u is bounded then $\mathscr{L}(u; a)=\emptyset$ for $a>\sup_R u$ and therefore $\omega(\mathscr{L}(u; a); z)=0$ for such a. This suggests that the asymptotic magnitude of $\omega(\mathscr{L}(u; a); z)$ as $a\to\infty$ can be used as a quasiboundedness criterion:

THEOREM. *For $u\in HP(R)$, $R\notin O_G$, the following conditions are equivalent by pairs:*

(a) $u\in HB'(R)$,

(b) $\lim_{a\to\infty} a\,\omega(\mathscr{L}(u; a); z)=0$ *for some (and hence for all) points* $z\in R$,

(c) $\liminf_{a\to\infty} a\,\omega(\mathscr{L}(u; a); z)=0$ *for some (and hence for all) points* $z\in R$.

It is clear that (b) implies (c). For this reason we have only to prove the implications (a)→(b) and (c)→(a). In both cases we may assume that u is unbounded on R and $R\notin O_{HP}$.

First we show that (a)→(b). Fix a point $z_0\in R$ and denote by G_a the component of $\{z\in R|u(z)<a\}$ $(a>u(z_0))$ containing the point z_0. Clearly $R=\bigcup_{a>u(z_0)} G_a$. Let $\varphi_a\in\mathbb{N}(R)$ such that $0\le\varphi_a\le 1$, $\varphi_a|\bar G_a=1$ and $\varphi_a|(R^*_{\mathbb{N}}-\bar G_{a+1})=0$. Then $\varphi_a u\in\mathbb{N}(R)$. For

$$v_a=\pi^{\mathbb{N}}_{K_a}(\varphi_a u)$$

with $K_a=\overline{(R^*_{\mathbb{N}}-\bar G_{a+1})}$ we have $0\le v_a\le v_{a'}\le u$ $(a<a')$ on R and $v_a=u$ on $\Delta_{\mathbb{N}}\cap\bar G_a$. Hence

$$v=\lim_{a\to\infty} v_a$$

exists on R and $v\le u$, which gives $v\in HB'(R)$. Obviously $v=u$ on $\Delta_{\mathbb{N}}$ except on $F_\infty=\{p\in\Delta_{\mathbb{N}}|u(p)=\infty\}$. Since $\mu_{\mathbb{N}}(F_\infty)=0$ (cf. 4 F)

$$\begin{aligned} u(z_0)&=\int_{\Delta_{\mathbb{N}}} u(p)\,d\mu_{\mathbb{N}}(p)=\int_{\Delta_{\mathbb{N}}-F_\infty} u(p)\,d\mu_{\mathbb{N}}(p)=\int_{\Delta_{\mathbb{N}}-F_\infty} v(p)\,d\mu_{\mathbb{N}}(p)\\ &=\int_{\Delta_{\mathbb{N}}} v(p)\,d\mu_{\mathbb{N}}(p)=v(z_0)\end{aligned}$$

where the center of $\mu_{\mathbb{N}}$ is z_0.

It follows that $v\equiv u$ on R, i.e. the function

$$w_a(z)=u(z)-v_a(z)$$

is superharmonic on R, $w_a|\mathscr{L}(u; a+1)=a+1$, and

$$\lim_{a\to\infty} w_a(z_0)=0. \tag{72}$$

We conclude that

$$\omega(\mathscr{L}(u; a+1); z_0)\le\frac{1}{a+1}\,w_a(z_0).$$

This with (72) implies (b).

We proceed to the implication (c)→(a). Choose a sequence $\{a_n\}$ of positive numbers such that $u(z_0)<a_1<a_2<\cdots$, $\lim_{n\to\infty} a_n=\infty$, and

$a_n \omega(\mathscr{L}(u;a_n);z_0) < 1/n^3$. Set $b_n = n\,a_n$,

$$w_n(z)|G_{a_n} = \omega(\mathscr{L}(u;a_n);z), \qquad w_n(z)|(R-G_{a_n}) = 1,$$

and

$$w(z) = \sum_{n=1}^{\infty} b_n w_n(z)$$

on R. The function $w(z)$ is positive superharmonic on R and continuous on R_{N}^*. The difference $h(z) = u(z) - (B'u)(z)$ is in $HP''(R)$; it is positive on R and vanishes on Δ_{N}. Thus

$$0 \le h(p) \le u(p) \le a_n = \frac{b_n}{n} \le \frac{w(p)}{n}$$

on $(\bar{G}_{a_n} \cap \Delta_{\mathrm{N}}) \cup \partial G_{a_n}$. By 3 J, $0 \le h(z) \le w(z)/n$ on G_{a_n}. On letting $n \to \infty$ we conclude that $h \equiv 0$, i.e. $u = B'u \in HB'(R)$.

This completes the proof.

We illustrate the theorem by an example. Consider the circular slit disk

$$R = \{z \mid 0 < |z| < 1\} - \bigcup_{n=1}^{\infty} \mathscr{G}_n$$

where $\mathscr{G}_n$ is a circular slit $\{r_n e^{i\theta} \mid -\alpha_n \le \theta \le \alpha_n\}$ with

$$1 > r_1 > r_2 > \cdots > r_n > \cdots, \qquad \lim_{n\to\infty} r_n = 0$$

and $0 \le \alpha_n < \pi$. Clearly $\log(1/|z|) \in HP(R)$.

Choose α_n so large that the harmonic measure of the complementary circular slit $\{r_n e^{i\theta} \mid \alpha_n \le \theta \le 2\pi - \alpha_n\}$ with respect to $|z| < 1$ at some $z_0 \in R$ is smaller than $\varepsilon_n/\log(1/r_n)$ with $\varepsilon_n \to 0$. Then condition (c) is satisfied and $\log 1/|z| \in HB'(R)$.

Next take α_n so small that the harmonic measure of $\{r e^{i\theta} \mid 0 \le \theta \le 2\pi\}$ with respect to R is not less than $\frac{1}{2}\log(1/r)$ for any $r \in (0,1)$. The extreme case is $\alpha_n = 0$. Then condition (b) is not satisfied and thus $\log 1/|z| \notin HB'(R)$.

6. Φ-Bounded Harmonic Functions

6 A. Φ-Boundedness. We have considered the classes HP', HP, HB', HB, HD and their intersections and also the corresponding null classes. To organize these systematically we shall consider a new class of harmonic functions, called Φ-bounded harmonic functions.

Throughout this no. the symbol $\Phi(t)$ denotes *a nonnegative finite real-valued function defined on the real half-line* $[0,\infty)$. We do not impose any additional conditions such as measurability, continuity, monotonicity or the like unless they are explicitly stated. The following quantities will play a fundamental role:

$$\bar{d}(\Phi)=\limsup_{t\to\infty}\frac{\Phi(t)}{t}, \qquad \underline{d}(\Phi)=\liminf_{t\to\infty}\frac{\Phi(t)}{t}. \tag{73}$$

Trivially $0\le\underline{d}(\Phi)\le\bar{d}(\Phi)\le\infty$.

A harmonic function u on R is called *Φ-bounded* if the composite function $\Phi(|u|)$ admits a harmonic majorant, i.e. if there exists an HP-function v such that

$$\Phi(|u|)\le v \tag{74}$$

on R. We denote by $H\Phi(R)$ the family of Φ-bounded harmonic functions on R and by $O_{H\Phi}$ the corresponding null class of Riemann surfaces, i.e. the totality of Riemann surfaces R for which $H\Phi(R)$ consists of only constants.

6 B. Determination of $O_{H\Phi}$. Despite the generality of the concept of Φ-boundedness the corresponding null class $O_{H\Phi}$ turns out to be quite specific: it coincides with either O_{HB} or O_{HP}. To see this we first prove the following general inclusion relations:

THEOREM. *If $\bar{d}(\Phi)<\infty$ then $O_{H\Phi}\subset O_{HP}$. If $\bar{d}(\Phi)=\infty$ then $O_{H\Phi}\supset O_{HB}$.*

We start with the case $\bar{d}(\Phi)<\infty$. There exists a point $t_0\in[0,\infty)$ such that $\Phi(t)\le C\,t$ $(t_0\le t<\infty)$ with a finite constant C. Assume the existence of a nonconstant $u\in HP(R)$. For $v=u+t_0\in HP(R)$ we have $\Phi(|v|)=\Phi(v)\le C\,v$ on R and conclude that v is a nonconstant $H\Phi$-function on R.

Consider next the case $\bar{d}(\Phi)=\infty$. Assume that there exists a nonconstant $H\Phi$-function u on R and a fortiori an HP-function h on R such that $\Phi(|u|)\le h$ on R; we may postulate that u is unbounded. We have to show that $R\notin O_{HB}$. Suppose this were not so:

$$R\in O_{HB}. \tag{75}$$

Since $\bar{d}(\Phi)=\infty$ we can find a strictly increasing sequence $\{r_n\}_1^\infty$ of positive numbers r_n such that $\lim_{n\to\infty} r_n=\infty$, $\Phi(r_n)>0$,

$$G_n=\{z\in R\,|\,|u(z)|<r_n\}\neq\emptyset,$$

and

$$\lim_{n\to\infty} a_n=0 \tag{76}$$

with $a_n=r_n/\Phi(r_n)$. Clearly $G_1\subset G_2\subset\cdots$ and $R=\bigcup_{n=1}^\infty G_n$.

First we show that $G_n\notin SO_{HB}$ for all sufficiently large n. Since $a_n h-|u|$ is superharmonic, bounded from below on G_n, and

$$a_n h-|u|\ge a_n\Phi(|u|)-|u|=a_n\Phi(r_n)-r_n=0$$

on ∂G_n, $G_n \in SO_{HB}$ would imply that $a_n h - |u| \geq 0$ on G_n. Since $a_n \to 0$ we would have $u \equiv 0$ on R, which contradicts the nonconstancy of u. Hence by choosing a suitable subsequence of $\{r_n\}$ if necessary we may assume that

$$G_n \notin SO_{HB} \qquad (n=1, 2, \ldots). \tag{77}$$

If $G_n - \bar{G}_1 \notin SO_{HB}$ for some $n>1$ then by the two region test (3 M), $G_1 \notin SO_{HB}$ and $G_n - \bar{G}_1 \notin SO_{HB}$ imply that $R \notin O_{HB}$, which contradicts (75). From this we infer that

$$G_n - \bar{G}_1 \in SO_{HB} \qquad (n=2, 3, \ldots). \tag{78}$$

The function

$$w_n = a_n h + r_1 - |u|$$

is superharmonic and bounded from below on G_n and also on $G_n - \bar{G}_1$. As before we see that $w_n \geq a_n h - |u| \geq 0$ on ∂G_n. Clearly $w_n \geq r_1 - |u| = 0$ on ∂G_1. Since $G_n - \bar{G}_1 \in SO_{HB}$ we now conclude that $w_n \geq 0$ on G_n, or $|u| \leq a_n h + r_1$ there. Hence (76) implies $|u| \leq r_1$ on R. This contradicts (75) and we must have $R \notin O_{HB}$.

We have proved the theorem.

The function $\Phi(t)$ is said to be *completely unbounded* on $[0, \infty)$ if $\Phi(t)$ is not bounded in any neighborhood of any point in $[0, \infty)$. In the problem of Φ-boundedness a bounded Φ and a completely unbounded Φ are of little interest, the class $O_{H\Phi}$ being determined as follows:

COROLLARY. (a) *If Φ is bounded on $[0, \infty)$ then $O_{H\Phi}$ consists of closed Riemann surfaces.*

(b) *If Φ is completely unbounded on $[0, \infty)$ then $O_{H\Phi}$ consists of all Riemann surfaces.*

(c) *If Φ is neither bounded nor completely unbounded then $O_{H\Phi} = O_{HP}$ or O_{HB} according as $\bar{d}(\Phi) < \infty$ or $\bar{d}(\Phi) = \infty$.*

If Φ is bounded then every nonconstant harmonic function is a nonconstant $H\Phi$-function. Thus $O_{H\Phi}$ consists of Riemann surfaces carrying no nonconstant harmonic functions, but these are the closed Riemann surfaces.

If Φ is completely unbounded then for any nonconstant harmonic function u on R, $\Phi(|u|)$ is again completely unbounded on R. Thus there exist no nonconstant $H\Phi$-functions on any Riemann surface, i.e. $O_{H\Phi}$ consists of all Riemann surfaces.

Now assume that Φ is neither bounded nor completely unbounded. Let $\bar{d}(\Phi) = \infty$ and suppose there exists a nonconstant HB-function u on R. We can find an interval (a, b) on which $\Phi(t) < c = \text{const}$. The function $v = (a+b)/2 + ((b-a)/2)(\sup_R |u|)^{-1} u$ is nonconstant and bounded and $\Phi(|v|) = \Phi(v) < c$ on R. Thus $O_{H\Phi} \subset O_{HB}$. This with Theorem 6 B gives $O_{H\Phi} = O_{HB}$.

Next let $\bar{d}(\Phi)<\infty$. We know that $O_{HP}\supset O_{H\Phi}$. Suppose there exists an $R\in O_{HP}-O_{H\Phi}$. Let u be a nonconstant $H\Phi$-function on R. Then $\Phi(|u|)\le c=\text{const}$ on R. Since Φ is unbounded and $|u|(R)$ is connected in $[0,\infty)$, u is bounded on R. Then $\sup_R|u|+u$ is a nonconstant HP-function on R, and this contradicts $R\in O_{HP}$. Thus $O_{H\Phi}=O_{HP}$.

6 C. Harmonic Measures on the Disk. We shall construct examples of harmonic functions on the disk which show that $H\Phi$ does not coincide with HP' or HB' even for a simple Φ. To this end we first give some preliminaries.

Let U be the unit disk $|z|<1$. Take an arc X in the unit circumference ∂U: $|z|=1$. Denote by $w(z;X)$ the harmonic measure of X taken at $z\in U$ with respect to U:

$$w(z;X)=\frac{1}{2\pi}\int_X \frac{1-r^2}{1-2r\cos(\theta-\varphi)+r^2}\,d\varphi,\qquad z=r\,e^{i\theta}. \tag{79}$$

Let α be the length of X and β the angle subtended by the arc X at $z\in U$. Then

$$w(z;X)=\frac{2\beta-\alpha}{2\pi}. \tag{80}$$

To see this let $\zeta=e^{i\varphi}$. Then (79) takes the form

$$w(z;X)=\frac{1}{2\pi}\int_X \frac{1-|z|^2}{|z-\zeta|^2}\,d\varphi. \tag{81}$$

Using $d\varphi$ define $d\sigma$ as indicated in the infinitesimal scheme of Fig. 11. By the similarity of the infinitesimal triangles $\Delta z\,AB$ and $\Delta z\,CD$ we obtain

$$d\varphi : d\sigma=|z-\zeta| : \frac{1-|z|^2}{|z-\zeta|},$$

that is

$$\frac{1-|z|^2}{|z-\zeta|^2}\,d\varphi=d\sigma. \tag{82}$$

From $\Delta z\,B\,C$ we see that the infinitesimal angle $\angle A\,z\,B=d\beta$ is $\frac{1}{2}(d\varphi+d\sigma)$. Therefore

$$\frac{1-|z|^2}{|z-\zeta|^2}\,d\varphi=2\,d\beta-d\varphi.$$

This with (81) gives

$$w(z;X)=\frac{1}{2\pi}\int_X (2\,d\beta-d\varphi)=\frac{2\beta-\alpha}{2\pi}.$$

Let L_X be the line segment connecting the end points of X. It follows from (80) that

$$w(0;X)=\frac{\alpha}{2\pi}, \tag{83}$$

$$w(z;X)|L_X=1-\frac{\alpha}{2\pi}. \tag{84}$$

Denote by A_j the arc in ∂U with end points 1 and $e^{i\alpha_j}$ $(j=1,2)$ where $0<\alpha_1<\alpha_2\le\pi/2$. Let $\tilde{A}_j$ be the arc with end points 1 and $e^{-i\alpha_j}$ and set

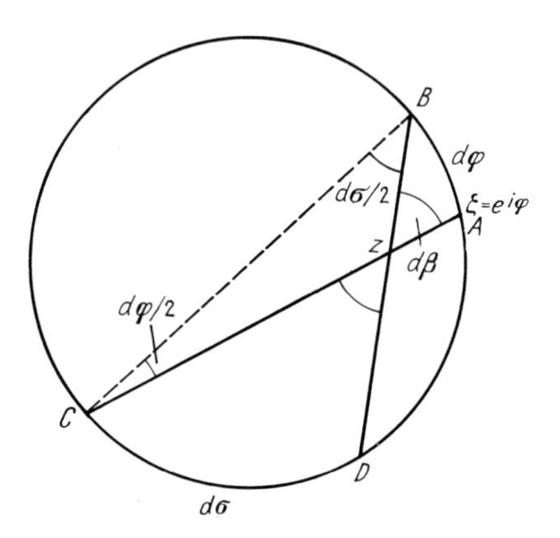

Fig. 11

$A_j'=A_j\cup\tilde{A}_j$. Write L_2 for the line segment $L_{A_2'}$ connecting the end points of A_2'.

We shall prove that

$$|w(z;A_1)-w(z;\tilde{A}_1)||L_2<\frac{\pi^4}{4}\cdot\frac{\alpha_1^2}{(\alpha_2^2-\alpha_1^2)^2}. \tag{85}$$

To this end denote the points $e^{i\alpha_1}$, $\frac{1}{2}(e^{i\alpha_1}+e^{-i\alpha_1})$, $e^{i\alpha_1}$, 1, $\frac{1}{2}(e^{i\alpha_2}+e^{-i\alpha_2})$, and $z\in L_2$ with $\mathrm{Im}(z)\ge 0$ by A, B, C, D, E, and P. Set $\overline{PA}=a$, $\overline{PB}=b$, $\overline{PC}=c$, $\overline{AB}=\overline{BC}=d$, and $\overline{EB}=k$. By (80)

$$w(z;A_1)-w(z;\tilde{A}_1)=\frac{\angle APD-\angle DPC}{\pi}.$$

Let $\angle APB=\theta_1$ and $\angle BPC=\theta_2$. Then $\angle APD\le\theta_1$ and $\angle DPC\ge\theta_2$. Consequently

$$0\le w(z;A_1)-w(z;\tilde{A}_1)\le\frac{\theta_1-\theta_2}{\pi}.$$

By the cosine theorem applied to d^2 in the triangles APB and BPC we obtain

$$2ab\cos\theta_1 = a^2+b^2-d^2, \qquad 2bc\cos\theta_2 = b^2+c^2-d^2,$$

which give

$$2abc(\cos\theta_2-\cos\theta_1)=(c-a)(ac-b^2+d^2).$$

Pappus' identity applied to the triangle APC yields $a^2+c^2=2(b^2+d^2)$. Thus

$$\sin\tfrac{1}{2}(\theta_1-\theta_2)=(c-a)\frac{4d^2-(a-c)^2}{8abc\sin\frac{1}{2}(\theta_1+\theta_2)}.$$

Here we have

$$ac\sin\tfrac{1}{2}(\theta_1+\theta_2)\geq ac\sin\tfrac{1}{2}(\theta_1+\theta_2)\cos\tfrac{1}{2}(\theta_1+\theta_2)$$

$$=\frac{ac}{2}\sin\angle APC=\Delta APC=\Delta AEC=dk.$$

The triangle inequality applied to APC gives $c-a<2d$. In view of this and $b\geq k$ we conclude that $\sin\frac{1}{2}(\theta_1-\theta_2)<d^2k^{-2}$. Since $\sin\theta\geq(2/\pi)\theta$ for $0\leq\theta<\frac{1}{2}\pi$,

$$\theta_1-\theta_2<\pi d^2k^{-2}.$$

Observe that $d=\sin\alpha_1<\alpha_1$ and

$$k=\cos\alpha_1-\cos\alpha_2=2\sin\tfrac{1}{2}(\alpha_1+\alpha_2)\sin\tfrac{1}{2}(\alpha_2-\alpha_1)>(2/\pi^2)(\alpha_2^2-\alpha_1^2).$$

Hence we obtain (85).

We shall use (85) in the following particular case:

LEMMA. *Let $0<\alpha_1<\alpha_2/\sqrt{2}$ and $\alpha_2\leq\pi/2$. There exists a universal constant s $(\leq\pi^4)$ such that*

$$|w(z;A_1)-w(z;\tilde{A}_1)||L_2<s\cdot\frac{\alpha_1^2}{\alpha_2^4}. \tag{86}$$

6 D. Convergence on the Boundary. Let (R,γ) be a bordered surface with border γ. Consider a sequence $\{u_n\}_1^\infty\subset HP(R)$ such that

$$u=\lim_{n\to\infty}u_n \tag{87}$$

exists on R.

LEMMA. *Suppose u_n is continuous on $R\cup\gamma$, $u_n\in HP(R)$, and $\{u_n|\gamma\}_1^\infty$ converges uniformly on every compact subset of γ. Then u is continuous on $R\cup\gamma$ and $u|\gamma=\lim_n u_n|\gamma$.*

Take an arbitrary relatively compact open arc $\alpha\subset\gamma$. We have only to show that u is continuous on $R\cup\alpha$ and $u|\alpha=\lim_n u_n|\alpha$. Join the end points of α by a simple analytic arc $\beta\subset R$ so that the subregion F of R

bounded by $\alpha \cup \bar{\beta}$ is simply connected. By the Riemann mapping theorem we may choose $F = U = \{|z| < 1\}$ and $\alpha \cup \bar{\beta} = \partial U = \{|z| = 1\}$. Hereafter we shall consider u_n and u only on U. Let

$$v_n(z) = \frac{1}{2\pi} \int_\alpha \frac{1-|z|^2}{|z-\zeta|^2} u_n(\zeta) |d\zeta|$$

and

$$v(z) = \frac{1}{2\pi} \int_\alpha \frac{1-|z|^2}{|z-\zeta|^2} (\lim_n u_n(\zeta)) |d\zeta|.$$

Clearly v, v_n are continuous on $U \cup \alpha$ and $v|\alpha = (\lim_n u_n)|\alpha$, $v_n|\alpha = u_n|\alpha$. Moreover $v = \lim_n v_n$ on $U \cup \alpha$. By replacing u_n, u by $u_n - v_n$, $u - v$ we may assume that $u_n|\alpha = 0$. Thus we must show that u has continuous boundary values zero at α.

Define a regular Borel measure μ_n on ∂U by

$$d\mu_n(\zeta) = \frac{1}{2\pi} u_n(\zeta) |d\zeta|. \tag{88}$$

By Poisson's formula

$$u_n(z) = \int_{\partial U} \frac{1-|z|^2}{|z-\zeta|^2} d\mu_n(\zeta). \tag{89}$$

Since $\mu_n(\partial U) = u_n(0)$ is bounded, by the selection theorem we may assume on choosing a suitable subsequence that there exists a regular Borel measure μ with

$$\lim_n \int_{\partial U} \lambda(\zeta)\, d\mu_n(\zeta) = \int_{\partial U} \lambda(\zeta)\, d\mu(\zeta) \tag{90}$$

for any real-valued continuous function λ on ∂U. In particular for $\lambda(\zeta) = (1-|z|^2)/|z-\zeta|^2$ we conclude by (87) that

$$u(z) = \int_{\partial U} \frac{1-|z|^2}{|z-\zeta|^2} d\mu(\zeta). \tag{91}$$

By (88) and $u_n|\alpha = 0$, $\mu_n(\alpha) = 0$. Therefore

$$\mu(\alpha) = 0. \tag{92}$$

Take a point $z_1 \in \bar{\alpha}$ and let ρ be the distance between z_1 and $\partial U - \alpha$. From (91) and (92) it follows that

$$0 \le u(z) \le 4\rho^{-2} \mu(\partial U)(1-|z|^2)$$

for $z \in U$, $|z - z_1| < \rho/2$. This shows that u has continuous boundary values 0 at z_1 and thus at α.

6 E. First Example. We are now ready to give an example of a function Φ which is continuous and increasing and has $\bar{d}(\Phi)=\infty$, $\underline{d}(\Phi)=0$. We shall also furnish an $H\Phi$-function u on the disk U: $|z|<1$ which is not an HP'-function.

EXAMPLE. *Let p be a constant such that $0<p<\min(4^{-1}, 4^{-s})$ where s is the constant of Lemma 6 C, and let $\{p_n\}_1^\infty$ be the sequence of numbers $p_1=p$ and*

$$p_n=(p^{4^\nu})^{2^\nu+\mu} \tag{93}$$

for $n=2^\nu+\mu$ ($\nu=0, 1, 2, \dots; \mu=1, \dots, 2^\nu$). On the unit circle consider the arcs

$$A_n=\left\{e^{i\theta} \,|\, 0\le\theta\le 2p_n\frac{\pi}{n}\right\}, \qquad \tilde{A}_n=\left\{e^{i\theta} \,|\, -2p_n\frac{\pi}{n}\le\theta\le 0\right\}. \tag{94}$$

Let $\{r_\nu\}_1^\infty$ and $\{b_\nu\}_1^\infty$ be the sequences of positive numbers

$$r_\nu=2/(p^{4^{\nu-1}})^{2^\nu}, \qquad b_\nu=2^{\nu/2}\, r_\nu. \tag{95}$$

Define $\Phi(t)$ on $[0,\infty)$ by

$$\Phi(t)=\begin{cases} 0 & \text{for } t\in[0,r_1],\\ b_1(t-r_1) & \text{for } t\in[r_1, r_1+1],\\ b_\nu & \text{for } t\in[r_\nu+1, r_{\nu+1}] \qquad (\nu=1,2,\dots),\\ b_\nu+(b_{\nu+1}-b_\nu)(t-r_{\nu+1}) & \text{for } t\in[r_{\nu+1}, r_{\nu+1}+1] \qquad (\nu=1,2,\dots).\end{cases} \tag{96}$$

On U consider

$$u(z)=\sum_{n=1}^{n}\left(w(z;A_n)-w(z;\tilde{A}_n)\right)/p_n. \tag{97}$$

Then

(a) *$\Phi(t)$ is continuous and increasing, with $\bar{d}(\Phi)=\infty$ and $\underline{d}(\Phi)=0$,*
(b) *$u(z)$ is well-defined and harmonic on U,*
(c) *$u\in H\Phi(U)$,*
(d) *$u\notin HP'(U)$.*

Proof of (a). By definition $\Phi(t)$ is continuous and increasing. Since

$$\Phi(r_\nu+1)/(r_\nu+1)=b_\nu/(r_\nu+1)=2^{\nu/2}\, r_\nu/(r_\nu+1)>2^{\nu/2-1}$$

and $r_\nu+1\to\infty$ $(\nu\to\infty)$ we see that $\bar{d}(\Phi)=\infty$. Similarly since

$$\Phi(r_{\nu+1})/r_{\nu+1}=b_\nu/r_{\nu+1}=2^{\nu/2}\, r_\nu/r_{\nu+1}=2^{\nu/2}\, p^{4^{\nu-1}\cdot 2^\nu\cdot 7}$$

and $r_{\nu+1}\to\infty$ $(\nu\to\infty)$ we conclude by $p<4^{-1}$ that $\underline{d}(\Phi)=0$.

Proof of (b). For each $n=1, 2, \dots$ set

$$v_n(z)=w(z;A_n)-w(z;\tilde{A}_n), \qquad u_n(z)=\sum_{k=1}^{n} v_k(z)/p_k.$$

Observe that v_n and u_n are harmonic on U, positive on the upper half of U, $v_n(\bar{z}) = -v_n(z)$, $u_n(\bar{z}) = -u_n(z)$ on U, and $v_n = u_n = 0$ on the real axis in U. To show that the series (97) is convergent on U and defines a harmonic function there we have only to prove that $\{u_n(i/2)\}_1^\infty$ converges. By (86)

$$0 < v_n(i/2) < s(2p_n\pi/n)^2/(\pi/2)^4 = s' p_n^2,$$

where s' is a constant independent of $n > 1$. Thus

$$0 < u_{n+m}(i/2) - u_n(i/2) = \sum_{k=n+1}^{n+m} v_k(i/2)/p_k$$

$$< s' \sum_{k=n+1}^{n+m} p_k < s' p^n/(1-p).$$

This shows the convergence of $\{u_n(i/2)\}_1^\infty$.

Proof of (c). For each $\nu = 1, 2, \ldots$ we denote by L_ν the line segment $L_{A'_{2^\nu}}$ connecting the end points of $A'_{2^\nu} = A_{2^\nu} \cup \tilde{A}_{2^\nu}$. Since $|v_k(z)| < 1$ on U we have

$$|v_k(z)/p_k| \le 1/p_k \le 1/(p^{4^{\nu-1}})^k \qquad (1 \le k \le 2^\nu)$$

on U and a fortiori on L_ν. Next for $k = 2^\nu + \mu$ ($\mu = 1, 2, \ldots$) and $z \in L_\nu$ we infer by (86) that

$$\begin{aligned} |v_k(z)/p_k| &\le s(2p_k\pi/k)^2/((2p_{2^\nu}\pi/2^\nu)^4 p_k) \\ &= s(2^{4\nu}/4\pi^2k^2)(p_k/p_{2^\nu}^4) \\ &= s(2^{4\nu}/4\pi^2k^2)\left[(p^{4^\nu})^k/((p^{4^{\nu-1}})^{2^\nu})^4\right] \\ &= s(2^{4\nu}/4\pi^2k^2)\, p^{4^\nu\mu} \le p^{4^\nu(\mu-1)}. \end{aligned}$$

Hence for $z \in L_\nu$

$$\begin{aligned} |u(z)| &\le \sum_{k=1}^{2^\nu} |v_k(z)/p_k| + \sum_{k=2^\nu+1}^{\infty} |v_k(z)/p_k| \\ &< \sum_{k=1}^{2^\nu} 1/(p^{4^{\nu-1}})^k + \sum_{\mu=1}^{\infty} p^{4^\nu(\mu-1)} \\ &< 2/(p^{4^{\nu-1}})^{2^\nu} = r_\nu. \end{aligned}$$

Since $u(z)$ is quasibounded on the upper half of U and on the lower half of U we obtain for $e^{i\theta} \in \partial U - A'_{2^\nu}$

$$\begin{aligned} |u(e^{i\theta})| &= \sum_{k=1}^{2^\nu} |v_n(e^{i\theta})/p_k| \le \sum_{k=1}^{2^\nu} 1/p_k \\ &\le \sum_{k=1}^{2^\nu} 1/(p^{4^{\nu-1}})^k < r_\nu. \end{aligned}$$

From this and the maximum principle we obtain $0<u(z)<r_\nu$ on that part of the upper half of U which lies to the left of L_ν. Hence

$$|u(z)|<r_\nu$$

to the left of L_ν on U.

By (84) $w(z;A'_{2^\nu})=1-p_{2^\nu}/2^{\nu-1}$ on L_ν and

$$w(z;A'_{2^\nu})\geq 1-p_{2^\nu}/2^{\nu-1}$$

to the right of L_ν on U. Therefore if z lies between L_ν and $L_{\nu+1}$ in U

$$b_\nu w(z;A'_{2^\nu})\geq b_\nu-2^{-\nu/2+2}\geq \Phi(|u(z)|)-2^{-\nu/2+2},$$

that is,

$$\Phi(|u(z)|)\leq b_\nu w(z;A'_{2^\nu})+2^{-\nu/2+2}$$

since $\Phi(t)\leq b_\nu$ for $t\leq r_{\nu+1}$. On the other hand

$$2\pi b_\nu w(0;A'_{2^\nu})=b_\nu(4p_{2^\nu}\pi/2^\nu)=8\pi\, 2^{-\nu/2}.$$

Hence if we set

$$w(z)=\sum_{\nu=1}^{\infty}\left(b_\nu w(z;A'_{2^\nu})+2^{-\nu/2+2}\right)$$

then $w(0)=8\sum_{\nu=1}^{\infty}2^{-\nu/2}<\infty$ and therefore $w\in HP(U)$. Thus

$$\Phi(|u(z)|)\leq b_\nu w(z;A'_{2^\nu})+2^{-\nu/2+2}<w(z)$$

between L_ν and $L_{\nu+1}$ on U. Since ν is arbitrary we obtain $\Phi(|u(z)|)\leq w(z)$ on L_1 and to the right of it on U. If z lies to the left of L_1 on U then $|u(z)|\leq r_1$ and consequently $\Phi(|u(z)|)\leq w(z)$ there as well. This shows that $u\in H\Phi(U)$.

Proof of (d). Contrary to the assertion assume that $u\in HP'(U)$. Then $|u(z)|$ has a harmonic majorant $h(z)$ on U. Since $u(z)$, $u_n(z)$, and $v_n(z)$ are positive on the upper half of U and antisymmetric with respect to the real line, i.e. $u(z)=-u(\bar z)$ etc., we infer that $h\geq|u|\geq|u_n|$ on U. Clearly

$$|u_n(z)|=\sum_{k=1}^{n}|w(z;A_k)-w(z;\tilde A_k)|/p_k$$

and the least harmonic majorant of the subharmonic function $|u_n(z)|$ is $\sum_{k=1}^{n}w(z;A'_k)/p_k$, where $A'_k=A_k\cup\tilde A_k$ as before. Hence

$$\sum_{k=1}^{n}w(z;A'_k)/p_k\leq h(z)$$

on U for any $n=1, 2, \ldots$. Thus in particular

$$\sum_{k=1}^{n} w(0; A_k')/p_k \leq h(0)$$

which gives the following contradiction:

$$\infty = 2 \sum_{k=1}^{\infty} 1/k = \frac{1}{2\pi} \sum_{k=1}^{\infty} 4 p_k \pi/k\, p_k \leq h(0).$$

6 F. Second Example. Consider the functions

$$\Phi(t) = \log^+ t = (\log t) \cup 0$$

for $t \in [0, \infty)$ and

$$u(z) = (\cos\theta)/r$$

for $z = r e^{i\theta} \in U_0 = \{z \mid 0 < |z| < 1\}$. Then $\Phi(t)$ is unbounded, increasing, continuous, and $\bar{d}(\Phi) = \underline{d}(\Phi) = 0$. Clearly $u \in H\Phi(U_0)$ but $u \notin HP'(U_0)$.

This example depends essentially on the weakness of the boundary point 0 of U_0. Without using such a special boundary property we can easily construct an example of this kind on the open unit disk $U: |z| < 1$ by employing Example 6 E:

Example. *Let $\Phi(t)$ and $u(z)$ be as in Example 6 E. Let*

$$\Phi_a(t) = \Phi(t) \cap a\,t \tag{98}$$

with a constant $a \in (0, \infty)$. Then

(a) *Φ_a is unbounded, increasing, and continuous on $[0, \infty)$, and $\bar{d}(\Phi_a) = a < \infty$, $\underline{d}(\Phi_a) = 0$,*

(b) *$u \in H\Phi_a(U)$,*

(c) *$u \notin HP'(U)$.*

Since $\Phi_a(t)/t \leq a\,t/t = a$, $\bar{d}(\Phi_a) \leq a$. Conversely, since $\bar{d}(\Phi) = \infty$, we can find an increasing sequence $\{t_n\}_1^\infty \subset (0, \infty)$ with $\lim_n t_n = \infty$ and $\Phi(t_n) > a\,t_n$. Hence $\Phi_a(t_n)/t_n = a$, i.e. $\bar{d}(\Phi_a) \geq a$ and therefore $\bar{d}(\Phi_a) = a$. By $\underline{d}(\Phi_a) \leq \underline{d}(\Phi)$ we see that $\underline{d}(\Phi_a) = 0$. The remainder of (a) is clear.

Assertion (b) follows from $\Phi_a(|u|) \leq \Phi(|u|)$ on U and $u \in H\Phi(U)$.

The last assertion (c) is nothing more than (d) of Example 6 E.

6 G. The Inclusion $H\Phi \subset HP'$. The concept of Φ-boundedness is so general that even the inclusion $H\Phi \subset HP'$ does not necessarily hold (Examples 6 E and 6 F). In view of this it is of interest to determine the weakest possible condition on Φ to assure that $H\Phi \subset HP'$:

Theorem. *In order that $H\Phi(R) \subset HP'(R)$ for every R it is sufficient that $\underline{d}(\Phi) > 0$. If $\underline{d}(\Phi) = 0$, $H\Phi(R) \subset HP'(R)$ is not necessarily true.*

The second half is seen by Examples 6 E and 6 F.

For the first half assume that $\underline{d}(\Phi)=2c>0$. Let $t_0\in(0,\infty)$ be such that $\Phi(t)>ct$ for $t>t_0$. Then for every $t\in[0,\infty)$

$$\Phi(t)+c\,t_0\geq c\,t.$$

Let $u\in H\Phi(R)$ and denote by h the harmonic majorant of $\Phi(|u|)$ on R. Then

$$h+c\,t_0\geq\Phi(|u|)+c\,t_0\geq c\,|u|$$

on R. Thus $|u|$ possesses a harmonic majorant $(h+c\,t_0)/c$ on R, i.e. $u\in HP'(R)$.

6 H. The Inclusion $H\Phi\cap HP'\subset HB'$. We turn to the relation between Φ-boundedness and B'-boundedness. We first prove:

THEOREM. *If $\bar{d}(\Phi)=\infty$ then $H\Phi(R)\cap HP'(R)\subset HB'(R)$.*

Let $u\in H\Phi(R)\cap HP'(R)$. Take an HP-function h on R such that $\Phi(|u|)\leq h$. Set $M\,u=u\vee 0+(-u)\vee 0$. Clearly $M\,u\geq|u|$ on R. To show that $B'u=u$ it is sufficient to prove $B'\,M\,u=M\,u$.

By the assumption $\bar{d}(\Phi)=\infty$ there exists an increasing sequence $\{r_n\}_1^\infty$ of positive numbers tending to ∞ such that $\Phi(r_n)>0$ and

$$\lim_{n\to\infty}a_n=0,\qquad a_n=\frac{r_n}{\Phi(r_n)}. \tag{99}$$

Set $G_n=\{z\in R\,|\,|u(z)|<r_n\}$ $(n=1,2,\ldots)$. Clearly $\{G_n\}_1^\infty$ is increasing and $R=\bigcup_{n=1}^\infty G_n$. Take an exhaustion $\{R_m\}_1^\infty$ of R and let w_m be the harmonic function on $R_m\cap G_n$ with boundary values $w_m|(\partial R_m)\cap G_n=\min(M\,u-B'\,M\,u,r_n)$ and $w_m|(\partial G_n)\cap R_m=0$. On setting $w_m|(R_m-G_n)=0$ we obtain a subharmonic w_m on R_m. Since $\min(M\,u-B'\,M\,u,r_n)$ is superharmonic on R

$$w_m\geq w_{m+1} \tag{100}$$

on R_m.

Let w'_m be harmonic on R_m with boundary values $w'_m|(\partial R_m)\cap G_n=\min(M\,u-B'\,M\,u,r_n)$ and $w'_m|(\partial R_m-\bar{G}_n)=0$. Clearly $\{w'_m\}_{m=1}^\infty$ is a bounded sequence and $0\leq w'_m\leq M\,u-B'\,M\,u$ for every m. Let w' be a limiting function of some convergent subsequence of $\{w'_m\}_1^\infty$. Then $0\leq w'\leq M\,u-B'\,M\,u$ and we obtain

$$0\leq B'\,w'\leq B'(M\,u-B'\,M\,u)=B'\,M\,u-B'^2\,M\,u=B'\,M\,u-B'\,M\,u=0.$$

Since w' is bounded $w'=B'\,w'=0$ on R. Therefore

$$\lim_{m\to\infty}w'_m=0 \tag{101}$$

on R. In view of $w'_m \ge w_m \ge 0$ on R_m this implies

$$\lim_{m\to\infty} w_m = 0 \tag{102}$$

on R.

Observe that on $(\partial R_m)\cap G_n$, $|u|<r_n$ and $|u|\le M u = B' M u + (M u - B' M u)$, that is $|u| - B' M u \le M u - B' M u$. Hence on $(\partial R_m)\cap G_n$, $|u| - B' M u \le \min(M u - B' M u, r_n)$ or what amounts to the same, $|u| \le B' M u + w_m$. On ∂G_n we have $|u| = r_n = a_n \Phi(|u|) \le a_n h$. Therefore

$$|u| \le a_n h + B' M u + w_m \tag{103}$$

on $\partial(R_m \cap G_n)$. Since $|u|$ is subharmonic and $a_n h + B' M u + w_m$ is harmonic on $R_m \cap G_n$, (103) is valid on $R_m \cap G_n$. On letting $m\to\infty$ and then $n\to\infty$ we infer that

$$|u| \le B' M u \tag{104}$$

on R. Note that $M u = u \vee 0 + (-u) \vee 0$ is the least harmonic majorant of $|u|$ on R. Thus (104) implies $M u \le B' M u$. The reverse inequality is trivial and $B' M u = M u$ as claimed.

COROLLARY. *Let $\bar{d}(\Phi)=\infty$. In order that $H\Phi(R)\subset HB'(R)$ for any R it is sufficient that $\underline{d}(\Phi)>0$. If $\underline{d}(\Phi)=0$, this is not necessarily true.*

Because of $HB'(R)\subset HP'(R)$, Example 6 E gives the second half. The first half follows from Theorems 6 G and 6 H.

6 I. The Relative Class $SO_{H\Phi}$. Let $SO_{H\Phi}$ be the class of bordered Riemann surfaces (R,γ) such that every $H\Phi$-function u which vanishes continuously on γ vanishes on R. Clearly $SO_{H\Phi}$ consists of all bordered compact surfaces if Φ is bounded, and all bordered surfaces if Φ is not bounded in any neighborhood of $t=0$ on $[0,\infty)$. To exclude these cases of little interest we make in 6 I and 6 J the following assumption:

$\Phi(t)$ is bounded in a neighborhood of $t=0$ and unbounded on $[0,\infty)$.

By an obvious modification of the argument in 6 B we obtain:

THEOREM. *If $\bar{d}(\Phi)=\infty$ then $SO_{H\Phi}=SO_{HB}$.*

6 J. The Class $SO_{H\Phi}$ for $\bar{d}(\Phi)<\infty$. Here we pause to determine the class SO_{HP} of bordered surfaces (R,γ) on which every HP-function vanishing continuously on γ vanishes on R. The following result gives a complete solution:

THEOREM. *The class SO_{HP} consists of all bordered compact surfaces.*

Clearly all bordered compact surfaces belong to SO_{HP}. Let (R,γ) be noncompact. Take the Green's function $g(z,\zeta)$ on R with pole $\zeta\in R$, and fixed point $z_0 \in R$. Since $R \cup \gamma$ is noncompact there exists a sequence $\{\zeta_n\} \subset R$ which does not accumulate in $R \cup \gamma$. Let

$$u_n(z) = \frac{g(z, \zeta_n)}{g(z_0, \zeta_n)}.$$

By virtue of $u_n(z_0) = 1$ and $u_n > 0$ there exists a convergent subsequence of $\{u_n\}$ on R. On choosing a subsequence we may assume that $\{u_n\}$ itself is convergent. Set

$$u = \lim_n u_n.$$

Then $u(z_0) = 1$ and thus $u > 0$ on R. By Lemma 6 D, u vanishes identically on γ, i.e. $(R, \gamma) \notin SO_{HP}$.

COROLLARY. *If $\bar{d}(\Phi) < \infty$ then $SO_{H\Phi} = SO_{HP}$, and $SO_{H\Phi}$ consists of all bordered compact surfaces.*

In fact $\bar{d}(\Phi) < \infty$ means that $\Phi(t) \leq c_1 + c_2 t$ on $[0, \infty)$ for some constants $c_1, c_2 \geq 0$. Hence every HP-function is an $H\Phi$-function, and the assertion follows.

6 K. Increasing Convex Φ. Hereafter in 6 we only consider functions $\Phi(t)$ which are *increasing* and *convex* on $[0, \infty)$. It is elementary that

$$\bar{d}(\Phi) = \underline{d}(\Phi) = d(\Phi) > 0. \tag{105}$$

From Corollary 6 B we know that

$$O_{H\Phi} = O_{HP} \tag{106}$$

if $d(\Phi) < \infty$, and

$$O_{H\Phi} = O_{HB} \tag{107}$$

if $d(\Phi) = \infty$ (PARREAU [4]). From 6 G and 6 H it follows that

$$H\Phi(R) \subset HP'(R) \tag{108}$$

in general and

$$H\Phi(R) \subset HB'(R) \tag{109}$$

if $d(\Phi) = \infty$ (PARREAU [4]).

Particularly important in this case is the fact that $\Phi(|u|)$ is subharmonic if u is harmonic as is seen from Jensen's inequality (e.g. HARDY-PÓLYA-LITTLEWOOD [1, p. 152]). Thus if $u \in H\Phi(R)$ then $\Phi(|u|)$ admits a least harmonic majorant, which we denote by $\hat{u}_\Phi$.

THEOREM. *If $d(\Phi) = \infty$ and $u \in H\Phi(R)$ then*

$$\hat{u}_\Phi(z) = \int_{\Delta_N} P_N(z; p)\, \Phi(|u(p)|)\, d\mu_N(p). \tag{110}$$

We know from (109) that $u \in HB'(R)$. For $u_n = (u \wedge n) \vee (-n)$ we thus have $u = \lim_n u_n$ and $\Phi(|u_1|) \leq \Phi(|u_2|) \leq \cdots$, $\Phi(|u|) = \lim_n \Phi(|u_n|)$. From this it follows that

$$\hat{u}_\Phi = \lim_{n \to \infty} \widehat{(u_n)}_\Phi. \tag{111}$$

To prove (110) we may therefore assume that $u \in HB$. Then $\hat{u}_\Phi$ is the bounded harmonic function on R which coincides with $\Phi(|u|)$ on $\Delta_{\mathbb{N}}$ (see 3 J), and (110) ensues.

6 L. Φ-Mean Boundedness. A harmonic function u on R is called *Φ-mean bounded* if there exists a constant $C < \infty$ such that

$$\Big| \int_{\partial\Omega} \Phi(|u(z)|) * d_z g_\Omega(z, \zeta) \Big| < C \tag{112}$$

for every regular subregion Ω of R with $\zeta \in \Omega$, where g_Ω is the Green's function on Ω and ζ is a fixed point in R. We claim (Parreau [4], Rudin [1]):

Theorem. *For increasing and convex Φ, $u \in H\Phi(R)$ if and only if u is Φ-mean bounded on R.*

Clearly $u \in H\Phi(R)$ implies (112) with $C = 2\pi\, h(\zeta)$, where h is a harmonic majorant of $\Phi(|u|)$ on R. Suppose (112) is valid and set

$$v_\Omega(z) = -\frac{1}{2\pi} \int_{\partial\Omega} \Phi(|u(\zeta)|) * d_\zeta g_\Omega(\zeta, z).$$

Since v_Ω increases with Ω and $v_\Omega \geq \Phi(|u|)$ on Ω we have $\hat{u}_\Phi = \lim_\Omega v_\Omega$.

6 M. A Relation to *HD*. Consider the special case

$$\Phi(t) = t^2$$

and take a $u \in HD(R)$. We shall show that $u \in H\Phi(R)$. To this end we may assume that $u(\zeta) = 0$ and $R \notin O_G$. By Green's formula

$$\int_{\partial\Omega} (g_\Omega * d(u^2) - u^2 * dg_\Omega) = \int_\Omega (g_\Omega \Delta u^2 - u^2 \Delta g_\Omega)\, dx\, dy. \tag{113}$$

Since $\Delta u^2 = 2u \Delta u + 2\,|\mathrm{grad}\, u|^2 = 2\,|\mathrm{grad}\, u|^2$, (113) implies

$$\Big| \int_{\partial\Omega} u^2(z) * dg_\Omega(z, \zeta) \Big| = 2 \int_\Omega g_\Omega(z, \zeta)\, |\mathrm{grad}\, u(z)|^2\, dx\, dy. \tag{114}$$

Let g_R be the Green's function on R. Then the right-hand side of (114) is dominated by $2 \int_R g_R(z, \zeta)\, |\mathrm{grad}\, u(z)|^2\, dx\, dy$, which is finite since $u \in HD$ and g_R is bounded outside of a neighborhood of ζ in which the integrand is integrable. Thus (112) is valid and we obtain (Parreau [4]):

THEOREM. *For* $\Phi(t)=t^2$, $HD(R)\subset H\Phi(R)$.

In this case $d(\Phi)=\infty$ and therefore $H\Phi(R)\subset HB'(R)$. Thus $HD(R)\subset HB'(R)$ and $O_{HB}\subset O_{HD}$, i.e. we have given one more alternative proof of $O_{HB}\subset O_{HD}$.

Remark. The Φ-bounded harmonic functions on Riemann surfaces and the corresponding null classes were first systematically investigated by PARREAU [4] for an increasing and convex Φ.

Recently YAMASHITA [1] showed that $HB'(R)=\bigcup H\Phi(R)$ where the union is taken with respect to every increasing convex Φ with $d(\Phi)=\infty$.

§ 3. Lindelöfian Meromorphic Functions

In connection with O_{HB} we again return to a class of analytic functions. A meromorphic function f on the unit disk U is, by definition, of bounded characteristic if it is of the form φ_1/φ_2 with φ_i, $i=1,2$, analytic functions such that $|\varphi_i|<1$ on U. The condition is equivalent to the requirement that $\log^+|f|$ admit a superharmonic majorant on U. This is the property we shall take for the definition of the class $MB^*(R)$ of meromorphic functions on an arbitrarily given Riemann surface R. Thus the corresponding null class O_{MB^*} is a subclass of O_{AB} and we shall see that the class has a close connection with O_{HB}. It is for this reason that we include its study in the present chapter.

A function $f\in MB^*$ is continuous on $R_{\mathfrak{N}}^*$ and we have the inclusion $U_{HB}\subset O_{MB^*}$. These topics are discussed in 7. Covering properties of $f(R)$ with $f\in MB^*(R)$ over the Riemann sphere will be studied in 8. Various strict inclusion relations concerning O_{MB^*} are given in 9.

7. Inclusion Relations

7 A. Boundary Theorem of Riesz-Lusin-Privaloff Type. First we state a counterpart of Theorem III. 5 H, which will play a fundamental role in the classification problem of this no.

THEOREM. *Let E be a subset of $\Gamma_{\mathfrak{N}}$ with $\mu_{\mathfrak{N}}(E)>0$, and let G be a normal subregion of R such that $\bar{G}$ is a neighborhood of E in $R_{\mathfrak{N}}^*$. A meromorphic function f on G with continuous boundary values zero at each point of E vanishes identically on G.*

Contrary to the assertion assume that $f\not\equiv 0$. The set $F=\{z\in G\,|\,|f(z)|<1\}$ is normal and open in R, with $\bar{F}$ a neighborhood of E in $R_{\mathfrak{N}}^*$. We may suppose that E is compact and $E\subset\Delta_{\mathfrak{N}}$. Moreover we may assume that E is both open and closed in $\Delta_{\mathfrak{N}}$ (Corollary 2 in 4 D). By Theorem 3 E we can find a $u\in HB(F)$ such that $u=0$ on ∂F, $u=1$ on E, $u=0$ on

$\Delta_{\mathbb{N}} \cap \bar{F} - E$, and $0 \le u \le 1$. From this and Theorem 3 J we conclude that $u(z) \le \varepsilon \log |1/f(z)|$ on F for any $\varepsilon > 0$. Therefore u must vanish identically, a contradiction.

7 B. Lindelöfian Meromorphic Functions. Denote by $M(R)$ the totality of meromorphic functions on R. A function $f \in M(R)$ is called *Lindelöfian* or *of bounded characteristic* if there exists a superharmonic function s such that

$$\log^+ |f| \le s \tag{115}$$

on R. The family of functions $f \in M(R)$ with (115) will be denoted by $MB^*(R)$ and the corresponding null class by O_{MB^*}. Since $AB(R) \subset MB^*(R)$,

$$O_{MB^*} \subset O_{AB}. \tag{116}$$

Clearly $\log^+ |f|$ is subharmonic on $R - f^{-1}(\infty)$. Hence by (115) there exists a least harmonic majorant u_f of $\log^+ |f|$ on $R - f^{-1}(\infty)$.

Assume f is not constant and $f^{-1}(\infty) = \{a_1, \ldots, a_n, \ldots\} \ne \emptyset$. Then by (115), $R \notin O_G$, and the Green's function $g(\cdot, a_n) = \lim_{\Omega \to R} g_\Omega(\cdot, a_n)$ exists on R. Let $\nu_n = \nu(a_n)$ be the order of the pole a_n. For $a_n \in \Omega$, $u_f - \nu_n g_\Omega(\cdot, a_n)$ is regular at a_n and >0 on Ω and a fortiori $u_f - \nu_n g(\cdot, a_n) > 0$ on R. We conclude that

$$\tilde{u}_f = u_f - \sum_{a_n \in f^{-1}(\infty)} \nu_n g(\cdot, a_n) \in HP(R). \tag{117}$$

Let $f^{-1}(0) = \{b_1, \ldots, b_n, \ldots\}$. Then

$$w_f = u_f - \log |f| \tag{118}$$

is nonnegative and harmonic on $R - f^{-1}(0)$. Let $\mu_n = \mu(b_n)$ be the order of the zero b_n. Since u_f is regular at b_n it follows that

$$\tilde{w}_f = w_f - \sum_{b_n \in f^{-1}(0)} \mu_n g(\cdot, b_n) \in HP(R). \tag{119}$$

Observe that w_f is superharmonic and nonnegative on R, and that $u_f > 0$. Thus (118) implies that w_f qualifies as an s in (115) for $1/f$ in place of f in (115). We conclude:

The class $MB^*(R)$ is a field.

7 C. The Inclusion $O_G \subset O_{MB^*}$. Assume that $R \in O_G$. We wish to show that every $f \in MB^*(R)$ is constant. Suppose this were not the case. Then by (117) and (119) we would have $f^{-1}(0) \cup f^{-1}(\infty) = \emptyset$ and by (118), $\log |f| \in HP'(R)$. Thus $R \notin O_{HP}$ and the proof of

$$O_G \subset O_{MB^*} \tag{120}$$

will be complete, by contradiction, if we establish

$$O_G \subset O_{HP}. \tag{121}$$

Suppose a nonconstant $u \in HP(R)$ exists on $R \in O_G$ and choose a disk $U \subset R$. By 3 J, $u|(R-U)$ takes its minimum on ∂U and $u|\bar{U}$ takes its minimum on ∂U. Therefore $u|R$ takes its minimum on ∂U, a contradiction.

7 D. The Decomposition Theorem. For a nonconstant $f \in M(R)$, $R \notin O_G$, let $f^{-1}(\infty) = \{a_1, \ldots, a_n, \ldots\}$ and $f^{-1}(0) = \{b_1, \ldots, b_n, \ldots\}$. Denote by ν_n the order of a_n, by μ_n the order of b_n, and set

$$\begin{aligned} G_1(z; f) &= \sum_{a_n \in f^{-1}(\infty)} \nu_n g(z, a_n) \leq \infty, \\ G_2(z; f) &= \sum_{b_n \in f^{-1}(0)} \mu_n g(z, b_n) \leq \infty \end{aligned} \tag{122}$$

where g is the Green's function on R. We then have (Heins [4], Parreau [4], and others):

Theorem. *Let $R \notin O_G$ and let $f \in M(R)$ be nonconstant. Then $f \in MB^*(R)$ if and only if $G_1(z; f) < \infty$ on $R - f^{-1}(\infty)$, $G_2(z; f) < \infty$ on $R - f^{-1}(0)$, and there exists a $u(z; f-a) \in HP'(R)$ such that*

$$\log|f(z) - a| = G_1(z; f) - G_2(z; f-a) + u(z; f-a) \tag{123}$$

on R for every complex number a.

Clearly (115) follows from (123). The converse is obtained by (117), (118), and (119).

7 E. Continuity on $R_{\mathfrak{N}}^*$. We denote by $M_{\mathfrak{N}}(R)$ the class of functions $f \in M(R)$ which are continuous on $R_{\mathfrak{N}}^*$, i.e. f: $R \to \{|z| \leq \infty\}$ can be extended (uniquely) to a continuous mapping f: $R_{\mathfrak{N}}^* \to \{|z| \leq \infty\}$.

Theorem. *Every $f \in MB^*(R)$ is continuous on $R_{\mathfrak{N}}^*$, i.e. $MB^*(R) \subset M_{\mathfrak{N}}(R)$. Moreover $\log|f| \in \mathscr{W}(R)$.*

The right-hand member of (123) belongs to $\mathscr{W}(R)$, and therefore $\log|f-a| \in \mathscr{W}(R)$. Hence $|f-a|$ is continuous on $R_{\mathfrak{N}}^*$ for every complex a. From this it follows easily that f is continuous on $R_{\mathfrak{N}}^*$.

7 F. Properties of $B'u(z; f-a)$. In (123) it is clear that $G_1|\Delta_{\mathfrak{N}} = G_2|\Delta_{\mathfrak{N}} = 0$ (cf. the argument in 3 I). Hence by 4 E

$$\log|f(p) - a|\,|\Delta_{\mathfrak{N}} = u(p; f-a)\,|\Delta_{\mathfrak{N}} = B'u(p; f-a)|\,\Delta_{\mathfrak{N}} \tag{124}$$

or equivalently:

THEOREM. *The following representation is valid for nonconstant* $f \in MB^*(R)$:

$$B' u(z; f-a) = \int_{\Delta_{\mathrm{N}}} P(z, p) \log |f(p)-a| \, d\mu_{\mathrm{N}}(p). \tag{125}$$

As a corollary we conclude (RAO [1]):

COROLLARY 1. $B' u(z; f-a)$ *is nonconstant except for at most one* a.

Assume that $B' u(z; f-a_j) \equiv c_j$ (const) for $j=1,2$ with $a_1 \neq a_2$. By (124), $\log |f-a_j| = c_j$ on Δ_{N}. It follows that $f(\Delta_{\mathrm{N}}) \subset C_j = \{w \mid |w-a_j| = e^{c_j}\}$ $(j=1,2)$ and therefore $f(\Delta_{\mathrm{N}}) \subset C_1 \cap C_2$. Let $C_1 \cap C_2 = \{\alpha, \beta\}$. Then $f^{-1}(\alpha) \cap \Delta_{\mathrm{N}}$ and $f^{-1}(\beta) \cap \Delta_{\mathrm{N}}$ are both open and closed. Thus at least one, say $f^{-1}(\alpha) \cap \Delta_{\mathrm{N}}$, has positive μ_{N}-measure. By Theorem 7 A we conclude that $f \equiv \alpha$ on R, a contradiction.

COROLLARY 2. *The following inclusion relations hold:*

$$O_{HB} \subset O_{MB^*} \subset O_{AB}. \tag{126}$$

This result is due to HEINS [4].

7 G. The Inclusion $U_{HB} \subset O_{MB^*}$. Let $R \in U_{HB}$. Then Δ_{N} contains a point with positive μ_{N}-measure (cf. 3 K). Thus Theorems 7 E and 7 A imply that $R \in O_{MB^*}$ (CONSTANTINESCU-CORNEA [3]):

$$U_{HB} \subset O_{MB^*} \subset O_{AB}. \tag{127}$$

If $R \in O_{HB} - O_G$ then $R - K \in U_{HB}$ for any compact set K (cf. 3 K), or more generally for K with $(R-K, \partial(R-K)) \notin SO_{HB}$ (cf. 3 L). It follows that $R - K \in O_{MB^*} \subset O_{AB}$ and therefore (KURAMOCHI [1]):

THEOREM. *Let U be a parametric region of a Riemann surface $R \in O_{HB}$. The ideal boundary of R is such that $R - \bar{U} \in O_{AB}$ despite the strong border ∂U.*

8. Covering Properties

8 A. Capacity of a Plane Set. Let K be a compact set. We say that K has *capacity zero* if $\{|z| \leq \infty\} - K$ is connected and belongs to O_G. Otherwise K is said to have positive capacity (cf. II. 10 C, V. 7 F):

THEOREM. *If $K \subset \{|z| \leq \infty\}$ has capacity zero and U is a disk with $\bar{U} \cap K = \emptyset$ then there exists a nonnegative continuous function s on $\{|z| \leq \infty\}$ such that $s \mid \bar{U} = 0$, $s \mid K = \infty$, and s is a finitely continuous superharmonic function on $\{|z| \leq \infty\} - K \cup \bar{U}$.*

Let $\{R_n\}_0^\infty$ be an exhaustion of $R = \{|z| \leq \infty\} - K$ with $R_0 = U$. Let w_n be continuous on $\{|z| \leq \infty\}$ such that $w_n \mid \bar{U} = 0$, $w_n \mid (\{|z| \leq \infty\} - R_n) = 1$, and $w_n|(R_n-\bar{U}) \in H(R_n-\bar{U})$. Then $w_1 \geq w_2 \geq \cdots$ on $\{|z| \leq \infty\}$ and $\lim_n w_n = 0$ on R. By choosing a suitable subsequence of $\{R_n\}$, if necessary, we may assume that

$$s = \sum_1^\infty w_n$$

is convergent on R. Then s has the required properties.

8 B. Capacity of $f(\Delta_{\mathrm{N}})$. From 7 A it is clear that $f(\Delta_{\mathrm{N}})$ cannot be a finite set for any nonconstant $f \in MB^*(R)$. Moreover:

Theorem. *For every nonconstant $f \in MB^*(R)$, $f(\Delta_{\mathrm{N}})$ has positive capacity.*

Contrary to the assertion assume that the compact set $f(\Delta_{\mathrm{N}})$ has capacity zero. Let U be a disk with $\bar{U} \cap f(\Delta_{\mathrm{N}}) = \emptyset$. The function s of Theorem 8 A for $K = f(\Delta_{\mathrm{N}})$ is superharmonic on $F = \{|z| \leq \infty\} - \bar{U}$, and $G = f^{-1}(F)$ is a normal open set in R such that $\bar{G} \supset \Delta_{\mathrm{N}}$. The function

$$S = s \circ f$$

is superharmonic on G, vanishes on ∂G, and $\lim_{z \in G, z \to p} S(z) = \infty$ for every $p \in \Delta_{\mathrm{N}}$. Hence by 3 J, $(G, \partial G) \in SO_{HB}$, which contradicts 3 L.

8 C. Meromorphic Functions on O_{MB^*}-Surfaces. Assume that $R \in O_{MB^*}$. Then for nonconstant $f \in M(R)$, $\log^+ |f|$ never satisfies (115). This suggests smallness of $\{|z| \leq \infty\} - f(R)$. More precisely:

Theorem. *If $R \in O_{MB^*}$ and $f \in M(R)$ is not constant then $K_f = \{|z| \leq \infty\} - f(R)$ is of capacity zero, i.e. $f(R) \in O_G$.*

We may assume that the closure of the unit disk U is contained in $f(R)$. Suppose $f(R) \notin O_G$ and let $g(w, 0)$ be the Green's function on $f(R)$ with pole 0. Observe that

$$\log^+ \frac{1}{|w|} \leq g(w, 0)$$

on $f(R)$ and $g(f(z), 0)$ is superharmonic on R. Therefore

$$\log^+ \left| \frac{1}{f(z)} \right| \leq g(f(z), 0),$$

i.e. $1/f \in MB^*(R)$ and therefore $f \in MB^*(R)$.

The theorem of course remains valid for any null class which is a subclass of O_{MB^*}. In particular:

Corollary 1. *If $R \in O_G$, O_{HP}, O_{HB}, or U_{HB} and $f \in M(R)$ is not constant then $K_f = \{|z| \leq \infty\} - f(R)$ has capacity zero.*

COROLLARY 2. *If $R \in O_{HB}$ and $f \in M(R)$ is not constant then any component of $f^{-1}(|w-w_0|<\rho)$ belongs to SO_{HB} for every w_0 and $\rho>0$.*

Suppose a component $G \notin SO_{HB}$. By 3 K and 5 C, $G \in U_{HB}$. Thus $f(G) \subset (|w-w_0|<\rho)$ belongs to O_G, a contradiction.

8 D. Parabolic Ends. By an *end* R we mean the interior of a bordered Riemann surface (R, γ_R) with compact border $\gamma_R = \bigcup_1^n \gamma_R^{(j)}$, the $\gamma_R^{(j)}$ being the components of γ_R, such that the Stoïlow boundary $\Gamma_S = R_S^* - R$ consists of $n+1$ points. Hence for an end R we can always find a Riemann surface $\tilde{R}$ with only one Stoïlow ideal boundary point such that $\tilde{R}_0 = \tilde{R} - R - \gamma_R$ is a regular subregion. Let $\{\tilde{R}_n\}_0^\infty$ be an exhaustion of $\tilde{R}$. Then $R - \tilde{R}_n$ is connected for every n.

We call R a *parabolic end* if R is an end for which $\tilde{R} \in O_G$. Clearly the parabolicity of R does not depend on the choice of $\tilde{R}$.

The simplest example of a parabolic end R is U_0: $0<|z|<1$. Here γ_U: $|z|=1$. We know that every $f \in AB(U_0)$ has limit at $z=0$. This can be generalized to an arbitrary parabolic end R by noticing that every $f \in AB(R)$ satisfies (115).

Before stating the theorem (HEINS [2], see also VI. 1) we observe that for an end R Alexandroff's ideal boundary point a_∞ of the associated $\tilde{R}$ can be identified with Stoïlow's ideal boundary point of R, which we denote by $a_\infty = a_\infty(R)$. Hence $R \cup \gamma_R \cup a_\infty(R)$ is compact.

THEOREM. *If $f \in AB(R)$ on a parabolic end R then*

$$\lim_{z \in R,\, z \to a_\infty(R)} f(z) = f\big(a_\infty(R)\big) \tag{128}$$

always exists.

We continue denoting by $\tilde{R}$ the surface associated with R and by $\{\tilde{R}_n\}_0^\infty$ an exhaustion of $\tilde{R}$ with $\tilde{R}_0 = \tilde{R} - R - \gamma_R$. We may assume that f is nonconstant analytic on $R \cup \gamma_R = R \cup \partial R$ and $r_f = \sup_R |f| < 1$ there. Let $n(w)$ be the number of zeros of $f-w$ on R with multiplicities taken into account. Clearly $n(w)$ is lower semicontinuous on $|w|<\infty$. We shall prove that it is bounded.

Let $g(z, z_0)$ be the Green's function on R with pole $z_0 \in R$. The function is symmetric, i.e. $g(z, z_0) = g(z_0, z)$, as is easily seen by applying the Green's formula to $g_\Omega(z_0, z)$ and $g_\Omega(z, z_0)$ on Ω less small disks shrinking to z_0 and z, and then letting $\Omega \to R$. Since $|f(z)|<1$ on R, $\log\big|(1-\bar{w} f(z))/(f(z)-w)\big|$ is positive harmonic on $R - f^{-1}(w)$ and at each point $a_i \in f^{-1}(w)$ it has a logarithmic singularity of the form $\nu(a_i) \log|z-a_i|^{-1}$ where $\nu(a_i)$ is the multiplicity of f at a_i. In the same manner as in 7 B we conclude that

$$\sum_{a_i \in f^{-1}(w)} \nu(a_i)\, g(z, a_i) < \log\left|\frac{1-\bar{w} f(z)}{f(z)-w}\right| \tag{129}$$

on R.

Let $R_n = \tilde{R} - \tilde{R}_n - \partial\tilde{R}_n$. Since $\tilde{R} \in O_G$ we infer by 3 J that

$$m_1(z) = \inf_{\zeta \in R_1} g(z, \zeta) = \inf_{\zeta \in R_1} g(\zeta, z) > 0 \tag{130}$$

for a given $z \in R$. Take an arbitrary w_0 with $|w_0| \le r_f < 1$ and a neighborhood V_{w_0} of w_0 such that $\overline{V}_{w_0} \subset \{|w| < 1\}$ and $f(R) - \overline{V}_{w_0} \neq \emptyset$. Clearly

$$c'_{w_0} = \sup_{w \in V_{w_0}} \sum_{a_i \in f^{-1}(w) \cap (R \cup \partial R - R_1)} \nu(a_i) < \infty . \tag{131}$$

Let $z_0 \in R$ such that $f(z_0) \in f(R) - \overline{V}_{w_0}$. Then $\sup_{w \in V_{w_0}} \log|(1 - \overline{w} f(z_0))/(f(z_0) - w)| = c'''_{w_0} < \infty$, $c''_{w_0} = c'''_{w_0}/m_1(z_0) < \infty$, and by (129)

$$\sup_{w \in V_{w_0}} \sum_{a_i \in f^{-1}(w) \cap R_1} \nu(a_i) \le c''_{w_0} . \tag{132}$$

On setting $c_{w_0} = c'_{w_0} + c''_{w_0}$ we obtain

$$\sup_{w \in V_{w_0}} \sum_{a_i \in f^{-1}(w)} \nu(a_i) \le c_{w_0} < \infty . \tag{133}$$

In view of $n(w) = \sum_{a_i \in f^{-1}(w) \cap R} \nu(a_i)$, (133) implies the boundedness of $n(w)$ on V_{w_0}. Since $|w| \le r_f$ is covered by a finite number of such V_{w_0} we conclude that $n(w)$ is bounded on $|w| \le r_f$. Obviously $n(w) = 0$ for $|w| > r_f$ and consequently $n(w)$ is bounded on $|w| < \infty$.

The set $K = \bigcap_0^\infty \overline{f(R_n)}$ is a continuum. If it degenerates to a one point set then (128) follows. We shall assume that K is a nondegenerate continuum. One easily sees that there exists a point $w_1 \in K \subset \{|w| < 1\}$ such that $n(w_1) = \max_{w \in K} n(w)$. Since $n(w)$ is lower semicontinuous $n|K$ is continuous at w_1, i.e. n is a constant in a neighborhood of w_1 on K. Hence we can find an R_m and a nondegenerate continuum K_1 with $w_1 \in K_1 \subset K$ and $f(R_m \cup \partial R_m) \cap K_1 = \emptyset$.

By 3 J, $|f| \,|\, R_m \cup \partial R_m$ attains its maximum on ∂R_m. Hence $\sup_{K_1} |w| \le \sup_K |w| \le \sup_{f(\partial R_m)} |w|$. Thus $f(R_m \cup \partial R_m)$ is contained in the component Ω of $\{|z| \le \infty\} - K_1$ containing ∞. Let $g_\Omega(w, \infty)$ be the Green's function on Ω with pole $\infty \in \Omega$. Since $f(R_m \cup \partial R_m) \cap K_1 = \emptyset$ and $g_\Omega(w, \infty)$ vanishes only on K_1, $g_\Omega(f(z), \infty) > 0$ on $R_m \cup \partial R_m$. Thus by 3 J

$$\inf_{R_m} g_\Omega(f(z), \infty) = \min_{\partial R_m} g_\Omega(f(z), \infty) > 0 .$$

On the other hand $\overline{f(R_m)} \supset K \supset K_1$ implies the existence of a sequence $z_n \in R_m$ such that $f(z_n) \to K_1$. Thus $g_\Omega(f(z_n), \infty) \to 0$ and

$$\inf_{R_m} g_\Omega(f(z), \infty) = 0 ,$$

a contradiction.

8 E. Bounded Valence. Let G be a parabolic end of an open Riemann surface R. We maintain (HEINS [7]):

THEOREM. *Every $f \in MB^*(R)$ has bounded valence on G.*

Let $g(z, z_0)$ be the Green's function on R with pole $z_0 \notin G$. Observe that by 3 J, $m = \min_{z \in G \cup \partial G} g(z, z_0) = \min_{z \in G \cup \partial G} g(z_0, z) > 0$. By Theorem 7 D

$$\sum_{b_i \in f^{-1}(\alpha) \cap G} \mu_i\, g(z_0, b_i) < \infty$$

for every α with $|\alpha| \leq \infty$. Since $m > 0$

$$n_{f|G}(\alpha) = \sum_{b_i \in f^{-1}(\alpha) \cap G} \mu_i \leq \frac{1}{m} \sum_{b_i \in f^{-1}(\alpha) \cap G} \mu_i\, g(z_0, b_i) < \infty .$$

By the lower semicontinuity of $n_{f|G}(\alpha)$ for $|\alpha| \leq \infty$, $n_{f|G}(\alpha)$ is bounded.

COROLLARY. *Let G be a parabolic end. For every $f \in MB^*(R)$,* (128) *exists for G and f.*

Let $\{G_n\}_1^\infty$ be a decreasing sequence of subends of G with $\bigcap_1^\infty G_n = \emptyset$. Since f is of bounded valence, $E_n = \{|w| \leq \infty\} - f(G_n)$ is compact in $\{|w| \leq \infty\}$, $\{E_n\}_1^\infty$ is increasing, and $\bigcup_1^\infty E_n = \{|w| \leq \infty\}$. Since $\{|w| \leq \infty\}$ is a complete metric space Baire's theorem states that $\bigcup_1^\infty E_n$ is not of the first category. In other words some E_n has interior points, i.e. $\overline{f(G_n)} \neq \{|w| \leq \infty\}$. Thus f can be considered to be bounded on G_n, and Theorem 8 D implies our assertion.

9. Examples Concerning O_{MB^*}

9 A. Lindelöfian Analytic Functions. Consider the class $AB^*(R) = A(R) \cap MB^*(R)$. Obviously $AB(R) \subset AB^*(R) \subset MB^*(R)$, that is

$$O_{MB^*} \subset O_{AB^*} \subset O_{AB} .$$

We shall prove the strictness of these inclusions. First we show (HEINS [4]):

$$O_{AB^*} < O_{AB} . \tag{134}$$

There exists a compact linear set K of linear measure zero and positive capacity; this will be shown by means of a generalized Cantor set in V. 11 (cf. Remark I. 6 A). Let $R = \{|z| < \infty\} - K$. By Theorem II. 10 A, $R \in O_{AB}$. On the other hand $f(z) = z \in AB^*(R)$. To see this let the distance between 0 and K be greater than 1. For the Green's function $g(z, 0)$ on R with pole 0 we clearly have $\log(1/|z|) \leq g(z, 0)$ on R and thus $\log^+ |1/f| \leq g(\cdot, 0)$ on R, i.e. $1/f \in MB^*(R)$ and therefore $f \in AB^*(R)$. Hence $R \notin O_{AB^*}$ and (134) follows.

Next we claim (HEINS [4]):

$$O_{MB^*} < O_{AB^*}. \tag{135}$$

Take two sequences $\{a_k\}_1^\infty$, $\{b_k\}_1^\infty$ of real numbers such that

$$-1 < \cdots < b_{k+1} < a_k < b_k < \cdots < a_1 < b_1 < -\tfrac{1}{2}$$

and $\lim_k a_k = \lim_k b_k = -1$. Let

$$S = \{|z|<1\} \cap \{\operatorname{Re} z < -\tfrac{1}{2}\} - \bigcup_{k=1}^{\infty} [a_k, b_k]$$

and

$$K = \left\{ -\frac{1}{2} + it \,\middle|\, -\frac{\sqrt{3}}{2} < t < \frac{\sqrt{3}}{2} \right\}.$$

Denote by u the harmonic measure of $K \cup \bigcup_1^\infty [a_k, b_k]$ with respect to the region S: $u \in H(S)$ has boundary values $u | (\partial S) - \overline{K \cup (\bigcup_1^\infty [a_k, b_k])} = 0$ and $u | K \cup (\bigcup_1^\infty [a_k, b_k]) = 1$. Let

$$\sigma(x) = \max\{u(x+iy) \,|\, |y| \le \sqrt{1-x^2}\},$$

$$\sigma_k = \min\{\sigma(x) \,|\, b_{k+1} < x < a_k\}.$$

By choosing the $[a_k, b_k]$ sufficiently small and the $[b_{k+1}, a_k]$ sufficiently large we may assume that

$$\sum_{k=1}^{\infty} \sigma_k < \infty. \tag{136}$$

Next take two sequences $\{c_k\}_1^\infty$ and $\{d_k\}_1^\infty$ of real numbers such that

$$0 < \cdots < d_{k+1} < c_k < d_k < \cdots < c_1 < d_1 < 1$$

and $\lim_k c_k = \lim_k d_k = 0$. Let

$$T_1 = \{0<|z|<1\} - \bigcup_1^\infty [c_k, d_k],$$

$$T_2 = \{0<|z|<1\} - \left(\bigcup_1^\infty [a_k, b_k]\right) \cup \left(\bigcup_1^\infty [c_k, d_k]\right),$$

and

$$T_3^k = \{|z| \le \infty\} - [a_k, b_k] \qquad (k = 1, 2, \ldots).$$

Consider the surface R obtained as follows: join T_1 and T_2 crosswise along $\bigcup_1^\infty [c_k, d_k]$, and to the surface so created join every T_3^k crosswise along $[a_k, b_k]$, $k = 1, 2, \ldots$. The resulting covering surface of $\{|z| \le \infty\}$ will be denoted by R.

Take the projection map f_0 of R onto the extended plane and a complex number α and a $\zeta_k \in T_3^k$ with $f_0(\zeta_k) = \alpha$. Let $z_0 \in T_1$ and let $g(z, z_0)$

be the Green's function on R with pole z_0. Consider $S \subset T_2$ and let $m = \max_{z \in K} g(z, z_0) < \infty$. Then since $g(\cdot, z_0)|S \leq m\,u$ on S,

$$g(z, z_0) \leq m\,\sigma_{k-1}$$

on T_3^k, and in particular

$$g(\zeta_k, z_0) \leq m\,\sigma_{k-1}.$$

Let $\{z_i\}_1^\infty = f_0^{-1}(\alpha)$ and denote by v_i the multiplicity of z_i as an α-point of f_0. Set

$$s(z) = \sum_{i=1}^{\infty} v_i\, g(z, z_i).$$

We take $|\alpha| > 1$; then all $v_i = 1$ and $s(z) = \sum_{k=1}^{\infty} g(z, \zeta_k)$. Since $s(z_0) < \infty$, s is a positive superharmonic function on R. Observe that

$$\log \frac{|\alpha| - 1}{|f_0 - \alpha|} \leq s$$

on R and therefore $f_0 \in MB^*(R)$, i.e. $R \notin O_{MB^*}$.

Take an arbitrary $f \in MB^*(R)$ and denote by F the component of $f_0^{-1}(|w| < \frac{1}{2})$ which is not relatively compact in R. Then F is a 2-sheeted covering surface of $|w| < \frac{1}{2}$ and consequently a parabolic end. By Corollary 8 E, $\lim_{z \to 0} f(z)$ exists. Thus we may assume $f \in AB(F)$. Let $\varphi(w)$ be the square of the difference of two values of f on $f_0^{-1}(w)$ for $|w| < \frac{1}{2}$. Clearly $\varphi(w) \in AB(|w| < \frac{1}{2})$ and it vanishes on $(\{c_k\}_1^\infty \cup \{d_k\}_1^\infty) \cap \{|w| < \frac{1}{2}\}$ which accumulates at 0. Thus $\varphi \equiv 0$ on $|w| < \frac{1}{2}$.

It follows that f takes on the same value on $T_1 \cup T_2$ at two points with the same projection. By an analytic continuation of $f \circ f_0^{-1}$ to the projection of the slit $[a_k, b_k]$ we conclude that $f \in MB^*(R)$ is constant on $f_0^{-1}(w)$ for every w. Thus we can find a rational function r such that $f = r \circ f_0$, i.e. every $f \in MB^*(R)$ has a pole unless it is constant. Therefore $R \in O_{AB^*}$.

9 B. Picard's End. We are going to construct a parabolic surface of infinite genus with only one Stoïlow boundary point and with every end Ω having the following Picard property (HEINS [2]):

Every nonconstant meromorphic function on Ω takes on all values infinitely often with the exception of at most two.

Let $\{a_n\}_1^\infty$ be a sequence of real numbers $0 < a_1 < \cdots < a_n < \cdots$ such that $a = \inf_n a_{n+1}/a_n > 1$. Subdivide each $[a_{2n+1}, a_{2n+2}]$ into an odd number $(> 1 + a_{2n+2})$ of subsegments. Set:

$T_1 = \{|z| < \infty\}$ less the slits $[a_{2n}, a_{2n+1}]$ and every alternate subsegment of $[a_{2n+1}, a_{2n+2}]$ starting with the second for all n;

$T_2 =$ the region T_1 less the slits $[-a_{2n+2}, -a_{2n+1}]$ for all n;

$T_3^n = \{|z| \leq \infty\}$ slit along the single interval $[-a_{2n+2}, -a_{2n+1}]$.

Now join T_1 and T_2 crosswise along their common slits. To the resulting surface join every T_3^n crosswise along $[-a_{2n+2}, -a_{2n+1}]$ for every n so as to obtain a covering surface R of the extended plane.

We first show that $R \in O_G$. Let A_n be the 2-sheeted covering of $a_{2n} < |z| < a_{2n+1}$ lying in the surface obtained by joining T_1 and T_2. Then A_{n+1} separates A_n from the ideal boundary of R. Let $w_n \in H(\bar{A}_n)$ such that $w_n = 1$ over $|z| = a_{2n}$ and $w_n = 0$ over $|z| = a_{2n+1}$. Let $w_n' \in H(a_{2n} < |z| < a_{2n+1})$ such that $w_n'(z) = w_n(f_0^{-1}(z))$ where f_0 is the projection of the surface obtained by joining T_1 and T_2 into $\{|z| \le \infty\}$. Then

$$D(w_n') = \frac{2\pi}{\operatorname{mod}(a_{2n} < |z| < a_{2n+1})} = \frac{2\pi}{\log(a_{2n+1}/a_{2n})} \le \frac{2\pi}{\log a}.$$

Since $D_{A_n}(w_n) = 2D(w_n')$ we have

$$D_R(w_n) \le \frac{4\pi}{\log a}$$

where w_n is extended to R by setting $w_n = 1$ on the relatively compact part of $R - \bar{A}_n$ and $w_n = 0$ elsewhere. Clearly $w_n \in \mathrm{MI}_\Delta(R)$ and $\lim_n w_n = 1$. Thus by Theorem III.1 G, $1 \in \mathrm{MI}_\Delta(R)$, and by Theorem III.2 F, $R \in O_G$.

Let R_n be the subregion of R bounded by the dividing cycle γ_n which lies over $|z| = \sqrt{a_{2n} a_{2n+1}}$ in the surface obtained by joining T_1 and T_2. Then $\{R_n\}_1^\infty$ defines an exhaustion of R. Clearly R has only one Stoïlow boundary point.

The set $\Omega_n = R - \bar{R}_n$ is an end. We have only to show that Ω_n has the asserted property. Suppose to the contrary that there exists a nonconstant meromorphic function f on Ω_n for which there exist three distinct values taken on at most a finite number of times. Without loss of generality we may suppose that $f \in M(\Omega_n \cup \partial\Omega_n)$ omits the values 0, 1, ∞. This can be accomplished by replacing Ω_n by some Ω_{n+p}.

We shall now deduce that there exists a nonconstant AB-function on Ω_n. If $\overline{f(\Omega_n)} \neq \{|w| \le \infty\}$ then there is nothing to prove. Thus we have only to show that this condition is satisfied. If this were not the case, i.e. $\overline{f(\Omega_n)} = \{|w| \le \infty\}$, then f would have to take arbitrarily large and arbitrarily small values on $\gamma_m = \partial R_m \subset \Omega_n$ for sufficiently large $m > n$ and hence also values of modulus 1. On applying Schottky's theorem (e.g. Tsuji [5, p. 268]) to f on A_m, the 2-sheeted covering of $a_{2m} < |z| < a_{2m+1}$, we see that

$$\sup_{m>n} \max_{\gamma_m} |f| < \infty.$$

Here we made essential use of the condition $a = \inf_m a_{2m+1}/a_{2m} > 1$ which implies that every A_m is conformally equivalent to some annulus containing a fixed annulus to the boundary of which the image of γ_m has a distance greater than a fixed positive number. Thus $\overline{f(\Omega_n)} \neq \{|w| \le \infty\}$.

We may now suppose that $f \in AB(\bar{\Omega}_n)$. Let G_n be the covering surface over $\operatorname{Re} z > a_{2n+1}$ in the surface obtained by joining T_1 and T_2. It is 2-sheeted over $\operatorname{Re} z > a_{2n+1}$ and a part of Ω_n. By the same argument as in 9 A (also see Myrberg's example in I. 10 B) we conclude that f takes on the same value on G_n at two points with a common projection. The same is true for Ω_n on T_1 and T_2. Thus by means of an analytic continuation to the projection of the slit $[-a_{2n+2}, -a_{2n+1}]$ along which T_2 and T_3^n are joined, $f \circ f_0^{-1}$ can be continued to all of $\{|z| \leq \infty\}$ so as to be in $AB(|z| \leq \infty)$. We conclude that f is constant, a contradiction.

9 C. General O_{MB^*}. In the general case we have the following inclusion relations (Heins [4]):

$$O_{HB} < O_{MB^*} < O_{AB^*} < O_{AB}. \tag{137}$$

In view of (126), (134), and (135), all we have to show is the strictness of $O_{HB} \subset O_{MB^*}$. Take an end Ω of the surface R in 9 B. Clearly $\Omega \notin O_{HB}$. Choose an arbitrary nonconstant $f \in M(\Omega)$. We have seen in 9 B that f is not of bounded valence on Ω or on any subend of Ω. On the other hand if $f \in MB^*(\Omega)$ then by 8 E, f is of bounded valence on any subend Ω' with $\bar{\Omega}' \subset \Omega$. Thus $f \notin MB^*(\Omega)$, i.e. $\Omega \in O_{MB^*}$.

9 D. O_{MB^*} for Finite Genus. We know from 9 A that $O_{AB^*} < O_{AB}$ even for plane regions and thus for surfaces of finite genus. We shall now show that $O_{AB^*} \subset O_{HB}$ for finite genus. To see this take an $R \notin O_{HB}$ of finite genus. Then R can be embedded in a closed Riemann surface R'.

Since $R' - R$ contains infinitely many points we can construct, by suitably modifying the process used in I. 9 A, a nonconstant meromorphic function f on R' such that $f \in A(R)$. We may assume that the compact set $f(\partial R)$ does not meet the unit disk $|w| < 1$. Let $g(z, \zeta_i)$ be the Green's functions on R with poles ζ_i where $R \cap f^{-1}(0) = \{\zeta_i\}$, a finite set. These functions exist since $O_G \subset O_{HB}$. Take an exhaustion $\{R_n\}$ of R and let $\{g_n(z, \zeta_i)\}$ be the corresponding Green's functions. For sufficiently large n, $f(\partial R_n)$ does not meet $|w| < 1$ and thus $\log 1/|f| - \sum_i g_n(\cdot, \zeta_i) \leq 0$ on ∂R_n and hence on R_n. Consequently

$$\log^+ \left| \frac{1}{f} \right| \leq \sum_i g(\cdot, \zeta_i)$$

on R, i.e. $1/f \in MB^*(R)$. Hence $f \in AB^*(R)$, i.e. $R \notin O_{AB^*}$.

From this and from (137) we conclude that

$$O_{HB} = O_{MB^*} = O_{AB^*} < O_{AB} \tag{138}$$

for surfaces of finite genus, and also for plane regions.

§ 4. Invariance under Deformation

In this chapter we have studied mainly the class O_{HB}, using Wiener's algebra $\mathbb{N}(R)$ and Wiener's compactification $R_{\mathbb{N}}^*$. The question arises: which properties of $\mathbb{N}(R)$ and $R_{\mathbb{N}}^*$ are essential in determining whether or not R belongs to the class O_{HB}. We close the chapter with this rather short section devoted to the question; the section is thus a counterpart of III. § 3.

The algebraic and topological structures of $\mathbb{N}(R)$ and $R_{\mathbb{N}}^*$ respectively are considered in 10, and each will be shown to be characteristic for what we call Wiener's structure. In 11 we shall prove that the classes O_G, O_{HB}, and U_{HB} are invariant under mappings preserving this structure.

10. Wiener's Structure

10 A. Wiener's Mapping. Let T be a topological mapping of a Riemann surface R_1 onto another R_2. We call T a *Wiener mapping* provided $f \circ T \in \mathbb{N}(R_1)$ if and only if $f \in \mathbb{N}(R_2)$. The class of Riemann surfaces between any two of which there exists a Wiener mapping will be called *Wiener's structure.* Similarly we define the *quasiconformal structure* and *Royden's structure.* We saw in III. § 3 that the algebraic structure of $\mathbb{M}(R)$ is characteristic of the quasiconformal structure of R and that the topological structure of $R_{\mathbb{M}}^*$ is characteristic of Royden's structure of R.

10 B. Algebraic Structure. First we show that the algebraic structure of $\mathbb{N}(R)$ is characteristic of Wiener's structure of R:

THEOREM. *Every Wiener mapping T of R_1 onto R_2 induces (and is induced by) an algebraic isomorphism $\sigma\colon f \to f^\sigma$ of $\mathbb{N}(R_1)$ onto $\mathbb{N}(R_2)$ satisfying the relation $f^\sigma = f \circ T^{-1}$ for every $f \in \mathbb{N}(R_1)$.*

We have only to prove that for a given σ the required T exists. Let $p_2 \in R_{2\mathbb{N}}^*$. It defines a character X_{p_2} on $\mathbb{N}(R_2)$ by $X_{p_2}(f) = f(p_2)$ for every $f \in \mathbb{N}(R_2)$. Let

$$X(f) = X_{p_2}(f^\sigma)$$

for $f \in \mathbb{N}(R_1)$. Clearly X is a character on $\mathbb{N}(R_1)$. Hence by Corollary 2 B there exists a unique point $p_1 \in R_{1\mathbb{N}}^*$ such that

$$X(f) = f(p_1)$$

for every $f \in \mathbb{N}(R_1)$. By $p_1 = S^*(p_2)$, S^* defines a mapping of $R_{2\mathbb{N}}^*$ into $R_{1\mathbb{N}}^*$ such that

$$f^\sigma(p) = f(S^*(p)) \tag{139}$$

for every $p \in R_{2\mathbb{N}}^*$ and every $f \in \mathbb{N}(R_1)$.

Similarly we can find a mapping T^* of $R^*_{1\mathbb{N}}$ into $R^*_{2\mathbb{N}}$ such that

$$f^\tau(p)=f\big(T^*(p)\big) \tag{140}$$

for every $p\in R^*_{1\mathbb{N}}$ and every $f\in\mathbb{N}(R_2)$ where τ is the inverse of σ. Since $\mathbb{N}(R_i)=B(R^*_{i\mathbb{N}})$ $(i=1,2)$ we infer by (139) and (140) that both S^* and T^* are continuous. Again from (139) and (140) it follows easily that both $S^*\circ T^*$ and $T^*\circ S^*$ are identities, i.e. $(T^*)^{-1}=S^*$ and T^* is a topological mapping of $R^*_{1\mathbb{N}}$ onto $R^*_{2\mathbb{N}}$.

By 2 G, $T^*p\in R_2$ if and only if T^*p is G_δ in $R^*_{2\mathbb{N}}$. Again T^*p has the latter property if and only if p has it in $R^*_{1\mathbb{N}}$, i.e. $p\in R_1$. Hence $T=T^*|R_1$ gives a topological mapping of R_1 onto R_2. From (139) it follows that

$$f^\sigma=f\circ T^{-1} \tag{141}$$

which shows that $f\in\mathbb{N}(R_1)$ if and only if $f\circ T^{-1}\in\mathbb{N}(R_2)$, i.e. T^{-1} and hence T are Wiener's mappings.

10 C. Topological Structure. Next we show that the topological structure of $R^*_{\mathbb{N}}$ is also characteristic of Wiener's structure of R:

THEOREM. *Every Wiener mapping T of R_1 onto R_2 can be extended to a homeomorphism T^* of $R^*_{1\mathbb{N}}$ onto $R^*_{2\mathbb{N}}$. Conversely for every homeomorphism T^* of $R^*_{1\mathbb{N}}$ onto $R^*_{2\mathbb{N}}$, $T=T^*|R_1$ is a Wiener mapping of R_1 onto R_2.*

Let T be a Wiener mapping. Then $f\to f^\sigma=f\circ T^{-1}$ induces an algebraic isomorphism of $\mathbb{N}(R_1)$ onto $\mathbb{N}(R_2)$. In the proof of 10 B we saw that there exists a homeomorphism T^* of $R^*_{1\mathbb{N}}$ onto $R^*_{2\mathbb{N}}$ such that $f^\sigma=f\circ T^{*-1}$. Therefore $T^*|R_1=T$.

Conversely let T^* be a homeomorphism of $R^*_{1\mathbb{N}}$ onto $R^*_{2\mathbb{N}}$. By 2 G, $T=T^*|R_1$ is a homeomorphism of R_1 onto R_2. Clearly f is continuous on $R^*_{1\mathbb{N}}$ if and only if $f\circ(T^*)^{-1}$ is continuous on $R^*_{2\mathbb{N}}$. Since $\mathbb{N}(R_i)=B(R^*_{i\mathbb{N}})$, $f\in\mathbb{N}(R_1)$ if and only if $f\circ T^{-1}\in\mathbb{N}(R_2)$, i.e. T is a Wiener mapping.

11. Boundary Behavior

11 A. Invariance of Harmonic Boundary. We know that the harmonic boundary $\Delta_{\mathbb{N}}$ determines the structure of the class $HB(R)$. It is of interest that the Wiener structure suffices to determine $\Delta_{\mathbb{N}}$:

THEOREM. *For every homeomorphism T^* of $R^*_{1\mathbb{N}}$ onto $R^*_{2\mathbb{N}}$*

$$T^*(\Delta_{1\mathbb{N}})=\Delta_{2\mathbb{N}}. \tag{142}$$

Let $p_1\in\Delta_{1\mathbb{N}}$. Clearly $p_2=T^*p_1\in\Gamma_{2\mathbb{N}}$. We have to show that $p_2\in\Delta_{2\mathbb{N}}$. Contrary to the assertion assume that $p_2\in\Gamma_{2\mathbb{N}}-\Delta_{2\mathbb{N}}$. Since $\Delta_{2\mathbb{N}}$ is compact we can find two open neighborhoods F_2^* and G_2^* of p_2 in $R_{2\mathbb{N}}^*$ such that $F_2^* \supset \overline{G_2^*}$ and $\overline{F_2^*} \cap \Delta_{2\mathbb{N}} = \emptyset$. Moreover we may assume that the relative boundaries of $F_2 = F_2^* \cap R_2$ and $G_2 = G_2^* \cap R_2$ consist of piecewise analytic simple curves which do not accumulate in R_2.

Set $F_1^* = T^{*-1}(F_2^*)$, $G_1^* = T^{*-1}(G_2^*)$, $F_1 = T^{-1}(F_2) = F_1^* \cap R_1$, and $G_1 = T^{-1}(G_2) = G_1^* \cap R_1$ where $T = T^*|R_1$. Since $p_1 \notin \Delta_{1\mathbb{N}} \cap \overline{R_1 - G_1}$, $\Delta_{1\mathbb{N}} \not\subset \overline{R_1 - G_1}$, or equivalently, $\bar{G}_1 - \partial G_1$ contains a point in $\Delta_{1\mathbb{N}}$. By 3 L we can find a component G_1' of G_1 such that $G_1' \notin SO_{HB}$. Let $G_2' = T(G_1')$.

Take an exhaustion $\{R_2^{(n)}\}_0^\infty$ of R_2 such that $\partial R_2^{(n)} \cap G_2' \neq \emptyset$. Choose an $f_2 \in B(R_2)$ such that $0 \leq f_2 \leq 1$ on R_2,

$$f_2 \Big| \bigcup_{n=0}^{\infty} (\partial R_2^{(2n+1)} \cap \overline{F_2^*} \cap R_2) \cup (R_2 - F_2) = 0, \tag{143}$$

$$f_2 \Big| \bigcup_{n=0}^{\infty} (\partial R_2^{(2n)} \cap \overline{G_2^*} \cap R_2) = 1. \tag{144}$$

We shall show that $f_2 \in \mathbb{N}(R_2)$. If $R_2 \in O_G$ this is trivially true by 1 B. If $R \notin O_G$ let $K = \overline{F_2^*} \cap \Gamma_{2\mathbb{N}} \subset \Gamma_{2\mathbb{N}} - \Delta_{2\mathbb{N}}$. By 3 I there exists a superharmonic function s on R_2 which is positive and finitely continuous on R_2, $s|\Delta_{2\mathbb{N}} = 0$, and $s = \infty$ on K. For every $\varepsilon > 0$ we can find some $R_2^{(n)}$ such that

$$\varepsilon s | \overline{F_2^*} \cap (R_2 - \overline{R_2^{(n)}}) > 1.$$

Thus by (143), $0 \leq f_2 \leq \varepsilon s$ on $R_2 - \overline{R_2^{(n)}}$, i.e., $\varepsilon s \in \bar{U}(R_2, f_2)$. Therefore $\overline{W}_{f_2}^{R_2} = 0$. Since $\underline{W}_{f_2}^{R_2} \geq 0$, $W_{f_2}^{R_2}$ exists.

For any normal subregion $R_2' \subset R_2$ the existence of $W_{f_2}^{R_2'}$ can be shown as follows. Let $u_n \in H(R_2' \cap R_2^{(n)})$ with $u_n | \partial(R_2' \cap R_2^{(n)}) = f_2$. We may assume that the sequence $\{u_n\}$ converges to a harmonic function $u \in H(R_2')$ with $u|\partial R_2' = f_2$. For some $a > 0$, $f_2 < as$ on R_2. Thus by setting $u = f_2$ on $R_2 - R_2'$, $0 \leq u \leq (as) \cap 1$ on R_2. In particular u vanishes identically on $\Delta_{2\mathbb{N}}$. Clearly $(u + \varepsilon s)|R_2' \in \bar{U}(R_2', f_2)$, i.e.

$$\overline{W}_{f_2}^{R_2'} \leq u. \tag{145}$$

Let $u^*(p) = \liminf_{z \in R_2,\, z \to p} u(z)$ for every $p \in \Gamma_{2\mathbb{N}}$. Then $K_\varepsilon = \{p \in \Gamma_{2\mathbb{N}} | u^*(p) \geq \varepsilon\}$ is compact in $\Gamma_{2\mathbb{N}}$ and $K_\varepsilon \cap \Delta_{2\mathbb{N}} = \emptyset$. Let s_ε be the function s for K_ε. Since $(u - \varepsilon' s_\varepsilon - \varepsilon' s - \varepsilon)|R_2' \in \underline{U}(R_2', f_2)$ for every $\varepsilon' > 0$ we have $u - \varepsilon \leq \underline{W}_{f_2}^{R_2'}$ for every ε and thus

$$u \leq \underline{W}_{f_2}^{R_2'}.$$

This with (145) shows that $W_{f_2}^{R_2'} = u$. Therefore $f_2 \in \mathbb{N}(R_2)$ and, since $T^*|R_1 = T$ is a Wiener mapping,

$$f_1 \in \mathbb{N}(R_1) \tag{146}$$

where $f_1 = f_2 \circ T$.

Let the decomposition of f_1 in 3 F.(33) be $f_1 = v + \varphi$, $v \in HB(R_1)$, and $\varphi | \Delta_{\mathrm{N}} = 0$. Let $v_n \in H(R_1^{(n)})$ with boundary values f_1 at $\partial R_1^{(n)}$ where $R_1^{(n)} = T^{-1}(R_2^{(n)})$. Similarly define φ_n to be harmonic on $R_1^{(n)}$ with boundary values φ at $\partial R_1^{(n)}$. Then clearly $v_n = v + \varphi_n$. For $s \in \overline{U}(R_1, |\varphi|)$ we have $|\varphi_n| \leq s$ for sufficiently large n and thus $\limsup_n |\varphi_n| \leq s$ for every s. Since $\inf s = 0$ we see that $\lim_n \varphi_n = 0$. Therefore

$$\lim_n v_n = v \tag{147}$$

on R_1.

Because of $G_1' \notin SO_{HB}$ there exists a harmonic function $w \in H(G_1')$ such that $0 < w < 1$ on G_1' and w has continuous boundary values zero at $\partial G_1'$. By (144), $f_1 | \partial R_1^{(2n)} \cap G' = 1$ and thus $v_{2n} | \partial R_1^{(2n)} \cap G' = 1$. Since $f_2 \geq 0$, $v_{2n} | \partial G' \cap R_1^{(2n)} \geq 0$. Therefore $v_{2n} \geq w$ on $R_1^{(2n)} \cap G_1'$ and

$$\lim_n v_{2n} \geq w > 0 \tag{148}$$

on G_1'. By (143), $f_2 | \partial R_1^{(2n+1)} = 0$. A fortiori $v_{2n+1} \equiv 0$ on $R_1^{(2n+1)}$, i.e.

$$\lim_n v_{2n+1} = 0 \tag{149}$$

on R_1. In view of (148) this violates (147).

11 B. Absolute Continuity. Property (142) of Wiener's mappings is a counterpart of III.(125) for Royden's mappings. However the following more detailed invariance can be established in the present case:

THEOREM. *Let T^* be a homeomorphism of $R_{1\mathrm{N}}^*$ onto $R_{2\mathrm{N}}^*$. A set $X \subset \Gamma_{1\mathrm{N}}$ has harmonic measure zero if and only if $T^*(X) \subset \Gamma_{2\mathrm{N}}$ has this property.*

We have only to show that $\mu_{1\mathrm{N}}(X) = 0$ implies $\mu_{2\mathrm{N}}(T^*(X)) = 0$. To this end we may assume without loss of generality that X is a compact subset of $\Delta_{1\mathrm{N}}$. Then $X_2 = T^*(X)$ is also a compact set in $\Delta_{2\mathrm{N}}$. Suppose that $\mu_{2\mathrm{N}}(X_2) > 0$. By Corollary 2 in 4 D

$$\mu_{2\mathrm{N}}(X_2) = 1 - \mu_{2\mathrm{N}}(\Delta_{2\mathrm{N}} - X_2) = 1 - \mu_{2\mathrm{N}}(\overline{\Delta_{2\mathrm{N}} - X_2}) = \mu_{2\mathrm{N}}(X_2')$$

where $X_2' = \Delta_{2\mathrm{N}} - \overline{\Delta_{2\mathrm{N}} - X_2}$ is open in $\Delta_{2\mathrm{N}}$. Since $\mu_{2\mathrm{N}}(X_2') > 0$, $X_2' \neq \emptyset$ and $X_2' \subset X_2$. Therefore $X' = T^{*-1} X_2'$ is open in $\Delta_{1\mathrm{N}}$ and $X' \subset X$. By $(\mu.5)$ in 4 A, $\mu_{1\mathrm{N}}(X') > 0$, which contradicts $\mu_{1\mathrm{N}}(X) = 0$.

11 C. Invariant Classes. We have seen that if R_1 and R_2 have the same Wiener structure then $\Delta_{1\mathrm{N}}$ and $\Delta_{2\mathrm{N}}$ are homeomorphic. The classes O_G (cf. 2 E), O_{HB}, O_{HB}^n $(1 \leq n \leq \infty)$, and U_{HB} (cf. 3 K, 3 N) are characterized only in terms of the topology of Δ_{N}. Thus we are now able to conclude:

THEOREM. *The classes O_G, O_{HB}, O_{HB}^n, and U_{HB} are invariant under Wiener's mappings.*

It is not known whether or not O_{HB}, O_{HB}^n, and U_{HB} are quasiconformally invariant.

11 D. Boundary Property. Suppose R_1 and R_2 are open Riemann surfaces with conformally equivalent neighborhoods V_1 and V_2 of their ideal boundaries. We may assume that the ∂V_i are analytic and $\bar{V}_1$, $\bar{V}_2$ conformally equivalent. Theorem 5 C ensures the existence of an absolutely continuous homeomorphism between $\Delta_{\mathfrak{N}}(R_1)$ and $\Delta_{\mathfrak{N}}(R_2)$. By 3 K we see that R_1 belongs to O_G (resp. O_{HB}, O_{HB}^n, U_{HB}) if and only if R_2 does.

Next assume that there exists a nonconstant HP-function u_2 on R_2. If u_2 is bounded then $R_2 \notin O_{HB}$ and consequently $R_1 \notin O_{HB}$, i.e. there exists a nonconstant HB-function which may be considered positive by adding a constant. If u_2 is unbounded then since $\int_{\partial V_2} * du_2 = 0$ we also have an unbounded HP-function v_1 on $\bar{V}_1$ with $\int_{\partial V_1} * dv_1 = 0$.

Let u_1 be the principal function with respect to (v_1, L, V_1). Here L may be any normal operator (cf. I.7) but for definiteness take $L = L_1$. Since $u_1 - v_1$ is bounded on V_1, u_1 is unbounded but it is bounded from below and we may assume that it is positive. Therefore $R_1 \in O_{HP}$ if and only if $R_2 \in O_{HP}$.

We combine the above with III. 8 I:

THEOREM. *Membership in any one of the classes O_G, O_{HP}, O_{HB}, O_{HD}, O_{HB}^n, O_{HD}^n, U_{HB}, U_{HD}, and $U_{\widetilde{HD}}$ is a property of the ideal boundary.*

Chapter V

Functions with Logarithmic Singularities

In the preceding chapters we have discussed regular harmonic and analytic functions. In connection with extremal and mapping problems we have lightly touched on functions with singularities of the form $(z-\zeta)^{-n}$. We have also encountered the singularity $-\log|z-\zeta|$ which arose in the definition of the Green's function. It is to this logarithmic singularity that we now direct our attention in more detailed study.

We start with the rudiments of the theory of capacity functions. In § 1 we construct capacity functions on an arbitrary Riemann surface. They will provide us with useful conformally invariant measures of not only the ideal boundary but also ideal boundary components. The latter will then be classified as weak, unstable, and strong, whereas the capacity of the boundary gives in § 2 the classes of parabolic and hyperbolic surfaces.

Parabolicity can be characterized in several ways: in terms of Green's functions, potentials, and function spaces. These characterizations, i.e. parabolicity tests, and planar parabolic surfaces will be the main topics of § 2. The Green's function leads to the important concept of Green's potential. In particular Green's potentials of Evans' type are characteristic of parabolicity and will be discussed extensively in § 3.

§ 1. Capacity Functions

We introduce in 1 capacity functions and capacities for ideal boundaries. By means of an extension of Schwarz's lemma to Riemann surfaces we solve in 2 the problem of minimizing the supremum of analytic functions with single-valued moduli on arbitrary Riemann surfaces. The minimum is the reciprocal of the capacity.

Capacity functions and capacities for boundary components are then discussed in 3. The class of surfaces with absolutely disconnected boundaries is introduced in 4 and shown to coincide with $O_{SB}=O_{SD}$ discussed in Chapter II. Null classes corresponding to functions with single-valued moduli are also shown in 4 to coincide with the class of surfaces with boundaries of vanishing capacity.

Stoïlow's ideal boundary points of plane regions can be realized as plane continua. These continua can potentially become either points or nondegenerate continua under conformal mappings of the region. If they are stable points or nondegenerate continua we call them weak or strong respectively. Sufficient conditions for weakness and strength will be given in 5 in terms of capacities. These are closely related to radial slit disks and circular slit disks. In connection with these we shall discuss maximal disks and minimal disks and then rigid radii in 6. The last topic is closely related to *AD*-removable sets discussed in Chapter I.

1. Capacity of the Boundary

1 A. An Inequality. On the unit disk $\bar{U}: |z| \leq 1$ consider the family of functions s of the form

$$s(z) = \log|z| + \varphi(z), \qquad \varphi(0) = 0$$

with $\varphi \in H(\bar{U})$. Let α_0: $|z| = 1/e$ and α_1: $|z| = 1$. The following simple inequality is valid for every function s in this class:

$$\min_{z \in \alpha_1} s(z) - \min_{z \in \alpha_0} s(z) \leq 1. \tag{1}$$

In fact by the maximum principle $\min_{z \in \alpha_0} \varphi(z) \geq \min_{z \in \alpha_1} \varphi(z)$. Since $\log|z| \,|\, \alpha_1 = 0$ and $\log|z| \,|\, \alpha_0 = -1$,

$$\begin{aligned} \min_{z \in \alpha_1} s(z) - \min_{z \in \alpha_0} s(z) &= \min_{z \in \alpha_1} \varphi(z) - \min_{z \in \alpha_0} \varphi(z) + 1 \\ &= 1 - \left(\min_{z \in \alpha_0} \varphi(z) - \min_{z \in \alpha_1} \varphi(z)\right) \leq 1. \end{aligned}$$

1 B. Minimum Property. Let R be an open Riemann surface and β its ideal boundary. Consider the class $\{S\}$ of (multivalued) functions S on R of the form

$$S(z) = s(z) + i\, s^*(z) = \log z + \Phi(z) \tag{2}$$

on a parametric disk $\bar{U}: |z| \leq 1$ and with a single-valued real part s on R. Here $\Phi(z) = \varphi(z) + i\, \varphi^*(z)$ is a holomorphic function on $\bar{U}$ with $\varphi(0) = 0$. The functions S are assumed to have no singularities on R other than $z = 0$ in U.

We use the conventional notation

$$\int_\beta s * ds = \int_\beta s\, ds^* = \lim_{\Omega \to R} \int_{\partial\Omega} s * ds,$$

with Ω in the class of regular subregions of R. Since $\int_{\partial\Omega} s * ds = \int_{\partial U} s * ds + D_{\Omega - \bar{U}}(s)$ for $\Omega \supset \bar{U}$, $\int_\beta s * ds \leq \infty$ always exists. The capacity function s_β will be introduced in 1 E by means of the following extremal property:

THEOREM. *There exists a function $S_\beta = s_\beta + i s_\beta^* \in \{S\}$ such that*

$$\min_{s=\operatorname{Re} S,\, S\in\{S\}} \int_\beta s*ds = \int_\beta s_\beta * ds_\beta \equiv 2\pi k_\beta. \tag{3}$$

The deviation of $\int_\beta s*ds$ from the minimum, if it is finite, is given by

$$\int_\beta s*ds - \int_\beta s_\beta * ds_\beta = D_R(s-s_\beta). \tag{4}$$

First suppose that (R,β) is a compact bordered surface with border β. By I.7 A there exists a function $s_\beta = \operatorname{Re} S_\beta$ with $S_\beta \in \{S\}$ such that

$$s_\beta | \beta = k_\beta \tag{5}$$

where k_β is a finite constant. Clearly s_β is unique. For $h = s - s_\beta$ we have

$$\int_{\partial\Omega} s*ds - \int_{\partial\Omega} s_\beta * ds_\beta = \int_{\partial\Omega} h*ds_\beta + \int_{\partial\Omega} s_\beta * dh + D_\Omega(h).$$

Let $\Omega - \Omega_t = \{s_\beta < t\}$. From

$$\Big(\int_{\partial\Omega_t} h*ds_\beta - \int_{\partial\Omega_c} h*ds_\beta\Big) - \Big(\int_{\partial\Omega_t} s_\beta * dh - \int_{\partial\Omega_c} s_\beta * dh\Big) = 0$$

for $c<t$ it follows that $\int_{\partial\Omega_t} h*ds_\beta \to 2\pi h(0) = 0$ as $c\to-\infty$, and thus the above formula takes the form

$$\int_{\partial\Omega_t} s*ds - \int_{\partial\Omega_t} s_\beta * ds_\beta = D_{\Omega_t}(h).$$

On letting $t\to k_\beta$ we obtain (4) and consequently (3) for a compact bordered surface.

1 C. Convergence Proof. We turn to an arbitrary open R. Let $\{R_n\}_1^\infty$ be a regular exhaustion of R with $\bar U \subset R_1$. Let $\beta_n = \partial R_n$ and denote by s_n the unique minimizing function for the bordered surface (R_n, β_n). Set

$$s_n(z) = \log|z| + \varphi_n(z)$$

on $\bar U$ and use the notation α_0 and α_1 of 1 A.

Take an arbitrary positive integer m, fixed for the time being, and set $K_m = \bar R_m - U$. By the maximum principle and the relation $\int_{\beta_n} * ds_n = 2\pi > 0$ we obtain for $n > m$

$$d_n = \min_{\alpha_0} s_n < \min_{\alpha_1} s_n = \min_{K_m} s_n = \min_{\alpha_1} \varphi_n \le 0.$$

On the other hand

$$\max_{K_m} s_n > \max_{\alpha_1} s_n = \max_{\alpha_1} \varphi_n \ge 0$$

and therefore

$$|s_n| < \max_{K_m} (s_n - d_n) \tag{6}$$

on K_m. Moreover $(s_n - d_n)|(R_n - \{|z| \le 1/e\}) \in HP(R_n - \{|z| \le 1/e\})$. Thus by Harnack's inequality

$$\max_{K_m} (s_n - d_n) < M_m \min_{\alpha_1} (s_n - d_n) \tag{7}$$

with a finite constant M_m. From (6), (7), and (1) we conclude that

$$|s_n| < M_m \tag{8}$$

on K_m. On choosing a subsequence we may suppose that

$$s_\beta = \lim_{n \to \infty} s_n$$

exists on R since m was arbitrary.

1 D. Deviation Formula. It remains to prove (3) and (4) for s_β. Using (4) for s_n and R_n and observing that $D_{R_{n+1} - \bar{R}_n}(s_{n+1}) \ge 0$ we obtain

$$2\pi k_n = \int_{\beta_n} s_n * ds_n \le \int_{\beta_n} s_{n+1} * ds_{n+1} \le \int_{\beta_{n+1}} s_{n+1} * ds_{n+1} = 2\pi k_{n+1}$$

where k_n is the constant value of s_n on β_n. Hence we have the existence of

$$-\infty < k_\beta = \lim_{n \to \infty} k_n \le \infty.$$

Again by virtue of (4) for R_n and s_n, the fact that $D_{R - R_n}(s) \ge 0$, and the definition of $\int_\beta s * ds$ we obtain

$$2\pi k_n = \int_{\beta_n} s_n * ds_n \le \int_{\beta_n} s * ds \le \int_\beta s * ds.$$

Since this is true for every n and $s \in \{s\}$

$$2\pi k_\beta \le \inf_{\{s\}} \int_\beta s * ds \le \int_\beta s_\beta * ds_\beta.$$

On the other hand for $m > n$, $\int_{\beta_n} s_m * ds_m \le \int_{\beta_m} s_m * ds_m = 2\pi k_m$. Thus

$$\int_\beta s_\beta * ds_\beta = \lim_{n \to \infty} \int_{\beta_n} s_\beta * ds_\beta = \lim_{n \to \infty} \lim_{m \to \infty} \int_{\beta_n} s_m * ds_m \le \lim_{n \to \infty} \lim_{m \to \infty} 2\pi k_m = 2\pi k_\beta.$$

Therefore

$$\min_{\{s\}} \int_\beta s * ds = \int_\beta s_\beta * ds_\beta = 2\pi k_\beta. \tag{9}$$

Let $s - s_\beta = h$ and $s_\varepsilon = s_\beta + \varepsilon h$ with a real constant ε. Then

$$\int_\beta s_\varepsilon * ds_\varepsilon = \int_\beta s_\beta * ds_\beta + \varepsilon \int_\beta (s_\beta * dh + h * ds_\beta) + \varepsilon^2 D_R(h).$$

We may suppose, in order to establish (4) that $\int_\beta s * ds < \infty$ or equivalently $D_R(h) < \infty$. Then in the quadratic form above, (9) implies that $\int_\beta (s_\beta * dh + h * ds_\beta) = 0$. On choosing $\varepsilon = 1$ we obtain (4).

Eq. (4) also gives the uniqueness of s_β if $k_\beta < \infty$.

1 E. Capacity c_β. The function s_β given by (3) is called the *capacity function* of β on R. The quantity

$$c_\beta = e^{-k_\beta}, \qquad k_\beta = \frac{1}{2\pi} \int_\beta s_\beta * ds_\beta, \tag{10}$$

is, by definition, the *capacity* of β. The capacity function is unique if $c_\beta > 0$ or equivalently $k_\beta < \infty$. If $c_\beta = 0$ we shall see later that there are infinitely many capacity functions. Among them we shall single out the important subclass of Evans-Selberg potentials (see 13).

2. The Class of Functions W

2 A. Minimum of $D(W)$. We have studied the class $\{S\}$ and the associated class $\{s\}$ of real parts of $\{S\}$. Next we consider the class $\{W\}$ of regular (multivalued) analytic functions W on R with

$$\lim_{z \to 0} \left| \frac{W(z)}{z} \right| = 1 \tag{11}$$

on a fixed parametric disk U: $|z| < 1$ and a single-valued modulus $|W|$ on R. The correspondence $S \to W = e^S$ gives a one-to-one and onto mapping between $\{S\}$ and $\{W\}$. Observe that the Dirichlet integral of W is given by

$$D_R(W) = \int_\beta |W|^2 \, d \arg W.$$

THEOREM. *The function $W_\beta = e^{S_\beta}$ gives the minimum of $D_R(W)$ among functions in $\{W\}$:*

$$\min_{\{W\}} D_R(W) = D_R(W_\beta) = \frac{2\pi}{c_\beta^2}. \tag{12}$$

We retain the notation of 1. Let $W = e^S = e^{s + is^*}$. Then

$$D_{R_n}(W) = \int_{\beta_n} |W|^2 \, d \arg W = \int_{\beta_n} e^{2s} * ds.$$

For $W_n = e^{S_n} = e^{s_n + is_n^*}$ on $\bar{R}_n$ and $s - s_n = h$

$$\begin{aligned} D_{R_n}(W) - D_{R_n}(W_n) &= \int_{\beta_n} (e^{2(s_n + h)} * d(s_n + h) - e^{2s_n} * ds_n) \\ &= e^{2k_n} \Big(\int_{\beta_n} (e^{2h} - 1) * ds_n + \int_{\beta_n} e^{2h} * dh \Big). \end{aligned}$$

As in 1, $\int_{\beta_n}(h*ds_n - s_n*dh)=0$ and thus

$$\int_{\beta_n}(e^{2h}-1)*ds_n \geq 2\int_{\beta_n} h*ds_n = 2\int_{\beta_n}(h*ds_n - s_n*dh)=0.$$

Furthermore on setting $H=h+ih^*$ we obtain

$$\int_{\beta_n} e^{2h}*dh = D_{R_n}(e^H)\geq 0$$

and conclude that

$$D_{R_n}(W)\geq D_{R_n}(W_n)=2\pi\, e^{2k_{\beta_n}}=\frac{2\pi}{c_{\beta_n}^2}. \tag{13}$$

The convergence of S_n to S_β implies that of $W_n=e^{S_n}$ to $W_\beta=e^{S_\beta}\in\{W\}$ on R. Since $k_n\leq k_{n+1}$

$$\lim_{n\to\infty} D_{R_n}(W_n)=\lim_{n\to\infty}\frac{2\pi}{c_{\beta_n}^2}=\frac{2\pi}{c_\beta^2}.$$

In view of $D_R(W)\geq D_{R_n}(W)$ and (13) it follows that $D_R(W)\geq 2\pi/c_\beta^2$ and a fortiori

$$D_R(W_\beta)\geq \inf_{\{W\}} D_R(W)\geq\frac{2\pi}{c_\beta^2}.$$

On the other hand for $m>n$, $D_{R_n}(W_m)\leq D_{R_m}(W_m)$. Therefore

$$D_R(W_\beta)=\lim_{n\to\infty} D_{R_n}(W_\beta)=\lim_{n\to\infty}\lim_{m\to\infty} D_{R_n}(W_m)\leq \lim_{n\to\infty}\lim_{m\to\infty} D_{R_m}(W_m)=\frac{2\pi}{c_\beta^2}$$

and we obtain (12).

2 B. Schwarz's Lemma. In the case of R: $|z|<1$ and β: $|z|=1$ we have $s_\beta=\log|z|$, $k_\beta=0$, $c_\beta=1$, and $W_\beta=z$. Thus the following theorem is a generalization of Schwarz's classical lemma:

THEOREM. *Let (R,β) be a compact bordered surface with compact border β. Consider the class $\{f\}$ of analytic functions f on R with single-valued moduli $|f|$ on R such that $f(0)=0$ in a fixed parametric disk $U=\{|z|<1\}\subset R$. If*

$$|f|\leq c_\beta^{-1} \tag{14}$$

on R then

$$|f|\leq|W_\beta|,\qquad \lim_{z\to 0}\left|\frac{f}{z}\right|\leq 1, \tag{15}$$

with equalities if and only if $|f|\equiv|W_\beta|$ on R.

Consider the function $y=\log|f/W_\beta|$. It is harmonic on R except for those points in R at which $|f|=0$. At each of these exceptional points y

has a negative logarithmic singularity. Hence by the maximum principle applied to the subharmonic function y we see that

$$y \leq \limsup_{z\in R,\, z\to\beta} y(z) \leq \log \frac{c_\beta^{-1}}{e^{k_\beta}} = 0$$

on R. Here y can assume the value 0 on R if and only if $y \equiv 0$ on R.

2 C. Minimax Property. We have established minimum property (12) for the class $\{W\}$. Using the above Schwarz lemma we obtain another minimum property:

THEOREM. *On an arbitrary open Riemann surface R*

$$\min_{\{W\}} (\sup_R |W|) = \sup_R |W_\beta| = \frac{1}{c_\beta}.$$

By Theorem 2 B we have for any fixed $W \in \{W\}$ on R

$$\sup_{R_n} |W| \geq \sup_{R_n} |W_n| = \frac{1}{c_{\beta_n}}. \tag{16}$$

In fact let $a = \sup_{R_n} |W|$ and $f = W/a\, c_{\beta_n}$. Then (14) is valid and thus by (15), $|f| \leq |W_n|$, i.e. $(1/c_{\beta_n})|W| \leq a\,|W_n|$. In view of the normalization $\lim_{z\to 0} |W/z| = \lim_{z\to 0} |W_n/z| = 1$ we obtain (16).

We pass to the limit $n \to \infty$ in (16):

$$\sup_R |W| \geq \frac{1}{c_\beta}.$$

We have to show that $\sup_R |W_\beta| = 1/c_\beta$ or equivalently $\sup_R s_\beta = k_\beta$. Since $s_n \leq k_{\beta_n} \leq k_\beta$ clearly $\sup_R s_\beta \leq k_\beta$. If $\sup_R s_\beta < k_\beta$ then $\int_\beta s_\beta * ds_\beta < 2\pi\, k_\beta$ which contradicts (3).

3. Capacity of an Ideal Boundary Point

3 A. Minimum Property. Let γ be a Stoïlow ideal boundary point (cf. IV.5 D) of an open Riemann surface R. Take a regular exhaustion $\{R_n\}_0^\infty$ of R with $R_0 = U$. Let β_{ni} be the relative boundary of a component of $R - \bar{R}_n$. Then $\partial R_n = \beta_n = \bigcup_{i=1}^{\nu(n)} \beta_{ni}$. Denote by γ_n that β_{ni} which is the relative boundary of the component of $R - \bar{R}_n$ belonging to the defining sequence of γ.

We now consider the subclass $\{T\} \subset \{S\}$ of functions $T = t + i\,t^*$ with the property $\int_{\gamma_n} * dt = \int_{\gamma_n} dt^* = 2\pi$ for all n and $\int_\alpha * dt = 0$ across every dividing cycle α not separating γ from R_0. Clearly the definition is independent of the exhaustion and we write symbolically $\int_\gamma * dt = 2\pi$. The following minimum property is valid:

THEOREM. *There exists a function* $T_\gamma = t_\gamma + i\, t_\gamma^* \in \{T\}$ *such that*

$$\min_{t=\operatorname{Re} T,\; T\in\{T\}} \int_\beta t * dt = \int_\beta t_\gamma * dt_\gamma = 2\pi\, k_\gamma. \tag{17}$$

The deviation from the minimum is given by

$$\int_\beta t * dt - \int_\beta t_\gamma * dt_\gamma = D_R(t - t_\gamma). \tag{18}$$

For an admissible t we have

$$\int_{\gamma_n} * dt = 2\pi, \qquad \int_{\beta_{ni} \neq \gamma_n} * dt = 0. \tag{19}$$

Let t_n be the function in $\{t\} = \{\operatorname{Re} T\}$ for R_n determined by the conditions

$$t_n | \beta_{ni} = k_{ni} \tag{20}$$

for $i = 1, \ldots, \nu(n)$ with constants k_{ni}. Such a t_n is of course unique. If $\nu(n) = 1$ then s_n qualifies as t_n. If $\nu(n) > 1$ then let $\beta_{n\nu(n)} = \gamma_n$. Take $\omega_i \in H(\bar{R}_n)$ such that $\omega_i | \beta_{ni} = 1$ and $\omega_i | \beta_n - \beta_{ni} = 0$ $(i = 1, \ldots, \nu(n) - 1)$. Set $a_{ij} = \int_{\beta_{nj}} * d\omega_i$ and $b_j = \int_{\beta_{nj}} * ds_n$ $(j = 1, \ldots, \nu(n) - 1)$. Clearly the solvability of

$$\sum_{i=1}^{\nu(n)-1} a_{ij}\, \xi_i = b_j \tag{21}$$

with respect to ξ_i, $i = 1, \ldots, \nu(n) - 1$ is equivalent to showing that $\sum_{i=1}^{\nu(n)-1} a_{ij} \xi_i = 0$ implies $\xi_i = 0$. Set $\omega = \sum_{i=1}^{\nu(n)-1} \xi_i \omega_i$. Then $\int_{\beta_{nj}} * d\omega = \sum_{i=1}^{\nu(n)-1} a_{ij}\, \xi_i = 0$ for all $j = 1, \ldots, \nu(n) - 1$ and thus also for $j = \nu(n)$. Since ω is constant on each β_{nj}, $D_{R_n}(\omega) = \sum_{i=1}^{\nu(n)-1} \xi_i \int_{\beta_{ni}} * d\omega = 0$. From $\omega | \gamma_n = 0$ we obtain $\omega \equiv 0$ on R_n. In particular $\xi_i = \omega | \beta_{ni} = 0$.

Let ξ_i be as in (21) and set $t_n = s_n - \sum_{i=1}^{\nu(n)-1} \xi_i \omega_i$. Clearly t_n satisfies (20); (21) implies the latter half of (19); the former half is a consequence of $\int_{\partial R_0} * dt_n = 2\pi$.

For $t - t_n = h$ observe that

$$\int_{\beta_n} t * dt = \int_{\beta_n} t_n * dt_n + \int_{\beta_n} h * dt_n + D_{R_n}(h).$$

As in 1 B we have

$$\int_{\beta_n} h * dt_n = \int_{\beta_n} (h * dt_n - t_n * dh) = \int_{t_n = c} (h * dt_n - t_n * dh) \to 0$$

for $c \to -\infty$. Thus

$$\int_{\beta_n} t * dt \geq \int_{\beta_n} t_n * dt_n = 2\pi\, k_{\gamma_n} \tag{22}$$

where k_{γ_n} stands for $k_{n\nu(n)}$. Since $\int_{\beta_{n+1}} t_{n+1} * dt_{n+1} - \int_{\beta_n} t_{n+1} * dt_{n+1} = D(t_{n+1}) > 0$, with D taken over $R_{n+1} - R_n$, (22) implies

$$k_{\gamma_n} \leq k_{\gamma_{n+1}}. \tag{23}$$

Thus we can define

$$k_\gamma = \lim_{n \to \infty} k_{\gamma_n}.$$

By means of (19) the compactness of $\{t_n\}$ can be shown in a manner similar to that in 1 C. A subsequence, again denoted by $\{t_n\}$, then converges to a function $t_\gamma \in \{t\}$ on R. The argument in 1 D used to prove (3) can be applied to establish (17) if we replace $s_n, k_{\beta_n}, s_\beta, k_\beta$ by $t_n, k_{\gamma_n}, t_\gamma, k_\gamma$ respectively. We may assume $\int_\beta t_\gamma * dt_\gamma < \infty$, and (18) can again be deduced using the argument in 1 D.

3 B. Capacity c_γ. We have thus obtained a measure characterizing the magnitude of a boundary point γ. As a counterpart of 1 E we call t_γ in (17) the *capacity function* of γ on R and refer to the quantity

$$c_\gamma = e^{-k_\gamma}, \qquad k_\gamma = \frac{1}{2\pi} \int_\beta t_\gamma * dt_\gamma, \tag{24}$$

as the *capacity* of γ. The boundary β of R is said to be *absolutely disconnected* if $c_\gamma = 0$ for every Stoïlow boundary point γ.

3 C. The Class $\{V\}$. We consider the subclass $\{V\} \subset \{W\}$ of functions V on R such that $\int_\gamma d \arg V = 2\pi$ for a fixed Stoïlow boundary point γ and $\int_\alpha d \arg V = 0$ for every dividing cycle α not separating γ from $R_0 = U = \{|z| < 1\}$. The one-to-one mapping in 2 A of $\{S\}$ onto $\{W\}$ takes $\{T\}$ onto $\{V\}$, i.e. $T \to V = e^T$ is a one-to-one mapping of $\{T\}$ onto $\{V\}$.

We now restrict our attention to a planar R and prove:

THEOREM. *Let R be a planar Riemann surface. There exists a function $V_\gamma = e^{T_\gamma} \in \{V\}$ such that among all univalent functions $V \in \{V\}$ on R*

$$\min D_R(V) = D_R(V_\gamma) = \frac{2\pi}{c_\gamma^2}. \tag{25}$$

Suppose first that (R, β) is a compact bordered surface with border β. The function $V_\gamma = e^{T_\gamma}$ maps R in a univalent manner onto a circular slit disk of radius e^{k_γ}. In the T_γ-plane the image is a vertical slit strip of width 2π to the left of $t_\gamma = k_\gamma$. For a univalent V the corresponding T is also univalent. It contributes to the area the quantity

$$\int_\beta t * dt - \int_\beta t_\gamma * dt_\gamma,$$

adding a certain area to the right of $t=k_\gamma$ and subtracting an area from the left of it. By (18) the former is at least as large as the latter. The same is true a fortiori for the corresponding areas in the V-plane since e^T enlarges the areas to the right of $t=k_\gamma$ more than those to the left of it. Thus (25) follows.

For an arbitrary planar R the assertion is then proved by the same limiting process as in 2 A.

3 D. Subboundaries. We have defined capacity functions and capacities both for the entire boundary $\beta=R_S^*-R$ and for one point γ of R_S^*-R. Between these two extreme cases we can also define the capacity function and the capacity for a compact subset K of R_S^*-R. To this end we merely replace γ by K and the defining sequence of γ by a decreasing sequence of neighborhoods of K tending to it and given by an exhaustion $\{R_n\}$ of R.

For further generalizations of capacities and capacity functions reference is here made to KURAMOCHI [23], MARDEN-RODIN [1], and SARIO-OIKAWA [27].

4. Surface Classes C_β and C_γ

4 A. Weak Boundaries. We denote by $\mathfrak{M}$ the class of analytic functions with single-valued moduli. The meaning of $O_{\mathfrak{M}D}$, $O_{\mathfrak{M}BH}$, $O_{\mathfrak{M}B}$ is clear; for example $O_{\mathfrak{M}BD}$ is the class of Riemann surfaces on which there is no nonconstant bounded Dirichlet finite analytic function with a single-valued modulus.

Let C_β be the class of open Riemann surfaces R whose ideal boundaries have vanishing capacity, i.e. $c_\beta=0$. For convenience we add closed surfaces to C_β. We have:

THEOREM. *The following equalities hold:*

$$C_\beta=O_{\mathfrak{M}D}=O_{\mathfrak{M}BD}=O_{\mathfrak{M}B}. \tag{26}$$

By (12) we infer that $C_\beta=O_{\mathfrak{M}D}$. The minimax relation (Theorem 2 C) furnishes $C_\beta=O_{\mathfrak{M}B}$. Both (12) and Theorem 2 C then give $C_\beta=O_{\mathfrak{M}BD}$.

4 B. Absolutely Disconnected Boundaries. Let C_γ be the class of open Riemann surfaces R with absolutely disconnected boundaries, i.e. such that $c_\gamma=0$ for every $\gamma\in R_S^*-R$. As in the case of C_β we include closed surfaces in C_γ.

Since $\{t\}\subset\{s\}$, (3) and (17) imply

$$k_\beta\le k_\gamma, \qquad c_\beta\ge c_\gamma$$

for every $\gamma\in R_S^*-R$. Therefore $C_\beta\subset C_\gamma$. Actually

$$C_\beta< C_\gamma \tag{27}$$

as will be seen by the example of the Cantor set of positive capacity to be discussed in 11.

The Ahlfors-Beurling equality $O_{SD}=O_{SB}$ is proved here again in the following setting:

THEOREM. *For planar surfaces*

$$C_{\gamma}=O_{SD}=O_{SB}. \tag{28}$$

Equality (25) implies $C_{\gamma}=O_{SD}$. Next suppose that $c_{\gamma}>0$ for some $\gamma\in R_S^*-R$. Since $V_n=e^{T_n}$ converges uniformly to $V_{\gamma}=e^{T_{\gamma}}$ on each compact set and $\max|V_n|=e^{k_{\gamma_n}}=1/c_{\gamma_n}$ we obtain $|V_{\gamma}|\leq 1/c_{\gamma}$ and thus $V_{\gamma}\in SB(R)$. It follows that $O_{SB}\subset C_{\gamma}$. Conversely if $R\notin O_{SB}$ then $SB(R)\subset SD(R)$ gives $R\notin O_{SD}$. We infer that $c_{\gamma}>0$ and therefore $C_{\gamma}\subset O_{SB}$.

5. Strong and Weak Components

5 A. Boundary Components under Conformal Mappings. Let R be a region in $|z|\leq\infty$. If it is simply connected then its ideal boundary $\beta=R_S^*-R$ consists of only one point and is realized as a continuum in $|z|\leq\infty$. A univalent conformal mapping φ of R is in a natural manner a homeomorphism of $R\cup\beta$ onto $\varphi(R)\cup\varphi(\beta)$ where $\varphi(\beta)=(\varphi(R))_S^*-\varphi(R)$. According as $R\in O_G$ or not, β is a point or a proper continuum. The same is true of $\varphi(R)$ and $\varphi(\beta)$. Therefore if β is a point, so is $\varphi(\beta)$ for every φ. The same is true of proper continua.

The situation is strikingly different for plane regions R of infinite connectivity. Since a defining sequence on R corresponds to a defining sequence on $\varphi(R)$ in a one-to-one and onto fashion, φ is a topological mapping of $R\cup\beta$ onto $\varphi(R)\cup\varphi(\beta)$. We shall denote the image of $\gamma\in\beta=R_S^*-R$ under φ by $\varphi(\gamma)$. There can exist a γ which is realized as a point whereas $\varphi(\gamma)$ is realized as a proper continuum.

As an example take the region $R=\{|z|<\infty\}-\bigcup_{n=0}^{\infty}E^2(p_1^n p_2^n\ldots)$ constructed in I.6C. Each point $\gamma\in\beta=R_S^*-R$ is realized as a point in $\bigcup_{n=0}^{\infty}E^2(p_1^n p_2^n\ldots)\cup\{\infty\}$. Since R has finite area, $z\in SD(R)$. By (28) there exists a $\varphi\in SB(R)$. The point $\infty\in\beta$ is then mapped by φ to a proper continuum $\varphi(\infty)$.

We call $\gamma\in\beta=R_S^*-R$ *weak* if $\varphi(\gamma)$ is a point under every univalent conformal mapping φ of R. In contrast $\gamma\in\beta$ is *strong* if $\varphi(\gamma)$ is a proper continuum for every φ. If γ is weak or strong it is said to be *stable*, otherwise it is *unstable*. The component ∞ in the above example is unstable.

These concepts were further studied by SAVAGE [1], OIKAWA [3], AKAZA [1], and AKAZA-OIKAWA [4] (see also the monograph SARIO-OIKAWA [27]).

5 B. Weak Boundary Components. Let R be a plane region and γ a Stoïlow boundary component, i.e. $\gamma \in R_S^* - R$.

THEOREM. *If $c_\gamma = 0$, then γ is weak.*

Suppose that γ is not weak. Then by deforming R conformally we may assume that γ is a proper continuum. Again on applying a conformal mapping we may assume that R is a subset of $|z|<1$, $0 \in R$, and $\gamma: |z|=1$. Let $\tilde{t}(z) = \log|z|$. Then clearly $\tilde{t} \in \{t\} = \{\operatorname{Re} T\}$ of 3 A and $\int_\beta \tilde{t} * d\tilde{t} = D_{R-\bar{R}_0}(\tilde{t}) - 2\pi\rho < D_{e^{-\rho}<|z|<1}(\tilde{t}) = 2\pi\rho < \infty$ where $\bar{R}_0 = \{|z| \le e^{-\rho}\} \subset R$. Thus by (17), $k_\gamma < \infty$ and $c_\gamma = e^{-k_\gamma} > 0$.

5 C. Tests. We shall discuss tests for $c_\beta = 0$ in 10. By natural modifications we obtain tests for $c_\gamma = 0$ (SAVAGE [1]), and these in turn give the weakness of γ.

5 D. Capacity c^γ. Before turning to strong components we define another kind of capacity of an ideal boundary component of an arbitrary open Riemann surface R.

We use the same notation as in 3 A. In particular $\{t\}$ is the family of real parts of functions in $\{T\}$. We now consider the subfamily $\{p\} \subset \{t\}$ on R_n of functions p such that $p|\gamma_n = k(p)$, a constant. In $\{p\}$ there exists a unique function t^n such that $*dt^n = 0$ on $\beta_n - \gamma_n$. We set $k(t^n) = k^{\gamma_n}$,

$$k^\gamma = \limsup_{n\to\infty} k^{\gamma_n}, \tag{29}$$

and introduce another capacity

$$c^\gamma = e^{-k^\gamma} \tag{30}$$

of γ.

The function t^n may be constructed for example by using the double of R_n about $\beta_n - \gamma_n$.

The geometric significance of t^n is as follows. The function $V^n = e^{T^n}$ with $T^n = t^n + i t^{n*} \in \{T\}$ on R_n maps R_n conformally onto a radial slit disk of radius $1/c^{\gamma_n} = e^{k^{\gamma_n}}$ with $\delta_n = \beta_n - \gamma_n$ going to the radial slits.

LEMMA. *The function t^n maximizes the functional*

$$2\pi k(p) - \int_{\delta_n} p * dp \tag{31}$$

in the class $\{p\}$ on R_n. The maximizing function is unique.

By an application of Green's formula to $D_{R_n}(p - t^n)$ as in 1 B we can easily prove the deviation formula

$$2\pi k(p) - \int_{\delta_n} p * dp = 2\pi k^{\gamma_n} - D_{R_n}(p - t^n) \tag{32}$$

from which our assertion follows.

If we take $p=t_n$ (cf. 3 A) we see that

$$k_{\gamma_n} \leq k^{\gamma_n}. \tag{33}$$

Therefore $c_{\gamma_n} \geq c^{\gamma_n}$ and consequently

$$c^{\gamma} \leq c_{\gamma}. \tag{34}$$

5 E. Strong Boundary Components. We again consider a plane region R and take a $\gamma \in R_S^* - R$. We claim:

THEOREM. *If $c^{\gamma} > 0$ then γ is strong.*

If γ is not strong, then we may assume it is the point ∞. We may also suppose that $0 \in R$. Take an arbitrarily large $M > 0$. For sufficiently large n, γ_n lies in $|z| > M$ and separates ∞ from 0. Let φ be the conformal mapping of the simply connected region bounded by γ_n onto the disk $|w| < \rho$ with $\varphi(0)=0$ and $\varphi'(0)=1$. The subharmonic function $|\varphi(z)/z|$ takes its maximum $\rho/\min_{\gamma_n}|z|$ on γ_n. This value dominates $\lim_{z \to 0}|\varphi(z)/z| = 1$. Thus $\rho \geq \min_{\gamma_n}|z|$ and therefore

$$\rho > M. \tag{35}$$

Since $\log|\varphi| \in \{p\}$ on R_n we have by Lemma 5 D

$$2\pi \log \rho - \int_{\delta_n} \log|\varphi(z)| \, d \arg \varphi(z) \leq 2\pi k^{\gamma_n}.$$

Here $-\int_{\delta_n} \log|\varphi(z)| \, d\arg\varphi(z)$, the logarithmic area enclosed by $\varphi(\delta_n)$, is nonnegative and consequently $\log \rho \leq k^{\gamma_n}$ or $c^{\gamma_n} \leq 1/\rho$. Hence by (35), $c^{\gamma} \leq 1/M$ for every $M > 0$, that is $c^{\gamma} = 0$.

5 F. Minimum Property. For the function $t_n \in \{p\} \subset \{t\}$ on R_n (cf. 3 A) we have the following counterpart of Lemma 5 D:

LEMMA. *The function t_n minimizes the functional*

$$2\pi k(p) + \int_{\delta_n} p * dp \tag{36}$$

in $\{p\}$ on R_n. The minimizing function is unique.

The proof is immediate by means of the deviation formula

$$2\pi k(p) + \int_{\delta_n} p * dp = 2\pi k_{\gamma_n} + D_{R_n}(p - t_n). \tag{37}$$

5 G. Reduction Theorem. We append here the following reduction principle which will be used in 6.

Let R be an arbitrary open Riemann surface and $\{R_n\}_1^{\infty}$ an exhaustion. Consider a class $\mathscr{F}$ (resp. $\mathscr{F}_n$) of functions on R (resp. R_n) such that

$$\mathscr{F} | R_n \subset \mathscr{F}_n, \qquad \mathscr{F}_{n+p} | R_n \subset \mathscr{F}_n. \tag{38}$$

Let a functional $m_n(\cdot)$ be given on $\mathscr{F}_n$ with the property

$$\lim_{\sup_{R_n} |f-f_0| \to 0} m_n(f) = m_n(f_0) \tag{39}$$

for $f, f_0 \in \mathscr{F}_n$. Moreover the existence of

$$\lim_{n\to\infty} m_n(f) = m(f) \tag{40}$$

is postulated for every $f \in \mathscr{F}$ so that $m(\cdot)$ is a functional on $\mathscr{F}$. We retain the notation C-lim of III.1 C in reference to convergence on compact subsets.

THEOREM. *Suppose that*

(a) *there exists an* $f_n \in \mathscr{F}_n$ *with* $m_n(f_n) = \min_{\mathscr{F}_n} m_n(f)$,
(b) $m_n(f) \le m_{n+p}(f)$ *for every* $f \in \mathscr{F}_{n+p}$,
(c) $C\text{-}\lim_n f_n = f_\infty \in \mathscr{F}$ *on* R.

Then

$$\min_{\mathscr{F}} m(f) = m(f_\infty) = \lim_{n\to\infty} m_n(f_n). \tag{41}$$

The proof starts with the relations

$$m(f_\infty) = \lim_n m_n(f_\infty) = \lim_n \big(\lim_k m_n(f_k)\big) \le \liminf_k m_k(f_k) \tag{42}$$

which are a consequence of $m_n(f_k) \le m_k(f_k)$ for $k \ge n$. On the other hand for $f \in \mathscr{F}$

$$m_n(f_n) \le m_n(f) \le m(f).$$

Since this is true for every n and every $f \in \mathscr{F}$

$$\limsup_n m_n(f_n) \le \inf_{\mathscr{F}} m(f) \le m(f_\infty). \tag{43}$$

On combining (42) and (43) we obtain (41).

6. Univalent Functions

6 A. The Class $\mathscr{F}$. Hereafter in this section we consider plane regions R of a particular shape. Let δ be a compact set in $\{|z|<1\}$ such that $R = \{|z|<1\} - \delta$ is connected and $0 \in R$. Choose $\gamma = \{|z|=1\}$.

Let $\mathscr{F}$ be the class of univalent analytic functions F on $R \cup \gamma$ with $F(0)=0$, $F'(0)=1$, and $|F(z)|\,|\,\gamma = r(F)$, a constant.

Take an "exhaustion" $\{R_n\}_1^\infty$ of R such that $0 \in R_1$ and γ is a component of ∂R_n. The logarithmic area enclosed by the image of $\delta_n = \partial R_n - \gamma$ is

$$\mathfrak{A}_n(F) = -\int_{\delta_n} \log|F(z)|\, d\arg F(z). \tag{44}$$

For $\lim_{n\to\infty} \mathfrak{A}_n(F)$ we write

$$\mathfrak{A}(F) = -\int_{\delta} \log|F(z)|\, d\arg F(z). \tag{45}$$

To study mappings of R we first prove:

THEOREM. *There is a unique $F_0 \in \mathscr{F}$ such that*

$$\max_{\mathscr{F}} \left(2\pi \log r(F) + \mathfrak{A}(F)\right) = 2\pi \log r(F_0). \tag{46}$$

For any $F \in \mathscr{F}$ the deviation from this maximum is $D_R(\log|F/F_0|)$. The area of the image δ_{F_0} of δ under F_0 vanishes. Each component of δ_{F_0} is a radial slit or a point.

For $\mathscr{F}_n$ we take $\mathscr{F}$ defined for R_n. The approximating functional m_n is $-2\pi \log r(F) - \mathfrak{A}_n(F)$. Clearly (38), (39), and (40) are satisfied. Condition (a) in 5 G is met by Lemma 5 D, and (b) is obvious. Let F_{0n} be the extremal function in $\mathscr{F}_n$. Then

$$2\pi \log r(F_{0,\,n+p}) \le 2\pi \log r(F_{0,\,n+p}) + \mathfrak{A}_n(F_{0,\,n+p}) \le 2\pi \log r(F_{0n})$$

and therefore the $r(F_{0n})$ are bounded. Since $|F_{0n}| < r(F_{0n})$ on R_n we may assume, on choosing a subsequence, that $\{F_{0n}\}$ converges to an $F_0 \in \mathscr{F}$ uniformly on compacta of $R \cup \gamma$; condition (c) in 5 G is fulfilled.

By Theorem 5 G, F_0 maximizes $2\pi \log r(F) + \mathfrak{A}(F)$ in $\mathscr{F}$ and the value of the maximum is

$$2\pi \lim_{n\to\infty} \log r(F_{0n}) = 2\pi \log r(F_0). \tag{47}$$

The area of the image δ_{F_0} of δ under F_0 vanishes since

$$2\pi \log r(F_0) + \mathfrak{A}(F_0) = 2\pi \log r(F_0).$$

Set $h = \log|F/F_0|$ on R and $p_\varepsilon = \log|F_0| + \varepsilon h$. The corresponding value of the functional of Lemma 5 D for R_n is

$$\Psi(\varepsilon) = 2\pi \log r(F_0) + \mathfrak{A}_n(F_0) + \varepsilon[\ ] - \varepsilon^2 \int_{\delta_n} h * dh$$

where the quantity in brackets is immaterial. In fact all other terms of Ψ, and Ψ itself, have unique limits as $n \to \infty$. Consequently so does the bracketed quantity. On passing to the limit we equate to zero the ε-derivative of $\Psi(\varepsilon)$ for $\varepsilon = 0$ and obtain $[\ ] = 0$. The value $\varepsilon = 1$ then gives the deviation formula

$$2\pi \log r(F) + \mathfrak{A}(F) = 2\pi \log r(F_0) - D_R\left(\log\left|\frac{F}{F_0}\right|\right). \tag{48}$$

From this the uniqueness of F_0 follows at once.

To prove that each component δ_{iF_0} of δ_{F_0} is a radial slit or a point suppose this were not the case. Then we could map the doubly connected region R_0 bounded by γ_{F_0} and δ_{iF_0} conformally onto a radial slit disk by a function $\varphi \in \mathscr{F}$ on R_0. Since δ_{iF_0} can always be thought of as being the "image" under some F of an analytic Jordan curve we conclude by Lemma 5 D that the radius of γ_φ is strictly greater than that of γ_{F_0}. The area of δ remains 0 under φ, and the restriction of φ to R increases the quantity $2\pi \log r + \mathfrak{A}$, a contradiction.

6 B. Maximal Disks. We now state two immediate consequences of the preceding theorem.

COROLLARY 1. *The function $F_0 \in \mathscr{F}$ has the extremal property*

$$\max_{\mathscr{F}} r(F) = r(F_0). \tag{49}$$

This is obvious since $2\pi \log r + \mathfrak{A} \geq 2\pi \log r$.

Next let $\mathscr{G} (\supset \mathscr{F})$ be the class of functions G on R which differs from $\mathscr{F}$ only in that the condition $|F(z)| \,|\, \gamma = \text{const}$ is dropped.

COROLLARY 2. *The function $F_0 \in \mathscr{G}$ gives*

$$\max_{\mathscr{G}} \min_{\gamma} |G(z)| = r(F_0). \tag{50}$$

Suppose there is a $G \in \mathscr{G}$ such that $\min_\gamma |G(z)| = r' > r(F_0)$. Then we map the simply connected finite region R' bounded by the "image" γ_G of γ under G onto a disk R'_φ by a function $\varphi \in \mathscr{F}$ on R'. The radius ρ of R'_φ is not less than r'. Thus $\rho > r(F_0)$ and the restriction of $\varphi \circ G$ to R gives a contradiction.

6 C. Minimal Disks. For an analogue of Theorem 6 A we consider the corresponding minimum problem and circular slit disks:

THEOREM. *There is a unique function $F_1 \in \mathscr{F}$ on R with*

$$\min_{\mathscr{F}} \big(2\pi \log r(F) - \mathfrak{A}(F)\big) = 2\pi \log r(F_1). \tag{51}$$

The deviation from this minimum is given by $D_R(\log |F/F_1|)$. The image δ_{F_1} of δ under F_1 vanishes. Each component of δ_{F_1} is either a circular slit or a point.

Condition 5 G.(a) is met by Lemma 5 F, and 5 G.(b) is obvious. By Lemma 5 D, $r(F_{1n}) \leq r(F_{0n})$ where F_{1n} is the minimal function in $\mathscr{F}$ on R_n. We have seen in 6 A that the $r(F_{0n})$ are bounded. Thus we may suppose that 5 G.(c) is also satisfied, i.e. $F_1 = C\text{-}\lim_n F_{1n}$ exists. The rest is the same as in the proof of Theorem 6 A. The deviation formula now takes the form

$$2\pi \log r(F) - \mathfrak{A}(F) = 2\pi \log r(F_1) + D_R\left(\log\left|\frac{F}{F_1}\right|\right). \tag{52}$$

As in 6 B we deduce:

COROLLARY 1. *The function F_1 gives*

$$\min_{\mathscr{F}} r(F) = r(F_1). \tag{53}$$

COROLLARY 2. *The function F_1 has the minimax property*

$$\min_{\mathscr{G}} \max_{\gamma} |G(z)| = r(F_1). \tag{54}$$

6 D. Rigid Disks. For $R = \{|z|<1\} - \delta$, $\gamma = \{|z|=1\}$ we have shown that

$$r(F_1) \le r(F) \le r(F_0) \tag{55}$$

for every $F \in \mathscr{F}$ on $R \cup \gamma$. If the range $[r(F_1), r(F_0)]$ of r reduces to a point then we say that R has a *rigid radius*. An immediate problem is: What condition on δ ensures that R has a rigid radius? If R has this property then by (55) and Theorems 6 A, 6 C we see that $F_1 \equiv F_0 \equiv z$ and δ is totally disconnected. A complete characterization is given as follows:

THEOREM. *A necessary and sufficient condition for $R = \{|z|<1\} - \delta$ to have a rigid radius is that δ be an AD-null set, i.e. $\{|z| \le \infty\} - \delta \in O_{AD}$.*

Suppose $\{|z| \le \infty\} - \delta \in O_{AD}$. Since $\mathscr{F} \subset AD(R)$ we infer by the characterization of O_{AD}-regions in I.8 that every $F \in \mathscr{F}$ can be analytically continued to all of $|z| \le 1$. Since $u(z) = \log|F(z)/z| \in H(|z| \le 1)$, $u|\gamma$ is constant, and $u(0) = 0$ we infer that $F(z) \equiv z$ for every $F \in \mathscr{F}$. Thus R has a rigid radius.

Assume $\{|z| \le \infty\} - \delta \notin O_{AD}$. Again by the characterization of O_{AD}-regions in I.8 there exists a conformal mapping $f(z)$ of $\{|z| \le \infty\} - \delta$ such that the image δ_f of δ under f has positive area. Let φ map the simply connected region bounded by $\gamma_f = f(\gamma)$ onto a disk R' of radius ρ. Then the image δ_φ of δ_f under φ continues to have positive area, hence also positive logarithmic area $\mathfrak{A}$. We assume φ is suitably normalized so that $\varphi \circ f \in \mathscr{F}$. Then by (46)

$$2\pi \log r(\varphi \circ f) + \mathfrak{A} \le 2\pi \log r(F_0),$$

and $\mathfrak{A} > 0$ implies $\rho = r(\varphi \circ f) < r(F_0)$.

Remark. For a comprehensive discussion of capacity functions and of weak, strong, and unstable boundary components we refer the reader to the monograph SARIO-OIKAWA [27].

§ 2. Parabolic and Hyperbolic Surfaces

The class C_β will be seen to coincide with the class O_G of parabolic Riemann surfaces. The parabolicity of a surface is also characterized by the nonexistence of nonconstant positive superharmonic functions. This leads once again to the inclusion relation $O_G \subset O_{HP}$. That this inclusion is strict will also be shown. These questions together with some further characterizations of parabolicity will be the main theme of 7.

Hyperbolic surfaces, i.e. surfaces not in O_G, are surfaces with Green's functions. If we vary poles, the Green's functions give rise to a kernel function which is symmetric, jointly continuous, and positive. The Green's kernel will also be extended to Royden's compactifications. These topics will be discussed in 8.

With respect to the Green's kernel we shall introduce in 9 transfinite diameters, Tchebycheff's constants, minimum energies, and Dirichlet's constants. Using Green's potentials we shall show that these quantities coincide for regular sets. This result remains valid of course in a much more general setting. However we shall treat the simplest situation necessary for later application in § 3: the construction of Evans' harmonic functions.

In 10 several parabolicity tests will be given. These constitute the counterparts of O_{AD}-tests discussed in Chapter I.

In the final no. 11 of this section we specialize to plane regions. A plane compact set with parabolic complement is said to be of capacity zero. These sets are important in that they are HB- and HD-removable. Robin's constants are also considered and used in a criterion for vanishing capacity in the case of generalized Cantor sets. To this end a brief discussion is included on logarithmic potentials.

7. Parabolicity

7 A. Review. We have called a Riemann surface R *parabolic* if it is closed or if the harmonic measure of its ideal boundary vanishes; the class of parabolic surfaces is denoted by O_G (see III. 2 F). In contrast an $R \notin O_G$ is called *hyperbolic*. We also say that $R \in O_G$ has a *null boundary* and $R \notin O_G$ a *positive boundary*.

We have obtained various characterizations of $R \in O_G$ in III. 2 F, IV. 1 B, and IV. 2 E. The equivalence of $R \in O_G$, $1 \in \mathrm{M}_\Delta(R)$, and $\Delta_{\mathrm{M}} = \emptyset$ is particularly useful.

7 B. Positive Superharmonic Functions. We denote by $\mathscr{S}(R)$ the class of nonnegative superharmonic functions on R and by $O_{\mathscr{S}}$ the class of Riemann surfaces R such that $\mathscr{S}(R)$ consists of only constants. Clearly closed surfaces are in $O_{\mathscr{S}}$ by the minimum principle for superharmonic functions. Moreover (OHTSUKA [1], AHLFORS [8]):

THEOREM. *The following identity is valid:*

$$O_G = O_{\mathscr{S}}. \tag{56}$$

Suppose $R \notin O_G$. Then the harmonic measure $\omega(z) = \omega(z; R, R_0)$ of the ideal boundary of R is not constant (cf. III.2 F). Let $s = 1 - \omega$. Clearly $s \in \mathscr{S}(R)$ and thus $R \notin O_{\mathscr{S}}$.

Conversely assume that $R \notin O_{\mathscr{S}}$. Take a nonconstant $s \in \mathscr{S}(R)$ and a constant c such that $\inf_R s < c < \sup_R s$. Then $v = (s \cap c)/c$ is in $\mathscr{S}(R)$ and does not reduce to a constant. By the lower semicontinuity of s there exists a disk $R_0 \subset R$ such that $v | \bar{R}_0 = 1$. Take an exhaustion $\{R_n\}_0^\infty$ of R. Let $w_n(z) = w(z; R_n, R_0)$ $(n \geq 1)$ be continuous on R and such that $w_n | \bar{R}_0 = 1$, $w_n | R_n - \bar{R}_0 \in H(R_n - \bar{R}_0)$, and $w_n | R - R_n = 0$. Since $v \geq w_n$ on R and $\{w_n\}_1^\infty$ is increasing, $w(z) = \lim_{n \to \infty} w_n(z)$ exists and $v \geq w$, $w | \bar{R}_0 = 1$, $w \in H(R - \bar{R}_0)$, and w is continuous on R. By the fact that $0 \leq \inf_R w \leq \inf_R v < 1$ and $w | \bar{R}_0 = 1$, w is not constant and the same is true of $1 - w$. Since $1 - w_n = \omega_n$ (cf. III.2 F), $1 - w = \omega$, which implies $R \notin O_G$.

7 C. Strict Inclusion $O_G < O_{HP}$. Since $\mathscr{S}(R) \supset HP(R)$ we obviously have $O_{\mathscr{S}} \subset O_{HP}$ and by (56), $O_G \subset O_{HP}$ (cf. also IV.7 C). To see that actually

$$O_G < O_{HP} \tag{57}$$

we shall now construct a hyperbolic Riemann surface R on which there is no nonconstant HP-function. The approach is stimulated by the ingenious example of TÔKI [2] to prove $O_{HB} < O_{HD}$ (see III.4 H) and by the pioneering work of AHLFORS [6].

Let $\{q_m\}_1^\infty$ be the sequence of odd primes, 3, 5, 7, ...; then $(m, n) \to \mu = q_m 2^n$ is a one-to-one mapping of the set of all pairs of positive integers into the set of positive integers. With each μ we associate a positive integer $\lambda = \lambda(\mu)$ to be specified later.

Set $r_i = 1 - 2^{-i}$. In the unit disk $|z| < 1$ consider the radial slits

$$S_{mn}^{\nu} = \left\{ z = r\, e^{i\theta} \,|\, r_{2\mu} \leq r \leq r_{2\mu+1},\ \theta = \nu \cdot \frac{2\pi}{2^{m+\lambda(\mu)}} \right\} \tag{58}$$

where for each fixed (m, n) or equivalently $\mu = q_m 2^n$, $\nu = 1, \ldots, 2^{m+\lambda(\mu)}$. Thus for each $m = 1, 2, \ldots$ we have an infinite sequence $n = 1, 2, \ldots$ of collections of slits $\nu = 1, \ldots, 2^{m+\lambda}$.

For each fixed m consider the sector

$$\Sigma_{mk}=\left\{z=r\,e^{i\theta}|0\leq r<1,\ (k-1)\frac{2\pi}{2^{m-1}}\leq\theta\leq k\frac{2\pi}{2^{m-1}}\right\},\tag{59}$$

$k=1,\ldots,2^{m-1}$. Identify by pairs those edges of S_{mn}^{ν} that lie in the same sector Σ_{mk} and are symmetrically located with respect to the bisecting half-ray d_{mk} of Σ_{mk}. The edges facing d_{mk} are here identified, as are the edges away from d_{mk}. In particular the edges of a slit on d_{mk} are mutually identified, and the left edge of a slit on $\theta=(k-1)\,2\pi/2^{m-1}$ is identified with the right edge of the corresponding slit on $\theta=k\,2\pi/2^{m-1}$.

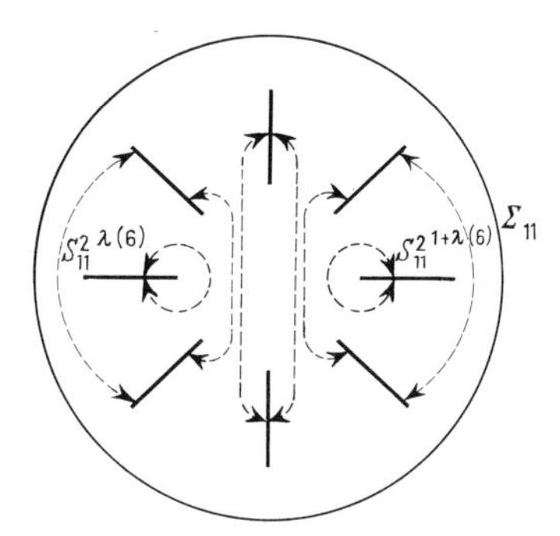

Fig. 12

The first step of this procedure is indicated in Fig. 12: if $m=1$ then $k=1$ and there is only one sector $\Sigma_{11}=\{z|0\leq\theta\leq 2\pi\}=\{|z|<1\}$, and the bisecting half ray is $d_{11}=\{z|\theta=\pi\}$, A topological model of the surface resulting from the identifications shown in Fig. 12 is a surface of genus 3 and with one boundary component.

We proceed with the general construction. The points identified above will be denoted by p and $p_m=p_m(p)$. For an end point p of a slit on the boundary of Σ_{mk} there are 2^{m-1} identical points $p_m^i(p)$ where $i=1,\ldots,2^{m-1}$.

When this identification is carried out for each $m=1,2,\ldots,$ a surface R of infinite genus is obtained.

We now make R into a Riemann surface as follows. Let $z=z(p)$ be the natural projection on $|z|<1$ of $p\in R$; it is 2- or more valued on the slits. If $p_0\in R$ is not on a slit let $V=V_{p_0}$ be a disk with center p_0 not touching any slit, and let $h(p)=h_{p_0}(p)=z(p)$. If p_0 is on the edge of a slit but is not one of its end points then V_{p_0} is to consist of two half-disks with equal diameters, one centered at p_0, the other at $p_m(p_0)$, neither reaching the end points or touching other slits. The half-disks are then transferred by appropriate rigid rotations $\tilde{p}(p)$ about $z=0$ so as to form a connected full disk; the mapping h is $h(p)=h_{p_0}(p)=z(\tilde{p}(p))$.

Next let p_0 be an end point of a slit S^{ν}_{mn} that does not lie on the boundary of Σ_{mk}. The neighborhood V_{p_0} of p_0 shall consist of two slit disks of equal radius, one centered at p_0, the other at $p_m(p_0)$, neither disk reaching the other end points or touching other slits. By suitable rigid rotations $\tilde{p}(p)$ about $z=0$ the two slit disks are transferred so as to form a connected doubly covered disk. The mapping $(z(\tilde{p}(p)))^{\frac{1}{2}}$ now serves as $h=h_{p_0}$.

Finally if p_0 is an end point of a slit on the boundary of Σ_{mk}, V_{p_0} shall consist of 2^{m-1} slit disks centered at the points $p^i_m(p_0)$, $i=1, \ldots, 2^{m-1}$. By rigid rotations $\tilde{p}(p)$ about $z=0$ the slit disks are again transferred so as to form a connected 2^{m-1}-fold disk. The mapping $h=h_{p_0}$ of V is now $(z(\tilde{p}(p)))^{2^{-m+1}}$.

By $\{(V, h)\}$, R gains a conformal structure, as is easily verified.

The natural projection $z=z(p)$ is not continuous on R and not even single-valued: $z(p) \neq z(p_m)$. However $|z(p)|$ is continuous on R and the same is true of $s(p)=\log(1/|z(p)|)$. By the definition of conformal structure, $s(p)$ is harmonic on R except where $z(p)=0$ and at the end points of the slits. Since it is bounded in a neighborhood of each such end point, $s(p) \in H(R-0)$, whereas $s(0)=\infty$. Thus s is in $\mathscr{S}(R)$ and does not reduce to a constant. We conclude by (56) that

$$R \notin O_G. \tag{60}$$

Actually s is the Green's function on R with pole 0. The proof of $R \in O_{HP}$ will be given below.

7 D. *HP*-Symmetry. We have seen that for $m=1$ the sectors Σ_{mk} reduce to the single sector $\Sigma_{11}=\{z=r\,e^{i\theta} | 0 \leq r < 1,\ 0 \leq \theta \leq 2\pi\}$, and the bisecting half-ray d_{11} is the negative real axis. For every $p \in R$ we denote by $T_1(p)$ the symmetric point of p with respect to d_{11}.

We first verify that T_1 is well-defined. If p lies on a slit so does $T_1(p)$, the edges corresponding in an obvious manner. In particular for $p \in S^{\nu}_{1n}$, $T_1(p)$ is the point $p_1(p)$ used in the construction. If p is on S^{ν}_{mn} with $m>1$ but is not an end point of an S^{ν}_{mn} on the boundary of Σ_{mk} then p is identical with $p_m(p)$, and T_1 leads to "two" points $T_1(p)$ and $T_1(p_m(p))$. Since the Σ_{mk} are by pairs symmetrically placed about the real axis these two points are identified by $p_m(p)$, and the operation T_1 becomes unique.

Finally if p is an end point of an S^{ν}_{mn} that lies on the boundary of Σ_{mk} then p is one of the 2^{m-1} identical points $p^i_m(p)$, $i=1, \ldots, 2^{m-1}$. The points $T_1(p^i_m(p))$ are, in a different order, identical with the $p^i_m(p)$, and the operation T_1 reduces to $p=T_1(p)$.

Thus T_1 is a well-defined sense reversing conformal self-mapping of R.

Now let $u \in HP(R)$. We shall use t for a generic point of R and also for a local parameter. In order to prove that u reduces to a constant we first show that $u(t) \equiv u(T_1(t))$. Let t_0 be the point corresponding to $z=0$. We may assume that $u(t_0)=1$. The function

$$u_1(t)=u(t)-u(T_1(t))$$

is again harmonic on R with $u_1(t)=0$ on S_{1n}^{ν}. Let c_r be a disk $r=\text{const}<1$. Then

$$\int_{c_r} |u_1(t(z))|\, d\theta \le \int_{c_r} u(t(z))\, d\theta + \int_{c_r} u(T_1(t(z)))\, d\theta = 4\pi \tag{61}$$

since the value of each integral on the right is $2\pi u(t_0)=2\pi$.

Next consider the annulus

$$A_\mu\colon\ r_{2\mu-1} \le r \le r_{2\mu+2} \tag{62}$$

for $m=1$ ($\mu=3\cdot 2^n$). The function $u_1(t(z))$ vanishes on the slits in A_μ but is not necessarily harmonic on them. Let $v_\mu(z)$ be a harmonic function on A_μ defined by $v_\mu(z)|\beta_\mu = |u_1(t(z))|$ with $\beta_\mu = \partial A_\mu$. Then

$$|u_1(t(z))| \le v_\mu(z) \tag{63}$$

on A_μ. Let $g(\zeta, z)$ be the Green's function on A_μ, with pole z, and consider the annulus

$$B_\mu\colon\ r_{2\mu} \le r \le r_{2\mu+1}. \tag{64}$$

If C_i is the circle $r=r_i$ then by (61), $\int v_\mu(z)\, d\theta \le 4\pi$ along $C_{2\mu-1}$ and $C_{2\mu+2}$. Hence for $z \in B_\mu$

$$\begin{aligned} v_\mu(z) &= \frac{1}{2\pi} \int_{\beta_\mu} v_\mu(\zeta) \frac{\partial g(\zeta, z)}{\partial n}\, ds \\ &\le 4 \max \left(\frac{\partial g(\zeta, z)}{\partial n} \middle| (\zeta, z) \in \beta_\mu \times B_\mu \right) = M_\mu . \end{aligned} \tag{65}$$

In view of (63) we conclude that $|u_1| \le M_\mu$ on B_μ. It is important to observe that M_μ is independent of u and λ.

If we let the number $2^{1+\lambda}$ of slits in B_μ tend to infinity the width of the sectors Γ_{1n}^{ν} bounded by S_{1n}^{ν}, $S_{1n}^{\nu+1}$, $C_{2\mu}$, and $C_{2\mu+1}$ tends to zero. Consider the harmonic function $w(z)$ on Γ_{1n}^{ν} with $w|S_{1n}^{\nu} \cup S_{1n}^{\nu+1}=0$ and $w|C_{2\mu} \cup C_{2\mu+1} = M_\mu$. As the last step choose $\lambda=\lambda(\mu)$ as the smallest positive integer for which

$$w(z) \le 2^{-\mu} \tag{66}$$

on the circle D_μ: $r=\frac{1}{2}(r_{2\mu}+r_{2\mu+1})$.

Since $u_1|S_{1n}^\nu \cup S_{1n}^{\nu+1}=0$ and $|u_1||C_{2\mu}\cup C_{2\mu+1}\le M_\mu$ we have $|u_1|\le w$ on B_μ and consequently

$$|u_1|\le 2^{-\mu}$$

on D_μ. On letting $\mu\to\infty$ we obtain $u_1\equiv 0$ on R. Thus $u(t)$ is symmetric about the real axis, and $\partial u/\partial\theta=0$ on the slits S_{1n}^ν and at the points on the real axis not located on any slit.

We are ready to prove, by induction, the symmetry of $u(t)$ on Σ_{mk} about d_{mk} for any m. Suppose it has been established for d_{ik} with $i=1,\ldots,m-1$ and $k=1,\ldots,2^{i-1}$; furthermore assume $\partial u/\partial\theta=0$ on S_{in}^ν and at the points on d_{ik} that do not lie on any slit.

For $p\in\Sigma_{mk}$ let $T_m(p)$ be the symmetric point about d_{mk}. If p is interior to Σ_{mk}, $T_m(p)$ is on a slit S_{hn}^ν if and only if p is. As in the case of T_1 one then shows that T_m is well-defined and a sense reversing conformal self-mapping of the interior of Σ_{mk}. Moreover $u(T_m(t))$ is harmonic in the interior of Σ_{mk} and the same is true if p is on an edge of S_{jn}^ν, $j\ge m$, on the boundary of Σ_{mk}.

The radial part of the boundary of Σ_{mk} consists of some d_{in} with $i<m$. If p is on this part of the boundary but not on an S_{jn}^ν with $j\ge m$ then a sufficiently small neighborhood U of p is transformed by $T_m(p)$ onto two half-disks. The diameters of the latter lie again on some d_{in} with $i<m$ and are located either on some slits S_{in}^ν with $i<m$ or not on any slit. But at such points $\partial u/\partial\theta=0$ and we conclude by the symmetry of $u(t)$ about any d_{in} with $i<m$ that $u(T_m(t))$ is harmonic on U. Thus the function

$$u_m(t)=u(t)-u(T_m(t))$$

is harmonic on all of R.

In the same fashion as for $m=1$ we now infer that, with the $\lambda=\lambda(\mu)$ properly chosen, $u_m(t)\equiv 0$ on R and $u(t)|\Sigma_{mk}$ is symmetric about d_{mk}. Furthermore $\partial u/\partial\theta=0$ on S_{mn}^ν and at the points on d_{mk} that do not lie on any slit.

Now take a circle C_r: $|z|=r$ without points in common with any S_{mn}^ν. Since $\partial u/\partial\theta=0$ at $C_r\cap d_{mk}$ we find on letting $m\to\infty$ that $\partial u/\partial\theta=0$ on all of C_r. Consequently $u\equiv$ const, i.e.

$$R\in O_{HP}. \tag{67}$$

This completes the proof of the strictness of the inclusion relation (57).

We remark in passing that the above example can be slightly modified to give a somewhat simpler proof of $O_{HB}<O_{HD}$.

7 E. Characterization by Maximum Principle. Let R again be an arbitrary Riemann surface and R_0 a regular subregion. Take a $u\in HB(R-R_0)$. If $R\in O_G$ then by IV.3 J

$$\max_{R-R_0} u=\max_{\partial R_0} u. \tag{68}$$

In general if (68) is valid for any u under consideration then we say that R admits the *maximum principle.* If $R \notin O_G$ we can construct a $u \in HB(R-R_0)$ such that $u|\Delta_{\mathrm{N}}=1$, $u|\partial R_0=0$. Again by IV.3 J, (68) is not valid for this u. Therefore (AHLFORS [8]):

THEOREM. *A surface R is parabolic if and only if it admits the maximum principle.*

This can also be shown in a more elementary manner. Let $R \in O_G$ and $u \in HB(R-R_0)$. Assume that R is open and set $m'=\sup_{R-R_0} u$, $m=\max_{\partial R_0} u$. For the harmonic measures $\omega_n(z)=\omega_n(z; R_n, R_0)$ we have $u \leq m+m'\omega_n$ on R_n-R_0. Since $\omega_n \to 0$ as $n \to \infty$ we infer that $u \leq m$ on $R-R_0$, i.e. $m' \leq m$ and thus $m'=m$. If $R \notin O_G$, $\omega(z)=\omega(z; R, R_0)$ does not satisfy (68).

7 F. The Identity $C_\beta = O_G$. Suppose the capacity $c_\beta = e^{-k_\beta}$ of the ideal boundary β of R is positive. Then by Theorem 2 C

$$\sup_{R-R_0} s_\beta = \sup_R s_\beta = k_\beta < \infty$$

where s_β is the capacity function whose pole is in R_0. Thus $s_\beta|R-R_0 \in HB(R-R_0)$ violates (68) and we have $R \notin O_G$, i.e. $O_G \subset C_\beta$.

Conversely suppose $R \in O_G$ and $k_\beta < \infty$. Then $s_\beta|R-R_0 \in HB(R-R_0)$ must satisfy (68), contradicting Theorem 2 C. We have obtained the identity

$$C_\beta = O_G. \tag{69}$$

7 G. The Space $\mathcal{N}(R)$. We retain the notation $\mathcal{N}(R)$ for the class of meromorphic differentials α on R having at most a finite number of poles and such that $\int \alpha \wedge * \bar{\alpha} < \infty$ over the complement of a neighborhood of the poles (cf. II.16 A). We also continue using $\mathcal{N}_1(R)$ for the subclass consisting of semiexact differentials $\alpha \in \mathcal{N}(R)$, i.e. those for which $\int_\gamma \alpha = 0$ for every dividing cycle γ not separating poles of α. We shall prove the following result anticipated in II.16 I:

THEOREM. *A surface R is parabolic if and only if $\mathcal{N}(R)=\mathcal{N}_1(R)$.*

Let V be a subregion of $R \in O_G$ such that $\gamma = \partial V$ is a dividing cycle of R and $\alpha \in \mathcal{N}(R)$ is regular on $V \cup \gamma$. By taking a slightly smaller region if necessary we may assume that γ is analytic. Take an "exhaustion" $\{V_n\}$ of V such that $\partial V_n \supset \gamma$ and let $w_n \in H(\bar{V}_n)$ with $w_n|\gamma=1$ and $w_n|\partial V_n - \gamma = 0$. By the maximum principle III.2 H, $w = B\text{-}\lim_n w_n$ is identically 1 and thus $D\text{-}\lim_n w_n = 0$. By Stokes' formula we obtain

$$\left|\int_\gamma \alpha\right|^2 = \left|\int_{\partial V_n} w_n \alpha\right|^2 \leq \left|\int_V \alpha \wedge * \bar{\alpha}\right| D_{V_n}(w_n)$$

and consequently $\int_\gamma \alpha = 0$, i.e. $\alpha \in \mathcal{N}_1(R)$.

Suppose $R \notin O_G$ and let $g(\cdot, \zeta)$ be the Green's function on R with pole $\zeta \in R$. The differential $\alpha = dg(\cdot, \zeta) + i * dg(\cdot, \zeta)$ belongs to $\mathcal{N}(R)$ and has the pole $-(z-\zeta)^{-1} dz$ at ζ as its only singularity. Let γ be a small circle about ζ. Then $\int_\gamma \alpha = -2\pi i \neq 0$, i.e. $\mathcal{N}_1(R) < \mathcal{N}(R)$.

7 H. The Class *L*. In analogy with $\log|f|$ for meromorphic functions f we introduce the class $L(R)$ of functions, harmonic on R except for a set of isolated logarithmic singularities with integral coefficients. The class of nonnegative members in $L(R)$ will be denoted by $LP(R)$. As a counterpart of $\log|f|$ with $f \in MB^*$ (cf. IV. § 3) we also consider the class $LB^*(R)$ of functions u in $L(R)$ such that $u \cup 0$ admits a superharmonic majorant on R.

Since the Green's functions are in LP and $LP \subset \mathcal{S}$ (cf. 7 B) we see that

$$O_G = O_{LP}.$$

By Theorem IV. 7 D we deduce at once that

$$O_{MB^*} = O_{LB^*}.$$

It is clear that $O_{LP} \subset O_{LB^*}$ and thus we again have $O_G \subset O_{MB^*}$.

7 I. Relative Classes. Let us fix a Riemann surface R and consider bordered subregions $(G, \partial G)$ of R with $R - G$ and ∂G compact. For such bordered surfaces $(G, \partial G)$ we have:

$(G, \partial G) \in SO_{HB}$, $(G, \partial G) \in SO_{HD}$, and $R \in O_G$ are equivalent.

In fact $R \in O_G$ is characterized by $\Delta_{\mathrm{MI}} = \emptyset$ and also by $\Delta_{\mathrm{N}} = \emptyset$. The conclusion follows easily from Theorem IV. 3 L.

In the general case we have the following complete string of strict inclusion relations:

$$\begin{array}{ccccc} SO_{HP} & < & SO_{HB} & < & SO_{HD} \\ & & \wedge & & \wedge \\ & & SO_{AB} & < & SO_{AD}. \end{array}$$

In fact SO_{HP} consists of all compact bordered surfaces (cf. IV. 6 J) and thus $SO_{HP} \subset SO_{HB}$ is trivial. If $(G, \partial G)$ is a compact bordered surface and p a point in G then $(G-p, \partial G)$ gives the strictness of the inclusion.

The relation $SO_{HB} \subset SO_{HD}$ is given by $SO_{HBD} = SO_{HD}$ which was proved in IV. 3 L. The strictness of the inclusion is again seen by means of Tôki's example R (see III. 4 H). Let U be the part of R over $|z| < 1/\sqrt{e}$ and let $\gamma = \partial U$, $G = R - \bar{U}$. Take the harmonic measure w of $|z| = 1$ with respect to $1/\sqrt{e} < |z| < 1$. The natural lifting of w to G proves that $(G, \gamma) \notin SO_{HB}$. As in III. 4 H we see that any function $u \in H^0B(G, \gamma)$ takes the same value at all points which lie over the same base point. This in turn shows that $(G, \gamma) \in SO_{HBD} = SO_{HD}$.

The inclusions $SO_{HB} \subset SO_{AB}$ and $SO_{HD} \subset SO_{AD}$ are trivial. To see their strictness consider the unit disk V and the interval $I = [-\frac{1}{2}, \frac{1}{2}]$. On I construct a Cantor set E of linear measure zero but positive capacity. By II. 10 we have $(V-E, \partial V) \in SO_{AB} \cap SO_{AD}$. Since the plane less E does not belong to O_G, $(V-E, \partial V)$ is in neither SO_{HB} nor SO_{HD} (cf. Theorem 11 B).

We recall from II. 2 C that $SO_{AB} \subset SO_{AD}$. By II. 11 D there exists an AD-removable but not AB-removable set F in I, and $(V-F, \partial V)$ serves to show the strictness of the inclusion relation.

7 J. The Class O_{HP}^n. Here we append a discussion of the class O_{HP}^n of surfaces R such that $\dim HP'(R) \leq n$ with the convention $O_{HP}^0 = O_G$. Every HP-function u on $0<|z|<1$ has the asymptotic behavior $-a \log|z|$ at $z=0$ with some constant $a \in [0, \infty)$ (cf. proof of Lemma 13 D). Let R be the surface of 7 C. Remove n points from R and let the resulting surface be $R^{(n)}$. Then

$$R^{(n)} \in O_{HP}^n - O_{HP}^{n-1}$$

for every $n = 1, 2, \ldots$. If we remove a countably infinite set of isolated points from R then the resulting surface $R^{(\infty)}$ belongs to the class O_{HP}^∞ of surfaces R of at most countably infinite $\dim(HP'(R))$ but not to O_{HP}^n $(n<\infty)$. Clearly $O_{HP}^n \subset O_{HB}^n$, and $R^{(n+1)} \in O_{HB} \subset O_{HB}^n$ gives the strictness of the inclusion. Let $E \subset R$ be a Cantor set of capacity zero and write $R^E = R - E$. Then $R^E \notin O_{HP}^\infty$ but $R^E \in O_{HB}^\infty$.

In the same fashion as for HB- and HD-functions we can define HP-minimality as follows: a function $u \in HP(R)$ is minimal if $u>0$ and $u \geq v \in HP(R)$ implies the existence of a constant c_v with $v = c_v u$. Here we impose the convention that $HP(R) = \{0\}$ if $R \in O_G$. We shall denote by U_{HP} the class of surfaces on which there exist HP-minimal functions. Clearly $R^{(1)}$ is in U_{HP} but not in U_{HB}. Moreover the existence of HB-minimal functions implies that of HP-minimal functions and therefore $U_{HP} > U_{HB}$. Recall that $R^E \notin O_{HP}^\infty$ but $R^E \in U_{HP}$.

We summarize our results obtained thus far (cf. III. 5 J and IV. 3 N):

$$\begin{array}{ccccccccccc}
O_G < O_{HP} & = & O_{HP}^1 & < & O_{HP}^2 & < \cdots < & O_{HP}^\infty & < & U_{HP} \cup O_G \\
\wedge & & \wedge & & \wedge & & \wedge & & \vee \\
O_{HB} & = & O_{HB}^1 & < & O_{HB}^2 & < \cdots < & O_{HB}^\infty & < & U_{HB} \cup O_G \\
\wedge & & \wedge & & \wedge & & \wedge & & \wedge \\
O_{HD} & = & O_{HD}^1 & < & O_{HD}^2 & < \cdots < & O_{HD}^\infty & < & U_{\widetilde{HD}} \cup O_G .
\end{array}$$

Let $R \in O_{HB} - O_G$ (resp. $O_{HD} - O_G$) and remove a closed disk $\bar{U}$ from R. Then $R - \bar{U}$ is in U_{HB} (resp. $U_{\widetilde{HD}}$) but not in O_{HB}^∞ (resp. O_{HD}^∞). Tôki's example in III. 4 H shows that $O_{HB}^\infty < O_{HD}^\infty$.

Remark. Membership in any one of the classes U_{HB}, U_{HD}, and $U_{\widetilde{HD}}$ means that the ideal boundary of the surface in question is in a sense of an intricate nature. In contrast the U_{HP}-property is quite normal. In fact MARTIN [1] proved that $HP(R)$, $R \notin O_G$, is always "spanned" by HP-minimal functions (see also CONSTANTINESCU-CORNEA [17]). Therefore $U_{HP} \cup O_G$ includes every Riemann surface.

Closely related to these questions are the so-called Martin compactification and Kuramochi compactification. These are important for potential theory using kernels but do not appear convenient for the construction of harmonic functions with various boundedness properties and thus for classification theory. For this reason we shall not discuss them in this book. The interested reader is referred to the authoritative monograph CONSTANTINESCU-CORNEA [17].

8. Green's Kernel

8 A. Green's Function. We have already defined a Green's function in III.6 A and used the fact that $R \notin O_G$ can be characterized by its existence. Once more we give its definition but this time in an axiomatic fashion. We also discuss it in connection with capacity functions introduced in §1.

Let R be a Riemann surface and suppose there exists on R a function $g(z, \zeta)$ such that

(g.1) $g(\cdot, \zeta) \in HP(R-\zeta)$,
(g.2) $g(z, \zeta) - \log(1/|z-\zeta|)$ *is harmonically extendable to* ζ,
(g.3) *every* $u \in HB(R)$ *with* $g(\cdot, \zeta) \geq u \geq 0$ *vanishes identically.*

We call $g(\cdot, \zeta)$ the *Green's function* on R with pole ζ. Clearly $g(\cdot, \zeta) \in \mathscr{S}(R)$ and thus the existence of a Green's function on R implies $R \notin O_G$.

Take an exhaustion $\{R_n\}$ of R. The capacity function $s_{\beta_n} = s_n$ on R_n with $\partial R_n = \beta_n$ and with pole $\zeta \in R_n$ takes on the constant value $k_{\beta_n} = k_n$ on β_n. Therefore $g_n(\cdot, \zeta) = k_n - s_n(\cdot)$ satisfies (g.1) and (g.2). Since $g_n(\cdot, \zeta)|\beta_n = 0$, (g.3) is also met. Thus $g_n(\cdot, \zeta)$ is the Green's function on R_n. For any h with properties (g.1) and (g.2) on R_n, $h - g_n$ is in $H(R)$ and its "boundary values" are nonnegative on β_n. Thus $h \geq g_n$. This shows the uniqueness of the Green's function on R_n.

Now suppose $R \notin O_G = C_\beta$. Then the relations

$$g_n(z, \zeta) = k_n - s_n(z) \tag{70}$$

and $\lim_n k_n = k_\beta < \infty$ imply for $g(z, \zeta) = \lim_n g_n(z, \zeta)$ that

$$g(z, \zeta) = k_\beta - s_\beta(z). \tag{71}$$

Assume $g(\cdot,\zeta)\geq u(\cdot)\geq 0$ on R with $u\in HB(R)$. This gives $g(\cdot,\zeta)-g_n(\cdot,\zeta)\geq u(\cdot)\geq 0$ on β_n and hence on R_n. On letting $n\to\infty$ we conclude that $0\geq u\geq 0$ on R, i.e. g satisfies not only (g.1) and (g.2) but also (g.3). We have shown the existence of a Green's function on R if $R\notin O_G$.

Again let h have properties (g.1) and (g.2). Then $h-g_n\in HP(R_n)$ and thus $h\geq g$ on R which shows the uniqueness of the Green's function on R. We conclude:

THEOREM. *There exists a unique Green's function $g(\cdot,\zeta)$ on a Riemann surface R with pole $\zeta\in R$ if and only if $R\notin O_G$. In this case $g(\cdot,\zeta)$ is the smallest function satisfying* (g.1) *and* (g.2).

8 B. Behavior on Royden's Compactification. Let $U_c=\{z\in R|g(z,\zeta)>c\}$. On R_n-V, with V a disk about ζ, g_n takes its maximum on ∂V. Therefore $g|R-V$ attains its maximum on ∂V. Fix a $c>\max_{\partial V}g$. Then U_c is a disk in V, and clearly $U_c^n=\{z\in R|g_n(z,\zeta)>c\}$ is a disk in U_c. Observe that

$$D_{R_n}(g_n\cap c)=\int_{\partial R_n-\partial U_c^n} g_n * dg_n=2\pi c$$

and B-$\lim_{n\to\infty} g_n\cap c=g\cap c$ on R, where we have extended g_n to R by setting $g_n=0$ on $R-R_n$. Clearly $g_n\cap c\in \mathbb{M}_\Delta(R)$. Hence by Theorem III.1 G, $g\cap c\in \mathbb{M}_\Delta(R)$ and the same is true for every $c>0$.

THEOREM. *Suppose $R\notin O_G$. Then the Green's function $g(z,\zeta)$ on R has the property $g(\cdot,\zeta)\cap c\in\mathbb{M}_\Delta(R)$ for every $c>0$. Moreover $g(\cdot,\zeta)$ is continuous on Royden's compactification $R_{\mathbb{M}}^*$ and $g(\cdot,\zeta)|\Delta_{\mathbb{M}}=0$. The capacity function s_β is continuous on $R_{\mathbb{M}}^*$ and $s_\beta(\cdot)|\Delta_{\mathbb{M}}=k_\beta$.*

By using the fiber space $(R_{\mathbb{N}}^*, R_{\mathbb{M}}^*, \rho)$ in IV.2 F we can here replace $\mathbb{M}$ by $\mathbb{N}$.

8 C. Kernels. In general a function $k(\cdot,\cdot)$ defined on the product space $\Omega\times\Omega$ of a space (set) Ω is called a *kernel* or kernel function on Ω.

Thus far we fixed (arbitrarily) $\zeta\in R$ in the discussion of $g(z,\zeta)$. If we vary ζ on R then $g(z,\zeta)$ can be considered as a function on $R\times R$. In this case we call $g(z,\zeta)$ the *Green's kernel* on $R\notin O_G$.

Take $a,b\in R$ and choose R_n with $a,b\in R_n$. For $a\neq b$ we have by Green's formula

$$\int_{\partial R_n-C_a-C_b}\big(g_n(z,b)*dg_n(z,a)-g_n(z,a)*dg_n(z,b)\big)=0$$

where C_a and C_b are small circles about a and b with radii ε. On letting $\varepsilon\to 0$ we obtain $2\pi g_n(a,b)-2\pi g_n(b,a)=0$, i.e.

$$g_n(a,b)=g_n(b,a).$$

This is also true for $a=b$ if we set $g_n(a,a)=\infty$. On passing to the limit we conclude that

$$g(z,\zeta)=g(\zeta,z) \tag{72}$$

on $R\times R$, i.e. $g(\cdot,\cdot)$ is a *symmetric* kernel on R.

8 D. Joint Continuity. Let U be a bounded plane region and X a topological space. Let $f(z,x)$ be a real-valued function on $U\times X$ such that

(a) $f(\cdot,x)\in H(U)$,
(b) $f(z,\cdot)$ is finitely continuous on X,
(c) $\inf_{U\times X} f>-\infty$.

Then the function $f(\cdot,\cdot)$ is finitely jointly continuous on $U\times X$.

In fact let $(z_0,x_0)\in U\times X$. Denote by k the Harnack function on $U\times U$ (III.4 C). By (c) we may assume without loss of generality that $f>0$ on $U\times X$. Then for any $(z,x)\in U\times X$

$$|f(z,x)-f(z_0,x_0)|\leq k(z,z_0)\,|f(z_0,x)-f(z_0,x_0)|+\big(k(z,z_0)-1\big)\,f(z_0,x_0).$$

Hence $\lim_{(z,x)\to(z_0,x_0)} f(z,x)=f(z_0,x_0)$.

8 E. Local Behavior. Let U: $|z|<1$ be a disk on $R\notin O_G$ and set

$$g(z,\zeta)=\log\frac{1}{|z-\zeta|}+u(z,\zeta) \tag{73}$$

for $(z,\zeta)\in U\times U$. We shall now show that $u(\cdot,\cdot)$ is finitely jointly continuous on $U\times U$, and consequently $g(\cdot,\cdot)$ is jointly continuous on $R\times R$.

Take U_r: $|z|<r$ $(0<r<1)$. Clearly $u(\cdot,\cdot)$ satisfies (a) and (b) of 8 D with $X=U_r$. By virtue of $\inf_{(z,\zeta)\in\partial U\times U_r} g(z,\zeta)\geq 0$ and $\sup_{\partial U\times U_r}\log(1/|z-\zeta|)=\log\big(1/(1-r)\big)$,

$$c=\inf_{\partial U\times U_r} u(z,\zeta)>-\infty.$$

Since $u(\cdot,\zeta)\in H(\bar{U})$ we have $u(z,\zeta)\geq c$ for every $(z,\zeta)\in U\times U_r$. Hence (c) in 8 D is also satisfied by $u(\cdot,\cdot)$. This shows that $u(\cdot,\cdot)$ is continuous on $U\times U_r$ for every $0<r<1$ and consequently on $U\times U$.

8 F. Extension to Royden's Compactification. By Theorem 8 B and (72) we can always define

$$g(z,p)=\lim_{w\in R,\,w\to p} g(z,w) \tag{74}$$

for $(z,p)\in R\times R^*_{\mathrm{MI}}$. Obviously $g(z,p)|R\times R$ is the Green's kernel on R.

THEOREM. *For every* $p\in R^*_{\mathrm{MI}}$, $g(\cdot,p)\in HP(R-p)$. *Moreover* $g(\cdot,\cdot)$ *is continuous on* $R\times R^*_{\mathrm{MI}}$, *finitely continuous on* $R\times\Gamma_{\mathrm{MI}}$ *with* $g(\cdot,p)\cap c\in \mathrm{MI}_\Delta(R)$, *and*

$$D_R\big(g(\cdot,p)\cap c\big)\leq 2\pi c \tag{75}$$

for every $c>0$.

The function $g(\cdot,\cdot)$ is continuous on $R\times R$ and finitely continuous outside of the diagonal set. To prove the finite continuity of $g(\cdot,\cdot)$ at $(z_0, p_0)\in R\times\Gamma_{\mathrm{MI}}$ let $\varepsilon>0$ be arbitrary and choose an open neighborhood W of p_0 in R^*_{MI} such that $z_0\notin\overline{W}$ and

$$|g(z_0, p)-g(z_0, p_0)|<\frac{\varepsilon}{2} \tag{76}$$

for every $p\in W$. Take a small disk U about z_0 such that $\overline{U}\subset R^*_{\mathrm{MI}}-\overline{W}$. The quantity $d=\sup_{w\in R-U} g(z_0, w)=\sup_{p\in R^*_{\mathrm{MI}}-U} g(z_0, p)$ is finite. Using the Harnack function k on $U\times U$ we obtain

$$k(z, z_0)^{-1} g(z_0, \zeta)\leq g(z, \zeta)\leq k(z, z_0)\, g(z_0, \zeta)$$

for every $(z, \zeta)\in U\times(R-U)$. We can find a smaller disk $V\subset\overline{V}\subset U$ such that $k(z, z_0)<1+\varepsilon/2d$ and $k(z, z_0)^{-1}>1-\varepsilon/2d$ for all $z\in V$. Then

$$|g(z, \zeta)-g(z_0, \zeta)|\leq\frac{\varepsilon}{2}$$

for every $(z, p)\in V\times(W\cap R)$. On passing to the limit $\zeta\to p$ we obtain for every $(z, p)\in V\times W$

$$|g(z, p)-g(z_0, p)|\leq\frac{\varepsilon}{2}. \tag{77}$$

From (76) and (77) it follows that

$$|g(z, p)-g(z_0, p_0)|<\varepsilon$$

for all $(z, p)\in V\times W$, a neighborhood of (z_0, p_0).

Since

$$g(z_0, \zeta)=\frac{1}{2\pi}\int_{|z-z_0|=1} g(z, \zeta)\, d\theta,$$

where z is a parameter in $\{|z-z_0|\leq 1\}\subset R-\zeta$ and $d\arg(z-z_0)=d\theta$, we obtain on letting $\zeta\to p$

$$g(z_0, p)=\frac{1}{2\pi}\int_{|z-z_0|=1} g(z, p)\, d\theta$$

and therefore $g(z, p)\in HP(R-p)$. Here we have used the fact, a consequence of the finite continuity of $g(\cdot,\cdot)$ on $R\times\Gamma_{\mathrm{MI}}$, that $g(\cdot, \zeta)$ converges uniformly to $g(\cdot, p)$ as $\zeta\to p$.

Since by Green's formula $D_{R_n}(g_n\cap c)=2\pi c$ we have $D_R(g(\cdot, \zeta)\cap c)\leq 2\pi c$ for $\zeta\in R$ and obtain (75) for every $c>0$ on letting $\zeta\to p\in R^*_{\mathrm{MI}}$.

Choose an arbitrary neighborhood W of $p_0\in\Gamma_{\mathrm{MI}}$. For an exhaustion $\{R_n\}$ of R set $W_n=W-R_n$. Since $C\text{-}\lim_{\zeta\in R, \zeta\to p_0} g(\cdot, \zeta)=g(\cdot, p_0)$ on R we can find a $\zeta_n \in W_n$ such that

$$C\text{-}\lim_{n\to\infty} g(\cdot, \zeta_n) = g(\cdot, p_0)$$

where $\{\zeta_n\}$ accumulates in $\overline{W} \cap \Gamma_{\mathrm{MI}}$. From $D_R(g(\cdot, \zeta_n) \cap c) \leq 2\pi c$ and $C\text{-}\lim_n g(\cdot, \zeta_n) \cap c = g(\cdot, p_0) \cap c$ we then conclude by Theorem III.1 G that $g(\cdot, p) \cap c \in \mathrm{MI}_\Delta(R)$.

8 G. Green's Kernel on R^*_{MI}. We have seen that

$$g(\cdot, p) = \lim_{w\in R,\, w\to p} g(\cdot, w)$$

is continuous on R^*_{MI}.

Definition. *The Green's kernel $G(p, q)$ on R^*_{MI} with $R \notin O_G$ is the function defined for $(p, q) \in R^*_{\mathrm{MI}} \times R^*_{\mathrm{MI}}$ by the double limit*

$$G(p, q) = \lim_{z\in R,\, z\to p} \; \lim_{w\in R,\, w\to q} g(z, w). \tag{78}$$

In addition to the harmonic boundary Δ_{MI} we distinguish an important part of $\Gamma_{\mathrm{MI}} = R^*_{\mathrm{MI}} - R$ relative to the Green's kernel:

$$\Xi = \{p \in \Gamma_{\mathrm{MI}} \mid g(p, z) > 0,\ z \in R\}. \tag{79}$$

Since $g(p, z) = g(z, p) = G(z, p) = G(p, z)$ for $(z, p) \in R \times \Gamma_{\mathrm{MI}}$, (79) can be defined, e.g. in terms of $G(z, p)$ instead of $g(p, z)$. By 8 B, $g(p, z) = 0$ for $(p, z) \in \Delta_{\mathrm{MI}} \times R$ and we infer that

$$\Gamma_{\mathrm{MI}} - \Xi \supset \Delta_{\mathrm{MI}}. \tag{80}$$

We call Ξ the *essentially irregular boundary* of R.

8 H. Properties of $G(\cdot, \cdot)$. For later reference we state here:

Theorem. *The Green's kernel $G(\cdot, \cdot)$ on R^*_{MI} enjoys the following properties:*

(G.1) $G(z, w) = g(z, w)$ *for* $(z, w) \in R \times R$,

(G.2) $G(z, p) = G(p, z) = g(z, p) = g(p, z)$ *for* $(z, p) \in R \times R^*_{\mathrm{MI}}$,

(G.3) $G(\cdot, q)$ *is continuous on* R^*_{MI} *and finitely continuous on* $R^*_{\mathrm{MI}} - q$,

(G.4) $G(\cdot, \cdot)$ *is continuous on* $R \times R^*_{\mathrm{MI}}$ *and finitely continuous on* $R \times R^*_{\mathrm{MI}}$ *outside of the diagonal, in particular on* $R \times \Gamma_{\mathrm{MI}}$,

(G.5) $G(\cdot, p) \in HP(R - p)$,

(G.6) $D_R(G(\cdot, p) \cap c) \leq 2\pi c$ *for any* $c > 0$ *and* $p \in R^*_{\mathrm{MI}}$,

(G.7) $G(\cdot, \cdot) \mid \Delta_{\mathrm{MI}} \times R^*_{\mathrm{MI}} = 0$,

(G.8) $G(\cdot, \cdot) \mid (R \cup \Xi) \times (R \cup \Xi) > 0$,

(G.9) $G(\cdot, \cdot) \mid R^*_{\mathrm{MI}} \times (\Gamma_{\mathrm{MI}} - \Xi) = 0$.

Properties (G.1) and (G.2) are given by the definition; (G.3)–(G.7) by Theorem 8 F; and (G.9) by (79). To prove (G.8) it suffices to treat the case $(p, q) \in \Xi \times \Xi$. Let U be a disk in R and fix a point $w \in U$. Clearly $G(\cdot, q) = g(\cdot, q) = g(q, \cdot) \in HP(R)$ and $\min_{z \in \partial U} G(z, q) > 0$. Thus we can find an $a > 0$ with $aG(\cdot, q) \geq g(\cdot, w)$ on ∂U. Furthermore $aG(\cdot, q) = g(\cdot, w) = 0$ on Δ_{M} and $aG(\cdot, q) - g(\cdot, w)$ is bounded from below on $R - \bar{U}$. Thus by Theorem III.3 I, $aG(\cdot, q) \geq g(\cdot, w)$ on $R^*_{\mathrm{M}} - \bar{U}$, and in particular $aG(p, q) \geq g(p, w) > 0$ for $p \in \Xi$.

9. Green's Potentials

9 A. Transfinite Diameter ρ. Throughout 9 we shall assume that R is an open Riemann surface with $R \notin O_G$ and we denote by $G(\cdot, \cdot)$ the Green's kernel on Royden's compactification R^*_{M} of R.

For each set $X \subset R^*_{\mathrm{M}}$ we write

$$\binom{n}{2} \rho_n(X) = \inf_{p_1, \ldots, p_n \in X} \sum_{i<j}^{1, \ldots, n} G(p_i, p_j) \tag{81}$$

if $X \neq \emptyset$ and $\rho_n(\emptyset) = \infty$.

Let $p_1, \ldots, p_{n+1}$ be arbitrary points in X. Then

$$\sum_{i<j}^{1, \ldots, n+1} G(p_i, p_j) = \sum_{i=1}^{k-1} G(p_i, p_k) + \sum_{j=k+1}^{n+1} G(p_k, p_j) + \sum_{i<j;\, i, j \neq k}^{1, \ldots, n+1} G(p_i, p_j)$$

and therefore

$$\sum_{i<j}^{1, \ldots, n+1} G(p_i, p_j) \geq \sum_{i=1}^{k-1} G(p_i, p_k) + \sum_{j=k+1}^{n+1} G(p_k, p_j) + \binom{n}{2} \rho_n(X)$$

for each $k = 1, \ldots, n+1$. On summing these $n+1$ inequalities we obtain

$$(n+1) \sum_{i<j}^{1, \ldots, n+1} G(p_i, p_j) \geq 2 \sum_{i<j}^{1, \ldots, n+1} G(p_i, p_j) + (n+1) \binom{n}{2} \rho_n(X)$$

or equivalently

$$(n-1) \sum_{i<j}^{1, \ldots, n+1} G(p_i, p_j) \geq (n+1) \binom{n}{2} \rho_n(X).$$

Thus $(n-1) \binom{n+1}{2} \rho_{n+1}(X) \geq (n+1) \binom{n}{2} \rho_n(X)$ and a fortiori

$$0 \leq \rho_n(X) \leq \rho_{n+1}(X) \leq \infty. \tag{82}$$

It follows that we can define

$$\rho(X) = \rho(X; R^*_{\mathrm{M}}, G) = \lim_{n \to \infty} \rho_n(X), \tag{83}$$

the *transfinite diameter* of X with respect to R^*_{M} and G.

9 B. Tchebycheff's Constant τ. Again for $X \subset R_{\mathfrak{M}}^*$ set

$$n\,\tau_n(X) = \sup_{p_1,\ldots,p_n \in X} \left(\inf_{p \in X} \sum_{i=1}^{n} G(p, p_i) \right) \tag{84}$$

if $X \neq \emptyset$, and $\tau_n(\emptyset) = \infty$. Let $p_1, \ldots, p_{n+m}$ be arbitrary points in X. Then

$$\sum_{i=1}^{n+m} G(p, p_i) = \sum_{i=1}^{n} G(p, p_i) + \sum_{i=n+1}^{n+m} G(p, p_i)$$

and therefore

$$\inf_{p \in X} \sum_{i=1}^{n+m} G(p, p_i) \geq \inf_{p \in X} \sum_{i=1}^{n} G(p, p_i) + \inf_{p \in X} \sum_{i=n+1}^{n+m} G(p, p_i).$$

Thus it follows from (84) that

$$(n+m)\,\tau_{n+m}(X) \geq n\,\tau_n(X) + m\,\tau_m(X) \tag{85}$$

for every $n, m = 1, 2, \ldots$.

Let $\alpha = \sup_n \tau_n(X)$. For any $\beta < \alpha$ we can find an m such that $\tau_m(X) > \beta$. Any positive integer n has a unique representation $n = qm + r$ with nonnegative integers q and r, $0 \leq r < m$. By (85), $\tau_{qm}(X) \geq \tau_m(X)$ for a positive integer q, and again by (85)

$$n\,\tau_n(X) = (q\,m + r)\,\tau_{qm+r}(X) \geq q\,m\,\tau_{qm}(X) + r\,\tau_r(X) \geq q\,m\,\tau_m(X).$$

Therefore

$$\tau_n(X) \geq \frac{q\,m}{q\,m + r}\,\tau_m(X) > \frac{q\,m}{q\,m + r}\,\beta.$$

Since $q \to \infty$ as $n \to \infty$ we infer that

$$\alpha \geq \limsup_{n \to \infty} \tau_n(X) \geq \liminf_{n \to \infty} \tau_n(X) \geq \beta.$$

On letting $\beta \to \alpha$ we deduce the existence of

$$\tau(X) = \tau(X; R_{\mathfrak{M}}^*, G) = \lim_{n \to \infty} \tau_n(X), \tag{86}$$

the (modified) *Tchebycheff constant* of X with respect to $R_{\mathfrak{M}}^*$ and G.

9 C. An Estimate. Both ρ and τ give certain magnitudes of sets in $R_{\mathfrak{M}}^*$. The larger $\rho(X)$ and $\tau(X)$ the smaller in general X. Actually for a finite set $X \subset R_{\mathfrak{M}}^*$, $\rho(X) = \tau(X) = \infty$. We claim:

THEOREM. *For an arbitrary set* $X \subset R_{\mathfrak{M}}^*$

$$\tau(X) \geq \rho(X). \tag{87}$$

If X is finite then $\tau(X)=\rho(X)=\infty$. Thus we may assume that X is infinite. Let $n>1$ and $r=1/(n-1)$. We can choose n points $p_n, \ldots, p_1 \in X$ inductively so as to satisfy

$$\sum_{j=n-i+1}^{n} G(p_{n-i}, p_j) \leq \inf_{p \in X} \sum_{j=n-i+1}^{n} G(p, p_j) + r \tag{88}$$

for $i=1, \ldots, n-1$. In fact let $p_n \in X$ be arbitrary. Assume that $p_n, \ldots, p_{n-i+1}$ $(i<n-1)$ have already been chosen. Consider

$$f(p) = \sum_{j=n-i+1}^{n} G(p, p_j)$$

on X. Since $f(p)$ is nonnegative on X we can find a point $p_{n-i} \in X$ such that

$$f(p_{n-i}) \leq \inf_{p \in X} f(p) + r.$$

This is in fact (88), and the induction is complete.

By the definition of $\tau_i(X)$ in (84) we have

$$\inf_{p \in X} \sum_{j=n-i+1}^{n} G(p, p_j) \leq i\, \tau_i(X).$$

In view of (88) we conclude that

$$\sum_{j=n-i+1}^{n} G(p_{n-i}, p_j) \leq i\, \tau_i(X) + r$$

for every $i=1, \ldots, n-1$. On summing these $n-1$ inequalities we obtain

$$\sum_{i<j}^{1, \ldots, n} G(p_i, p_j) \leq \sum_{i=1}^{n-1} i\, \tau_i(X) + (n-1)\, r.$$

Therefore by (81)

$$\binom{n}{2} \rho_n(X) \leq \sum_{i=1}^{n-1} i\, \tau_i(X) + 1$$

or equivalently

$$\rho_n(X) \leq \frac{\tau_1(X) + 2\tau_2(X) + \cdots + (n-1)\, \tau_{n-1}(X)}{\binom{n}{2}} + \frac{1}{\binom{n}{2}}. \tag{89}$$

Since $\tau_n(X) \to \tau(X)$ as $n \to \infty$ it is readily seen that

$$\lim_{n \to \infty} \frac{\tau_1(X) + 2\tau_2(X) + \cdots + (n-1)\, \tau_{n-1}(X)}{\binom{n}{2}} = \tau(X).$$

Hence (89) gives (87).

9 D. Minimum Energy ε. Thus far we have considered general sets $X \subset R^*_{\mathbb{M}}$. Hereafter in 9 we restrict our attention to sets $K \subset R$ consisting of finitely many compact analytic arcs.

Consider the family m_K of positive regular unit Borel measures μ with supports $S_\mu \subset K$. For any positive regular Borel measures μ and ν we write

$$(\mu, \nu) = \iint G(z, w)\, d\mu(z)\, d\nu(w) \tag{90}$$

and call it the *mutual energy* of μ and ν. In particular

$$(\mu, \mu) = \iint G(z, w)\, d\mu(z)\, d\mu(w) \tag{91}$$

is the *energy* of μ.

Now we define a third quantity, the *minimum energy* $\varepsilon(K)$ of K with respect to R and G, as

$$\varepsilon(K) = \varepsilon(K; R, G) = \inf_{\mu \in m_K} (\mu, \mu) \tag{92}$$

if $K \neq \emptyset$, and $\varepsilon(\emptyset) = \infty$. Again the larger $\varepsilon(K)$ the smaller K.

For a Borel measure μ we call

$$G_\mu(z) = \int G(z, w)\, d\mu(w) \tag{93}$$

the *Green's potential* of μ.

THEOREM. *Let K consist of a finite number of compact analytic arcs contained in R. Then there exists a unique measure $\mu_K \in m_K$ such that*

$$\varepsilon(K) = (\mu_K, \mu_K) < \infty. \tag{94}$$

The Green's potential G_{μ_K} is continuous on $R^*_{\mathbb{M}}$, belongs to $HD(R-K)$, and

$$G_{\mu_K} | K = \varepsilon(K), \qquad G_{\mu_K} | R^*_{\mathbb{M}} \leq \varepsilon(K). \tag{95}$$

Moreover $G_{\mu_K} \in \mathbb{M}_\Delta(R)$ and

$$D_R(G_{\mu_K}) = 2\pi\, \varepsilon(K). \tag{96}$$

The proof will extend from 9 E through 9 J. The essence of the theorem is due to FROSTMAN [1].

9 E. Maximum Principle. Let μ be a positive regular Borel measure with compact support $S_\mu \subset R$. Clearly the Green's potential G_μ belongs to $H(R - S_\mu)$. Since $G^n_\mu(z) = \int G(z, \zeta) \cap n\, d\mu(\zeta)$ is continuous on R, increases with n, and tends to $G_\mu(z)$ as $n \to \infty$, G_μ is lower semicontinuous on R. It is also easily seen that G_μ is nonnegative and superharmonic on R. By 8 H, G_μ is finitely continuous on $R^*_{\mathbb{M}} - S_\mu$ and vanishes on $\Delta_{\mathbb{M}}$.

THEOREM. *For every $z_0 \in \partial S_\mu$*

$$\limsup_{z \in R - S_\mu,\ z \to z_0} G_\mu(z) \leq \limsup_{z \in S_\mu,\ z \to z_0} G_\mu(z). \tag{97}$$

Take a disk U about z_0. Since $G_\mu = G_{\mu-\mu_{\bar U}} + G_{\mu_{\bar U}}$ and $G_{\mu-\mu_{\bar U}}$ is finitely continuous at z_0, where $\mu_{\bar U}(\cdot) = \mu(\cdot \cap \bar U)$, we may assume that $S_\mu \subset \bar U$. Moreover by (73) we have only to prove (97) for

$$L_\mu(z) = \int \log \frac{1}{|z-\zeta|} \, d\mu(\zeta)$$

under the assumption $\mu(z_0) = 0$.

Let $\varepsilon > 0$. We can find a $0 < \rho < 1$ with $\mu(\bar U_\rho) < \varepsilon$ where $U_\rho : |z - z_0| < \rho$. For each $z \in \bar U_\rho$ choose a point $\zeta(z) \in S_\mu \cap \bar U_\rho$ such that

$$|z - \zeta(z)| = \inf_{\zeta \in S_\mu \cap \bar U_\rho} |z - \zeta|.$$

Then for $\zeta \in S_\mu \cap \bar U_\rho$

$$|\zeta - \zeta(z)| \le |\zeta - z| + |z - \zeta(z)| \le 2\,|\zeta - z|$$

and thus

$$\log \frac{1}{|\zeta - z|} \le \log \frac{1}{|\zeta(z) - \zeta|} + \log 2.$$

It follows that

$$\begin{aligned} L_\mu(z) &= \int_{\bar U_\rho} \log \frac{1}{|z-\zeta|} \, d\mu(\zeta) + \int_{\zeta \notin \bar U_\rho} \log \frac{1}{|z-\zeta|} \, d\mu(\zeta) \\ &\le \int_{\bar U_\rho} \log \frac{1}{|\zeta(z)-\zeta|} \, d\mu(\zeta) + \varepsilon \log 2 + \int_{\zeta \notin \bar U_\rho} \log \frac{1}{|z-\zeta|} \, d\mu(\zeta) \\ &= L_\mu(\zeta(z)) + \varepsilon \log 2 + \int_{\zeta \notin \bar U_\rho} \left(\log \frac{1}{|z-\zeta|} - \log \frac{1}{|\zeta(z)-\zeta|} \right) d\mu(\zeta). \end{aligned}$$

Clearly $z \to z_0$ implies $\zeta(z) \to z_0$. Hence

$$\limsup_{z \to z_0} L_\mu(z) \le \limsup_{z \in S_\mu,\, z \to z_0} L_\mu(z) + \varepsilon \log 2$$

and on letting $\varepsilon \to 0$ we obtain (97) for L_μ.

COROLLARY 1 (CONTINUITY PRINCIPLE). *If $G_\mu | S_\mu$ is continuous on S_μ then $G_\mu | R$ is continuous.*

In fact since G_μ is lower semicontinuous on R we have for $z_0 \in S_\mu$

$$G_\mu(z_0) \le \liminf_{z \to z_0} G_\mu(z) \le \limsup_{z \to z_0} G_\mu(z) \le \limsup_{z \in S_\mu,\, z \to z_0} G_\mu(z) = G_\mu(z_0).$$

Thus G_μ is continuous at any $z_0 \in S_\mu$ and clearly at $z_0 \in R - S_\mu$.

COROLLARY 2 (FROSTMAN'S MAXIMUM PRINCIPLE). *If $G_\mu | S_\mu \le c$ then $G_\mu | R \le c$.*

Since $G_\mu \in H(R - S_\mu)$, $G_\mu | R^*_{\mathbf{M}} - S_\mu$ is continuous, and $G_\mu | \Delta_{\mathbf{M}} = 0$, we see by (97) and Theorem III.3 I that $G_\mu | R - S_\mu \le c$.

9 F. Measure ξ. Let K be as in 9 D and let $R-K$ consist of k components $R^1, \ldots, R^k$. Clearly $\bigcup_{i=1}^k \partial R^i = K$. For any finitely continuous function f on ∂R^i we can find a function $H^i(f)$ harmonic on R^i, continuous on $\bar{R}^i$, with the closure taken in R^*_{MI}, such that $H^i(f)|\partial R^i = f$ and $H^i(f)|\Delta_{\mathrm{MI}} \cap \bar{R}^i = 0$. The existence and uniqueness follow from III. 3 B. Fix $z^i \in R^i$. Then $H^i(f)(z^i)$ is a bounded linear functional on $B(\partial R^i)$. Thus there exists a unique positive regular Borel measure ξ^i on ∂R^i such that $H^i(f)(z^i) = \int f d\xi^i$. We set $\xi^i|K-\partial R^i \equiv 0$ and conclude that

$$\xi = \sum_{i=1}^{k} \xi^i \tag{98}$$

is a regular Borel measure on K.

Lemma. *Let F be a subset of K with $\xi(F)=0$. Then there exists a positive harmonic function h on $R-K$ such that*

$$\lim_{z\in R-K,\, z\to z_0} h(z) = \infty \tag{99}$$

for every $z_0 \in F$.

The proof is analogous to that of Theorem IV. 4 F. In fact we can find a nonnegative lower semicontinuous f^i on ∂R^i such that $f^i|F \cap \partial R^i = \infty$ and $\int f^i d\xi^i < \infty$. Then $h|R^i = \lim_{n\to\infty} H^i(f^i \cap n)$ is convergent and provides us with the required function.

9 G. Gauss' Variation. Given a bounded continuous function f on K we set for every regular Borel measure μ on K

$$V_f(\mu) = \iint G(z,\zeta)\, d\mu(z)\, d\mu(\zeta) - 2\int f(z)\, d\mu(z). \tag{100}$$

Gauss' variational problem is to find a μ which minimizes $V_f(\mu)$.

Theorem. *There exists a regular Borel measure ν on K such that*

$$V_f(\nu) = \min_{\mu} V_f(\mu), \tag{101}$$

$G_\nu|S_\nu \le f$, and $G_\nu|K \ge f$ ξ-a.e.

Let $a = \inf_\mu V_f(\mu)$, $a_1 = \inf_{(z,\zeta)\in K\times K} G(z,\zeta) > 0$, and $a_2 = \max_K f < \infty$. Then $V_f(\mu) \ge a_1 \mu^2(K) - 2a_2 \mu(K) \ge -a_2^2/a_1$, i.e. $a \ge -a_2^2/a_1 > -\infty$. Let $\{\nu_n\}_1^\infty$ be a sequence of regular Borel measures on K such that $V_f(\nu_n) \to a$ as $n\to\infty$. Since $a_1 \nu_n^2(K) - 2a_2 \nu_n(K) \le V_f(\nu_n)$ is bounded we see that $\{\nu_n(K)\}_1^\infty$ is a bounded sequence. Thus there exists a regular Borel measure ν on K such that

$$\int \varphi d\nu = \lim_{n\to\infty} \int \varphi d\nu_n \tag{102}$$

for every bounded continuous φ on K. Set

$$\varphi_n^c(z)=\int (G(z,\zeta)\cap c)\,d\nu_n(\zeta), \qquad \varphi^c(z)=\int (G(z,\zeta)\cap c)\,d\nu(\zeta).$$

By the continuity of $G(\cdot,\cdot)|K\times K$ and by (102) we conclude that $U\text{-}\lim_{n\to\infty}\varphi_n^c=\varphi^c$ on K. Therefore, again by (102), $\int \varphi_n^c\,d\nu_n\to\int\varphi^c\,d\nu$ as $n\to\infty$. Thus

$$\begin{aligned}(\nu,\nu)&=\lim_{c\to\infty}\iint (G(z,\zeta)\cap c)\,d\nu(z)\,d\nu(\zeta)=\lim_{c\to\infty}\lim_{n}\int\varphi_n^c(\zeta)\,d\nu_n(\zeta)\\&\le\liminf_n\iint G(z,\zeta)\,d\nu_n(z)\,d\nu_n(\zeta).\end{aligned}$$

We infer that

$$a\le V_f(\nu)\le\lim_n V_f(\nu_n)=a$$

i.e. (101) is satisfied.

Now suppose $f(z_0)<G_\nu(z_0)$ for some $z_0\in K$. In a neighborhood U of z_0, $f<G_\nu$. Let $\mu=\nu|U$ and $0<\varepsilon<1$. Then $\nu-\varepsilon\mu\ge 0$ and from (101) it follows that $V_f(\nu)\le V_f(\nu-\varepsilon\mu)$. Moreover since $G_\nu\ge G_\mu$

$$\int G_\mu\,d\mu\le\int G_\nu\,d\mu=\int G_\mu\,d\nu\le\int G_\nu\,d\nu<\infty.$$

Thus $V_f(\nu)=(\nu,\nu)-2\int f\,d\nu$ and

$$V_f(\nu)\le V_f(\nu-\varepsilon\mu)=(\nu,\nu)-2\varepsilon(\mu,\nu)+\varepsilon^2(\mu,\mu)-2\int f\,d\nu+2\varepsilon\int f\,d\mu$$

imply

$$\int (f-G_\nu)\,d\mu+\frac{\varepsilon}{2}\int G_\mu\,d\mu\ge 0.$$

In view of $(f-G_\nu)|U<0$ we conclude on letting $\varepsilon\to 0$ that $\mu(U)=\nu(U)=0$. Thus $z_0\notin S_\nu$, $G_\nu|S_\nu\le f$.

For $\varepsilon>0$ the set

$$K_\varepsilon=\{z\in K\,|\,f(z)\ge G_\nu(z)+\varepsilon\} \tag{103}$$

is compact. Suppose $\xi(K_\varepsilon)>0$. Then we can find an R^i such that $\xi^i(K_\varepsilon)>0$. For $\mu=\xi^i$ and $z\in R^i$

$$G_\mu(z)=\int G(z,\zeta)\,d\xi^i(\zeta)=H^i(G(z,\cdot))(z^i)\le G(z,z^i)$$

which in turn gives

$$\int G_\mu\,d\mu\le\int G(z,z^i)\,d\mu(z)<\infty, \qquad \int G_\mu\,d\nu\le\int G(z,z^i)\,d\nu(z)<\infty.$$

For $\eta>0$, $V_f(\nu)\le V_f(\nu+\eta\mu)$. On rewriting this as before we obtain

$$\int (G_\nu-f)\,d\mu+\frac{\eta}{2}\int G_\mu\,d\mu\ge 0.$$

We pass to the limit $\eta \to 0$ and deduce from (103) that

$$-\varepsilon\,\mu(K_\varepsilon) \geq \int (G_\nu - f)\,d\mu \geq 0$$

and consequently $\mu(K_\varepsilon) = \xi^i(K_\varepsilon) = 0$, a contradiction.

Thus $\xi(K_\varepsilon) = 0$ for every $\varepsilon > 0$, i.e. $G_\nu | K \geq f$ ξ-a.e.

9 H. Energy Principle. Mutual energy and energy are related by a Schwarz type inequality which is often referred to as the *energy principle:*

$$(\mu, \nu)^2 \leq (\mu, \mu)(\nu, \nu). \tag{104}$$

We prove this only for μ, ν with $S_\mu, S_\nu \subset K$. Let $\{W_n\}$ be a sequence of regular open sets such that $W_n \supset \overline{W}_{n+1} \supset W_{n+1} \supset K$ and $\bigcap W_n = K$; let h_n be continuous on R with $h_n | R - W_n = G_\mu$ and $h_n | W_n \in H(W_n)$. Then $h_n | K \in B(K)$ and $0 < h_n | K < G_\mu | K$. Since the h_n are superharmonic and $h = \lim_n h_n = G_\mu$ on $R - K$ we conclude by the vanishing of the Lebesgue measure of K that $h \equiv G_\mu$ on R. Thus

$$\lim_{n\to\infty} h_n | K = G_\mu. \tag{105}$$

Let v_n be continuous on $R^*_{\mathbf{M}}$ with $v_n \in H(R - K)$, $v_n | K = h_n | K$, and $v_n | \Delta_{\mathbf{M}} = 0$. Since $v_n \leq h_n$ on R, v_n is positive superharmonic on R. Let λ_n furnish the minimum (101) for $f = v_n$. Then $G_{\lambda_n} | S_{\lambda_n} \leq v_n | S_{\lambda_n}$ and we conclude by 9 E, the relation $v_n | \Delta_{\mathbf{M}} = 0$, and the superharmonicity of v_n on R that $G_{\lambda_n} \leq v_n$ on R. Let w be the function in (99) for

$$F = \{z \in S_{\lambda_n} | G_{\lambda_n}(z) < v_n(z)\}$$

which satisfies $\xi(F) = 0$. Then for any $\varepsilon > 0$, $(G_{\lambda_n} + \varepsilon w) | (R - K)$ has boundary values on K dominating those of $v_n | (R - K)$. Hence $G_{\lambda_n} + \varepsilon w \geq v_n$ on $R - K$ and on letting $\varepsilon \to 0$ we obtain $G_{\lambda_n} \geq v_n$ on $R - K$.

We conclude that $G_{\lambda_n} = v_n$ on $R - K$ and that these functions are superharmonic. Therefore $G_{\lambda_n} = v_n$ on R and

$$V_{v_n}(\lambda_n) = (\lambda_n, \lambda_n) - 2\int v_n\,d\lambda_n = (\lambda_n, \lambda_n) - 2\int G_{\lambda_n}\,d\lambda_n = -(\lambda_n, \lambda_n).$$

For $\eta > 0$, $V_{v_n}(\eta\,\nu) \geq V_{v_n}(\lambda_n) = -(\lambda_n, \lambda_n)$ and consequently

$$\eta^2(\nu, \nu) - 2\eta \int v_n\,d\nu + (\lambda_n, \lambda_n) \geq 0$$

for every $\eta > 0$ and trivially for $\eta \leq 0$. Thus by $v_n = h_n$ on K we have

$$\left(\int h_n\,d\nu\right)^2 \leq (\lambda_n, \lambda_n)(\nu, \nu). \tag{106}$$

Observe that $G_{\lambda_n} = v_n \leq h_n \leq G_\mu$. For this reason $(\lambda_n, \lambda_n) = \int G_{\lambda_n}\,d\lambda_n \leq \int G_\mu\,d\lambda_n = \int G_{\lambda_n}\,d\mu \leq \int G_\mu\,d\mu = (\mu, \mu)$. From (106) and (105) we obtain on letting $n \to \infty$

$$\left(\int G_\mu\,d\nu\right)^2 \leq (\mu, \mu)(\nu, \nu),$$

which is (104).

9 I. Proof of Theorem 9 D. We are now ready to prove (94)–(96). Let R_0 be a regular subregion of R with $R_0 \supset K$. Take $u \in \mathrm{M}_\Delta(R)$ such that $u|K=1$, $u|\Delta_{\mathrm{M}}=0$, and $u|R-K \in H(R-K)$. Clearly

$$D_R(u)=\int_{\partial(R-K)} u * du=\int_{\partial(R-K)} * du=-\int_{\partial R_0} * du \tag{107}$$

where $\int_{\partial(R-K)}$ is understood as $\lim_n \int_{\partial(R-W_n)}$, with W_n as in 9 H.

Let μ_1 give the minimum (101) for $f=u$. Since u is positive superharmonic on R we conclude as in the proof of $G_{\lambda_n}=v_n$ in 9 H that

$$u=G_{\mu_1}, \qquad S_{\mu_1} \subset K. \tag{108}$$

Obviously $\mu_1(K)>0$. For $\mu \in m_K$ we have by 9 H

$$1=\left(\int d\mu\right)^2=\left(\int G_{\mu_1} d\mu\right)^2=(\mu_1, \mu)^2 \leq (\mu_1, \mu_1)(\mu, \mu)$$

and thus

$$(\mu, \mu) \geq \frac{1}{(\mu_1, \mu_1)}.$$

On the other hand

$$\mu_1(K)=\int d\mu_1=\int G_{\mu_1} d\mu_1=(\mu_1, \mu_1).$$

Hence if we set $\mu_K=\mu_1/\mu_1(K)$ then $\mu_K \in m_K$ and

$$(\mu_K, \mu_K)=\frac{(\mu_1, \mu_1)}{\mu_1(K)^2}=\frac{(\mu_1, \mu_1)}{(\mu_1, \mu_1)^2}=\frac{1}{(\mu_1, \mu_1)}.$$

Thus $(\mu, \mu) \geq (\mu_K, \mu_K)$ for every $\mu \in m_K$ and we have (94).

Clearly $G_{\mu_K}=\varepsilon(K)\, G_{\mu_1}=\varepsilon(K) u$ satisfies (95). By (107)

$$\begin{aligned}-D_R(G_{\mu_K}) &= \varepsilon(K)^2 \int_{\partial R_0} * du=\varepsilon(K) \int_{\partial R_0} * dG_{\mu_K}(\cdot) \\ &= \varepsilon(K) \int_K \left(\int_{\partial R_0} * dG(\cdot, \zeta)\right) d\mu_K(\zeta).\end{aligned}$$

Since $\zeta \in K \subset R_0$, $\int_{\partial R_0} * dG(\cdot, \zeta)=-2\pi$. We have obtained (96).

9 J. Uniqueness. It remains to prove the uniqueness of μ_K in (94). Observe that for $f=G_{\mu_K}|K$

$$V_f(\mu_K)=-\varepsilon(K).$$

For any μ with $S_\mu \subset K$, $\mu'=\mu/\mu(K) \in m_K$ and thus

$$\begin{aligned}V_f(\mu) &= \mu(K)^2 (\mu', \mu')-2\varepsilon(K)\, \mu(K) \\ &\geq \mu(K)^2\, \varepsilon(K)-2\mu(K)\, \varepsilon(K) \geq -\varepsilon(K).\end{aligned}$$

Therefore μ_K is a solution of the minimum problem for $V_f(\mu)$.

Now suppose there exists a $\lambda \in m_K$ with $(\lambda, \lambda) = \varepsilon(K)$. Then

$$V_f(\lambda) = (\lambda, \lambda) - 2\int f d\lambda = \varepsilon(K) - 2\varepsilon(K) = -\varepsilon(K).$$

Thus λ gives $\min_\mu V_f(\mu)$. By the same method as in 9 H we see that $G_\lambda = G_{\mu_K}$.

For any $\varphi \in C^2(R_0)$ with support in R_0 we obtain by Green's formula

$$\varphi(z) = -\frac{1}{2\pi}\int G(z,\zeta)\,\Delta_\zeta\,\varphi(\zeta)\,dS_\zeta \tag{109}$$

where dS_ζ is the Euclidean area element with respect to ζ. We integrate both sides with respect to $d\mu_K(z)$ and $d\lambda(z)$:

$$\begin{aligned}\int \varphi\, d\mu_K &= -\frac{1}{2\pi}\int G_{\mu_K}(\zeta)\,\Delta_\zeta\,\varphi(\zeta)\,dS_\zeta,\\ \int \varphi\, d\lambda &= -\frac{1}{2\pi}\int G_\lambda(\zeta)\,\Delta_\zeta\,\varphi(\zeta)\,dS_\zeta.\end{aligned} \tag{110}$$

Since $G_{\mu_K} = G_\lambda$ we obtain

$$\int \varphi\, d\mu_K = \int \varphi\, d\lambda$$

first for $\varphi \in C^2(R_0)$ with compact support and then for $\varphi \in B(K)$. This shows that $\mu_K = \lambda$ on K.

The proof of Theorem 9 D is herewith complete.

9 K. Dirichlet's Constant δ. Let $u_K \in \mathrm{M}_\Delta(R)$ such that $u_K | K = 1$ and $u_K \in H(R-K)$. We call the quantity

$$\delta(K) = \frac{2\pi}{D_R(u_K)} \tag{111}$$

the *Dirichlet constant* of K with respect to R. Since $G_{\mu_K} = \varepsilon(K)\,u_K$, we see at once by (96) that

$$\varepsilon(K) = \delta(K). \tag{112}$$

9 L. Identities. We have considered the transfinite diameter ρ, the Tchebycheff constant τ, the minimum energy ε, and the Dirichlet constant δ. We shall now show that they are all equal:

THEOREM. *For any set $K \subset R \notin O_G$ consisting of a finite number of compact analytic arcs in R*

$$\tau(K) = \rho(K) = \varepsilon(K) = \delta(K). \tag{113}$$

By (87) and (112) we have only to show that $\rho(K) \geq \varepsilon(K)$ and $\varepsilon(K) \geq \tau(K)$. We start with the former.

For each n choose n points $p_1^{(n)}, \ldots, p_n^{(n)} \in K$ such that

$$\binom{n}{2} \rho_n(K) \geq \sum_{i<j}^{1,\ldots,n} G(p_i^{(n)}, p_j^{(n)}) - \frac{1}{n}. \tag{114}$$

Let $\mu_n \in m_K$ with $\mu_n(p_i^{(n)}) = 1/n$ $(i=1, \ldots, n)$. There exists a subsequence of $\{\mu_n\}$, say $\{\mu_{n_k}\}$, such that

$$\lim_{k\to\infty} \int \varphi \, d\mu_{n_k} = \int \varphi \, d\mu \tag{115}$$

for some $\mu \in m_K$ and for every $\varphi \in B(K)$. Set

$$\varphi_k^c(z) = \int \big(G(z, w) \cap c\big) \, d\mu_{n_k}(w), \qquad \varphi^c(z) = \int \big(G(z, w) \cap c\big) \, d\mu(w).$$

As in 9 G, U-$\lim_k \varphi_k^c = \varphi^c$ on K. Therefore

$$\lim_{k\to\infty} \int \varphi_k^c \, d\mu_{n_k} = \int \varphi^c \, d\mu$$

or equivalently

$$\lim_{k\to\infty} \iint \big(G(z, w) \cap c\big) \, d\mu_{n_k}(z) \, d\mu_{n_k}(w) = \iint \big(G(z, w) \cap c\big) \, d\mu(z) \, d\mu(w). \tag{116}$$

From (114) it follows that

$$\rho_{n_k}(K) + \frac{2}{n_k^3} \geq \frac{2}{n_k^2} \sum_{i<j}^{1,\ldots,n_k} G(p_i^{(n_k)}, p_j^{(n_k)}) \geq \frac{2}{n_k^2} \sum_{i<j}^{1,\ldots,n_k} G(p_i^{(n_k)}, p_j^{(n_k)}) \cap c$$

$$= \iint \big(G(z, w) \cap c\big) \, d\mu_{n_k}(z) \, d\mu_{n_k}(w) - \frac{c}{n_k^2}.$$

On passing to the limit $k \to \infty$ we conclude by (116) that

$$\rho(K) \geq \iint \big(G(z, w) \cap c\big) \, d\mu(z) \, d\mu(w).$$

Since $c > 0$ is arbitrary we obtain on letting $c \to \infty$

$$\rho(K) \geq \iint G \, d\mu \, d\mu = (\mu, \mu) \geq \varepsilon(K).$$

We proceed to the proof of $\varepsilon(K) \geq \tau(K)$. Let μ_K be the measure in (94) and choose arbitrarily n points $p_1, \ldots, p_n \in K$. Then by (95)

$$\varepsilon(K) = \frac{1}{n} \sum_{i=1}^{n} G_{\mu_K}(p_i) = \frac{1}{n} \int \sum_{i=1}^{n} G(z, p_i) \, d\mu_K(z) \geq \frac{1}{n} \inf_{p \in K} \sum_{i=1}^{n} G(p, p_i).$$

Therefore

$$\varepsilon(K) \geq \frac{1}{n} \sup_{p_1, \ldots, p_n \in K} \left(\inf_{p \in K} \sum_{i=1}^{n} G(p, p_i) \right) = \tau_n(K).$$

On letting $n \to \infty$ we conclude that $\varepsilon(K) \geq \tau(K)$.

Remark. Equalities (113) were proved by FÉKETE [1] and SZEGÖ [1] for an arbitrary compact set in the plane with respect to the logarithmic potential. Theorem 9 E is essentially due to MARIA [1].

If we replace the measure ξ in 9 F by Green's capacity the entire discussion in 9 D – 9 L applies mutatis mutandis to an arbitrary compact set instead of K.

The reader interested in potential theory with kernels is referred to CONSTANTINESCU-CORNEA [17], TSUJI [5], OHTSUKA [6], and BRELOT [4, 5], among others.

10. Parabolicity Tests

10 A. Modular Test. Let R be an open Riemann surface and $\{R_n\}_0^\infty$ an exhaustion of R. Denote by

$$v_n = \operatorname{mod}(R_n - \bar{R}_0, \partial R_0, \partial R_n) \tag{117}$$

the modulus of the configuration $(R_n - \bar{R}_0, \partial R_0, \partial R_n)$. First we prove (NEVANLINNA [3]):

THEOREM. *Necessary and sufficient for R to belong to O_G is that*

$$\lim_{n\to\infty} v_n = \infty. \tag{118}$$

Let $\omega_n(z) = \omega_n(z; R_n, R_0)$ be the harmonic measure of ∂R_n with respect to $R_n - \bar{R}_0$, i.e. $\omega_n \in \mathrm{MI}(R)$, $\omega_n | R_n - \bar{R}_0 \in H(R_n - \bar{R}_0)$, $\omega_n | \bar{R}_0 = 0$, and $\omega_n | R - R_n = 1$. Then

$$\log v_n = \frac{2\pi}{D_R(\omega_n)}. \tag{119}$$

Thus (118) is equivalent to $\lim_n D_R(\omega_n) = 0$ and this in turn to $\lim_n \omega_n = 0$, i.e. $R \in O_G$.

10 B. Divergent Modular Product. We denote by μ_n the modulus of the configuration $(R_n - \bar{R}_{n-1}, \partial R_{n-1}, \partial R_n)$:

$$\mu_n = \operatorname{mod}(R_n - \bar{R}_{n-1}, \partial R_{n-1}, \partial R_n). \tag{120}$$

As a counterpart of the modular O_{AD}-test in I.1 D we have the following condition the necessity of which is due to NOSHIRO [1]:

THEOREM. *In order that $R \in O_G$ it is necessary and sufficient that there exist an exhaustion $\{R_n\}_0^\infty$ of R with*

$$\prod_{n=1}^{\infty} \mu_n = \infty. \tag{121}$$

Let $w_n(z)$ be the harmonic measure of ∂R_n with respect to $R_n - \bar{R}_{n-1}$. Then

$$\log \mu_n = \frac{2\pi}{D_R(w_n)}. \tag{122}$$

Set $u_n = \omega_n / D_R(\omega_n)$ and $v_n = w_n / D_R(w_n)$. Clearly

$$\int_{\partial R_n} * du_n = \int_{\partial R_n} \omega_n * d\omega_n / D_R(\omega_n) = 1$$

and similarly $\int_{\partial R_n} * dv_n = 1$. By (119) and (122)

$$\log v_n = 2\pi D_R(u_n), \qquad \log \mu_n = 2\pi D_R(v_n). \tag{123}$$

Schwarz's inequality gives for $1 \le i \le n$

$$D_{R_i - \bar{R}_{i-1}}(u_n) \ge \frac{D^2_{R_i - \bar{R}_{i-1}}(v_i, u_n)}{D_{R_i - \bar{R}_{i-1}}(v_i)} = \frac{\left(\int\limits_{\partial R_i - \partial R_{i-1}} v_i * du_n \right)^2}{\int\limits_{\partial R_i - \partial R_{i-1}} v_i * dv_i}$$

$$= \frac{(1/(D_R(w_i)))^2}{1/(D_R(w_i))} = \frac{1}{D_R(w_i)} = \frac{1}{2\pi} \log \mu_i.$$

Therefore $D_R(u_n) = \sum_{i=1}^n D_{R_i - \bar{R}_{i-1}}(u_n) \ge (2\pi)^{-1} \sum_{i=1}^n \log \mu_i$. Since $D_R(u_n) = (2\pi)^{-1} \log v_n$ we obtain

$$v_n \ge \prod_{i=1}^n \mu_i.$$

By Theorem 10 A, (121) implies $R \in O_G$.

Conversely assume $R \in O_G$. Take regular regions R_0 and R_1 with $R_1 \supset \bar{R}_0$ and a regular region $R_2 \supset \bar{R}_1$ with $\mu_2 \ge 2$. This is possible by (118). Here R_2 can be arbitrarily large. By repeating this procedure we obtain an exhaustion $\{R_n\}_0^\infty$ of R with $\mu_i \ge i$ and thus with (121).

10 C. Conformal Metric Test. We use the same notation as in I.2 A and 2 B. Let $L(\rho)$ be the length of the curve $\Gamma(\rho)$ in the given conformal metric ds. Then we have (Ahlfors [2], Laasonen [1]; see also Kuroda [1]):

Theorem. *If the length $L(\rho)$ grows so slowly that*

$$\int_\varepsilon^\infty \frac{d\rho}{L(\rho)} = \infty \qquad (\varepsilon > 0) \tag{124}$$

then R belongs to O_G.

The proof is a direct analogue of that in I.2 B and is given here for comparison.

Let w_n be the harmonic measure of ∂R_n with respect to $R_n - \bar{R}_{n-1}$ and w_n^* a conjugate harmonic function of w_n on $R_n - \bar{R}_{n-1}$. Then the function

$$W_n = w_n + i\, w_n^*$$

maps $R_n - \bar{R}_{n-1}$, less suitable slits γ_j on which w_n^* is constant, conformally onto the horizontally sliced rectangle T: $0 \le w_n \le 1$, $0 \le w_n^* \le 2\pi/\log \mu_n$ in the W-plane.

The Euclidean length of $\Gamma(\rho)$ in T is at least $2\pi/\log \mu_n$:

$$\frac{2\pi}{\log \mu_n} \le \int_{\Gamma(\rho)} \left| \frac{dW_n}{dz} \right| |dz|.$$

By Schwarz's inequality

$$\left(\frac{2\pi}{\log \mu_n} \right)^2 \le \int_{\Gamma(\rho)} \left| \frac{dW_n}{dz} \right|^2 \frac{|dz|}{\lambda} \int_{\Gamma(\rho)} \lambda \, |dz|.$$

We denote by $A(\rho)$ the area of that part of T which lies to the left of the image of $\Gamma(\rho)$. Since on $\Gamma(\rho)$

$$\frac{|dz|}{\lambda} = |dz| \frac{|dz|}{ds} = |dz| \frac{|dz|}{d\rho}$$

we infer that

$$\int_{\Gamma(\rho)} \left| \frac{dW_n}{dz} \right|^2 \frac{|dz|}{\lambda} = \frac{d}{d\rho} \int_{\rho_{n-1}}^{\rho} \int_{\Gamma(\rho)} \left| \frac{dW_n}{dz} \right|^2 |dz| \frac{|dz|}{d\rho} \, d\rho = \frac{d}{d\rho} A(\rho).$$

Therefore

$$\left(\frac{2\pi}{\log \mu_n} \right)^2 \le \frac{dA(\rho)}{d\rho} L(\rho)$$

and

$$\int_{\rho_{n-1}}^{\rho_n} \frac{d\rho}{L(\rho)} \le \left(\frac{\log \mu_n}{2\pi} \right)^2 \int_{\rho_{n-1}}^{\rho_n} dA = \frac{1}{2\pi} \log \mu_n .$$

We conclude that

$$\frac{1}{2\pi} \log \prod_1^{\infty} \mu_n \ge \int_{\rho_0}^{\infty} \frac{d\rho}{L(\rho)} = \infty.$$

Condition (121) is satisfied and $R \in O_G$.

***10 D. Euclidean Metric Test.** We shall use the notation in I.2 C and 2 D. Let $l(r)$ be the length of $\Gamma(r)$. As a counterpart of Theorem I.2 D we prove (Laasonen [2]):

Theorem. *Let G be a Fuchsoid group and B_0 its fundamental polygon containing the origin. If the total Euclidean length $l(r)$ of the cycles on $|z| = r$ in B_0 grows so slowly that*

$$\int_{\varepsilon}^{1} \frac{dr}{l(r)} = \infty \qquad (\varepsilon > 0), \tag{125}$$

then the Riemann surface R corresponding to G belongs to O_G.

The proof is again essentially the same as that in I.2 D.

Let ρ be the hyperbolic length corresponding to r and let $L(\rho)$ be that of $\Gamma(r)$. Then $\rho=\frac{1}{2}\log((1+r)/(1-r))$, $d\rho/dr=1/(1-r^2)$, and

$$L(\rho)=\int_{\Gamma(r)} \frac{|dz|}{1-|z|^2}=\int_{\Gamma(r)} \frac{r\,d\theta}{1-r^2}=\frac{d\rho}{dr}\int_{\Gamma(r)} r\,d\theta=\frac{d\rho}{dr}\,l(r).$$

Hence $L(\rho)/l(r)=d\rho/dr$ and for $\varepsilon'=\tanh^{-1}\varepsilon$

$$\int_{\varepsilon'}^{\infty} \frac{d\rho}{L(\rho)}=\int_{\varepsilon}^{1} \frac{dr}{l(r)}.$$

Thus (125) implies (124) which gives $R\in O_G$.

10 E. Regular Chain Test. We use the notation in I.3 A and 3 B and set

$$l_n=\sum_{i=1}^{\nu_n} l_{ni}. \tag{126}$$

The following is a slight modification of a test due to NEVANLINNA [2]:

THEOREM. *If there exists a regular chain set for which the numbers l_n of disks in the chains have the property*

$$\sum_n \frac{1}{l_n}=\infty \tag{127}$$

then R belongs to O_G.

Let $\tilde{\mu}_n$ be the modulus of $\bigcup_{i=1}^{\nu_n} C_{ni}$ and $\tilde{\mu}_{ni}$ that of C_{ni}. In I.3 C we saw that

$$\frac{1}{\log\tilde{\mu}_{ni}}<\frac{N}{d\pi}\,l_{ni}. \tag{128}$$

We denote the harmonic measures on C_{ni} and $\bigcup_{i=1}^{\nu_n} C_{ni}$ by w_{ni} and w_n and deduce from $\log\tilde{\mu}_{ni}=2\pi/D_R(w_{ni})$, $\log\tilde{\mu}_n=2\pi/D_R(w_n)$, and $D_R(w_n)=\sum_{i=1}^{\nu_n} D_R(w_{ni})$ that

$$\frac{1}{\log\tilde{\mu}_n}=\sum_{i=1}^{\nu_n}\frac{1}{\log\tilde{\mu}_{ni}}. \tag{129}$$

From this and from (129), (128), and (126) we obtain

$$\log\tilde{\mu}_n>\frac{d\pi}{N}\,\frac{1}{l_n}$$

and consequently

$$\log\prod_{n=1}^{\infty}\tilde{\mu}_n>\frac{d\pi}{N}\sum_n\frac{1}{l_n}=\infty.$$

This implies the existence of an exhaustion with (121) and we have $R\in O_G$.

Remark. On the basis of the modular test (121) we can derive several other explicit tests such as the deep covering test and the test by triangulation (see AHLFORS-SARIO [12, pp. 234–238]). In particular, tests for a covering surface of the sphere to belong to O_G can be discussed in a manner parallel to that in I.12–16, the details being left to the reader. The tests in terms of line complexes are especially interesting in this case. ROYDEN'S [2] well-known hyperbolicity test is suggestive in this direction. For these and further topics on line complexes (topological trees, Speiser graphs) we refer the reader to PFLUGER'S monograph [6].

Historically parabolicity is the oldest notion in classification theory. It already appeared in RIEMANN [1] in connection with his mapping theorem. The so-called classical type problem was to determine, mainly on the basis of the distribution of branch points, if a given simply connected Riemann surface belonged to O_G (TEICHMÜLLER [1], AHLFORS [1, 2], KAKUTANI [1–4], KOBAYASHI [1–4]). KAKUTANI was the first to make effective use of quasiconformal mappings in classification theory. For a lucid survey of the classical type problem we refer to LE-VAN [1].

The concept of harmonic measure was introduced by SCHWARZ [1] and effectively used by BEURLING [1]. NEVANLINNA [1] coined the phrase "harmonic measure" and introduced the class of "nullbounded" surfaces characterized by the vanishing of the harmonic measure. That this class coincides with the class O_G of "parabolic" surfaces was shown by MYRBERG [2] for surfaces of finite genus.

In addition to the rather modern treatment of the class O_G thus far given in this book there are numerous other important results in the theory of this class; among these we mention here, because of its practical significance, the test concerning welding as given by STREBEL [4] and OIKAWA [10].

11. Plane Regions

11 A. Inclusion Relations. By 7 C, IV.3 C, III.4 H, and III.4 G, we have the strict inclusion relations

$$O_G < O_{HP} < O_{HB} < O_{HD} = O_{HBD} \tag{130}$$

for general Riemann surfaces. In contrast we know from III.5 G that

$$O_G = O_{HP} = O_{HB} = O_{HD} = O_{HBD} \tag{131}$$

for surfaces of almost finite genus and of course for surfaces of finite genus.

In particular (131) is true for plane regions. Hereafter in 11 we shall mainly discuss plane regions R. Without loss of generality we can assume that R contains the point at infinity. Then $\delta = \{|z| \leq \infty\} - R$ is a compact set. If δ contains a proper continuum then R can be conformally deformed into a bounded region and thus $R \notin O_G$. In questions concerning property (131) we can assume that δ is totally disconnected. Then $\delta = \beta = \partial R$.

We shall say that the compact set δ is of *capacity zero* if $R \in O_G$ (cf. 1 E, 7 F).

11 B. Removable Sets. A compact set δ is called *HX-removable* or *HX-null* if for some disk U with $\delta \subset U$ all HX-functions on $\bar{U} - \delta$ have harmonic extensions to $\bar{U}$, where $X = B'$, B, or D.

Let $H_0 X(\bar{U} - \delta)$ be the class of HX-functions on $\bar{U} - \delta$ with vanishing values at the boundary α of U. We first remark:

A compact set δ is HX-removable if and only if $H_0 X(\bar{U} - \delta) = \{0\}$ where $X = B', B$, or D.

If δ is HX-removable, then $H_0 X(\bar{U} - \delta) = H_0 X(\bar{U}) | \bar{U} - \delta$. By the maximum principle $H_0 X(\bar{U}) = \{0\}$ and therefore $H_0 X(\bar{U} - \delta) = \{0\}$. Conversely if $H_0 X(\bar{U} - \delta) = \{0\}$ then the fact that $u - \tilde{u} \in H_0 X(\bar{U} - \delta)$ for any $u \in HX(\bar{U} - \delta)$ and $\tilde{u} \in HX(\bar{U})$ with $\tilde{u} = u$ on α implies that $\tilde{u} | \bar{U} - \delta = u$.

THEOREM. *For a plane compact set δ the following are equivalent:*

(a) *δ has capacity zero,*
(b) *δ is HB'-removable,*
(c) *δ is HB-removable,*
(d) *δ is HD-removable.*

Let $R = \{|z| \leq \infty\} - \delta$, U be a disk containing δ, and $\alpha = \partial U$. Suppose (a) holds, i.e. $R \in O_G$. By IV.3 D we have $H_0 B'(U - \delta) = \{0\}$ and a fortiori $H_0 B(U - \delta) = \{0\}$. In view of III.2 I, $H_0 D(U - \delta) = \{0\}$. Properties (b)–(d) follow.

Since HB, $HD \subset HB'$, (b) implies (c) and (d). To complete the proof we have only to show the implications (c) $\Rightarrow$ (a), and (d) $\Rightarrow$ (a).

Let $u \in HB(R)$ (resp. $u \in HD(R)$). Then $u | U - \delta$ can be harmonically extended to all of U. Thus u is harmonic on $|z| \leq \infty$ and consequently constant. Hence $R \in O_{HB}$ (resp. O_{HD}). Now (131) implies that $R \in O_G$, i.e. δ has capacity zero.

The concept of a plane HP-removable set is meaningless: even a point ζ is not a removable singularity for $\log(1/|z - \zeta|) \in HP(0 < |z - \zeta| < 1)$. In this connection we observe that for a general $R \in O_{HP} - O_G$, $R - p \notin O_{HP}$, where p is an arbitrary point in R. In fact $G(\cdot, p) \in HP(R - p)$ (cf. 7 J).

Remark. The arguments in 11 A and 11 B are valid for compact sets δ in a closed Riemann surface. The details are left to the reader.

11 C. Robin's Constant. Let δ be a plane compact set with connected complement $R=\{|z|\leq\infty\}-\delta$. Take a sequence $\{R_n\}_0^\infty$ of regions such that $R_0\subset R_1\subset\cdots$, $R=\bigcup_0^\infty R_n$, $\infty\in R_0$, and $\beta_n=\partial R_n$ consists of a finite number of analytic arcs. The $\delta_n=\{|z|\leq\infty\}-R_n$ converge decreasingly to δ. The boundary β may have vanishing capacity.

Let $g_n(z,\infty)$ be the Green's function on R_n with pole ∞ and denote by $s_n(z)$ the capacity function on R_n with the same pole ∞. Then $s_n|\beta_n=k_{\beta_n}$, a constant, and $k_{\beta_n}-s_n=g_n(\cdot,\infty)$ (cf. 8 A). Thus

$$g_n(z,\infty)=\log|z|+k_{\beta_n}+h_n(z), \qquad h_n(\infty)=0, \tag{132}$$

where $h_n\in H(\bar{R}_0)$. The quantity

$$r(\delta_n)=k_{\beta_n} \tag{133}$$

is by definition the *Robin constant* of δ_n. If $R\notin O_G$ then by (132)

$$g(z,\infty)=\log|z|+k_\beta+h(z) \tag{134}$$

with $h\in H(R)$, and the Robin constant of δ is

$$r(\delta)=k_\beta. \tag{135}$$

If $R\in O_G$, (135) also gives the Robin constant of $\delta=\beta$, viz.

$$r(\delta)=r(\beta)=k_\beta=\infty. \tag{136}$$

In any case $r(\delta_1)\leq r(\delta_2)\leq\cdots$ and

$$\lim_{n\to\infty} r(\delta_n)=r(\delta). \tag{137}$$

The quantity

$$\operatorname{cap}(\delta)=e^{-r(\delta)}$$

is called the *logarithmic capacity* of δ. It coincides with the capacity $c_\beta=e^{-k_\beta}$. For a general set X the (inner) logarithmic capacity is defined as

$$\operatorname{cap}(X)=\sup_{\delta\subset X}\operatorname{cap}(\delta)$$

with compact δ.

11 D. Change under Conformal Mapping. Although the property $r(\delta)=\infty$ is invariant under conformal mappings of R the quantity $r(\delta)$ is in general not.

Let f be a conformal mapping of R onto R' such that

$$z=f^{-1}(z')=a z'+b+\frac{c}{z'}+\frac{d}{z'^2}+\cdots \qquad (a\neq 0) \tag{138}$$

at $z'=z=\infty$. Denote by δ' the compact complement of R'. Then

$$r(\delta')=r(\delta)+\log|a|. \tag{139}$$

In fact if $R \in O_G$ then $R' \in O_G$ and (139) is trivially true in the sense $\infty = \infty$. Hence we assume $R \notin O_G$ and consequently $R' \notin O_G$. By (134) we have

$$g_R(z, \infty) = \log|z| + r(\delta) + h_R(z),$$

$$g_{R'}(z', \infty) = \log|z'| + r(\delta') + h_{R'}(z').$$

Here

$$g_{R'}(z', \infty) = g_R(f^{-1}(z'), \infty) = \log|z'| + \log|a| + r(\delta) + \tilde{h}(z')$$

with $\tilde{h}(\infty) = 0$. Thus we obtain (139).

As an example let $\delta = \{|z| \leq 1\}$ and $\delta' = [0, 1]$. Then the function

$$z' = f(z) = \frac{1}{4}\left(z + \frac{1}{z}\right) + \frac{1}{2}$$

maps R conformally onto R'. Clearly

$$z = f^{-1}(z') = 4z' + \cdots$$

and thus

$$r(\delta') = r(\delta) + \log 4. \tag{140}$$

On the other hand $g_R(z, \infty) = \log|z|$. Therefore

$$r(\delta) = 0, \qquad r(\delta') = \log 4. \tag{141}$$

11 E. Logarithmic Potentials. For a positive regular unit Borel measure μ with compact support S_μ in $|z| < \infty$ the *logarithmic potential* L_μ of μ is defined as

$$L_\mu(z) = \int \log \frac{1}{|z - \zeta|}\, d\mu(\zeta).$$

In the same manner as in 9 E, $L_\mu(z)$ is superharmonic on $|z| < \infty$, harmonic on $\{|z| < \infty\} - S_\mu$, and

$$L_\mu(z) = -\log|z| + h_\mu(z), \qquad h_\mu(\infty) = 0, \tag{142}$$

where h_μ is harmonic at ∞. We have previously proved in 9 E that

$$\limsup_{z \notin S_\mu,\, z \to z_0} L_\mu(z) \leq \limsup_{z \in S_\mu,\, z \to z_0} L_\mu(z) \tag{143}$$

for every $z_0 \in \partial S_\mu$. As a consequence of (143), (142) we have the validity of the continuity principle and Frostman's maximum principle for L_μ (cf. Corollaries in 9 E).

For a compact set K consisting of a finite number of analytic arcs and for $f \in B(K)$ we can also consider the Gauss variation

$$V_{L,f}(\mu) = \iint \log \frac{1}{|z - \zeta|}\, d\mu(z)\, d\mu(\zeta) - 2\int f\, d\mu \tag{144}$$

and measures μ with $S_\mu \subset K$. By a repetition of the proof of Theorem 9 G we deduce the existence of a measure ν on K such that

$$V_{L,f}(\nu) = \min_\mu V_{L,f}(\mu), \tag{145}$$

$L_\nu | S_\nu \leq f$, and $L_\nu | K \geq f$ ξ-a.e.

If the diameter of K is at most 1 then for measures μ, ν with $S_\mu, S_\nu \subset K$ we also have the energy principle

$$(\mu, \nu)_L^2 \leq (\mu, \mu)_L (\nu, \nu)_L \tag{146}$$

where $(\mu, \nu)_L = -\iint \log|z-\zeta|\, d\mu(z)\, d\mu(\zeta)$. We replace superharmonic functions vanishing on Δ_{MI} by superharmonic functions with behavior (142) at ∞, and the proof of (104) qualifies with obvious changes as that of (146).

After these preparations we can prove as in 9 I and 9 J the following counterpart of Theorem 9 D, with δ_n as in 11 C:

THEOREM. *Let δ_n be contained in the closed unit disk. There exists a unique measure $\mu_{\beta_n} \in m_{\beta_n}$ such that*

$$r(\delta_n) = k_{\beta_n} = (\mu_{\beta_n}, \mu_{\beta_n})_L = \min_{\mu \in m_{\beta_n}} (\mu, \mu)_L. \tag{147}$$

Moreover if we set $s_n(z) = k_{\beta_n}$ on δ_n then

$$L_{\mu_{\beta_n}}(z) = s_n(z). \tag{148}$$

In addition we observe that for any $\mu \in m_{\beta_n}$ with continuous $L_\mu | \beta_n$

$$\min_{z \in \beta_n} L_\mu(z) \leq r(\delta_n) \leq \max_{z \in \beta_n} L_\mu(z). \tag{149}$$

In fact since $h = L_\mu - s_n$ can be harmonically extended to ∞ we have $h \in H(R_n)$ and $h(\infty) = 0$. Thus $h | \beta_n = L_\mu | \beta_n - r(\delta_n)$ must change sign, and (149) follows.

11 F. Capacity of the Cantor Set. We consider next the generalized 1-dimensional Cantor set $E(p_1 p_2 \ldots)$ obtained from the unit interval $[0, 1]$ (cf. I.6 A). We shall prove:

THEOREM. *The Cantor set $E(p_1 p_2 \ldots)$ with $\sup_n p_n < \infty$ is of capacity zero if and only if*

$$\prod_{\nu=1}^{\infty} \left(1 - \frac{1}{p_\nu}\right)^{2^{-\nu}} = 0. \tag{150}$$

Let r_n be the Robin constant of $E(p_1 \ldots p_n)$, and r that of $E(p_1 p_2 \ldots)$. Then by (137)

$$r = \lim_{n \to \infty} r_n. \tag{151}$$

We also denote by r_0 the Robin constant of $E=[0,1]$ and by $r_{n\nu}$ that of $E(p_\nu \ldots p_n)$ for $1 \le \nu \le n$. We shall successively seek relations between r_0 and r_{nn}; r_{nn} and $r_{n,n-1}$; $r_{n,n-1}$ and $r_{n,n-2}$; ...; r_{n2} and $r_{n1}=r_n$. We know from (141) that

$$r(E)=r_0=\log 4. \tag{152}$$

First let us compare r_0 and r_{nn}. Observe that $E(p_n)$ consists of two parts E_1 and E_2 which are symmetric about $\frac{1}{2}$. By translation E_1 coincides with E_2. Thus by (139), $r(E_1)=r(E_2)$. Since the length of E_1 is $\frac{1}{2}(1-1/p_n)$ the complement R' of E_1 is mapped conformally onto the complement R of E by

$$z=\frac{2z'}{1-\dfrac{1}{p_n}}$$

and in view of (139) we have $r(E_1)=r(E)+\log(2/(1-1/p_n))$. It follows from (152) that

$$r(E_1)=r(E_2)=\log 4+\log\frac{2}{1-\dfrac{1}{p_n}}. \tag{153}$$

Let μ_i be the measure in (147) for E_i $(i=1,2)$. Set $\mu=\frac{1}{2}(\mu_1+\mu_2)$. Then $L_{\mu_i}|E_i=r(E_i)$ and by (149)

$$\min_{z\in E(p_n)} L_\mu(z) \le r_{nn} \le \max_{z\in E(p_n)} L_\mu(z). \tag{154}$$

Note that

$$L_\mu(z)=\tfrac{1}{2}L_{\mu_1}(z)+\tfrac{1}{2}L_{\mu_2}(z).$$

For $z\in E_1$ and $\zeta\in E_2$ we have $0\le\log(1/|z-\zeta|)\le\log p_n$ and therefore

$$\begin{aligned}\tfrac{1}{2}r(E_1)\le L_\mu|E_1\le\tfrac{1}{2}r(E_1)+\tfrac{1}{2}\log p_n,\\ \tfrac{1}{2}r(E_2)\le L_\mu|E_2\le\tfrac{1}{2}r(E_2)+\tfrac{1}{2}\log p_n.\end{aligned} \tag{155}$$

From (153), (154), and (155) it now follows that

$$\log 4+\log\frac{2}{\left(1-\dfrac{1}{p_n}\right)}\le 2r_{nn}\le\log 4+\log\frac{2}{\left(1-\dfrac{1}{p_n}\right)}+\log p_n. \tag{156}$$

Similarly we can compare r_{nn} and $r_{n,n-1}$; $r_{n,n-1}$ and $r_{n,n-2}$; ...; r_{n2} and $r_{n1}=r_n$. For greater clarity let us consider explicitly the next step, i.e. comparison of r_{nn} and $r_{n,n-1}$.

Observe again that $E(p_{n-1}\,p_n)$ consists of two parts E_1 and E_2 symmetric with respect to $\frac{1}{2}$. By translation E_1 coincides with E_2, and by (139), $r(E_1)=r(E_2)$. Since E_1 is contained in the interval of length $\frac{1}{2}(1-1/p_{n-1})$

the complement R' of E_1 is mapped conformally onto the complement R of $E(p_n)$ by

$$z=\frac{2z'}{1-\dfrac{1}{p_{n-1}}}.$$

We infer by (139) that $r(E_1)=r\big(E(p_n)\big)+\log\big(2/(1-1/p_{n-1})\big)$. Hence

$$r(E_1)=r(E_2)=r_{nn}+\log\frac{2}{1-\dfrac{1}{p_{n-1}}}. \tag{157}$$

Let μ_i be the measure in (147) for E_i $(i=1,2)$. Set $\mu=\frac{1}{2}(\mu_1+\mu_2)$. Then $L_{\mu_i}|E_i=r(E_i)$ and by (149)

$$\min L_\mu|E(p_{n-1}\,p_n)\le r_{n,n-1}\le\max L_\mu|E(p_{n-1}\,p_n). \tag{158}$$

Again observe that $L_\mu=\frac{1}{2}L_{\mu_1}+\frac{1}{2}L_{\mu_2}$ and $0\le\log(1/|z-\zeta|)\le\log p_{n-1}$. As a consequence

$$\begin{aligned}\tfrac{1}{2}r(E_1)\le L_\mu|E_1\le\tfrac{1}{2}r(E_1)+\tfrac{1}{2}\log p_{n-1},\\ \tfrac{1}{2}r(E_2)\le L_\mu|E_2\le\tfrac{1}{2}r(E_2)+\tfrac{1}{2}\log p_{n-1}.\end{aligned} \tag{159}$$

It follows by (157), (158), and (159) that

$$r_{nn}+\log\frac{2}{1-\dfrac{1}{p_{n-1}}}\le 2r_{n,n-1}\le r_{nn}+\log\frac{2}{1-\dfrac{1}{p_{n-1}}}+\log p_{n-1}. \tag{160}$$

We may now write

$$\tfrac{1}{2}r_{n\nu}+\tfrac{1}{2}\log\frac{2}{1-\dfrac{1}{p_{\nu-1}}}\le r_{n,\nu-1}\le\tfrac{1}{2}r_{n\nu}+\tfrac{1}{2}\log\frac{2}{1-\dfrac{1}{p_{\nu-1}}}+\tfrac{1}{2}\log p_{\nu-1} \tag{161}$$

for $\nu=2,\ldots,n+1$, with $r_{n,n+1}=r_0=\log 4$.

We form the sum

$$(161)_{\nu=2}+\tfrac{1}{2}(161)_{\nu=3}+(\tfrac{1}{2})^2\,(161)_{\nu=4}+\cdots+(\tfrac{1}{2})^{n-1}\,(161)_{\nu=n+1}$$

and obtain

$$\begin{aligned}&\frac{\log 4}{2^n}+\log\left(1\Big/\prod_{\nu=1}^{n}(1-1/p_\nu)^{2^{-\nu}}\right)+\left(1-\frac{1}{2^n}\right)\log 2\\ &\le r_n\le\frac{\log 4}{2^n}+\log\left(1\Big/\prod_{\nu=1}^{n}(1-1/p_\nu)^{2^{-\nu}}\right)+\left(1-\frac{1}{2^n}\right)\log 2+\sum_{\nu=1}^{n}\frac{\log p_\nu}{2^\nu}.\end{aligned} \tag{162}$$

If (150) holds we conclude by (151) that $r=\infty$. Conversely if $r=\infty$ then by (151), $r_n\to\infty$ as $n\to\infty$. Thus (150) follows under the condition $\sum_1^\infty 2^{-\nu}\log p_\nu<\infty$ which is a consequence of $\sup_\nu p_\nu<\infty$.

COROLLARY. *If*

$$\prod_{\nu=1}^{\infty}\left(1-\frac{1}{p_\nu}\right)^{2^{-\nu}}=0 \tag{163}$$

then the Cantor set $E(p_1\, p_2\, \ldots)$ is of capacity zero. Conversely if $E(p_1\, p_2\, \ldots)$ is of capacity zero then

$$\prod_{\nu=1}^{\infty}\left(\frac{p_\nu-1}{p_\nu^2}\right)^{2^{-\nu}}=0. \tag{164}$$

§ 3. Existence of Kernels

Simply connected Riemann surfaces are classified into three types: the hyperbolic surface $|z|<1$, the parabolic surface $|z|<\infty$, and the elliptic surface $|z|\le\infty$.

The hyperbolic kernel $\log(|1-\bar{\zeta}z|/|z-\zeta|)$ is associated with the first type. It is characterized as the harmonic function with a single positive logarithmic pole and without a positive harmonic minorant. In § 2 we called functions with this property on a general Riemann surface Green's functions, and surfaces carrying such functions or their kernels, hyperbolic.

With the second type we shall now associate the parabolic kernel or the logarithmic kernel $\log(1/|z-\zeta|)$. It is characterized as the harmonic function with a single positive logarithmic pole and with ideal boundary values $-\infty$. Kernel functions on general surfaces with this property may be called Evans' kernels. For $|z|<\infty$ it is also characteristic that there exists a function with one negative logarithmic pole and with ideal boundary values ∞, e.g. the function $\log|z|$. Such a function on a general surface is called an Evans-Selberg potential. It is natural to consider surfaces with an Evans-Selberg potential and to call them parabolic. That this parabolicity is exactly the same as that discussed in § 2 is the main topic of this section.

In 12 we consider hyperbolic surfaces and show the existence of a harmonic function with boundary values ∞ at the essentially irregular boundary of the surface. This function is used in 13 to construct an Evans-Selberg potential on every open surface in O_G. We are also led to the construction of Evans' kernel.

The elliptic kernel $\log(1/[z,\zeta])$ is considered on the elliptic surface $|z|\le\infty$. Its natural generalization to an arbitrary surface, called the s-kernel, does not imply classification: we can construct the s-kernel on every surface, open or closed. This topic will also be discussed in 13.

12. The Hyperbolic Case

12 A. The Fundamental Theorem. Given $R \notin O_G$ let R^*_{MI} be Royden's compactification of R, and Ξ the essentially irregular boundary of R, i.e. the subset of Royden's boundary $\Gamma_{\mathrm{MI}} = R^*_{\mathrm{MI}} - R$ on which the Green's kernel does not vanish (cf. 8 G). We shall show that the Tchebycheff constant $\tau(\Xi)$ and the transfinite diameter $\rho(\Xi)$ of Ξ with respect to the Green's kernel on R^*_{MI} (cf. 9 A and 9 B) are in a sense almost infinite:

THEOREM. *The essentially irregular boundary Ξ of a hyperbolic Riemann surface R is so small that for any compact subset Ξ' of Ξ the Tchebycheff constant $\tau(\Xi')$ and the transfinite diameter $\rho(\Xi')$ are infinite:*

$$\tau(\Xi') = \rho(\Xi') = \infty. \tag{165}$$

This result will play a fundamental role in the present section. The proof will be given in 12 B through 12 E. If $\Xi = \emptyset$ then the assertion is trivially true. We shall assume $\Xi \neq \emptyset$ throughout the proof.

12 B. An Auxiliary Function. Choose a point $z_0 \in R$ and let $\{r_n\}_1^\infty$ be an arbitrary but then fixed sequence of positive numbers r_n such that

$$r_n > r_{n+1}, \qquad \lim_{n\to\infty} r_n = 0$$

with the additional properties that the set

$$U_n = \{z \in R \mid G(z, z_0) > r_n\}$$

is not relatively compact in R and $dG(z, z_0) \neq 0$ on ∂U_n. Clearly the set

$$\Xi_n = \bar{U}_n \cap \Gamma_{\mathrm{MI}} = \{q \in \Gamma_{\mathrm{MI}} \mid G(q, z_0) \geq r_n\}$$

is compact in R^*_{MI} and also in Γ_{MI}, and we have

$$\Xi = \bigcup_{n=1}^{\infty} \Xi_n. \tag{166}$$

We also fix a regular exhaustion $\{R_m\}_1^\infty$ of R such that $z_0 \in R_1$. Then ∂R_m meets ∂U_n at most at a finite number of points and the set

$$F_{nm} = \bar{U}_n \cap \partial R_m \tag{167}$$

consists of a finite number of disjoint analytic arcs.

The set U_n is connected, for if this were not the case there would exist a component not containing z_0. Then by III.2 H, $G(z, z_0)$ would be the constant r_n on this component, a contradiction.

First we prove:

Lemma. *There exists a unique function* $w_{nm} \in HP(U_{n+1} - \bar{R}_m)$ *continuous on* $\bar{U}_{n+1} - R_m - (\partial R_m) \cap (\partial U_{n+1})$ *with*

$$\begin{gathered} w_{nm} | \partial U_{n+1} - R_m - (\partial R_m) \cap (\partial U_{n+1}) = 0, \\ w_{nm} | F_{n+1,m} - (\partial R_m) \cap (\partial U_{n+1}) = 1, \qquad w_{nm} | \Xi_n \geq \sigma_n, \end{gathered} \tag{168}$$

where $\sigma_n > 0$ *is a positive constant independent of* $m = 1, 2, \ldots$.

Take $k > m+1$ and let u_k be harmonic on $R_k \cap U_{n+1} - \bar{R}_m$ with continuous boundary values $u_k = 1$ on $F_{n+1,m} - (\partial R_m) \cap (\partial U_{n+1})$ and $u_k = 0$ on $\partial(R_k \cap U_{n+1}) - \bar{R}_m$. Since $u_k < u_{k+p}$

$$w_{nm} = \lim_{k \to \infty} u_k \tag{169}$$

is a well-defined positive harmonic function on $U_{n+1} - \bar{R}_m$ with boundary values (168) except possibly for $w_{nm} | \Xi_n \geq \sigma_n$.

Observe that

$$D_{R_k \cap U_{n+1} - \bar{R}_{m+1}}(u_k) = \int_{\partial R_{m+1} \cap U_{n+1}} u_k * du_k.$$

Since the convergence in (169) is uniform on $\partial R_{m+1} \cap U_{n+1}$ we see that

$$\lim_{k \to \infty} \int_{\partial R_{m+1} \cap U_{n+1}} u_k * du_k = \int_{\partial R_{m+1} \cap U_{n+1}} w_{nm} * dw_{nm} < \infty.$$

By Fatou's lemma

$$\begin{aligned} D_{U_{n+1} - R_{m+1}}(w_{nm}) &\leq \liminf_{k \to \infty} D_{R_k \cap U_{n+1} - \bar{R}_{m+1}}(u_k) \\ &= \int_{\partial R_{m+1} \cap U_{n+1}} w_{nm} * dw_{nm} < \infty. \end{aligned}$$

Let $v \in H(R_{m+1} \cap U_{n+1} - \bar{R}_m)$ with boundary values $v | F_{n+1,m+1} = w_{nm}$ and $v | \partial(R_{m+1} \cap U_{n+1} - \bar{R}_m) - F_{n+1,m+1} = 0$. Then clearly $v \in HD(R_{m+1} \cap U_{n+1} - \bar{R}_m)$. Set $f = w_{nm}$ on $U_{n+1} - R_{m+1}$, $f = v$ on $R_{m+1} \cap U_{n+1} - \bar{R}_m$, and $f = 0$ on $R - (U_{n+1} - \bar{R}_m)$. Then it is easy to see that $f \in \mathrm{M}(R)$. Thus f is continuous on R^*_{M} and a fortiori $w_{nm} | \bar{U}_{n+1} - R_{m+1} = f | \bar{U}_{n+1} - R_{m+1}$ is continuous on $\bar{U}_{n+1} - R_{m+1}$. Therefore w_{nm} is continuous on $\bar{U}_{n+1} - R_m - (\partial R_m) \cap (\partial U_{n+1})$.

Take $b > 0$ so large that the closure of $R_0 = \{z \in R | G(z, z_0) > b\}$ is contained in $R_1 \cap U_{n+1}$. Set

$$w(z) = \frac{G(z, z_0) - r_{n+1}}{b - r_{n+1}}.$$

Then $w \in HP(U_{n+1} - z_0)$, $w | \partial U_{n+1} = 0$, and $w | U_{n+1} - \bar{R}_0 < 1$. Observe that $(w_{nm} - w) | F_{n+1,m} = 1 - w | F_{n+1,m} > 0$ and $(w_{nm} - w) | \partial U_{n+1} - \bar{R}_m = 0$. From this and III.2 H it follows that

$$w_{nm} \geq w$$

on $U_{n+1} - \bar{R}_m$, and in particular

$$w_{nm} | \Xi_n \geq \inf_{\Xi_n} w.$$

Here $\sigma_n = \inf_{\Xi_n} w = \left(\inf_{\Xi_n} G(\cdot, z_0) - r_{n+1}\right)/(b - r_{n+1}) \geq (r_n - r_{n+1})/(b - r_{n+1}) > 0$. Thus we obtain (168).

The uniqueness of a w_{nm} with the required properties is clear from III.2 H.

12 C. An Inequality. The auxiliary function w_{nm} obtained in 12 B is now used to prove the following inequality:

LEMMA. *The transfinite diameter of Ξ_n is related to that of the approximating set $F_{n+1,m}$ by*

$$\rho(\Xi_n) \geq \sigma_n^2 \, \rho(F_{n+1,m}) \tag{170}$$

for every $m = 1, 2, \ldots$.

We may assume that Ξ_n is infinite since otherwise (170) is trivially true.

Let $k \geq 4$ be a fixed integer and $p_1, \ldots, p_k$ arbitrary points in Ξ_n. We shall find k points $z_1, \ldots, z_k$ in $F_{n+1,m}$ such that

$$\begin{aligned} \sigma_n^2 \sum_{i<j}^{1,\ldots,\nu} G(z_i, z_j) + \sigma_n \sum_{i=1}^{\nu} \sum_{j=\nu+1}^{k} G(z_i, p_j) + \sum_{i<j}^{\nu+1,\ldots,k} G(p_i, p_j) \\ \leq \sum_{i<j}^{1,\ldots,k} G(p_i, p_j) \end{aligned} \tag{171}$$

for each $\nu = 1, \ldots, k$. It is understood that $\sum_{i<j}^{1,\ldots,\nu} = 0$ if $\nu = 1$; $\sum_{i<j}^{\nu+1,\ldots,k} = 0$ if $\nu = k-1$; $\sum_{j=\nu+1}^{k} = \sum_{i<j}^{\nu+1,\ldots,k} = 0$ if $\nu = k$.

We shall use induction. Suppose $z_1, \ldots, z_{h-1}$ $(1 \leq h \leq k-1)$ have been found with (171) for $\nu = 1, \ldots, h-1$ and set

$$u_h(z) = \sum_{j=h+1}^{k} G(z, p_j) + \sigma_n \sum_{i=1}^{h-1} G(z_i, z). \tag{172}$$

We are going to determine z_h. We set $\sum_{i=1}^{h-1} = 0$ for $h = 1$ and $\sum_{j=h+1}^{k} = 0$ for $h = k$.

Since u_h is continuous on $F_{n+1,m}$ there exists a $z_h \in F_{n+1,m}$ such that

$$u_h(z_h) = \min u_h | F_{n+1,m}. \tag{173}$$

By (168) the function $u_h - u_h(z_h) \, w_{nm}$ is nonnegative on $\partial(U_{n+1} - \bar{R}_m)$. Clearly it is harmonic on $U_{n+1} - \bar{R}_m$ and bounded from below. Thus by

Theorem III.2 H it is nonnegative on $U_{n+1}-\bar{R}_m$ and, by continuity, on $\bar{U}_{n+1}-R_m$. In view of (168) we have in particular

$$u_h(p_h)\geq u_h(z_h)\,w_{nm}(p_h)\geq\sigma_n\,u_h(z_h).$$

This means that

$$\sum_{j=h+1}^{k} G(p_h,p_j)+\sigma_n\sum_{i=1}^{h-1} G(z_i,p_h)\geq\sigma_n\sum_{j=h+1}^{k} G(z_h,p_j)+\sigma_n^2\sum_{i=1}^{h-1} G(z_i,z_h). \tag{174}$$

In the case $h=1$ we deduce (171) for $v=1$. If $h>1$ then by the induction assumption we have (171) for $v=h-1$:

$$\begin{aligned}\sum_{i<j}^{1,\dots,k} G(p_i,p_j)&\geq\sigma_n^2\sum_{i<j}^{1,\dots,h-1} G(z_i,z_j)+\sigma_n\sum_{i=1}^{h-1}\sum_{j=h}^{k} G(z_i,p_j)+\sum_{i<j}^{h,\dots,k} G(p_i,p_j)\\ &=\sigma_n^2\sum_{i<j}^{1,\dots,h-1} G(z_i,z_j)+\sigma_n\sum_{i=1}^{h-1}\sum_{j=h+1}^{k} G(z_i,p_j)+\sum_{i<j}^{h+1,\dots,k} G(p_i,p_j)\\ &\quad+\left\{\sigma_n\sum_{i=1}^{h-1} G(z_i,p_h)+\sum_{j=h+1}^{k} G(p_h,p_j)\right\}.\end{aligned}$$

On applying (174) to { } we obtain

$$\begin{aligned}\sum_{i<j}^{1,\dots,k} G(p_i,p_j)&\geq\sigma_n^2\sum_{i<j}^{1,\dots,h-1} G(z_i,z_j)+\sigma_n\sum_{i=1}^{h-1}\sum_{j=h+1}^{k} G(z_i,p_j)+\sum_{i<j}^{h+1,\dots,k} G(p_i,p_j)\\ &\quad+\left\{\sigma_n^2\sum_{i=1}^{h-1} G(z_i,z_h)+\sigma_n\sum_{j=h+1}^{k} G(z_h,p_j)\right\}\\ &=\sigma_n^2\sum_{i<j}^{1,\dots,h} G(z_i,z_j)+\sigma_n\sum_{i=1}^{h}\sum_{j=h+1}^{k} G(z_i,p_j)+\sum_{i<j}^{h+1,\dots,k} G(p_i,p_j).\end{aligned}$$

This is (171) for $v=h$.

The induction is herewith complete. In particular for $v=k$ in (171)

$$\sigma_n^2\sum_{i<j}^{1,\dots,k} G(z_i,z_j)\leq\sum_{i<j}^{1,\dots,k} G(p_i,p_j).$$

By (81) it now follows that

$$\sigma_n^2\binom{k}{2}\,\rho_k(F_{n+1,m})\leq\sum_{i<j}^{1,\dots,k} G(p_i,p_j).$$

Here $p_1,\dots,p_k\in\Xi_n$ are independent of the left-hand side and we have again by (81)

$$\sigma_n^2\binom{k}{2}\,\rho_k(F_{n+1,m})\leq\binom{k}{2}\,\rho_k(\Xi_n),$$

that is $\sigma_n^2\,\rho_k(F_{n+1,m})\leq\rho_k(\Xi_n)$. On passing to the limit $k\to\infty$ we obtain (170).

12 D. Dirichlet Constant of $F_{n+1,m}$. Let $u_{F_{n+1,m}} \in \mathrm{MI}_\Delta(R)$ with $u_{F_{n+1,m}}|F_{n+1,m}=1$ and $u_{F_{n+1,m}} \in H(R-F_{n+1,m})$. The Dirichlet constant of $F_{n+1,m}$ was defined (cf. 9 K) as

$$\delta(F_{n+1,m}) = \frac{2\pi}{D_R(u_{F_{n+1,m}})}. \tag{175}$$

We assert:

LEMMA. *The approximations $F_{n+1,m}$ of Ξ_n become so small that*

$$\lim_{m\to\infty} \delta(F_{n+1,m}) = \infty. \tag{176}$$

Let $v_m \in \mathrm{MI}_\Delta(R)$ such that $v_m|\bar{U}_{n+1}-R_m=1$ and $v_m \in H(R-(\bar{U}_{n+1}-R_m))$. Then by Theorem III.3 B

$$D_R(u_{F_{n+1,m}}) \le D_R(v_m).$$

In order to prove (176) it is thus sufficient to show that

$$\lim_{m\to\infty} D_R(v_m) = 0. \tag{177}$$

Again by III.3 B

$$D_R(v_m - v_{m+p}) = D_R(v_m) - D_R(v_{m+p}).$$

From III.3 I we have $v_m \ge v_{m+p} > 0$ which shows that $v = BD\text{-}\lim_{m\to\infty} v_m$ exists on R. Clearly $v \in HD(R)$ and since $v_m \in \mathrm{MI}_\Delta(R)$, $v \in \mathrm{MI}_\Delta(R)$. Therefore $v=0$ and we obtain (177).

12 E. Proof of $\tau(\Xi_n)=\rho(\Xi_n)=\infty$. By 9 C, 12 C, and 9 L we conclude that

$$\tau(\Xi_n) \ge \rho(\Xi_n) \ge \sigma_n^2 \rho(F_{n+1,m}) = \sigma_n^2 \delta(F_{n+1,m}).$$

From this and 12 D it follows that $\tau(\Xi_n)=\rho(\Xi_n)=\infty$.

Let Ξ' be any compact subset of Ξ. If $\Xi'=\emptyset$ then (165) is trivial. Suppose $\Xi' \neq \emptyset$. By $G(\cdot, z_0)|\Xi > 0$ we have $G(\cdot, z_0)|\Xi' > 0$. Moreover $G(\cdot, z_0)|\Xi'$ is continuous on Ξ' since $G(\cdot, z_0)$ is so on R^*_{MI}. Therefore $G(\cdot, z_0)|\Xi' \ge r_n$ for some n. As a consequence

$$\Xi' \subset \Xi_n$$

and by 9 A.(81)

$$\rho_k(\Xi') \ge \rho_k(\Xi_n)$$

for every $k=1, 2, \ldots$. We infer that $\rho(\Xi') \ge \rho(\Xi_n) = \infty$, and 9 C gives $\tau(\Xi') \ge \rho(\Xi') = \infty$.

The proof of Theorem 12 A is herewith complete.

12 F. Green's Potential on R_{MI}^*. In 9 we considered the Green's potential $G_\mu(z)=\int G(z,w)\,d\mu(w)$ of a measure μ with $S_\mu\subset R$. We shall now discuss the general case $S_\mu\subset R_{\mathrm{MI}}^*$. However G_μ will be defined primarily on R since for $p\in\Gamma_{\mathrm{MI}}$ we do not in general even know if $G(p,\cdot)|\Gamma_{\mathrm{MI}}$ is measurable.

We are ready to state our main result in this section:

THEOREM. *Let $R\notin O_G$ and suppose its essentially irregular boundary Ξ is nonvoid. Then there exists a positive regular unit Borel measure μ on Ξ with the following properties.*

There exists a sequence $\{q_i\}_{i=1}^\infty$ of points in Ξ and a sequence $\{t_i\}_{i=1}^\infty$ of numbers $t_i>0$ with $\sum_{i=1}^\infty t_i=1$ such that

$$\mu=\sum_{i=1}^{\infty} t_i\,\varepsilon_{q_i} \tag{178}$$

where ε_{q_i} is a unit point measure at q_i. The Green's potential G_μ of μ,

$$G_\mu(p)=\int G(p,q)\,d\mu(q)=\sum_{i=1}^{\infty} t_i\,G(p,q_i), \tag{179}$$

is positive harmonic on R and continuous on $R\cup\Xi\cup\Delta_{\mathrm{MI}}$ with

$$G_\mu|\Xi=\infty \tag{180}$$

and

$$G_\mu|\Delta_{\mathrm{MI}}=0. \tag{181}$$

The function $G_\mu|R\cup\Xi\cup\Delta_{\mathrm{MI}}$ can be continuously extended to all of R_{MI}^*. Moreover

$$D_R(G_\mu\cap c)\le 2\pi c \tag{182}$$

for every positive number $c>0$.

The integral in (179) is defined as the sum in (179). We shall use this conventional notation for measures of the form (178). At this point we remark that

$$\liminf_{p\neq p_0,\,p\to p_0} G_\mu(p)\ge G_\mu(p_0) \tag{183}$$

on R_{MI}^*. In fact let V be a neighborhood of p_0, and n a positive integer. Then $G(p,q_i)\ge \inf_{p\in V-p_0} G(p,q_i)$ on $V-p_0$ implies

$$\inf_{V-p_0}\sum_{i=1}^{\infty} t_i\,G(p,q_i)\ge\sum_{i=1}^{n} t_i\inf_{V-p_0} G(p,q_i).$$

Since $\lim_{V\to p_0}\inf_{V-p_0} G(p,q_i)=G(p_0,q_i)$

$$\liminf_{p\neq p_0,\,p\to p_0} G_\mu(p)\ge\sum_{i=1}^{n} t_i\,G(p_0,q_i).$$

On letting $n\to\infty$ we obtain (183).

This shows that G_μ *is lower semicontinuous* on R^*_{MI}. We shall prove that G_μ is continuous on $R^*_{\mathrm{MI}}-(\Gamma_{\mathrm{MI}}-\Xi-\Delta_{\mathrm{MI}})$. It is not continuous on $\Gamma_{\mathrm{MI}}-\Xi-\Delta_{\mathrm{MI}}$ in general but we shall see that $G_\mu|R\cup\Xi\cup\Delta_{\mathrm{MI}}$ has a continuous extension to R^*_{MI}.

The proof of the theorem will be given in 12 G through 12 I.

12 G. Evans' Potential. We use the notation of 12 B, in particular $\Xi_n=\{q\in\Gamma_{\mathrm{MI}}|G(q,z_0)\geq r_n\}$. Since Ξ_n is compact and contained in Ξ we have by the fundamental theorem 12 A, $\tau(\Xi_n)=\infty$. By virtue of $\tau(\Xi_n)=\lim_{m\to\infty}\tau_m(\Xi_n)$ (cf. 9 B) there exists an increasing sequence $\{m_k\}_{k=1}^\infty$ of positive integers such that $\tau_{m_k}(\Xi_n)>2^k$ for $k=1,2,\ldots$. Because of (84) we can find m_k points $p_{ki}^{(n)}$ $(i=1,\ldots,m_k)$ in Ξ_n such that

$$\inf_{p\in\Xi_n}\sum_{i=1}^{m_k}G(p,p_{ki}^{(n)})>2^k m_k. \tag{184}$$

For

$$\mu_{nk}=\frac{1}{2^k m_k}\sum_{i=1}^{m_k}\varepsilon_{p_{ki}^{(n)}} \tag{185}$$

it follows that

$$G_{\mu_{nk}}(p)>1 \tag{186}$$

for every $p\in\Xi_n$. Again set

$$\mu_n=\sum_{k=1}^\infty\mu_{nk}.$$

By (186) we have for every $p\in\Xi_n$

$$G_{\mu_n}(p)=\sum_{k=1}^\infty G_{\mu_{nk}}(p)=\infty. \tag{187}$$

Observe that μ_n has total measure 1.

Once more define a new measure μ by

$$\mu=\sum_{n=1}^\infty\frac{1}{2^n}\mu_n. \tag{188}$$

It also has total measure 1. By (187)

$$G_\mu(p)=\sum_{n=1}^\infty\frac{1}{2^n}G_{\mu_n}(p)=\infty \tag{189}$$

for $p\in\Xi$. Clearly μ can be written in the form (178). Then for $z\in R$

$$G_\mu(z)=\sum_{i=1}^\infty t_i\,G(z,q_i).$$

Since $G(z,q_i)\in HP(R)$ and $\sum_{i=1}^\infty t_i\,G(z_0,q_i)\leq\sum_{i=1}^\infty t_i\,c_0=c_0<\infty$ with $c_0=\sup_{\Gamma_{\mathrm{MI}}}G(z_0,q)<\infty$ (cf. 8 H) we conclude that $G_\mu|R\in HP(R)$.

Relations (183) and (189) show that G_μ is continuous on $R\cup\Xi$ and satisfies (180).

12 H. Reduction. Set

$$x_n(z) = \sum_{i=1}^{n} t_i\, G(z, q_i). \tag{190}$$

Then clearly

$$C\text{-}\lim_{n\to\infty} x_n(z) = G_\mu(z) \tag{191}$$

on R.

Suppose we can show that

$$D_R(x_n \cap c) \leq 2\pi c \tag{192}$$

for every $c>0$ and every $n=1,2,\ldots$. By 8 H, $x_n|\Delta_{\mathrm{MI}}=0$ and thus $x_n \cap c \in \mathrm{MI}_\Delta(R)$. From (191) we see that $C\text{-}\lim_n x_n \cap c = G_\mu \cap c$. In view of (192) we conclude by Theorem III.1 G that $G_\mu \cap c \in \mathrm{MI}_\Delta(R)$. Therefore $G_\mu|R$ can be continuously extended to R^*_{MI} and this continuous extension vanishes on Δ_{MI}. By 8 H, (181) is obviously valid and we infer that G_μ is continuous on $R \cup \Delta_{\mathrm{MI}}$ and hence on $R \cup \Xi \cup \Delta_{\mathrm{MI}}$. Fatou's lemma together with (192) gives (182).

To complete the proof of Theorem 12 F we have only to establish (192).

12 I. An Estimate. For simplicity we write $x = x_n$,

$$x(z) = \sum_{j=1}^{n} t_j\, G(z, q_j)$$

on R. To prove (192) we set $dy(z) = *dx(z)$ and use $x+iy$ as a local parameter at each point of R except for at most a countable number of isolated points where $dx(z)=0$. Let

$$L(\alpha) = \sum_{j=1}^{n} \int_{x=\alpha} |*dG(z, q_j)|$$

where $\alpha>0$ is such that $dx \neq 0$ on $\{z \in R \mid x(z) = \alpha\}$. Since

$$L(\alpha) = \int_{x=\alpha} \sum_{j=1}^{n} \left| \frac{\partial}{\partial x} G(z, q_j) \right| dy$$

Schwarz's inequality yields

$$\begin{aligned} L(\alpha)^2 &\leq \int_{x=\alpha} \sum_{j=1}^{n} \left(\frac{\partial}{\partial x} G(z, q_j) \right)^2 dy \cdot n \int_{x=\alpha} dy \\ &\leq \int_{x=\alpha} \sum_{j=1}^{n} |\operatorname{grad} G(z, q_j)|^2\, dy \cdot n \int_{x=\alpha} *dx. \end{aligned}$$

Note that except for at most a countable number of $\alpha>0$, $dx \neq 0$ on $\{z\in R|x(z)=\alpha\}$. By Theorem 8 H, $x \cap c \in \mathrm{MI}_\Delta(R)$ for every $c>0$.

On applying Theorem III.3 J to $K=\{p\in R^*_{\mathrm{MI}}|x(p)\geq\alpha\}$ and $u=(x\cap\alpha)/\alpha$ we obtain

$$\alpha^{-2} D_R(x\cap\alpha)=D_R((x\cap\alpha)/\alpha)=\int_{x=\alpha} *d(x/\alpha)$$

which implies

$$D_R(x\cap\alpha)=\alpha\int_{x=\alpha} *dx. \tag{193}$$

If $0<c<\alpha<c'$ with $dx\neq 0$ on $\{z\in R|x(z)=c'\}$ we thus have

$$\int_{x=\alpha} *dx=\alpha^{-1} D_R(x\cap\alpha)\leq c^{-1} D_R(x\cap c')<\infty.$$

Therefore

$$L(\alpha)^2\leq n\,c^{-1} D_R(x\cap c')\int_{x=\alpha}\sum_{j=1}^{n}|\operatorname{grad} G(z,q_j)|^2\,dy \tag{194}$$

for $c<\alpha<c'$. On the other hand

$$\begin{aligned}\int_c^{c'} d\alpha\int_{x=\alpha}\sum_{j=1}^{n}|\operatorname{grad} G(z,q_j)|^2\,dy &= \sum_{j=1}^{n}\iint_{c<x<c'}|\operatorname{grad} G(z,q_j)|^2\,dx\,dy \\ &\leq \sum_{j=1}^{n}\iint_{x<c'}|\operatorname{grad} G(z,q_j)|^2\,dx\,dy.\end{aligned} \tag{195}$$

Since $G(z,q_j)<c'\,t_j^{-1}$ on $\{z\in R|x(z)<c'\}$ for every $j=1,\ldots,n$, Theorem 8 H gives

$$\iint_{x<c'}|\operatorname{grad} G(z,q_j)|^2\,dx\,dy\leq D_R\big(G(z,q_j)\cap c'\,t_j^{-1}\big)\leq 2\pi\,c'\,t_j^{-1}.$$

From this together with (194) and (195) we obtain

$$\int_{c<x<c'} L(x)^2\,dx\leq n\,c^{-1} D_R(x\cap c')\sum_{j=1}^{n}\frac{2\pi\,c'}{t_j}<\infty. \tag{196}$$

Therefore $L(\alpha)<\infty$ and

$$\int_{x=\alpha}|*dG(z,q_j)|<\infty \tag{197}$$

for $j=1,\ldots,n$ and for almost every $c<\alpha<c'$ and hence for almost every $\alpha>0$.

For convenience we say that $\alpha>0$ is *regular* for $x(z)$ if $dx(z)\neq 0$ on $\{z\in R|x(z)=\alpha\}$ and (197) is valid. Every $\alpha>0$ is clearly regular except for those of a set of linear measure zero.

Let $c>0$ be regular for $x(z)$. By (193)

$$D_R(x \cap c) = c \sum_{j=1}^{n} t_j \int_{x=c} *dG(z, q_j). \tag{198}$$

Let $\alpha > c/t_j$ and $dG(z, q_j) \neq 0$ on $\{z \in R | G(z, q_j) = \alpha\}$. Then the interior of $K' = \{p \in R^*_{\mathrm{MI}} | x(p) \geq c\}$ contains $K = \{p \in R^*_{\mathrm{MI}} | G(p, q_j) \geq \alpha\}$. Thus we can again use Theorem III.3 J for K, K', and $(G(z, q_j) \cap \alpha)/\alpha$. Here (197) is valid with α replaced by c. In view of 8 H we now have

$$\int_{x=c} *d\big(\alpha^{-1} G(z, q_j)\big) = \int_{\partial K'} *d\big(\alpha^{-1} G(z, q_j)\big) = \int_{\partial K} *d\big(\alpha^{-1} G(z, q_j)\big)$$
$$= D_{R-K}\big(\alpha^{-1} G(z, q_j)\big) = D_R\big(\alpha^{-1}(G(z, q_j) \cap \alpha)\big) \leq \alpha^{-2} \cdot 2\pi\alpha$$

and therefore

$$\int_{x=c} *dG(z, q_j) \leq 2\pi.$$

This and (198) give $D_R(x \cap c) \leq 2\pi c \sum_{i=1}^{n} t_i \leq 2\pi c$:

$$D_R(x \cap c) \leq 2\pi c \tag{199}$$

for a regular $c>0$. For an arbitrary $c>0$ take regular c_n's with $c<c_n$ and $c_n \to c$ as $n \to \infty$. Then by (199)

$$D_R(x \cap c) \leq D_R(x \cap c_n) \leq 2\pi c_n.$$

On letting $n \to \infty$ we see that (192) holds for every $c>0$.

The proof of Theorem 12 F is herewith complete.

12 J. Irregular Hyperbolic Surfaces. Let R be a hyperbolic Riemann surface and let $g(z, z_0)$ be the Green's function on R with pole $z_0 \in R$. Denote by $\mathscr{T}(R)$ the totality of sequences $\{z_n\}_1^\infty \subset R$ not accumulating in R and with

$$\liminf_{n \to \infty} g(z_n, z_0) > 0. \tag{200}$$

Clearly for a sequence $\{z_n\}_1^\infty$ this condition does not depend on $z_0 \in R$. Therefore the class $\mathscr{T}(R)$ is determined by R.

We call $R \notin O_G$ *regular* if $\mathscr{T}(R) = \emptyset$. It is easy to see that R is regular if and only if $\{z \in R | g(z, z_0) = c\}$ is compact in R for every $c>0$. It is also evident that R is regular if and only if $\Xi = \emptyset$.

A surface $R \notin O_G$ which is not regular is called *irregular*, i.e. $R \notin O_G$ is irregular if and only if $\mathscr{T}(R) \neq \emptyset$.

For example $|z|<1$ is regular hyperbolic whereas $0<|z|<1$ is irregular hyperbolic.

We now state the essence of Theorem 12 F without referring to Royden's compactification:

THEOREM. *On any irregular hyperbolic Riemann surface R there exists a positive harmonic function $u(z)$ with the following properties:*

(u.1) *$u(z)$ is an Evans harmonic function on R, i.e. $u \in HP(R)$ and $\lim_{n\to\infty} u(z_n) = \infty$ for every $\{z_n\} \in \mathcal{T}(R)$,*

(u.2) *$u(z)$ is quasi-Dirichlet finite of the first order, i.e. there exists a constant K_u with $D_R(u \cap c) \leq K_u c$ for every $c > 0$,*

(u.3) *$u(z)$ is singular, i.e. the greatest harmonic minorant of $u \cap c$ on R is identically zero for every $c > 0$.*

Take an exhaustion $\{R_n\}_1^\infty$ of R, a positive number $a > 0$, and a point $\zeta \in R$. Set

$$V(\zeta, a, n) = \{z \in R \mid g(z, \zeta) > a\} - \bar{R}_n. \tag{201}$$

It is clear that property (u.1) is equivalent to the following:

(u.1′) $\lim_{n\to\infty} \inf_{z \in V(\zeta, a, n)} u(z) = \infty$ *for every (ζ, a) such that $V(\zeta, a, 1)$ is not relatively compact in R.*

For the proof of the theorem we take the Green's potential G_μ of Theorem 12 F. We shall show that $u(z) = G_\mu(z)$ is the required function. Clearly $u \in HP(R)$. Property (182) is merely (u.2) and we see that u is continuous on R^*_{MI}. From (181) we know that $u \mid \Delta_{\mathrm{MI}} = 0$ and by III.2 H we obtain (u.3). Property (u.1) follows from (180) and the continuity of $G_\mu \mid R \cup \Xi$.

12 K. Green's Star Region. As an application of Theorem 12 J we shall show that the capacity vanishes for the set on the unit circle which is the image of the end points of singular Green's lines in Green's star region.

Let $\mathbb{G}' = \mathbb{G}'(R, z_0)$ be the Green's star region of an irregular hyperbolic Riemann surface (cf. III.6 B). We now let $\mathbb{E} = \mathbb{E}(R, z_0)$ signify the set of singular Green's lines and use the notation $\mathbb{E}_0$ in an analogously modified sense. Since $\mathbb{G}'$ is a simply connected region it can be mapped onto $U\colon |z| < 1$ by a one-to-one conformal mapping Φ. Let $L \in \mathbb{G}$, i.e. a Green's line issuing from z_0. Then $\Phi(L)$ terminates at a point $p(L, \Phi)$ on $C\colon |z| = 1$. In fact let φ be a branch of $r(p)\, e^{i\theta(p)}$ on $\mathbb{G}'$. Since $\varphi(\Phi^{-1}(z)) \in AB(U)$ has the asymptotic value $(\sup_L r(p))\, e^{i\theta}$ along $\Phi(L)$, $L = L_\theta$, $\Phi(L)$ must by Koebe's theorem terminate at a point in C.

We set

$$E' = E'(R, z_0; \Phi) = \{p(L, \Phi) \in C \mid L \in \mathbb{E}\} \tag{202}$$

and claim:

THEOREM. *The capacity of $E'(R, z_0; \Phi)$ is zero.*

Let $X=\{p(L,\Phi)|L\in\mathbb{E}(R,z_0)-\mathbb{E}_0(R,z_0)\}$ where $\mathbb{E}_0$ is countable (cf. III.6 B). We must show that X is of capacity zero.

We shall make use of the harmonic function u of Theorem 12 J. Let $v(z)=u\circ\Phi^{-1}(z)$ and denote by v^* a conjugate of v on U. Set

$$f(z)=\frac{1}{1+v(z)+i\,v^*(z)}$$

on U. Then $f\in AB(U)$ and $f(z)$ has by (u.1) the asymptotic value zero along $\Phi(L)$, with $L\in\mathbb{E}-\mathbb{E}_0$. Hence by Lindelöf's theorem the set Y of points in C where $f(z)$ has angular limit zero contains X:

$$Y\supset X. \tag{203}$$

Let $\rho>0$, $\Delta(\rho)=\{z\in U\,|\,|f(z)|<\rho\}$ and set $\Delta_n=\Delta(2^{-n}\rho)-\Delta(2^{-n-1}\rho)$. Since $|v|\leq 1/|f|\leq 2^{n+1}/\rho$ on Δ_n we obtain by (u.2)

$$\begin{aligned}D_{\Delta_n}(f)&=\iint_{\Delta_n}|f(z)|^4\,|\mathrm{grad}\,v(z)|^2\,dx\,dy\leq(2^{-n}\rho)^4\,D_{\Delta_n}(v)\\&\leq 2^{-4n}\rho^4\,D_U(v\cap 2^{n+1}\rho^{-1})\leq 2^{-4n}\rho^4\,D_R(u\cap 2^{n+1}\rho^{-1})\\&\leq 2^{-4n}\rho^4\,K\,2^{n+1}\rho^{-1}=2^{-3n+1}K\rho^3.\end{aligned}$$

It follows that

$$D_{\Delta(\rho)}(f)=\sum_{n=0}^{\infty}D_{\Delta_n}(f)\leq 2K\rho^3\sum_{n=0}^{\infty}8^{-n}. \tag{204}$$

Since f is bounded we have for large $\rho>0$, $D_{\Delta(\rho)}(f)=D_U(f)<\infty$. Again (204) implies that

$$\lim_{\rho\to 0}\frac{D_{\Delta(\rho)}(f)}{\rho^2}=0, \tag{205}$$

i.e. 0 is an ordinary value of f in the sense of BEURLING [1]. Therefore by Beurling's theorem Y has capacity zero (see e.g. TSUJI [5; p. 349]) and by (203) the same is true of X.

13. The Parabolic Case

13 A. Evans-Selberg Potential. Let R be an open Riemann surface. It may a priori be parabolic or hyperbolic.

A function $h(z)$ with the following three properties is called an *Evans-Selberg potential* on R with pole $\zeta\in R$:

(a) *h is harmonic on R except at ζ,*

(b) *$h(z)-\log|z-\zeta|$ in a punctured neighborhood about ζ can be harmonically extended to ζ,*

(c) *$h(z)$ tends to ∞ as z tends to the Alexandroff ideal boundary point a_∞ of R.*

We shall now show that the existence of such a function is characteristic of parabolicity.

THEOREM. *An open Riemann surface R belongs to O_G if and only if there exists an Evans-Selberg potential on R.*

The proof will be given in 13 B and 13 C.

13 B. Capacity Functions with Compact Level Lines. Suppose there exists an Evans-Selberg potential $h(z)$ on R with pole $\zeta \in R$. For a sufficiently large M, the region $U = \{z \in R \mid h(z) < -M\}$ is simply connected and relatively compact. Let $R_n = \{z \in R \mid h(z) < n\}$ $(n = 1, 2, \ldots)$. Then by (c), $\{R_n\}_1^\infty$ is an exhaustion of R. Set

$$s_n = h \mid R_n .$$

If we use the local parameter $z = e^{h + ih^*}$ in U, with h^* the conjugate of h, then s_n is the capacity function on R_n (cf. 1) and

$$k_{\beta_n} = s_n \mid \beta_n = h \mid \beta_n = n$$

where $\beta_n = \partial R_n$. Thus

$$h = \lim_{n \to \infty} s_n$$

is the capacity function on R and

$$k_\beta = \lim_{n \to \infty} k_{\beta_n} = \infty .$$

Hence $c_\beta = e^{-k_\beta} = 0$, i.e. $R \in C_\beta = O_G$.

We remark in passing:

COROLLARY. *An Evans-Selberg potential is a capacity function with compact level lines.*

13 C. Existence Proof. Suppose $R \in O_G$. Take a disk U and a slightly larger disk R_0 with $\bar{U} \subset R_0$. Then $R' = R - \bar{U} \notin O_G$. Let $g'(z, \zeta')$ be a Green's function on R'. It is easily seen that $g'(\cdot, \zeta') \mid \partial U = 0$ and

$$\inf_{z \in R - R_0} g'(z, \zeta') > 0$$

for $\zeta' \in R'$. Thus $\mathscr{T}(R') \neq \emptyset$ and actually $\mathscr{T}(R')$ consists of all sequences $\{z_n\}_1^\infty \subset R'$ tending to the Alexandroff ideal boundary point a_∞ of R.

Let u be the function of Theorem 12 J for R':

$$\lim_{z \in R', z \to a_\infty} u(z) = \infty . \tag{206}$$

By Theorem III. 3 B there exists a unique $v \in \mathrm{MI}(R)$ such that $v=u$ on ∂R_0 and $v|R-\bar{R}_0 \in H(R-\bar{R}_0)$. Let

$$t=(u-v)|(R-R_0).$$

Clearly $t \in H(R-R_0)$, $t|\partial R_0=0$, and by virtue of the boundedness of v

$$\lim_{z\in R-R_0, z\to a_\infty} t(z)=\infty. \tag{207}$$

Since $t \in HP(R-\bar{R}_0)$ and t takes its minimum on ∂R_0 we have

$$\int_{\partial(R-\bar{R}_0)} *dt<0.$$

For a suitable constant $a>0$

$$\int_{\partial(R-\bar{R}_0)} *d(at)=-2\pi. \tag{208}$$

The Green's function $g(z,\zeta)$ on U with pole $\zeta \in U$ has the flux

$$\int_{\partial U} *dg(\cdot,\zeta)=-2\pi. \tag{209}$$

Consider the open Riemann surface $R-\zeta$. For a neighborhood V of the ideal boundary we choose $V=(R-R_0)\cup(\bar{U}-\zeta)$. On V we take the function

$$s|R-R_0=at, \quad s|\bar{U}-\zeta=-g(\cdot,\zeta).$$

It has vanishing flux,

$$\int_{\partial V} *ds=0.$$

By Theorem I. 7 A there exists a harmonic function h on $R-\zeta$ such that $L(h-s)=h-s$ on V. Here $(h-s)|V$ is bounded and h has the required properties (a), (b), and (c).

13 D. Positive Singularities. Throughout 13 D – 13 H we denote by R a parabolic open Riemann surface. Let $\tilde{R}$ be the Alexandroff compactification and a_∞ the point at infinity, i.e. $\tilde{R}=R\cup a_\infty$.

A (normalized) *positive singularity* l_q at $q\in\tilde{R}$ is a positive harmonic function in a punctured open neighborhood $V(l_q)\subset R$ (i.e. $V(l_q)\cup q$ is an open neighborhood of q in $\tilde{R}$) with the following properties:

$$\lim_{p\in V(l_q), p\to q} l_q(p)=\infty \tag{210}$$

and

$$\int_\alpha *dl_q=-2\pi \tag{211}$$

for some (and hence for every) analytic cycle $\alpha \subset V(l_q)$ which is the boundary of a neighborhood of q and is positively oriented with respect to this neighborhood.

Two singularities l_q and l'_q at $q \in \tilde{R}$ are said to be *equivalent* if $l_q - l'_q$ is bounded in a punctured neighborhood of q.

LEMMA. *There exists a positive singularity l_q for every $q \in \tilde{R}$. All l_q are equivalent for a fixed $q \in R$.*

For the proof let $q \in R$ and let $\{U, z\}$ be a parametric disk at q, i.e. U is a neighborhood of q and z a conformal mapping of U onto $|z|<1$ with $z(q)=0$. Then $l_q(p)=\log(1/|z(p)|)$ on $V(l_q)=U-q$ is a positive singularity at q.

Let l'_q be another positive singularity at q. Denote by $p=p(z)$ the inverse mapping of $z=z(p)$ and assume that $l'_q(p(z))$ is defined and positive on $0<|z|<r$ $(0<r<1)$. Let $*\,l'_q(p(z))$ be the multivalued conjugate of $l_q(p(z))$ on $0<|z|<r$ and consider $f(z)=\exp(-l'_q(p(z))-i*l'_q(p(z)))$. In view of (211), $\int_\alpha * dl'_q = -2\pi$ with $\alpha=p(|z|=r')$ and $0<r'<r$. Hence $f(z)$ is single-valued on $0<|z|<r$. It is also bounded since $l'_q(p(z))>0$. Therefore $f(z)$ can be continued to all of $|z|<r$ and there exists a bounded analytic function $\varphi(z)$ on $|z|<r$ with $\varphi(0) \neq 0$ and $f(z)=z^n \varphi(z)$ for some positive integer n. It follows that

$$l'_q(p(z)) = -\log|f(z)| = -n\log|z| - \log|\varphi(z)|.$$

Clearly $\log|\varphi(z)|$ is harmonic on some $|z|<r''$ $(0<r''<r)$, and thus (211) implies that $n=1$. Therefore $l'_q - l_q$ is bounded in a neighborhood of q.

The existence of a positive singularity l_{a_∞} at a_∞ follows from 13 A.

There can exist several nonequivalent singularities at a_∞. For example take $R=\{0<|z|<\infty\}$ and let $l^\lambda_{a_\infty}(z)=-\lambda\log|z|$ $(0<|z|<1)$ and $(1-\lambda)\log|z|$ $(|z|>1)$. Since $\{0<|z|<1\} \cup \{1<|z|<\infty\}$ is a neighborhood of the Alexandroff point a_∞ at infinity for R, all $l^\lambda_{a_\infty}$ $(0<\lambda<1)$ are positive singularities at a_∞, but the $l^\lambda_{a_\infty} - l^{\lambda'}_{a_\infty}$ are not bounded in any neighborhood of a_∞ if $\lambda \neq \lambda'$.

13 E. Evans' Kernel. The logarithmic kernel $\log(1/|z-\zeta|)$ on the plane $P: |z|<\infty$ is a harmonic function of z on $P-\zeta$. It possesses positive and negative singularities at ζ and ∞ respectively and is symmetric on $P \times P$. With this in mind we generalize the logarithmic kernel to an arbitrary open Riemann surface with null boundary as follows:

DEFINITION. *An Evans kernel $e(p,q)$ on R is a mapping of $R \times R$ onto $(-\infty, \infty]$ satisfying the following conditions:*

(a) *$e(p,q)$ is harmonic in p on $R-q$,*

(b) *$e(p,q)$, as a function of p, is a positive singularity at q,*

(c) $-e(p,q)$, *as a function of* p, *is a positive singularity at* a_∞, *and* $-e(p,q)$ *and* $-e(p,q')$ *are equivalent for every pair* $(q,q')\in R\times R$,

(d) $e(p,q)$ *is symmetric, i.e.* $e(p,q)=e(q,p)$ *on* $R\times R$.

Condition (b) means that there exists a positive singularity l_q at q such that

$$e(p,q)=l_q(p)+h_q(p) \tag{212}$$

in a punctured neighborhood $V(l_q)$ of q, where h_q is a harmonic function on $V(l_q)\cup q$. Since l_q is unique up to equivalence, (212) has a definite meaning. Condition (c) requires the existence of a positive singularity l_{a_∞} at a_∞ *independent of* q such that $\sup_{R-K_q} h_q(p)<\infty$ and

$$e(p,q)=-l_{a_\infty}(p)+h_q(p) \tag{213}$$

on the complement of a compact set $K_q\subset R$. Since there can exist two or more nonequivalent positive singularities, $e(p,q)$ depends essentially on l_{a_∞}. For this reason we prefer to call $e(p,q)$ an l_{a_∞}-*Evans kernel* to indicate the dependence on l_{a_∞}.

We are ready to state:

THEOREM. *On an arbitrary parabolic open Riemann surface* R *there exists an* l_{a_∞}-*Evans kernel which is unique up to an additive constant.*

The existence of a function $\mathfrak{P}$ satisfying (a), (b), and (c) was obtained in 13 A. Thus the problem is to find a suitable function $k(p)$ on R with

$$\mathfrak{P}(p,q)+k(q)=\mathfrak{P}(q,p)+k(p)$$

for every $p,q\in R$. Instead of seeking such a $k(p)$ however we shall in 13 F prove the theorem indirectly.

13 F. Proof. Let q_0 be an arbitrary but then fixed point in R. Consider open sets

$$R_n=\{p\in R\,|\,\mathfrak{P}(p,q_0)>-n\} \tag{214}$$

for each positive integer n. By (c) we conclude that $\bar{R}_n$ is compact in R. From the maximum principle we also infer that R_n is connected. Clearly the relative boundary ∂R_n of R_n consists of a finite number of piecewise analytic Jordan curves. The sequence $\{R_n\}_1^\infty$ is a regular exhaustion of R.

Let $g_n(p,q)$ be the Green's kernel on R_n (cf. 8). Consider the kernel $u_n(p,q)$ on R_n defined by

$$u_n(p,q)=g_n(p,q)-n. \tag{215}$$

Since $R\in O_G$ the increasing sequence $\{g_n(p,q)\}_1^\infty$ diverges to ∞. However we shall show that

$$e(p,q)=\lim_{n\to\infty} u_n(p,q) \tag{216}$$

exists on $R \times R$ and is an l_{a_∞}-Evans kernel; the convergence is uniform on $K \times \{q\}$ for every $q \in R$ and every compact set $K \subset R - q$.

Let $q \in R$. By (c) there exists an integer $n(q)$ such that

$$|\mathfrak{P}(p, q) - \mathfrak{P}(p, q_0)| < \tfrac{1}{2} c(q)$$

on $R - R_{n(q)}$, where $c(q)$ is a finite constant depending only on q. For $n \geq n(q)$, $\mathfrak{P}(p, q) - u_n(p, q)$ is harmonic on R_n and $\mathfrak{P}(p, q_0) - u_n(p, q) = 0$ on ∂R_n. Hence

$$|\mathfrak{P}(p, q) - u_n(p, q)| < \tfrac{1}{2} c(q) \tag{217}$$

for every $p \in R_n$ and a fortiori

$$|u_{n+m}(p, q) - u_n(p, q)| < c(q) \tag{218}$$

for every $p \in R_n$ with $n \geq n(q)$ and $m \geq 1$. We conclude that for an arbitrarily fixed $q \in R$ there exists a subsequence of $\{u_n(p, q)\}_1^\infty$ which is uniformly convergent on each compact subset of $R - q$.

Let D be a countable dense subset of R. Using Cantor's diagonal process we can find a subsequence $\{n_k\}_{k=1}^\infty$ of $\{n\}_1^\infty$ such that for each fixed $q \in D$, $\{u_{n_k}(p, q)\}_{k=1}^\infty$ converges to a harmonic function, say $e_q(p)$, uniformly on each compact subset of $R - q$.

Next fix p arbitrarily in R. By (215), $u_n(p, q) = u_n(q, p)$ and thus $|u_{n+m}(p, q) - u_n(p, q)| = |u_{n+m}(q, p) - u_n(q, p)|$. Hence we obtain by (218)

$$|u_{n+m}(p, q) - u_n(p, q)| < c(p)$$

for every $q \in R_n$ with $n \geq n(p)$ and $m \geq 1$. Fix k_0 with $n_{k_0} \geq n(p)$. Then $\{u_{n_k}(p, q) - u_{n_{k_0}}(p, q)\}_{k_0}^\infty$ is a uniformly bounded sequence of functions harmonic in q and converges on the dense subset D of R. Consequently by Harnack's convergence theorem $\{u_{n_k}(p, q) - u_{n_{k_0}}(p, q)\}_{k_0}^\infty$, and a fortiori $\{u_{n_k}(p, q)\}_{k_0}^\infty$, converges uniformly on each compact subset of $R - p$. The function

$$h_p(q) = \lim_{k \to \infty} u_{n_k}(p, q)$$

is harmonic on $R - \{p\}$.

We conclude that

$$e(p, q) = \lim_{k \to \infty} u_{n_k}(p, q) \tag{219}$$

exists for every $(p, q) \in R \times R$. Again by (218), $|e(p, q) - u_{n_k}(p, q)| < c(q)$. By Harnack's theorem the convergence in (219) is uniform on $K \times \{q\}$ for every $q \in R$ and every compact set $K \subset R - q$. Because of $u_n(p, q) = u_n(q, p)$, $e(p, q)$ satisfies (d). In view of (219), (a) is clearly met. From (217) it follows that

$$|\mathfrak{P}(p, q) - e(p, q)| < \tfrac{1}{2} c(q) \tag{220}$$

for every $p \in R$. Since $\mathfrak{P}(p, q)$ satisfies (212) and (213), $e(p, q)$ has properties (b) and (c). Therefore $e(p, q)$ is an l_{a_∞}-Evans kernel.

Finally we prove that (219) implies (216). Assume to the contrary that (216) is not valid. Let $\{v_k\}_{k=1}^{\infty}$ be the complementary subsequence of $\{n_k\}_{k=1}^{\infty}$, i.e. $\{v_k\}_{k=1}^{\infty} \cup \{n_k\}_{k=1}^{\infty} = \{n\}_1^{\infty}$. Since $\{u_{v_k}\}_{k=1}^{\infty}$ does not converge to $e(p, q)$ on $R \times R$ there exists a point $(p_1, q_1) \in R \times R$ and a subsequence $\{\mu_k\}_{k=1}^{\infty}$ of $\{v_k\}_{k=1}^{\infty}$ such that

$$\lim_{k \to \infty} u_{\mu_k}(p_1, q_1) \neq e(p_1, q_1) \tag{221}$$

exists. Now $\{u_{\mu_k}\}_{k=1}^{\infty}$ satisfies (217) and we can find in the same manner as above a subsequence $\{m_k\}_{k=1}^{\infty}$ of $\{\mu_k\}_{k=1}^{\infty}$ and an l_{a_∞}-Evans kernel $e'(p, q)$ on R such that (219) is valid for $e'(p, q)$ and $\{u_{m_k}\}_{k=1}^{\infty}$. By (221), $e(p_1, q_1) \neq e'(p_1, q_1)$.

On the other hand (220) is also true for $e'(p, q)$ and thus $|e(p, q) - e'(p, q)| < c(q)$ for every $p \in R$. Therefore $p \to e(p, q) - e'(p, q)$ is a bounded harmonic function on R and consequently a constant $a(q)$.

By symmetry, $q \to e(p, q) - e'(p, q) = a(q)$ is also a bounded harmonic function and $a(q)$ is in turn a constant a, i.e. $e(p, q) = e'(p, q) + a$ on $R \times R$. Since $u_n(p, q_0) = \mathfrak{P}(p, q_0)$, $e(p, q_0) = e'(p, q_0)$ on R and therefore $a = 0$. In particular $e(p_1, q_1) = e'(p_1, q_1)$, a contradiction.

To complete the proof of Theorem 13 E it remains to establish the uniqueness of the l_{a_∞}-Evans kernel up to an additive constant. Let $e(p, q)$ and $e'(p, q)$ be l_{a_∞}-Evans kernels. Consider the difference $E(p, q) = e(p, q) - e'(p, q)$. By (b) and (c), $p \to E(p, q)$ is a bounded harmonic function on R and so is $q \to E(p, q)$. As above we conclude that $E(p, q)$ is constant.

13 G. Joint Continuity of Evans' Kernel. From a potential-theoretic point of view it is important that the logarithmic kernel $\log(1/|z - \zeta|)$ be continuous on $P \times P = \{(z, \zeta) \,|\, |z|, |\zeta| < \infty\}$ in the extended sense. The joint continuity of Green's kernel was proved in 8 E. We can also deduce the corresponding property for Evans' kernel:

THEOREM. *Evans' kernel $e(p, q)$ on R is jointly continuous, i.e. e is a continuous mapping of $R \times R$ onto $(-\infty, \infty]$.*

Specifically, $e(p, q)$ is finitely continuous on $R \times R$ outside of the diagonal set, and for any relatively compact subregion $V \subset R$ the decomposition

$$e(p, q) = g_V(p, q) + v_V(p, q) \tag{222}$$

is valid on $V \times V$. Here g_V is the Green's kernel on V and v_V is a finitely continuous function on $V \times V$.

Let $q_0 \in R$ and take a relatively compact subregion V of R, $q_0 \in V$, such that the relative boundary ∂V of V consists of a finite number of piecewise analytic Jordan curves. For $q \in V$ set

$$d(q)=d(q; q_0, V)=\max_{p\in\partial V}|e(p,q)-e(p,q_0)|.$$

Observe that by (c), $e(p,q)-e(p,q_0)$ is a bounded harmonic function of p on $R-V$. Since $R\in O_G$, $e(p,q)-e(p,q_0)$ takes its maximum and minimum on ∂V and therefore

$$|e(p,q)-e(p,q_0)|\leq d(q) \tag{223}$$

for every $p\in R-V$.

First we show that

$$\lim_{q\to q_0} d(q)=0. \tag{224}$$

If this were not the case then there would exist a sequence $\{q_n\}_1^\infty\subset V$ such that $\lim_n q_n=q_0$, $d(q_n)>0$, and $\lim_n d(q_n)>0$. Let W be a subregion of R of the same type as V and such that $\{q_n\}_0^\infty\subset W\subset\overline{W}\subset V$. The function $v_n(p)=\big(e(p,q_n)-e(p,q_0)\big)/d(q_n)$ is harmonic on $R-W$ and, by (223), $|v_n(p)|\leq 1$ for $p\in R-V$. Let g_V be the Green's kernel on V and set $\varphi_n(p)=\big(g_V(p,q_n)-g_V(p,q_0)\big)/d(q_n)$. The functions $\varphi_n(p)+1-v_n(p)$ and $-\varphi_n(p)+1+v_n(p)$ are harmonic and nonnegative on V. Therefore

$$-1+\min_{\partial W}\varphi_n(p)\leq v_n(p)\leq 1+\max_{\partial W}\varphi_n(p) \tag{225}$$

on $V-W$ and consequently on $R-W$.

Since $g_V(p,q)$ is continuous on $V\times V$ and $\{1/d(q_n)\}_1^\infty$ is bounded the bounds of v_n in (225) tend to finite limits. It follows that $\{v_n\}_1^\infty$ is a sequence of uniformly bounded harmonic functions on $R-W$. Let $p\in R-W$ be arbitrary but fixed for the present. Since $e(p,q_n)\to e(p,q_0)$ $(n\to\infty)$ and $\{1/d(q_n)\}_1^\infty$ is bounded, $\lim_n v_n(p)=0$. Thus $\{v_n\}_1^\infty$ converges to zero uniformly on each compact subset of $R-W$ and in particular on ∂V. This is impossible since $\max_{p\in\partial V}|v_n(p)|=d(q_n)/d(q_n)=1$. Hence (224) must be valid.

Let $(p_0,q_0)\in R\times R$ with $p_0\neq q_0$ and choose V such that $p_0\notin\overline{V}$, $q_0\in V$. For $p\notin\overline{V}$ and $q\in V$ it follows from (223) that

$$|e(p,q)-e(p_0,q_0)|\leq|e(p,q_0)-e(p_0,q_0)|+d(q;q_0,V).$$

By (224) and $\lim_{p\to p_0} e(p,q_0)=e(p_0,q_0)$ we conclude that

$$\lim_{(p,q)\to(p_0,q_0)} e(p,q)=e(p_0,q_0),$$

i.e. $e(p,q)$ is finitely continuous on $R\times R$ outside of the diagonal set.

Let

$$v(p,q)=e(p,q)-g_V(p,q) \tag{226}$$

on $V\times V$. From what we have seen we deduce that $v(p,q)$ is finitely continuous on $V\times V$ outside of the diagonal set. Let $p_0\in V$ and denote by W an open neighborhood of p_0 with $\overline{W}\subset V$. Given $\varepsilon>0$ we can find an open neighborhood U of p_0 such that $\overline{U}\subset W$ and

$$v(p,p_0)-\varepsilon<v(p,q)<v(p,p_0)+\varepsilon \tag{227}$$

for every $(p,q)\in(\partial W)\times U$. For an arbitrarily fixed $q\in U$ the positive singularities cancel in (227), and the functions of p on $\overline{W}$ are harmonic. Since (227) is valid on ∂W the maximum principle gives it on $\overline{W}\times U$. Thus it holds in particular for every $U\times U$ and therefore

$$|v(p,q)-v(p_0,q_0)|\le|v(p,p_0)-v(p_0,p_0)|+2\varepsilon.$$

Since $\lim_{p\to p_0} v(p,p_0)=v(p_0,p_0)$, we have $\lim_{(p,q)\to(p_0,q_0)} v(p,q)=v(p_0,q_0)$. As a consequence e is the sum of a finitely continuous function v and the Green's kernel which is also continuous on $V\times V$.

Remark. The function $d(q)$ and its use similar to that in the above proof is due to HEINS [3].

13 H. Joint Uniform Convergence. As a statement complementary to 1 C, 13 B, and 13 F we shall prove:

THEOREM. *Let $e(p,q)$ be an Evans kernel on R, and $g_\lambda(p,q)$ the Green's kernel on $R_\lambda=\{p\,|\,p\in R,\ e(p,q_0)>-\lambda\}$ with a fixed $q_0\in R$. Then*

$$e(p,q)=\lim_{\lambda\to\infty}\big(g_\lambda(p,q)-\lambda\big) \tag{228}$$

uniformly on each compact subset of $R\times R$, i.e.

$$\lim_{\lambda\to\infty}\ \sup_{(p,q)\in K\times K}\big|e(p,q)-\big(g_\lambda(p,q)-\lambda\big)\big|=0$$

for compact sets $K\subset R$.

In a manner similar to that in the proof 13 F we can show that $e'(p,q)=\lim_{\lambda\to\infty}\big(g_\lambda(p,q)-\lambda\big)$ exists on $R\times R$ and $e'(p,q)$ is an Evans kernel such that $p\to e'(p,q)$ gives a positive singularity at q equivalent to that of $p\to e(p,q)$. Moreover the convergence is uniform on $K\times\{q\}$ with an arbitrary $q\in R$ and an arbitrary compact set $K\subset R-q$. Since $p\to e(p,q)-e'(p,q)$ is bounded and harmonic on R we see as in 13 F that $e(p,q)-e'(p,q)$ is constant on $R\times R$. Moreover $e(p,q_0)=g_\lambda(p,q_0)-\lambda=e'(p,q_0)$ on R_λ and we conclude that $e(p,q)=e'(p,q)$ on $R\times R$, i.e. identity (228) is valid.

To prove the uniform convergence let $w_\lambda(p,q)=e(p,q)-\big(g_\lambda(p,q)-\lambda\big)$ on R_λ. Fix an arbitrary $\lambda_0>0$ and let $\lambda>\lambda_0$. For an arbitrarily fixed $q\in\overline{R}_{\lambda_0}$, $p\to w_\lambda(p,q)$ is harmonic on $\overline{R}_\lambda$ and for $p\in\partial R_\lambda$, $w_\lambda(p,q)=e(p,q)-$

$(g_\lambda(p, q_0)-\lambda)=e(p, q)-e(p, q_0)$. Therefore $|w_\lambda(p, q)| \le \max_{p\in\partial R_\lambda} |e(p, q)-e(p, q_0)|$ for $p\in R_\lambda$ and a fortiori

$$|w_\lambda(p, q)| \le \max_{(p, q)\in(\partial R_\lambda)\times \bar{R}_{\lambda_0}} |e(p, q)-e(p, q_0)| \tag{229}$$

for every $(p, q)\in R_\lambda \times \bar{R}_{\lambda_0}$. By Theorem 13 G, $|e(p, q)-e(p, q_0)|$ is finitely continuous on $(\partial R_\lambda)\times \bar{R}_{\lambda_0}$ and thus

$$M_\lambda = \max_{(p, q)\in(\partial R_\lambda)\times \bar{R}_{\lambda_0}} |e(p, q)-e(p, q_0)| < \infty. \tag{230}$$

In view of 13 E.(c), $p\to e(p, q)-e(p, q_0)$ is a bounded harmonic function on $R-R_\lambda$ for each fixed $q\in \bar{R}_{\lambda_0}$. It follows from $R\in O_G$ that

$$|e(p, q)-e(p, q_0)| \le \max_{p\in\partial R_\lambda} |e(p, q)-e(p, q_0)| \le M_\lambda$$

for every $(p, q)\in(R-R_\lambda)\times \bar{R}_{\lambda_0}$. In particular

$$M_{\lambda'} \le M_\lambda \tag{231}$$

for all $\lambda'>\lambda$. Therefore by (229), (230), and (231) there exist a finite constant M and a $\lambda_1\in(\lambda_0, \infty)$ such that

$$|w_\lambda(p, q)| \le M \tag{232}$$

for every $(p, q)\in \bar{R}_{\lambda_0}\times \bar{R}_{\lambda_0}$ and $\lambda>\lambda_1$.

Set $f_\lambda(p, q)=w_\lambda(p, q)+M$. Then

$$0 \le f_\lambda(p, q) \le 2M \tag{233}$$

on $\bar{R}_{\lambda_0}\times \bar{R}_{\lambda_0}$. Hence $p\to f_\lambda(p, q)$ and $q\to f_\lambda(p, q)$ are nonnegative harmonic functions on R_{λ_0}. As a consequence

$$\begin{aligned} k(p, p')^{-1} f_\lambda(p', q) &\le f_\lambda(p, q) \le k(p, p')\, f_\lambda(p', q), \\ k(q, q')^{-1} f_\lambda(p', q') &\le f_\lambda(p', q) \le k(q, q')\, f_\lambda(p', q') \end{aligned} \tag{234}$$

for arbitrary points p, p', q, and q' in R_{λ_0}. Here for $(s, t)\in R_{\lambda_0}\times R_{\lambda_0}$, $k(s, t)$ is the Harnack function given in III.4 C:

$$k(s, t)=k_{R_{\lambda_0}}(s, t)=\inf\{c \mid c^{-1}h(s)\le h(t)\le c\,h(s) \text{ for every } h\in HP(R_{\lambda_0})\}.$$

Recall that

$$1\le k(s, t)<\infty, \qquad \lim_{s\to t} k(s, t)=1. \tag{235}$$

By (233) and (234) we obtain

$$|f_\lambda(p, q)-f_\lambda(p', q')| \le 2M\big(k(p, p')\,k(q, q')-1\big)$$

and therefore

$$|w_\lambda(p, q)-w_\lambda(p', q')| \le 2M\big(k(p, p')\,k(q, q')-1\big) \tag{236}$$

for every (p, q) and (p', q') in $R_{\lambda_0} \times R_{\lambda_0}$. From (235) and (236) it follows that the family $\{w_\lambda(p, q)\}_{\lambda > \lambda_1}$ is equicontinuous on $R_{\lambda_0} \times R_{\lambda_0}$. As a consequence the convergence $\lim_{\lambda \to \infty} w_\lambda(p, q) = 0$ on $R_{\lambda_0} \times R_{\lambda_0}$ outside of the diagonal implies the uniform convergence $\lim_{\lambda \to \infty} w_\lambda(p, q) = 0$ on $\bar{R}_{\lambda_0/2} \times \bar{R}_{\lambda_0/2}$.

13 I. The *s*-Kernel. The most important potential-theoretic kernel on the extended plane $\tilde{P}$: $|z| \leq \infty$ is the *elliptic kernel* $\log(1/[z, \zeta])$ where $[z, \zeta] = |z - \zeta| / (\sqrt{1 + |z|^2}\sqrt{1 + |\zeta|^2})$. For $s(z, \zeta) = \log(1/[z, \zeta])$ and $e(z, \zeta) = \log(1/|z - \zeta|)$ we observe that

$$s(z, \zeta) = \tfrac{1}{2} \log[(1 + e^{-2e(z, 0)})(1 + e^{-2e(\zeta, 0)})] + e(z, \zeta). \tag{237}$$

In view of this relation the most natural generalization of the elliptic kernel to an arbitrary closed surface R is

$$s(p, q) = \tfrac{1}{2} \log[(1 + e^{-2e(p, a)})(1 + e^{-2e(q, a)})] + e(p, q) \tag{238}$$

where $(p, q) \in R \times R$, a is an arbitrary but fixed point in R, a_∞ a point in R different from a, and $e(p, q)$ an Evans kernel on $R - a_\infty$.

For an open Riemann surface $R \in O_G$ the kernel $s(p, q)$ can also be defined by (238), with a_∞ taken as the Alexandroff point at infinity of R.

Even if $R \notin O_G$, we can maintain the form of (237):

$$s(p, q) = \tfrac{1}{2} \log[(1 + e^{-2g(p, a)})(1 + e^{-2g(q, a)})] + g(p, q), \tag{239}$$

where $g(p, q)$ is the Green's kernel on R.

The kernel $s(p, q)$ so defined on an arbitrary Riemann surface R possesses most of the important properties of the elliptic kernel and thus may be regarded as a generalization of the latter. It satisfies the following conditions:

(a) *$s(p, q)$ is bounded from below on $R \times R$,*

(b) *$s(p, q) = s(q, p)$ on $R \times R$,*

(c) *$\Delta_p s(p, q)$ exists on $R - \{a, q\}$, is continuously extendable to R, and the associated 2-form is independent of q,*

(d) *for every relatively compact subregion $\Omega \subset R$ with $\Omega \notin O_G$ there exists a finitely continuous function $v_\Omega(p, q)$ on $\Omega \times \Omega$ such that*

$$s(p, q) = g_\Omega(p, q) + v_\Omega(p, q) \tag{240}$$

on $\Omega \times \Omega$ where $g_\Omega(p, q)$ is the Green's kernel on Ω.

A function with properties (a)–(d) is called the *s-kernel* on R. We shall prove (a)–(d) in 13 J.

13 J. Proof. Properties (b) and (c) are direct consequences of (238) and (239). For an open R, (d) is implied by (222) and the very definition of $s(p, q)$. For a closed R it suffices to consider the case in which Ω is a punctured parametric disk at a_∞ and $a \notin \bar{\Omega}$. Observe that there is only one positive singularity $g_\Omega(p, a_\infty)$ at a_∞ up to equivalence.

Let $v(p,q) = e(p,q) + g_\Omega(p, a_\infty) + g_\Omega(q, a_\infty) - g_\Omega(p,q)$. Both $p \to v(p,q)$ and $q \to v(p,q)$ are harmonic on $\Omega \cup \partial\Omega$. Clearly $v(p,q) \geq \min_{(p,q) \in (\partial\Omega) \times (\partial\Omega)} v(p,q)$ for every $(p,q) \in \Omega \times \Omega$. Since $v(p,q) = e(p,q)$ $(> -\infty)$ is continuous on $(\partial\Omega) \times (\partial\Omega)$ there exists a constant c such that $v(p,q) \geq c > -\infty$. As in the proof of (236) we obtain

$$|v(p,q) - v(p',q')| \leq |v(p',q') - c| \left(k_\Omega(p,p')\, k_\Omega(q,q') - 1\right).$$

Thus $v(p,q)$ is finitely continuous on $\Omega \times \Omega$ and (240) follows.

It remains to prove (a). We deal only with an open $R \in O_G$. If R is closed then it suffices to consider $R - a_\infty$. For $R \notin O_G$ the same procedure as below with the replacement of $e(p,q)$ by $g(p,q)$ and with an obvious modification gives the proof.

Take a relatively compact subregion V of R containing a. Set $A_1 = \inf_{(p,q) \in \bar{V} \times \bar{V}} s(p,q)$, $A_2 = \inf_{(p,q) \in \bar{V} \times (R - \bar{V})} s(p,q) = \inf_{(p,q) \in (R - \bar{V}) \times \bar{V}} s(p,q)$, and $A_3 = \inf_{(p,q) \in (R - \bar{V}) \times (R - \bar{V})} s(p,q)$. We must show that $A_i > -\infty$ $(i = 1, 2, 3)$.

In general $s(p,q) > e(p,q) > -\infty$. Since $e(p,q)$ is continuous on $\bar{V} \times \bar{V}$ we have $A_1 \geq \min_{(p,q) \in \bar{V} \times \bar{V}} e(p,q) > -\infty$.

Next consider the case $(p,q) \in (R - \bar{V}) \times \bar{V}$. Clearly

$$s(p,q) > e(p,q) - e(p,a) \equiv w(p,q).$$

By 13 E. (c), $p \to w(p,q)$ is bounded and harmonic on $R - \bar{V}$. Since $R \in O_G$

$$w(p,q) \geq \min_{p \in \partial V} w(p,q)$$

for every $(p,q) \in (R - \bar{V}) \times \bar{V}$. The function $w(p,q)$ $(> -\infty)$ is continuous on $(\partial V) \times \bar{V}$ and thus

$$w(p,q) \geq \min_{p \in \partial V} w(p,q) \geq \min_{(p,q) \in (\partial V) \times \bar{V}} w(p,q) > -\infty$$

for all $(p,q) \in (R - \bar{V}) \times \bar{V}$. Therefore $A_2 \geq \min_{(p,q) \in (\partial V) \times \bar{V}} w(p,q) > -\infty$.

Finally let $(p,q) \in (R - \bar{V}) \times (R - \bar{V})$ and observe that

$$s(p,q) > e(p,q) - e(p,a) - e(q,a) \equiv v(p,q).$$

Because of 13 E. (c), $p \to v(p,q)$ is bounded in a punctured neighborhood of a_∞. Moreover it is harmonic on $R - \bar{V} - q$, and $v(q,q) = \infty$. By $R \in O_G$

we infer that

$$v(p, q) \geq \min_{p \in \partial V} v(p, q) \tag{241}$$

for every $(p, q) \in (R - \bar{V}) \times (R - \bar{V})$. Fix p arbitrarily in ∂V. In the same manner as above the minimum principle applied to the harmonic function $q \to v(p, q)$ gives

$$v(p, q) \geq \min_{q \in \partial V} v(p, q) \tag{242}$$

for every $(p, q) \in (\partial V) \times (R - \bar{V})$. From (241) and (242) it follows that

$$v(p, q) \geq \min_{(p, q) \in (\partial V) \times (\partial V)} v(p, q)$$

for all $(p, q) \in (R - \bar{V}) \times (R - \bar{V})$. Again since $v(p, q)$ $(> -\infty)$ is continuous on $(\partial V) \times (\partial V)$ we conclude that

$$A_3 \geq \min_{(p, q) \in (\partial V) \times (\partial V)} v(p, q) > -\infty.$$

Chapter VI

Functions with Iversen's Property

Once more we return to classes of analytic functions. We mainly consider those with Iversen's property. The most important consequence of this property is Stoïlow's principle: the cluster set at a Stoïlow ideal boundary point is either a point or total. Thus naturally Iversen's property closely relates to continuity on Stoïlow's compactification.

In this chapter we therefore consider Stoïlow's compactification as the fundamental space. In § 1 we discuss O_{A^0D} and O_{A^0B} from the viewpoint of continuity of analytic functions on Stoïlow's boundary.

We then study in § 2 relations between these classes and null classes discussed in connection with HB- and HD-minimal functions. The section closes with a complete string of strict inclusion relations for null classes considered in our book.

§ 1. Classes O_{A^0D} and O_{A^0B}

In 1 we shall prove the so-called Stoïlow principle for functions with Iversen's property. It will mainly be used in a neighborhood of each Stoïlow ideal boundary point. For this reason Iversen's property, and consequently Stoïlow's principle itself, will be presented in a localized form.

The class O_{A^0D} introduced in Chapter I is considered in 2. Every function in the new class MD^* of meromorphic functions with finite spherical Dirichlet integrals is seen to have the localizable Iversen property on O_{A^0D}-surfaces. In particular AD-functions are continuous in any neighborhood of Stoïlow's ideal boundary of such surfaces.

The class O_{A^0B} briefly discussed in Chapter II is further studied in 3. It will be characterized in terms of Lindelöf's maximum principle. On O_{A^0B}-surfaces every meromorphic function has the localizable Iversen property. In particular the exceptional set of any meromorphic function on such a surface is a countable union of Painlevé null sets. Hence all AB- and also MD^*-functions are continuous on every neighborhood of the Stoïlow ideal boundary.

Tests for O_{A^0D} were given in Chapter I. We now present O_{A^0B}-tests in 3. These are used to exhibit some examples related to O_{A^0D} and O_{A^0B}, and we determine the position of these classes in the classification scheme. Here again generalized Cantor sets play an important role.

1. Iversen's Property

1 A. Functions with Iversen's Property. Consider a noncompact bordered Riemann surface (R, α) with compact border α. Here and hereafter we include the case $\alpha = \emptyset$, i.e. we consider an open Riemann surface R as a bordered surface $(R, \emptyset)$ with degenerate border $\emptyset$.

A function f in the class $M(R \cup \alpha)$ of (single-valued) meromorphic functions on $R \cup \alpha$ is said to have *Iversen's property* with respect to the ideal boundary β of R if the following conditions are satisfied:

(a) *f is not constant,*

(b) *for an arbitrary disk U on $|w| \leq \infty$ with $f(\alpha) \cap U = \emptyset$ and $f(R) \cap U \neq \emptyset$, and for every component V of $f^{-1}(U)$, the set $U - f(V)$ is totally disconnected in U, i.e. $U - f(V)$ does not contain any nondegenerate continua.*

Let R' be a subregion of R such that (R', α'), $\alpha' = \partial R'$, is a noncompact bordered surface with compact border α', and the ideal boundary β' of R' is a subset of β. For $f \in M(R \cup \alpha)$ we denote by f' the restriction of f to $R' \cup \alpha'$, i.e. $f' \in M(R' \cup \alpha')$.

We shall say that $f \in M(R \cup \alpha)$ has the *localizable Iversen property* with respect to β if the following is true:

(c) *Not only does f have Iversen's property with respect to β but f' as well has this property with respect to β' for every R'.*

Observe that even if $f \in M(R \cup \alpha)$ has Iversen's property it need not have the localizable one. For example let W be a 2-sheeted covering surface of $|w| \leq \infty$ with a finite number of branch points and without relative boundary. Remove two disks U_1 and U_2 from W away from the branch points and over a disk U on $|w| \leq \infty$. On the resulting surface R consider the border α of U_1 as the border α of R, and the border β of U_2 as the ideal boundary β of R. We obtain a noncompact bordered surface (R, α) with compact border α. Let $f \in M(R \cup \alpha)$ be the restriction to $R \cup \alpha$ of the natural projection of the covering surface W onto $|w| \leq \infty$. It is easy to see that f has Iversen's property with respect to β but it does not have the localizable one.

If the border α of R is empty then it is readily seen that every $f \in M(R)$ with Iversen's property automatically has the localizable one.

1 B. Cluster Set at the Ideal Boundary. For a bordered surface (R, α) with compact border α there exists a surface $\tilde{R}$ which contains R as a subregion, and $\tilde{R}-R$ is compact with $\partial R=\alpha$. We denote by $R^*_{S,\alpha}$ the space $\tilde{R}^*_S-(\tilde{R}-R\cup\alpha)$ where $\tilde{R}^*_S$ is Stoïlow's compactification of $\tilde{R}$ (cf. IV.5 D). We call $R^*_{S,\alpha}$ the *relative Stoïlow compactification* of R. The boundary of the compact space $R^*_{S,\alpha}$ consists of two parts: the border α and the ideal boundary $\beta_S(R)$; the latter will be called *Stoïlow's ideal boundary* of R. It is clear that $R^*_{S,\alpha}$ does not depend on the choice of $\tilde{R}$ and is completely determined by (R, α).

Let $f\in M(R\cup\alpha)$ and $p\in\beta_S(R)$. A complex number w is called a *limiting value* of f at p if there exists a sequence $\{z_n\}\subset R$ such that

$$\lim_{n\to\infty} z_n=p \tag{1}$$

and

$$\lim_{n\to\infty} f(z_n)=w. \tag{2}$$

If condition (1) is replaced by

$$\overline{\{z_n\}}-\{z_n\}\subset\beta_S(R) \tag{3}$$

then w is called a limiting value of f at $\beta_S(R)$.

We denote by $C_R(f, p)$ (resp. $C_R(f, \beta_S)$) the totality of limiting values of f at p (resp. $\beta_S=\beta_S(R)$) and call it the *cluster set* of f at p (resp. β_S). Clearly $C_R(f, \beta_S)\supset C_R(f, p)$ and both sets are closed.

In this connection we also consider an *asymptotic value* w of f at p (resp. β_S) defined by a curve γ: $z=z(t)$, $0\le t<1$, called an *asymptotic path* for w in R, such that γ tends to p (resp. β_S) and

$$\lim_{t\to 1} f(z(t))=w. \tag{4}$$

The totality of asymptotic values of f at p (resp. β_S) will be denoted by $A_R(f, p)$ (resp. $A_R(f, \beta_S)$) and called the *asymptotic set* of f at p (resp. β_S). The following relations are immediate:

$$A_R(f, p)\subset C_R(f, p), \qquad A_R(f, \beta_S)\subset C_R(f, \beta_S). \tag{5}$$

We shall prove (Stoïlow [2, 4]):

Theorem. *If $f\in M(R\cup\alpha)$ has the localizable Iversen property then either the cluster set $C_R(f, p)$ of f at a Stoïlow ideal boundary point p consists of a single point or $C_R(f, p)=\{|w|\le\infty\}$.*

Let $\{G_n\}_1^\infty$ be a determining sequence of p (cf. IV.5 D) and $\alpha_n=\partial G_n$, i.e. $R\supset G_{n-1}\supset G_n\cup\alpha_n$ and $\bigcap_1^\infty \overline{G_n}=p$. Clearly

$$C_R(f, p)=\bigcap_1^\infty \overline{f(G_n)}. \tag{6}$$

Since $\overline{\{f(G_n)\}}_1^\infty$ is a decreasing sequence of continua, $C_R(f, p)$ is also a continuum in $|w| \leq \infty$. The set $\{|w| \leq \infty\} - C_R(f, p)$ is open and therefore consists of at most a countable number of plane regions W_i^∞.

We must show that either there does not exist any W_i^∞ or there exists only one W_i^∞ which is a punctured sphere.

The proof will be given in 1 C – 1 J.

1 C. Valence Function. As in IV. 8 D we denote by $n_f(w)$ the number of points in $f^{-1}(w) \cap R$ with the multiple points counted repeatedly. The function n_f may be called the *valence function* of $f|R$. The value of n_f is a nonnegative integer or ∞. It is lower semicontinuous in the sense that

$$\underline{n}_f(w_0) = \lim_{m \to \infty} \Big(\inf_{|w_0 - w| < \frac{1}{m}} n_f(w) \Big) = n_f(w_0) \tag{7}$$

for every $|w_0| \leq \infty$. Here of course $|w_0 - w| < 1/m$ is to be replaced by $|w| > m$ if $w_0 = \infty$. As a dual of (7) we set

$$\bar{n}_f(w_0) = \lim_{n \to \infty} \Big(\sup_{|w_0 - w| < \frac{1}{m}} n_f(w) \Big). \tag{8}$$

In general $\bar{n}_f \geq n_f = \underline{n}_f$ and $\bar{n}_f(w_0) = n_f(w_0) < \infty$ if and only if n_f is finitely continuous at w_0. A point w_0 with $\bar{n}_f(w_0) > n_f(w_0)$ (and consequently $n_f(w_0) < \infty$) is a point of discontinuity of n_f.

Set

$$\Theta = f(\alpha). \tag{9}$$

LEMMA. *Let $w_0 \notin \Theta$ be a point of discontinuity of n_f. If f has the Iversen property then $w_0 \in A_R(f, \beta_S)$.*

Set $n_f(w_0) = n$. Since $n < \bar{n}_f(w_0) \leq \infty$, n is a nonnegative integer.

First suppose $n = 0$. Take a small disk U about w_0 with $\bar{U} \cap \Theta = \emptyset$. By $\bar{n}_f(w_0) > 0$, $U \cap f(R) \neq \emptyset$. Take a component V of $f^{-1}(U)$. In view of 1 A.(b), $f(V)$ has a totally disconnected complement in U. Hence we can find a curve $\gamma_w \subset f(V)$ which starts from a point in $f(V)$ and tends to w_0. Since V is a complete covering surface of $f(V)$ with projection f there exists a curve γ: $z = z(t)$, $0 \leq t < 1$, $\gamma \subset V$ with $f(\gamma) = \gamma_w$. It issues from a point in V over the initial point of γ_w. By $\bar{U} \cap \Theta = \emptyset$, γ cannot tend to α as $t \to 1$. Similarly it cannot tend to a point in R since $n_f(w_0) = 0$. Therefore γ tends to β_S and f tends to w_0 along γ, i.e. $w_0 \in A_R(f, \beta_S)$.

Next consider the case $n > 0$. Let

$$f^{-1}(w_0) = \{z_1, \ldots, z_n\}$$

where multiple points are repeated. Take a disk U with center w_0 so small that $\bar{U} \cap \Theta = \emptyset$, the components

$$V_i = V(U, z_i) \qquad (i = 1, \ldots, n) \tag{10}$$

of $f^{-1}(U)$ which contain z_i are relatively compact in R, and

$$\bar{V}_i \cap \bar{V}_j = \emptyset \tag{11}$$

for $z_i \neq z_j$. Clearly every point in $\bar{U}$ is covered the same number n_i of times by $\bar{V}_i$.

In view of $\bar{n}_f(w_0) > n$, $f^{-1}(U)$ has a component V which is disjoint from $\bigcup_1^n V_i$. Since $U - f(V)$ is totally disconnected we can again find a curve $\gamma_w \subset f(V)$ starting from a point in $f(V)$ and tending to w_0; and a curve $\gamma \subset V$ with $f(\gamma) = \gamma_w$. It is easy to see that γ cannot tend to any point in $(R - \bigcup_1^n V_i) \cup \alpha\ (\supset V)$. Therefore γ tends to β_S and f tends to w_0 along γ, i.e. $w_0 \in A_R(f, \beta_S)$.

The intuitive meaning of the lemma is that less valence on R is compensated for by more valence on β_S for functions with Iversen's property.

1 D. Degree of a Component. Each component W_i of $\{|w| \leq \infty\} - C_R(f, \beta_S)$ is decomposed into W_{ij}'s by $\Theta = f(\alpha)$: $W_i - \Theta = \bigcup_j W_{ij}$. We now see that

$$n_f | W_{ij} = n_{ij}, \tag{12}$$

with n_{ij} a nonnegative integer which we call the *degree* of W_{ij}.

In fact let $w_0 \in W_{ij}$. It is neither a limiting value nor an asymptotic value of f at β_S. Thus $n_f(w_0) < \infty$ and by Lemma 1 C, $n_f(w_0) = \bar{n}_f(w_0)$. Hence $n_f | W_{ij}$ is continuous and finite on W_{ij}. Since $n_f(W_{ij})$ is connected and a set of integers we obtain (12).

1 E. Asymptotic Values. We also remark that

$$\{w \in C_R(f, \beta_S) \mid n_f(w) < \infty\} \subset A_R(f, \beta_S). \tag{13}$$

To see this let $w \in C_R(f, \beta_S)$ and $n_f(w) < \infty$. We must show that $w \in A_R(f, \beta_S)$. If we move α slightly toward β_S none of the above three properties is affected. Hence we may suppose that $w \notin \Theta$.

Let $f^{-1}(w) = \{z_1, \ldots, z_n\}$, $n = n_f(w)$ and take U and $V_i = V(U, z_i)$ as in (10). Choose $\{\zeta_m\}_1^\infty \subset R - \bigcup_1^n \bar{V}_i$ tending to β_S such that $\{f(\zeta_m)\}_1^\infty$ is contained in U and tends to w. Since each $f(\zeta_m)$ is taken in each V_i $(i = 1, \ldots, n)$ and $\zeta_m \notin \bigcup_1^n \bar{V}_i$ we infer that $n_f(f(\zeta_m)) \geq n+1$ and consequently $\bar{n}_f(w) \geq n+1 > n_f(w)$. By Lemma 1 C, $w \in A_R(f, \beta_S)$.

1 F. Deficiency on ∂W_{ij}. Let n_{ij} be the degree of W_{ij} (cf. 1 D) and $w_0 \in \partial W_{ij} - \Theta$. We shall prove that

$$n_f(w_0) \leq n_{ij} - 1. \tag{14}$$

Since $n_{ij} \geq \underline{n}_f(w_0) = n_f(w_0)$ it suffices to show that $n_f(w_0) \neq n_{ij}$. Set $n_{ij} = n$ and suppose to the contrary that $n_f(w_0) = n$. Let $f^{-1}(w_0) = \{z_1, \ldots, z_n\}$ and take U and $V_k = V(U, z_k)$ as in (10). In view of

$$w_0 \in \partial W_{ij} - \Theta \subset \partial W_i \subset C_R(f, \beta_S)$$

$f^{-1}(U)$ has a component V which is disjoint from $\bigcup_{k=1}^n V_k$. As before we can find a curve $\gamma_w \subset f(V)$ which starts from an $f(z_V)$, $z_V \in V$, and tends to a point $w \in U \cap W_{ij}$.

Let γ be a curve in V with $f(\gamma) = \gamma_w$. Since $w \in W_{ij}$ is taken $n = n_{ij}$ times by f on R and is already taken in each V_k $(k = 1, \ldots, n)$ the curve γ tends to β_S and $w \in A_R(f, \beta_S) \subset C_R(f, \beta_S)$. This contradicts $C_R(f, \beta_S) \cap W_{ij} = \emptyset$.

1 G. Totally Disconnected $C_R(f, \beta_S)$. We first consider the case in which the cluster set $C_R(f, \beta_S)$ is totally disconnected. Then the complement of $C_R(f, \beta_S)$ consists of only one component W_1, divided into a finite number of components $W_{11}, W_{12}, \ldots, W_{1j_1}$ by Θ.

For an arbitrary $w_0 \in C_R(f, \beta_S)$ we may slightly deform α so that $w_0 \notin \Theta$. By (14) and Lemma 1 C, $w_0 \in A_R(f, \beta_S)$, i.e.

$$C_R(f, \beta_S) = A_R(f, \beta_S) \tag{15}$$

if $C_R(f, \beta_S)$ is totally disconnected.

We know from 1 D that $n_f | W_{1j} = n_{1j}$, a constant, for every $j = 1, \ldots, j_1$. In particular if $\alpha = \emptyset$ then

$$n_f | W_1 = n_1, \qquad n_f | C_R(f, \beta_S) \leq n_1 - 1.$$

1 H. Total $C_R(f, \beta_S)$. Next suppose that $C_R(f, \beta_S)$ is not totally disconnected. We must show that $C_R(f, \beta_S)$ is *total*, i.e. $C_R(f, \beta_S) = \{|w| \leq \infty\}$.

Contrary to the assertion suppose that the boundary of some W_i contains a proper continuum K. By taking K sufficiently small and by suitably deforming α we may assume that $K \cap \Theta = \emptyset$ and $K \subset \partial W_{ij} - \Theta$. Let the degree n_{ij} of W_{ij} be n and suppose that $w_0 \in K$ satisfies

$$m = \max_{w \in K} n_f(w) = n_f(w_0). \tag{16}$$

Then by (14)

$$m \leq n - 1. \tag{17}$$

Choose a disk U about w_0 such that $\partial U \cap W_{ij} \neq \emptyset$, $\partial U \cap K \neq \emptyset$, and for $f^{-1}(w_0) = \{z_1, \ldots, z_m\}$ the sets $V_k = V(U, z_k)$ $(k = 1, \ldots, m)$ are as in (10) and (11).

Take a smaller concentric disk U' in U, and two distinct points a_1 and a_2 on $(\partial U') \cap W_{ij}$. Cover $\partial U'$ by a concentric circular annulus and divide it into two regions Ω_1 and Ω_2 by radial segments through a_1 and a_2 respectively.

Since $n_f(a_1)=n$ and a_1 is taken m times in $\bigcup_{k=1}^m V_k$ there exist $n-m\,(\geq 1)$ points in $R-\bigcup_{k=1}^m V_k$ at which f takes on a_1. Thus there exists a component V of $f^{-1}(U)$ disjoint from $\bigcup_{k=1}^m V_k$ and containing a point b_1 with $f(b_1)=a_1$. The set $U-f(V)$ being totally disconnected we can find curves $\gamma_\nu^w \subset \Omega_\nu$ $(\nu=1,2)$ issuing from a_1 and tending to a_2. Let γ_ν be the curves in V with $f(\gamma_\nu)=\gamma_\nu^w$. Since $\gamma_1^w \cup \gamma_2^w$ encloses w_0 it must meet K at some point, say w_1. Let $\zeta \in \gamma_1 \cup \gamma_2 \subset V$ with $f(\zeta)=w_1$. In each V_k $(k=1,\dots,m)$, w_1 is taken on by f. We conclude that

$$n_f(w_1) \geq m+1. \tag{18}$$

In view of $w_1 \in K$, (18) contradicts (16).

1 I. Stoïlow's Principle. We have proved in 1 G and 1 H (Stoïlow [2, 4]):

Stoïlow's Principle on Iversen's Property. *If $f \in M(R \cup \alpha)$ has the localizable Iversen property then one of the following alternatives must hold:*

(a) *$C_R(f, \beta_S)$ is total, i.e. $C_R(f, \beta_S)=\{|w|\leq\infty\}$;*

(b) *$C_R(f, \beta_S)$ is totally disconnected, $C_R(f, \beta_S)=A_R(f, \beta_S)$, and n_f is a finite constant on each component of $\{|w|\leq\infty\}-C_R(f,\beta_S)-f(\alpha)$.*

1 J. Proof of Theorem 1 B. For a determining sequence $\{G_n\}_1^\infty$ of $p \in \beta_S(R)$ with $\alpha_n=\partial G_n$ set $f_n=f\,|\,G_n \cup \alpha_n$. Clearly

$$C_R(f,p) \subset C_{G_n}(f_n, \beta_n) \subset \overline{f(G_n)} \tag{19}$$

where β_n is Stoïlow's ideal boundary of (G_n, α_n). Consequently by (6)

$$C_{G_n}(f_n, \beta_n) \supset C_{G_{n+1}}(f_{n+1}, \beta_{n+1}), \qquad C_R(f,p)=\bigcap_1^\infty C_{G_n}(f_n, \beta_n). \tag{20}$$

Stoïlow's principle applies to $C_{G_n}(f_n, \beta_n)$ and we conclude that $C_R(f,p)$ is either totally disconnected or total. In the former case, since $C_R(f,p)$ is a continuum, $C_R(f,p)$ consists of only one point.

This completes the proof of Theorem 1 B.

1 K. Continuity of AB-Functions. As a direct consequence of Theorem 1 B we obtain:

Theorem. *If $f \in AB(R \cup \alpha)$ has the localizable Iversen property then it is continuous on the relative Stoïlow compactification $R_{S,\alpha}^*$ of (R, α).*

By a method analogous to that used for $R_{S,\alpha}^*$ in 1 B we can also define the *relative Royden (resp. Wiener) compactification* $R_{\mathrm{M},\alpha}^*$ (resp.

$R^*_{\mathrm{N},\alpha}$) of a bordered surface (R,α) with compact border α. By Corollary IV. 5 D we see:

COROLLARY. *If $f\in AB(R\cup\alpha)$ has the localizable Iversen property then it is continuous on the relative Royden (resp. Wiener) compactification $R^*_{\mathrm{M},\alpha}$ (resp. $R^*_{\mathrm{N},\alpha}$) of (R,α) and constant on each component of $R^*_{\mathrm{M},\alpha}-R$ (resp. $R^*_{\mathrm{N},\alpha}-R$).*

1 L. Remark on Removable Sets. Let X be one of the classes HB, HD, HBD, AB, AD, and ABD. A compact set E in a closed Riemann surface R is referred to as an *X-removable* set or an X-null set if there exists a disk U in R with $U\supset E$ such that $X(\bar{U}-E)=X(\bar{U})|\bar{U}-E$ (cf. I. 8 D, II. 2 D, V. 11 B). We have seen that E is X-removable if and only if $R-E\in O_X$. Since the inclusion relations

$$(O_G=O_{HP}=)\ O_{HB}=O_{HD}=O_{HBD}<O_{AB}<O_{AD}=O_{ABD}$$

are valid for surfaces of finite genus the essentially distinct removable sets are those related to HB, AB, and AD.

For later reference we append here the following remark on the union of removable sets:

THEOREM. *If $\{E_n\}$ is a countable family, with compact union E, of X-removable sets in a closed Riemann surface R then E is again an X-removable set.*

This result is due to KAMETANI for $X=AB$. See also NOSHIRO [3; footnote 2 on p. 11], KURODA [2], and RUDIN [2].

The theorem implies in particular that if $\{E_n\}$ is a finite family then $E=\bigcup_n E_n$ is always X-removable. In this case the proof could be given by elementary considerations (cf. Remark I. 8 D).

The proof in the general case is by contradiction. Suppose E were not X-removable. Then there would exist a nonconstant function $f\in X(R-E)$. Here E is totally disconnected together with E_n, and consequently $R-E$ is a surface. Let e be the set of nonremovable singularities of f, i.e. the set of points $z\in R$ such that f cannot be continued to any neighborhood of z so as to be in X there. Clearly e is a compact subset of E.

Take a disk $U\subset R$ with $U\supset E$. Obviously $e_n=e\cap E_n$ is X-removable and $e=\bigcup_n e_n$. In terms of the induced metric of U, e becomes a complete metric space, and each e_n is a compact subset of e. Therefore by the Baire category theorem (see e.g. YOSIDA [1; p. 11]) there exists at least one e_n which contains a point z_0 such that an ε-ball $v(z_0)\subset e$ with its center at z_0 is contained in e_n. Observe that $v(z_0)=e\cap u(z_0)\subset e_n$ where $u(z_0)=\{z\,|\,|z-z_0|<\varepsilon\}\subset U$.

Since e_n is totally disconnected we can find a sequence $\{\gamma_m\}$ of analytic Jordan curves $\gamma_m \subset u(z_0)-e_n$ such that the simply connected regions $u_m(z_0)$ enclosed by γ_m exhaust $u(z_0)$. From the fact that $e_{nm}=e_n \cap u_m(z_0)$ is a compact X-removable subset of $u_m(z_0)$ and

$$f\,|\,\overline{u_m(z_0)}-e_{nm} \in X\big(\overline{u_m(z_0)}-e_{nm}\big)$$

we infer that this function can be extended so as to be in $X\big(\overline{u_m(z_0)}\big)$ for every m, and consequently $f\,|\,u(z_0)-e_n$ is extendable to a member of $X\big(u(z_0)\big)$. This contradicts the assumption $z_0 \in e$.

2. Meromorphic Functions on O_{A^0D}-Surfaces

2 A. Class MD^*. For $f \in M(R)$ the *spherical Dirichlet integral* $D_R^*(f)$ is defined as

$$D_R^*(f)=\iint\limits_R \frac{|f'(z)|^2}{(1+|f(z)|^2)^2}\, r\,dr\,d\theta, \qquad z=r\,e^{i\theta}. \tag{21}$$

In other words $D_R^*(f)$ is the spherical area of the Riemannian image of R under f over the extended w-plane.

We denote by $MD^*=MD^*(R)$ the class of functions $f \in M(R)$ with $D_R^*(f)<\infty$. We also consider $AD^*=A \cap MD^*$. The corresponding null classes are denoted by O_{MD^*} and O_{AD^*}. Clearly $D_R^*(f) \le D_R(f)$. Thus $AD \subset AD^* \subset MD^*$ and

$$O_{MD^*} \subset O_{AD^*} \subset O_{AD}.$$

Let $\tilde{\mathbb{P}}$: $|z| \le \infty$ and $\mathbb{P}$: $|z|<\infty$. Since $z \in MD^*(\tilde{\mathbb{P}})$ whereas $AD^*(\tilde{\mathbb{P}})$ consists of only constants, $O_{MD^*}<O_{AD^*}$. Similarly $z \in AD^*(\mathbb{P})$, $AD(\mathbb{P})$ contains only constants, and we have

$$O_{MD^*}<O_{AD^*}<O_{AD}. \tag{22}$$

The null class O_{MD^*} is essentially different from the others in that it does not contain any surfaces of finite genus, open or closed. In fact any such surface R is contained in a closed surface $\tilde{R}$, and $M(\tilde{R})=MD^*(\tilde{R})$. For this reason O_{MD^*} consists of some surfaces of infinite genus. We shall in fact see in 6 B that $U_{HD} \subset O_{MD^*}$ and thus $O_{MD^*} \neq \emptyset$.

2 B. MD^* on O_{A^0D}-Surfaces. Let $R \in O_{A^0D}$ (cf. I.1 G), i.e. every bordered subregion of R belongs to SO_{AD}. Consider a noncompact bordered subregion (G,α) of R with compact border α and nonempty Stoïlow's ideal boundary $\beta_S=\beta_S(G)$. We allow the case $R=G$ and $\alpha=\emptyset$.

The following result was obtained by Matsumoto [4] and Kuroda [6]:

Theorem. *Every function $f \in MD^*(G \cup \alpha)$, $G \cup \alpha \subset R \in O_{A^0D}$, has the localizable Iversen property.*

Since G is arbitrary together with (G, α) it suffices to prove that $f \in MD^*(G \cup \alpha)$ has Iversen's property. For this purpose take a disk U in $|w| \leq \infty$ with $f(\alpha) \cap U = \emptyset$ and $f(G) \cap U \neq \emptyset$ and choose a component V of $f^{-1}(U)$. We must show that $U - f(V)$ is totally disconnected. If the center of U is ∞ then we may map U onto a finite disk by the function $w \to 1/w$ since $1/f \in MD^*(G \cup \alpha)$. Thus we always assume that U is a finite disk.

First we shall show that $f(V)$ is dense in U. If this were not the case then we could find two concentric disks W_1 and W_2 such that $U - f(V) \supset \overline{W}_1 \supset W_1 \supset \overline{W}_2$. Obviously there exists a nonconstant $\tilde{g} \in AD(\overline{U} - \overline{W}_2)$ whose real part vanishes on ∂U. Set $g = \tilde{g} | \overline{U} - W_1$ and consider

$$h = g \circ f$$

on $V \cup \partial V$. Clearly $h \in A^0(V, \partial V)$, i.e. h is (single-valued) analytic on $V \cup \partial V$ with vanishing real part on ∂V. Observe that

$$D_V(h) = 2 \int_V |f'(z)|^2 \, |g'(f(z))|^2 \, dx\, dy.$$

Since $f(V) \subset \overline{U} - W_1$ and $g \in A(\overline{U} - W_1)$, $|g'(f(z))|^2$ is bounded on V. Let ρ be the radius of U. Because of $f(V) \subset U$, $|f|V| \leq \rho < \infty$. Hence

$$\iint_V |f'(z)|^2 \, dx\, dy \leq (1+\rho^2)^2 \iint_V \frac{|f'(z)|^2}{(1+|f(z)|^2)^2} \, dx\, dy \leq (1+\rho^2)^2 \, D_G^*(f) < \infty.$$

We conclude that $D_V(h) < \infty$ and $h \in A^0 D(V, \partial V)$. Therefore $(V, \partial V) \notin SO_{AD}$ which contradicts $R \in O_{A^0D}$.

Next suppose $U - f(V)$ contains a proper continuum K. Then we can choose a disk U_1 with $\overline{U}_1 \subset U$ such that $U_1 - K$ consists of at least two components. There exists a component V_1 of $f^{-1}(U_1)$ contained in V. Clearly $f(V_1)$ belongs to a component of $U_1 - K$ and thus $f(V_1)$ cannot be dense in U_1, a contradiction.

Remark. The above proof is due to MATSUMOTO [4].

2 C. AD on O_{A^0D}-Surfaces. Let $G \cup \alpha$ be as in 2 B. We now assert (MATSUMOTO [4]):

THEOREM. *Every function $f \in AD(G \cup \alpha)$ with $G \cup \alpha \subset R \in O_{A^0D}$ is bounded and continuous on the relative Stoïlow compactification $G^*_{S,\alpha}$ of (G, α).*

Let $f = u + iv$ on $G \cup \alpha$. Suppose there exists a point $z_0 \in G$ with $|u|\alpha| < |u(z_0)|$. After a multiplication by -1 if necessary we may suppose that $u|\alpha < u(z_0)$. Take a number c such that $u|\alpha < c < u(z_0)$ and a component V of $\{z \in G \,|\, u(z) > c\}$ containing z_0. Then $V \cup \partial V \subset G$ and $f - c \in A^0 D(V, \partial V)$, i.e. $(V, \partial V) \notin SO_{AD}$, a contradiction. Thus $|u|G| \leq$

$\max_\alpha |u|$. Similarly by considering $if \in AD(G \cup \alpha)$ we infer that $|v|G| \leq \max_\alpha |v|$. Hence $f \in ABD(G \cup \alpha)$.

By 2 B, f has the localizable Iversen property. Hence by Theorem 1 K, f is continuous on $G^*_{S,\alpha}$.

COROLLARY. *Every function $f \in AD(G \cup \alpha)$, with $G \cup \alpha \subset R \in O_{A^0D}$, is bounded and continuous on the relative Royden (resp. Wiener) compactification $G^*_{\mathrm{MI},\alpha}$ (resp. $G^*_{\mathrm{NI},\alpha}$) of (G, α) and constant on each component of $G^*_{\mathrm{MI},\alpha} - G$ (resp. $G^*_{\mathrm{NI},\alpha} - G$).*

2 D. The Inclusion $O_{HD} < O_{A^0D}$. We claim (KURODA [5, 6]):

$$O_{HD} < O_{A^0D} < O_{AD}. \tag{23}$$

The latter relation was established in I. 10 D. To prove the former we shall first show that $O_{HD} \subset O_{A^0D}$. Let $R \in O_{HD}$ and take a bordered subregion $(G, \partial G)$. If it is in SO_{HD} then in view of $SO_{HD} \subset SO_{AD}$ there is nothing to prove. If $(G, \partial G) \notin SO_{HD}$ we infer by III. 3 F, 5 D, and 4 N that $G \in U_{HD}$. Hence by III. 5 I, $G \in O_{AD}$, and consequently $(G, \partial G) \in SO_{AD}$, i.e. $O_{HD} \subset O_{A^0D}$.

To prove the strictness take the generalized Cantor set

$$E(p_1 p_2 \ldots)$$

(cf. I. 6 A, V. 11 F). In particular choose $p_\nu = p > 1$ for every $\nu = 1, 2, \ldots$. By I. 6 A the complement R of $E(p^\infty) = E(p\, p \ldots)$ with respect to $|z| \leq \infty$ belongs to $O_{A^0D} \subset O_{AD}$. On the other hand

$$\prod_{\nu=1}^{\infty} \left(1 - \frac{1}{p_\nu}\right)^{2^{-\nu}} = \left(1 - \frac{1}{p}\right)^{\sum_\nu 2^{-\nu}} = 1 - \frac{1}{p} > 0.$$

Thus because of Theorem V. 11 F, $E(p^\infty)$ has positive capacity. In view of V. 11 A we conclude that $R \notin O_G = O_{HD}$ for plane regions, and we have proved $O_{HD} < O_{A^0D}$ even for plane regions.

3. Meromorphic Functions on O_{A^0B}-Surfaces

3 A. The Class M on O_{A^0B}-Surfaces. As in II. 2 C we denote by SO_{AB} the class of bordered Riemann surfaces (R, α) for which the family $A^0B(R, \alpha)$ of AB-functions on $R \cup \alpha$ whose real parts vanish on α consists of only constants. The null class O_{A^0B} is made up of Riemann surfaces R such that every bordered subregion (G, α) belongs to SO_{AB}.

Consider a bordered subregion (G, α) of $R \in O_{A^0B}$. As before we include the case $G = R$ so that $\alpha = \emptyset$. For the class M of meromorphic functions we shall prove (KURODA [4]; see also A. MORI [1], YÛJÔBÔ [1], NOSHIRO [1]):

THEOREM. *Every* $f \in M(G \cup \alpha)$, *with* $G \cup \alpha \subset R \in O_{A^0B}$, *has the localizable Iversen property.*

Take a disk U in $|w| \le \infty$ such that $f(\alpha) \cap U = \emptyset$ and $f(G) \cap U \neq \emptyset$. Let V be a component of $f^{-1}(U)$. If $U - f(V)$ contains a proper continuum $K \subset U$ then there exists a nonconstant $g \in AB(\bar{U} - K)$ with $\operatorname{Re} g | \partial U = 0$. The function $h = g \circ f \in AB(V)$ is nonconstant and $\operatorname{Re} h | \partial V = 0$. Thus $(V, \partial V) \notin SO_{AB}$, in violation of $R \in O_{A^0B}$.

Since $G \subset R$ is arbitrary the property is clearly localizable.

3 B. A Characterization of O_{A^0B}. Let (G, α) be a bordered Riemann surface and suppose every $f \in AB(G \cup \alpha)$ satisfies the following *Lindelöf maximum principle:* $|f|\alpha| = c$, a constant, implies $|f|G| \le c$. Denote by $\mathscr{L}$ the family of such bordered Riemann surfaces (G, α).

The following characterization of O_{A^0B} is due to NOSHIRO [3]:

THEOREM. *In order that $R \in O_{A^0B}$ it is necessary and sufficient that every bordered subregion $(G, \partial G)$ of R belong to $\mathscr{L}$.*

Suppose that R satisfies the condition. Take an $f \in AB(G \cup \partial G)$ with $\operatorname{Re} f | \partial G = 0$. Then $e^{\pm f} \in AB(G \cup \partial G)$ and $|e^{\pm f}| = 1$ on ∂G. Thus by $(G, \partial G) \in \mathscr{L}$, $|e^{\pm f}| = e^{\pm \operatorname{Re} f} \le 1$ on G, or equivalently $\pm \operatorname{Re} f \le 0$. Hence $\operatorname{Re} f = 0$ on G, i.e. $(G, \partial G) \in SO_{AB}$, and we have $R \in O_{A^0B}$.

Conversely suppose $R \in O_{A^0B}$. Take an $f \in AB(G \cup \partial G)$ with $|f|\partial G| = c$, a constant. If there exists a point $z_0 \in G$ with $|f(z_0)| > c$ choose a constant d with $|f(z_0)| > d > c$ and let V be the component of $\{z \in G \,|\, |f(z)| > d\}$ containing z_0. Then $(V, \partial V)$ is a bordered surface, $|f| \,|\, \partial V = d > 0$, and $|f|V| > d$. Let $f | V \cup \partial V = g$ and $\sup_V |g| = k$. Consider a nonconstant function $h \in AB(d \le |w| \le k)$ with $\operatorname{Re} h | \{|w| = d\} = 0$. Then $F = h \circ g \in AB(V \cup \partial V)$, $\operatorname{Re} F | \partial V = 0$, and F is not constant, i.e. $(V, \partial V) \notin SO_{AB}$. This contradicts $R \in O_{A^0B}$.

Remark. It is also readily seen that if every bordered subregion $(G, \partial G)$ with compact border ∂G belongs to $\mathscr{L}$ then $R \in O_{A_0B}$.

3 C. Exceptional Sets. Let (G, α) be a bordered subregion of $R \in O_{A^0B}$. Consider an $f \in M(G \cup \alpha)$ and its valence function n_f, i.e. $n_f(w)$ is the number of points in $f^{-1}(w) \cap G$. Take a disk U in $|w| \le \infty$ and set

$$N_f^U = \sup_{w \in U} n_f(w), \tag{24}$$

$$E_f^U = \{w \in U \,|\, n_f(w) < N_f^U\}, \tag{25}$$

and

$$E_f^{U,n} = \{w \in U \,|\, n_f(w) \le n\} \tag{26}$$

for $0\le n<N_f^U$. Clearly $E_f^{U,n}$ is closed in U, $E_f^{U,n-1}\subset E_f^{U,n}$, and

$$E_f^U=\bigcup_{0\le n<N_f^U} E_f^{U,n}. \tag{27}$$

THEOREM. *If $R\in O_{A^0B}$ and $f\in M(G\cup\alpha)$ then every compact subset of E_f^U is a Painlevé null set.*

Let K be a compact subset of E_f^U and write $K_n=K\cap E_f^{U,n}$. Then $K_n\subset K_{n+1}$ and $K=\bigcup K_n$ for $0\le n<N_f^U$. Suppose there exists an n such that K_n is not a Painlevé null set whereas K_{n-1} is; here $K_{-1}=\emptyset$. Then there exists by Lindelöf's covering theorem a point $w_n\in K_n-K_{n-1}$ such that $K_\rho=K_n\cap\{|w-w_n|\le\rho\}$ is not a Painlevé null set for any $\rho>0$. Let $U_\rho=\{|w-w_n|<2\rho\}\subset U$. By Lemma 1 C, $f^{-1}(U_\rho)$ consists, for some ρ, of n relatively compact disks V_0 (k-sheeted disks are counted as k disks) and at least one relatively noncompact component V. Any point $w\in K_\rho$ is already covered n times by the V_0. Hence $U_\rho-f(V)\supset K_\rho$. Since K_ρ is not AB-removable there exists a nonconstant $g\in AB(\bar{U}_\rho-K_\rho)$ with $\operatorname{Re} g\,|\,\partial U_\rho=0$. Therefore $h=g\circ f\in AB(V\cup\partial V)$ is nonconstant and $\operatorname{Re} h\,|\,\partial V=0$. We infer that $(V,\partial V)\notin SO_{AB}$, in violation of $R\in O_{A^0B}$.

As a countable union of Painlevé null sets, K is a Painlevé null set (see 1 L).

COROLLARY. *The set E_f^U has 2-dimensional Lebesgue measure zero.*

Take an arbitrary compact set $K\subset E_f^U$. By the above, K is a Painlevé null set and consequently by II.2 E, K has Lebesgue measure zero. The same is true of E_f^U since K is arbitrary in E_f^U.

Actually E_f^U has "inner" Newtonian capacity zero and consequently "inner" $(1+\varepsilon)$-measure zero $(\varepsilon>0)$ (cf. II.2 H, 2 J).

3 D. $AB\subset MD^*$ for O_{A^0B}-Surfaces. Let (G,α) be a bordered subregion of an $R\in O_{A^0B}$. We have (MATSUMOTO [4]):

THEOREM. *Every $f\in MD^*(G\cup\alpha)$, $G\cup\alpha\subset R\in O_{A^0B}$, is continuous on the relative Stoïlow compactification $G_{S,\alpha}^*$ of (G,α). Moreover*

$$AB(G\cup\alpha)\subset MD^*(G\cup\alpha). \tag{28}$$

By Theorems 3 A and 1 I and Corollary 3 C we infer that (28) is valid and also that every $f\in MD^*(G\cup\alpha)$ is of finite valence. Hence by Theorem 1 B, $C_G(f,p)$ is a point for $p\in\beta_S(G)$, and therefore every $f\in MD^*(G\cup\alpha)$ is continuous on $G_{S,\alpha}^*$.

COROLLARY. *Every $f\in MD^*(G\cup\alpha)$, $G\cup\alpha\subset R\in O_{A^0B}$, is continuous on the relative Royden (Wiener) compactification $G_{\mathbf{M},\alpha}^*$ (resp. $G_{\mathbf{N},\alpha}^*$) of (G,α) and constant on each component of $G_{\mathbf{M},\alpha}^*-G$ (resp. $G_{\mathbf{N},\alpha}^*-G$).*

3 E. Integrated Form of O_{A^0B}-Test. Let $\{R_n\}_0^\infty$ be an exhaustion of an open R and let $R_n - \bar{R}_{n-1}$ consist of regions R_{nk}, $k=1, \ldots, k_n$. For R_0 we choose a disk $|z|<1$ in R. As in II.2 B we set

$$\begin{aligned} v_n &= \operatorname{mod}(R_n - \bar{R}_{n-1}, \partial R_{n-1}, \partial R_n), \\ \mu_{nk} &= \operatorname{mod}\big(R_{nk}, (\partial R_{n-1}) \cap (\partial R_{nk}), (\partial R_n) \cap (\partial R_{nk})\big), \end{aligned} \tag{29}$$

and

$$\mu_n = \min \mu_{nk} \tag{30}$$

with $k=1, \ldots, k_n$.

Define the function ξ on R as in II.2 B: $\xi | R_0 = |z|^2$ and $\xi | R_j - R_{j-1} = \tau_{j-1} + u_j$ with $\tau_0 = 1$, $\tau_j = \sum_{t=1}^{j} \log v_t + 1$; here u_j is the modulus function giving v_j.

For the modulus function u_{jk} corresponding to μ_{jk} we have

$$u_j | R_{jk} = c_{jk} u_{jk}, \qquad c_{jk} = \frac{\log v_j}{\log \mu_{jk}}. \tag{31}$$

In terms of the conformal metric $ds = |\operatorname{grad} \xi| |dz|$ we consider the curve $\Gamma(\rho)$ consisting of the points at distance ρ from $z_0 = 0$. It is made up of a finite number $M(\rho)$ of closed curves $\Gamma_i(\rho)$, $i=1, \ldots, M(\rho)$. Let $\Lambda_i(\rho)$ be the length of $\Gamma_i(\rho)$ and set

$$\Lambda(\rho) = \max \Lambda_i(\rho) \tag{32}$$

with $i=1, \ldots, M(\rho)$.

The following test is due to Kuroda (see also Kuroda [4, 6]):

Theorem. *If the maximum length $\Lambda(\rho)$ has the property*

$$\limsup_{j \to \infty} \int_{\tau_{j-1}}^{\tau_j} \exp\left(2\pi \int_{\varepsilon}^{\rho} \frac{d\rho}{\Lambda(\rho)}\right) d\rho = \infty \tag{33}$$

then R belongs to O_{A^0B}.

Remark. Condition (33) is also sufficient for $R \in O_{KB}$.

For the proof of the theorem let $(G, \partial G)$ be a noncompact bordered subregion of R and take $f \in AB(G, \partial G)$ with $\operatorname{Re} f | \partial G = 0$. We must show that f is constant. Suppose this were not the case.

We may assume that $\gamma(\rho) = \Gamma(\rho) \cap G \neq \emptyset$ $(\rho \geq \varepsilon)$. Denote by $\gamma_i(\rho)$ $(i=1, \ldots, m(\rho))$ the components of $\gamma(\rho)$ and by $D(\rho)$ the Dirichlet integral of f over $G_\rho = G \cap \{\xi < \rho\}$. Then

$$D(\rho) = \sum_{i=1}^{m(\rho)} \int_{\gamma_i(\rho)} h \frac{\partial h}{\partial \xi} * d\xi \tag{34}$$

where $f = h + i h^*$. Let $\lambda_i(\rho)$ be the length of $\gamma_i(\rho)$.

Suppose $\gamma_i(\rho)$ is a cross-cut, i.e. it issues and terminates at ∂G. Denote by η a conjugate ξ^* of ξ so that $d\eta = *d\xi = d\xi^*$. We may normalize η so that it increases from 0 to $\lambda_i(\rho)$ on $\gamma_i(\rho)$. Hence h can be considered as a function $h(\eta)$ on $[0, \lambda_i(\rho)]$, with $h(0) = h(\lambda_i(\rho)) = 0$. We extend $h(\eta)$ to $[-\lambda_i(\rho), \lambda_i(\rho)]$ antisymmetrically. Then the Fourier expansion of h is

$$h(\eta) = \sum_{k=1}^{\infty} b_k \sin \frac{\pi}{\lambda_i(\rho)} k\eta .$$

Therefore

$$\frac{d}{d\eta} h(\eta) = \sum_{k=1}^{\infty} \frac{\pi k b_k}{\lambda_i(\rho)} \cos \frac{\pi}{\lambda_i(\rho)} k\eta .$$

By Parseval's identity

$$\int_{\gamma_i(\rho)} h^2 * d\xi = \tfrac{1}{2} \lambda_i(\rho) \sum_{k=1}^{\infty} b_k^2 \tag{35}$$

and

$$\int_{\gamma_i(\rho)} \left(\frac{\partial h}{\partial \eta}\right)^2 * d\xi = \frac{\pi^2}{2\lambda_i(\rho)} \sum_{k=1}^{\infty} k^2 b_k^2 . \tag{36}$$

A comparison of the right-hand sides gives

$$\int_{\gamma_i(\rho)} h^2 * d\xi \leq \frac{\lambda_i(\rho)^2}{\pi^2} \int_{\gamma_i(\rho)} \left(\frac{\partial h}{\partial \eta}\right)^2 * d\xi .$$

Thus by Schwarz's inequality

$$\begin{aligned} \int_{\gamma_i(\rho)} h \frac{\partial h}{\partial \xi} * d\xi &\leq \frac{\lambda_i(\rho)}{\pi} \sqrt{\int_{\gamma_i(\rho)} \left(\frac{\partial h}{\partial \eta}\right)^2 * d\xi \int_{\gamma_i(\rho)} \left(\frac{\partial h}{\partial \xi}\right)^2 * d\xi} \\ &\leq \frac{\lambda_i(\rho)}{2\pi} \int_{\gamma_i(\rho)} |\operatorname{grad} h|^2 * d\xi . \end{aligned}$$

Since $\lambda_i(\rho) \leq \Lambda_j(\rho) \leq \Lambda(\rho)$ for some j we finally obtain

$$\int_{\gamma_i(\rho)} h \frac{\partial h}{\partial \xi} * d\xi \leq \frac{\Lambda(\rho)}{2\pi} \int_{\gamma_i(\rho)} |\operatorname{grad} h|^2 * d\xi . \tag{37}$$

Next suppose $\gamma_i(\rho)$ is a nondividing cycle of G. This time $h(\eta)$ is also defined on $[0, \lambda_i(\rho)]$ but $h(0) = h(\lambda_j(\rho))$ is not necessarily zero. However if we replace h by $h - h(0)$ the argument above applies and the counterpart of (37) holds:

$$\int_{\gamma_i(\rho)} (h - h(0)) \frac{\partial h}{\partial \xi} * d\xi \leq \frac{\Lambda(\rho)}{2\pi} \int_{\gamma_i(\rho)} |\operatorname{grad} h|^2 * d\xi . \tag{38}$$

Here

$$\int_{\gamma_i(\rho)} (h-h(0))\frac{\partial h}{\partial \xi} * d\xi = \int_{\gamma_i(\rho)} (h-h(0)) * dh = \int_{\gamma_i(\rho)} h * dh - h(0) \int_{\gamma_i(\rho)} * dh$$

$$= \int_{\gamma_i(\rho)} h * dh = \int_{\gamma_i(\rho)} h \frac{\partial h}{\partial \xi} * d\xi$$

since $f = h + i h^*$ is single-valued. Thus (37) holds in every case.

From (34) and (37) it follows that

$$D(\rho) \leq \frac{\Lambda(\rho)}{2\pi} \int_{\gamma(\rho)} |\operatorname{grad} h|^2 * d\xi. \tag{39}$$

Since

$$\frac{d}{d\rho} D(\rho) = \int_{\gamma(\rho)} |\operatorname{grad} h|^2 * d\xi$$

(39) takes the form

$$2\pi \frac{d\rho}{\Lambda(\rho)} \leq d \log D(\rho)$$

or equivalently

$$\exp\left(2\pi \int_{\varepsilon}^{\rho} \frac{d\rho}{\Lambda(\rho)}\right) \leq \frac{D(\rho)}{D(\varepsilon)}. \tag{40}$$

On the other hand

$$\frac{d}{d\rho} \int_{\gamma(\rho)} h^2 * d\xi = 2 \int_{\gamma(\rho)} h \frac{\partial h}{\partial \xi} * d\xi = 2 D(\rho)$$

and consequently

$$2 \int_{\tau_{j-1}}^{\tau_j} D(\rho)\, d\rho = \int_{\gamma(\tau_j)} h^2 * d\xi - \int_{\gamma(\tau_{j-1})} h^2 * d\xi.$$

Let $\sup_G |f| = c < \infty$. Then

$$\int_{\tau_{j-1}}^{\tau_j} D(\rho)\, d\rho \leq 2\pi c^2. \tag{41}$$

From this and (40) it follows that

$$\int_{\tau_{j-1}}^{\tau_j} \exp\left(2\pi \int_{\varepsilon}^{\rho} \frac{d\rho}{\Lambda(\rho)}\right) d\rho \leq \frac{2\pi c^2}{D(\varepsilon)} < \infty,$$

in violation of (33).

3 F. Modular O_{A^0B}-Test. We retain the notation used in 3 E and state the following modular O_{A^0B}-test of Kuroda (see also Kuroda [4, 6]):

Theorem. *If there exists an exhaustion of R and a subsequence $\{n_l\}$ with*

$$\mu_{n_l} \geq c > 1 \tag{42}$$

for every $l=1, 2, \ldots$ and

$$\limsup_{l\to\infty}\left(\prod_{j=1}^{n_l}\mu_j\right)\Big/ k_{n_l}=\infty \tag{43}$$

then R belongs to O_{A^0B}.

From II.2 B.(5) we obtain

$$\begin{aligned}
J_n &= \int_{\tau_{n-1}}^{\tau_n}\exp\left(2\pi\int_1^\rho \frac{d\rho}{\Lambda(\rho)}\right)d\rho\\
&\geq \int_{\tau_{n-1}}^{\tau_n}\left(\prod_1^{n-1}\mu_j\right)\exp\left(\frac{\log\mu_n}{\log\nu_n}(\rho-\tau_{n-1})\right)d\rho\\
&=\left(\prod_1^{n-1}\mu_j\right)\frac{\log\nu_n}{\log\mu_n}\left(\exp\left(\frac{\log\mu_n}{\log\nu_n}(\tau_n-\tau_{n-1})\right)-1\right)\\
&=\left(\prod_1^{n-1}\mu_j\right)\frac{\log\nu_n}{\log\mu_n}(\exp\log\mu_n-1)\\
&=\left(1-\frac{1}{\mu_n}\right)\left(\prod_1^{n}\mu_j\right)\frac{\log\nu_n}{\log\mu_n}.
\end{aligned}$$

On the other hand by (31)

$$2\pi c_{nk}=\int_{\Gamma(\tau_n)\cap\partial R_{nk}} *du_n$$

and

$$2\pi\sum_{k=1}^{k_n}c_{nk}=\int_{\Gamma(\tau_n)} *du_n=2\pi.$$

Thus we obtain

$$1=\sum_{k=1}^{k_n}\frac{\log\nu_n}{\log\mu_{nk}}\leq k_n\frac{\log\nu_n}{\log\mu_n}$$

and consequently

$$J_{n_l}\geq\left(1-\frac{1}{c}\right)\left(\prod_1^{n_l}\mu_j\right)\Big/ k_{n_l}.$$

Therefore (43) implies (33) and we have $R\in O_{A^0B}$.

Remark. Condition (43) is also sufficient for $R\in O_{KB}$.

3 G. The Inclusion $O_{HB}<O_{A^0B}$. The proof of $O_{HB}\subset O_{A^0B}$ parallels that of $O_{HD}\subset O_{A^0D}$. In fact take an $R\in O_{HB}$ and an arbitrary bordered subregion $(G, \partial G)$. If $(G,\partial G)\in SO_{HB}$ then $(G,\partial G)\in SO_{AB}$. If $(G,\partial G)\notin SO_{HB}$ then by IV.3 K and 3 L, $G\in U_{HB}$, and IV.7 G gives $G\in O_{MB^*}\subset O_{AB}$. Thus $(G,\partial G)\in SO_{AB}$ and consequently $R\in O_{A^0B}$.

We shall show that

$$O_{HB}<O_{A^0B} \tag{44}$$

is strict even for plane regions (Kuroda [4]).

Consider the generalized Cantor set $E(p^\infty)$ with $p=1/(1-2l), 0<l<\frac{1}{6}$. Let the center of each segment in $E(p^n)$ be ζ_{nk}, $k=1, \ldots, 2^n$. The annuli

$$A_{nk}=\{\zeta \mid l^n<|\zeta-\zeta_{nk}|<\tfrac{1}{2} l^{n-1}(1-l)\}$$

have the same modulus $\mu=(1-l)/2l$ for all n, k. For each n the $A_{n+1,k}$ $(k=1, \ldots, 2^{n+1})$ together separate $E(p^\infty)$ from the A_{nk} $(k=1, \ldots, 2^n)$. We suppose that each A_{nk} encloses $A_{n+1,2k-1}$ and $A_{n+1,2k}$. Let B_{nk} be the doubly connected region bounded by the inner contour of A_{nk} and the outer contours of $A_{n+1,2k-1}$ and $A_{n+1,2k}$ (the latter two are tangent to each other).

Consider the complement R of $E(p^\infty)$ with respect to $|\zeta| \leq \infty$. We know from 2 D that $R \notin O_G=O_{HB}=O_{HD}$ (for planar regions) and $R \in O_{A^0D}$. We shall also see that $R \in O_{A^0B}$.

Let R_{2n-1} be the unbounded region bounded by the outer contours of the A_{nk} $(k=1, \ldots, 2^n)$, and R_{2n} the unbounded region bounded by the inner contours of the A_{nk}. Then the minimum modulus μ_{2n} of $R_{2n}-\bar{R}_{2n-1}$ is μ, and $k_{2n}=2^n$. Hence

$$\left(\prod_{j=1}^{2n} \mu_j\right) \Big/ k_{2n} > \left(\prod_{i=1}^{n} \mu_{2i}\right) \Big/ k_{2n} = \mu^n/2^n$$

$$= \left(\frac{1-l}{4l}\right)^n > \left(\frac{5}{4}\right)^n$$

by virtue of $l<\frac{1}{6}$. Thus (42) and (43) are satisfied and $R \in O_{A^0B}$.

3 H. Inclusion Relations. We are now prepared to give the inclusion diagram.

THEOREM. *The following strict inclusion relations are valid:*

$$\begin{array}{ccccc} O_{HB} & < & O_{A^0B} & < & O_{AB} \\ \wedge & & \wedge & & \wedge \\ O_{HD} & < & O_{A^0D} & < & O_{AD} \end{array} \tag{45}$$

In view of I.10 D, II.2 C, II.11 D, III.4 H, (44), and (23) it only remains to prove the strictness of $O_{A^0B} \subset O_{A^0D}$. In III.4 H we gave the Tôki example of an $R \in O_{HD}$ as a complete covering surface of $|w|<1$ and with the ideal boundary point corresponding to $|w|=1$. The natural projection f of R onto $|w|<1$ is of course in $M(R)$. However it does not have Iversen's property, and by Theorem 3 A we conclude that $R \notin O_{A^0B}$. On the other hand $R \in O_{HD} \subset O_{A^0D}$.

Since Tôki's surface is not in O_{AB} we also have $O_{HD} \not\subset O_{A^0B}$ and $O_{HD} \not\subset O_{AB}$. The example in 3 G gives an $R \in O_{A^0B} \subset O_{AB}$ with $R \notin O_{HD}$. We infer that

$$O_{HD} \subset\!|\!\supset O_{A^0B}, \quad O_{HD} \subset\!|\!\supset O_{AB} \tag{46}$$

where $\subset\!|\!\supset$ indicates that there is no inclusion relation.

Myrberg's example I.10 B proves that $O_{AB} \not\subset O_{A^0D}$ and Tôki's example III.4 H implies $O_{A^0D} \not\subset O_{AB}$. Thus

$$O_{A^0D} \subset | \supset O_{AB}. \tag{47}$$

The analysis of diagram (45) is herewith complete.

At this point we append the following observation. By IV.9 D, $O_{MB^*} = O_{HB}$ for plane regions. Thus (45) shows that $O_{MB^*} \not\supset O_{A^0B}$. Take an $R' \in O_{HB} - O_G$ (cf. V.7 C) and let R be the surface obtained by removing a disk from R'. By IV.7 G, $R \in O_{MB^*}$. Denote by f' a nonconstant meromorphic function on R' and set $f = f'|R$. Then $f \in M(R)$ but clearly f does not have Iversen's property. Thus by 3 A, $R \notin O_{A^0B}$, i.e. $O_{MB^*} \not\subset O_{A^0B}$. We have shown that

$$O_{MB^*} \subset | \supset O_{A^0B}. \tag{48}$$

§ 2. Boundary Points of Positive Measure

Membership in any one of the "harmonic null classes" O_G, O_{HP}, O_{HB}, O_{HD}, O^n_{HB}, O^n_{HD}, and classes U_{HB}, U_{HD}, and $U_{\widetilde{HD}}$ is a property of the ideal boundary. Moreover the ideal boundary is in these cases relatively nonintricate in the following sense: for a closed disk K, or more generally a set K of positive capacity, we always have $R - K \notin O_X$ even if $R \in O_X$, where O_X is any one of the above harmonic null classes. This is in striking contrast with the cases of O_{AB} and O_{AD}. Myrberg's surface in Chapter I is an example, and an even more surprising one is Kuramochi's phenomenon discussed in Chapters III and IV: if $R \in O_{HD} - O_G$ (resp. $O_{HB} - O_G$), then $R - K \in O_{AD}$ (resp. O_{AB}). The underlying cause was shown to be the existence of points of positive harmonic measure on Royden's or Wiener's boundary.

We shall analyze this type of phenomenon in more detail in 4 where it will be shown to stem from the existence of points of positive harmonic measure on some abstract ideal boundary Γ_X, and the continuity of AB- or AD-functions on Γ_X. In 5 we shall prove this continuity in the case of Stoïlow's boundary and O_{A^0B} or O_{A^0D}-surfaces. Here Iversen's property will play a key role. Although these observations belong to the same line of reasoning as that in Chapters III and IV it will be shown by examples that HD- and HB-minimalities are sufficient but not necessary for these phenomena.

We then return in 6 once more to O_{MD^*} and establish the strictness of the inclusion $O_{HD} < O_{MD^*}$. This with $O_{HB} < O_{MB^*}$ proved in Chapter IV is another extension of Kuramochi's results. The theory is brought to a conclusion by completing the classification scheme for the main null classes.

4. General Compactifications

4 A. Relative X-Compactifications. Let R be an open Riemann surface. A *compactification* of R is a compact Hausdorff space R^* containing R as an open and dense subspace. Given two compactifications R_1^* and R_2^* of R assume there exists a continuous mapping π of R_1^* onto R_2^* such that $\pi|R$ is the identity and $\pi^{-1}(R)=R$. We say that (R_1^*, R_2^*, π) is a *fiber space* and π a *projection*. We indicate this relationship by writing $R_1^* \succ R_2^*$. Then $\succ$ defines a partial ordering of the set $\{R^*\}$ of all compactifications of R. The relation $R_1^* \succ R_2^*$ is equivalent to $B(R_1^*)|R \supset B(R_2^*)|R$ (cf. IV.2 F and 5 D). From this it follows easily that the Alexandroff compactification R_A^* is the smallest and the Čech compactification R_B^*, i.e. the one with $B(R_B^*)|R = B(R)$, is the largest:

$$R_A^* \prec R_S^* \prec R_{\mathbb{M}\mathbb{I}}^* \prec R_{\mathbb{N}}^* \prec R_B^*. \tag{49}$$

For a noncompact bordered surface (R, α) with compact border α there exists an open Riemann surface $\tilde{R}$ such that $R \cup \alpha$ is complex-analytically imbedded in $\tilde{R}$ and $\tilde{R} - R$ is a regular subregion with border α. For a compactification $\tilde{R}_X^*$ of $\tilde{R}$ let $R_{X,\alpha}^*$ be the compact set $\tilde{R}_X^* - (\tilde{R} - R \cup \alpha)$. We call $\tilde{R}_X^*$ the X-compactification of $\tilde{R}$, and $R_{X,\alpha}^*$ the *relative X-compactification* of (R, α). We consider only those relative X-compactifications which do not depend on the choice of $\tilde{R}$. These are exemplified by $R_{\mathbb{M}\mathbb{I},\alpha}^*$, $R_{\mathbb{N},\alpha}^*$, and $R_{S,\alpha}^*$.

4 B. Points of Positive Measure. For a relative X-compactification $R_{X,\alpha}^*$ of a noncompact bordered surface (R, α) with compact border $\alpha \neq \emptyset$ write

$$\Gamma_X = \Gamma_X(R, \alpha) = R_{X,\alpha}^* - R \cup \alpha.$$

Take a set $K \subset \Gamma_X$ and a family $\mathscr{F} = \mathscr{F}(R \cup \alpha, X; K)$ of nonnegative superharmonic functions s on R such that

$$\liminf_{z \in R,\, z \to p} s(z) \geq 1 \tag{50}$$

for every $p \in K$. We set

$$v_{X,\alpha}^K(z) = \inf_{s \in \mathscr{F}} s(z). \tag{51}$$

There are two alternatives: $v_{X,\alpha}^K > 0$ on R or $v_{X,\alpha}^K \equiv 0$ on R. In the latter case we say that K has *harmonic measure zero* with respect to $R_{X,\alpha}^*$. Suppose $\tilde{R} \notin O_G$. Then by III.5 A, 5 D, 5 E and IV.5 A, 5 C, 5 E

$$v_{X,\alpha}^K \equiv 0 \textit{ if and only if } \mu_X(K) = 0 \tag{52}$$

for a compact or Borel set K, and $X = S$, $\mathbb{M}\mathbb{I}$, and $\mathbb{N}$.

Given an open Riemann surface R let R_0 be a regular region with connected $R - \bar{R}_0$. Consider the class U_X of surfaces R such that

$(R-\bar{R}_0)^*_{X,\partial R_0}$ has a point on Γ_X of positive harmonic measure. By III.4 N, IV.3 K we see that

$$U_{\mathrm{MI}}=U_{\widetilde{HD}}, \qquad U_{\mathrm{N}}=U_{HB}.$$

Here we make the important observation that the property $R\in U_X$ does not depend on the choice of R_0. To see this take a regular exhaustion $\{R_n\}_0^\infty$ of R with connected $R-\bar{R}_0$ and $R-\bar{R}_1$. Denote by $HB_0(R-\bar{R}_0)$ and $HB_0(R-\bar{R}_1)$ the classes of HB-functions on $R-\bar{R}_0$ and $R-\bar{R}_1$ vanishing on ∂R_0 and ∂R_1 respectively.

For $u\in HB_0(R-\bar{R}_0)$ with $u\geq 0$ let $v_n\in HB(\bar{R}_n-R_1)$ and $h_n\in HB(\bar{R}_n-R_1)$ such that $v_n|\partial R_1=0$, $v_n|\partial R_n=u$, $h_n|\partial R_1=u$, and $h_n|\partial R_n=0$. Then both $\{v_n\}_2^\infty$ and $\{h_n\}_2^\infty$ are monotone and we conclude that

$$v=\lim_{n\to\infty} v_n, \qquad h=\lim_{n\to\infty} h_n$$

exist; $v|\partial R_1=0$; $h|\partial R_1=u$; $v,h\geq 0$; and $v+h=u$. We set $v=T_0\,u$ and $h=T_1\,u$. Since $HB_0(R-\bar{R}_0)=HP'B_0(R-\bar{R}_0)$, T_0 and T_1 can be extended to linear operators from $HB_0(R-\bar{R}_0)$ into $HB_0(R-R_1)$ and $HB(R-R_1)$ respectively.

If $u>0$ then $T_0\,u>0$. In fact if $T_0\,u=0$ then $u=T_1\,u$ on $R-R_1$. Since h_n takes its maximum on ∂R_1, so does $T_1\,u$. Hence $u|\partial R_0=0$ implies that u assumes its maximum at an interior point of ∂R_1, a contradiction.

As for a $p\in\Gamma_X$ suppose $v^p_{X,\partial R_0}>0$. Clearly $v^p_{X,\partial R_0}\in HB(R-\bar{R}_0)$. For any $s\in\mathscr{F}=\mathscr{F}((R-\bar{R}_1)\cup\partial R_1, X; p)$, $s+T_1\,v^p_{X,\partial R_0}\geq v^p_{X,\partial R_0}$ on $R-\bar{R}_1$. Thus $s\geq T_0\,v^p_{X,\partial R_0}$, i.e. $v^p_{X,\partial R_1}\geq T_0\,v^p_{X,\partial R_0}>0$.

For arbitrary R_0 and R_0' choose R_1 such that $R_1\supset\bar{R}_0,\bar{R}_0'$. On applying the above argument we conclude that $v^p_{X,\partial R_0}>0$ does not depend on R_0.

4 C. Localizable U_X. Suppose $R\in U_X$, i.e. there exists a point $p\in\Gamma_X=R^*_X-R$ of positive harmonic measure. Let G be an open subset of R such that $\bar{G}$ is a neighborhood of p and $\partial G\neq\emptyset$ is analytic. We can define $v^p_{X,\partial G}$ in the same manner as $v^p_{X,\alpha}$ in (51), i.e. in (50) replace the limit by

$$\liminf_{z\in G,\, z\to p} s(z)\geq 1,$$

where s is of course defined only on G. If every $R\in U_X$ has the property $v^p_{X,\partial G}>0$ for every admissible G then we say that U_X is *localizable.*

By III.5 D and IV.5 C, $U_{\mathrm{MI}}=U_{\widetilde{HD}}$ and $U_{\mathrm{N}}=U_{HB}$ are localizable.

4 D. The Class $C_{X,Y}$. Given a class $Y(R)\subset M(R)$ let O_Y be the corresponding null class. We are also interested in the class $C_{X,Y}$ of Riemann surfaces R such that every function in $Y(R-R_0)$ is continuous on $(R-\bar{R}_0)^*_{X,\partial R_0}$ for every regular subregion R_0 of R.

For example $C_{\mathrm{MI},AD}$, $C_{\mathrm{N},AB}$, and C_{N,MB^*} contain all open Riemann surfaces.

4 E. General Identity Theorem. We shall now establish the following central result:

THEOREM. *Assume that either* U_X *is localizable or* $Y \subset AB$. *Then* $R \in U_X \cap C_{X,Y}$ *implies* $R - K \in O_Y$ *for every compact subset* K *of* R *with connected complement.*

Suppose U_X is localizable, and $C_{X,Y}$ is *localizable* in the sense that for $R \in C_{X,Y}$ every function in $Y(G)$ is continuous at $\Gamma_X \cap \bar{G} - \overline{\partial G}$ for every subregion $G \subset R$. Then the compact set K in the theorem can be replaced by a closed set in R such that $\overline{R-K}$ is a neighborhood of a $p \in \Gamma_X$ of positive measure in R_X^*.

For the proof of the theorem let $f \in Y(R-K)$ and suppose to the contrary that f is not constant.

First assume that $Y \subset AB$. There exists a point $p \in \Gamma_X$ of positive measure. We may take $f(p)=0$ and $|f|<1$. Choose a regular region $R_0 \supset K$ with connected $R - \bar{R}_0$. The superharmonic function $s = -\log|f| > 0$ takes the value ∞ at p. Thus $\varepsilon s \in \mathscr{F}$ and therefore $v^p_{X,\partial R_0} = 0$, a contradiction.

Next suppose U_X is localizable. In this case take $G = \{z \in R - \bar{R}_0 \,|\, |f(z)| < 1\}$; then $\bar{G}$ is a neighborhood of p. In the same manner as above we infer that $v^p_{X,\partial G} = 0$ and consequently $v^p_{X,\partial R_0} = 0$, a contradiction.

This completes the proof.

Corollary III. 5 I and IV. 7 G are special cases of the above theorem including the methodological viewpoint.

In 5 we shall consider the case $X = S$, and $Y = AB$ or AD. The important facts to be observed are that $O_{A^0B} \subset C_{S,AB}$ and $O_{A^0D} \subset C_{S,AD}$. Thus the reasoning will essentially follow the same pattern as that in the proof of Theorem 4 E. However we shall repeat this somewhat sophisticated discussion in order to better prepare the reader for the subsequent development.

5. Identity Theorems on O_{A^0D}- and O_{A^0B}-Surfaces

5 A. The Class U_S. Consider the class U_S of open Riemann surfaces whose Stoïlow ideal boundary Γ_S contains a point of positive harmonic measure (cf. 4 B). The surfaces in the classes $U_{\widetilde{HD}}$ and U_{HB} were intricate in nature. However U_S contains quite simple surfaces. For example $R = \{|z|<1\} \in U_S$.

We insert here the remark that

$$U_{HB} < U_{\widetilde{HD}} < U_S. \tag{53}$$

By IV.(67) and (69) the inclusions $U_{HB} \subset U_{\widetilde{HD}} \subset U_S$ are obvious. The unit disk shows that $U_{\widetilde{HD}} < U_S$. The strictness of $U_{HB} < U_{\widetilde{HD}}$ was proved in IV. 3 N.

5 B. The Class $U_S \cap O_{A^0 D}$. We now state (MATSUMOTO [4]):

THEOREM. *Suppose $R \in U_S \cap O_{A^0 D}$ and let K be an arbitrary compact set in R with connected complement. Then $R - K \in O_{AD}$.*

Take a regular subregion $R_0 \subset R$ such that $R_0 \supset K$ and $R - \bar{R}_0$ is connected. Theorem 2 C shows that $AD(R - R_0) \subset AB(R - R_0)$ and every $f \in AD(R - R_0)$ is continuous on $R^*_{S,\alpha}$, $\alpha = \partial R_0$. We must prove that f is constant. Suppose the contrary is true. We can then assume that $|f| < 1$ and $f(p) = 0$, where $p \in \Gamma_S$ has positive measure. The function $s = -\log |f| > 0$ is superharmonic and $s(p) = \infty$. Hence $\varepsilon s \in \mathscr{F}$ for every $\varepsilon > 0$, i.e. $v^p_{S,\alpha} = 0$, a contradiction.

5 C. A Surface in $U_S \cap O_{A^0 \underline{D}} - U_{H\widetilde{D}}$. Remove a closed disk $\bar{U}$ from an $R' \in O_{HD} - O_G$. Then $R = R' - \bar{U} \in U_{H\widetilde{D}}$ but $R \notin O_{A^0 D}$, and therefore $U_{H\widetilde{D}} \not\subset U_S \cap O_{A^0 D}$. We shall show that there exists an $R \in U_S \cap O_{A^0 D} - U_{H\widetilde{D}}$, i.e.

$$U_S \cap O_{A^0 D} \subset | \supset U_{H\widetilde{D}}. \tag{54}$$

The following example is due to MATSUMOTO [4].

Let $E(p^\infty)$ be a Cantor set with $p = 1/(1 - 2l)$ and $0 < l < \frac{1}{2}$. We recall (2 D) that $E(p^\infty)$ has positive capacity, $E(p^0) = [0, 1]$, and $E(p^n)$ consists of 2^n intervals I_{nk} $(k = 1, \ldots, 2^n)$. We denote by ζ_{nk} the midpoint of I_{nk} and consider the annuli

$$A_{nk} = \{\zeta \mid l^n(1 - l) < |\zeta - \zeta_{nk}| < \tfrac{1}{2} l^{n-1}(1 - l)\}$$

in the complement W of $E(p^\infty)$ with respect to the extended plane. The $A_{n+1,k}$ $(k = 1, \ldots, 2^{n+1})$ together separate $E(p^\infty)$ from the A_{nk} $(k = 1, \ldots, 2^n)$. We suppose that each A_{nk} encloses $A_{n+1,2k-1}$ and $A_{n+1,2k}$ and we denote by B_{nk} the doubly connected region bounded by the inner contour of A_{nk} and the outer contours of $A_{n+1,2k-1}$ and $A_{n+1,2k}$. We fix an arbitrary slit S_{nk} in B_{nk}. The "smallness" of S_{nk} will be specified later.

Let $W_0 = W - \bigcup_{n,k} S_{nk}$. For each fixed $m \geq 1$ denote by W_m the set $W - \bigcup_{nk} S_{nk}$ with $n = 2^{m-1}(2i + 1)$, $i = 0, 1, \ldots$. Thus each slit S_{nk} appears exactly twice, once in W_0 and once in some W_m $(m \geq 1)$. Connect W_m with W_0 crosswise along each S_{nk} $(n = 2^{m-1}(2i + 1), i = 0, 1, \ldots; k = 1, \ldots, 2^n)$. The resulting covering surface R of W has only one Stoïlow ideal boundary point. Since $W \notin O_G$ we have $R \notin O_G$. This means that

$$R \in U_S. \tag{55}$$

Let A^m_{nk} be the annulus on W_m corresponding to A_{nk}. Then all A^m_{nk} $(m = 0, 1, \ldots; n \geq 1$ if $m = 0$; $n \geq 2^{m-1}$ if $m > 0)$ have the same modulus $\mu = 1/(2l) > 1$, and for each $n \geq 1$ the $A^m_{n+1,k}$ together separate the ideal boundary of R from the A^m_{nk}. For each $n \geq 1$ the minimum μ_n of the moduli of the A^m_{nk} is $\mu > 1$ and

$$\prod_1^\infty \mu_n = \lim_{n\to\infty} \mu^n = \infty .$$

On considering R_{2n-1}, R_{2n} as in 3 G we conclude by I.1 H that

$$R \in O_{A^0 D}. \tag{56}$$

Let D_{nk} be a concentric circular annulus separating S_{nk} from $E(p^\infty)$ in W_0. Without affecting (55) and (56) we may choose S_{nk} so small that D_{nk} can be taken to satisfy

$$\bar{D}_{nk} \cap \bar{D}_{n'k'} = \emptyset$$

if $(n, k) \neq (n', k')$ and

$$\sum_{n,k} \frac{1}{\log \operatorname{mod} D_{nk}} < \infty . \tag{57}$$

On R there is only one connected fragment $\tilde{D}_{nk}$ over D_{nk} which is 2-sheeted. Clearly $\log \operatorname{mod} \tilde{D}_{nk} = 2 \log \operatorname{mod} D_{nk}$. For this sequence $\{\tilde{D}_{nk}\}$ the conditions (α), (β), and (γ) in III.5 F are satisfied. Thus by Theorem III.5 F

$$R \notin U_{\widetilde{HD}}. \tag{58}$$

In view of (55), (56), and (58) we have $R \in U_S \cap O_{A^0 D} - U_{\widetilde{HD}}$.

5 D. The Class $U_S \cap O_{A^0 B}$. As a counterpart of Theorem 5 B we state the following result (Toda-Matsumoto [1]) the proof of which is now obvious.

Theorem. *Let $R \in U_S \cap O_{A^0 B}$. Then $R - K \in O_{AB}$ for every compact set $K \subset R$ with connected $R - K$.*

5 E. A Surface in $U_S \cap O_{A^0 B} - U_{HB}$. Remove a closed disk $\bar{U}$ from an $R' \in O_{HB} - O_G$. Then the resulting surface $R = R' - \bar{U}$ is in U_{HB} but not in $O_{A^0 B}$, and therefore $U_{HB} \not\subset U_S \cap O_{A^0 B}$. We shall now construct an $R \in U_S \cap O_{A^0 B} - U_{HB}$ to show that

$$U_S \cap O_{A^0 B} \subset\!|\!\supset U_{HB}. \tag{59}$$

The surface to be described is due to Toda-Matsumoto [1] and is similar to that in 5 C. The only differences are that $p = 1/(1-2l)$ with $0 < l < \frac{1}{6}$ and

$$A_{nk} = \{\zeta \mid l^n < |\zeta - \zeta_{nk}| < \tfrac{1}{2} l^{n-1}(1-l)\} .$$

Exactly as in 5 C we see that

$$R \in U_S. \tag{60}$$

In analogy with (58) we have

$$R \notin U_{HB}. \tag{61}$$

However this now follows directly from Corollary 1 in IV.8 C. In fact suppose $R \in U_{HB}$. The natural projection f of R onto W belongs to $M(R)$. Since $\{|z| \leq \infty\} - f(R)$ is of positive capacity we have a contradiction.

As for the counterpart of (56) we observe that the annuli A_{nk}^m have the same modulus $\mu = (1-l)/2l > \frac{5}{2}$ and consequently the minimum modulus $\mu_n > \frac{5}{2}$. The number k_n of these annuli satisfies $k_n \leq 2^n n$. Therefore

$$\frac{1}{k_n} \prod_1^n \mu_j > \frac{1}{n} \left(\frac{5}{4}\right)^n \to \infty$$

as $n \to \infty$, and by 3 F

$$R \in O_{A^0 B}. \tag{62}$$

6. The Class O_{MD^*}

6 A. Continuity on $R_{\mathbb{M}}^*$. Once more we return to the class MD^* (cf. 2 A). We first show:

THEOREM. *Every function $f \in MD^*(R)$ is continuous on the Royden compactification $R_{\mathbb{M}}^*$ of R.*

Clearly $|f| \cap n \in \mathbb{M}(R)$ for every $n > 0$, and consequently $|f|$ is continuous on $R_{\mathbb{M}}^*$ (cf. III.3 A). Let $G_n = \{z \in R \,|\, |f(z)| > n\}$. Then $K = \bigcap \bar{G}_n$ is compact in $R_{\mathbb{M}}^*$ and $\bar{G}_n$ is a neighborhood of K since $|f|$ is continuous. Clearly f is continuous on $R - \bar{G}_n$ for every n and $f = \infty$ continuously at each point of K.

6 B. MD^* on U_{HD}-Surfaces. We can now prove an analogue of Theorem IV.7 G:

THEOREM. *The following inclusion relation holds:*

$$U_{HD} \subset O_{MD^*}. \tag{63}$$

Suppose $R \in U_{HD}$ and $R \notin O_{MD^*}$. Since $MD^*(R)$ is a field we may assume for a nonconstant $f \in MD^*(R)$ that $f(p) = 0$ where p is a point in $\Gamma_{\mathbb{M}}$ with $\mu_{\mathbb{M}}(p) > 0$. By Theorem III.5 H, $f \equiv 0$, a contradiction.

6 C. M on U_{HB}-Surfaces. First we remark that

$$O_{HB}^{\infty} < O_{A^0 B}. \tag{64}$$

Let $R \in O_{HB}^{\infty}$ and consider a nonconstant $f \in M(R)$. Take a disk U in $|w| \leq \infty$ and let V be a component of $f^{-1}(U)$. Suppose $(V, \partial V) \notin SO_{HB} \subset SO_{AB}$; then $\bar{V} \cap \Delta_{\mathbb{N}} - \overline{\partial V}$ contains a point (cf. IV.3 L). Since $\Delta_{\mathbb{N}}$ is the closure of the set of isolated points in $\Delta_{\mathbb{N}}$ if and only if $R \in O_{HB}^{\infty}$, the set $\bar{V} \cap \Delta_{\mathbb{N}} - \partial V$ contains an isolated point in $\Delta_{\mathbb{N}}$. By Theorems IV.5 C and 7 A, $V \in O_{AB}$ and thus $(V, \partial V) \in SO_{AB}$. Therefore $O_{HB}^{\infty} \subset O_{A^0B}$. The example in 3 G serves to show that $O_{HB}^{\infty} < O_{A^0B}$.

We have seen that for $R \in O_{HB}^{\infty}$ every nonconstant $f \in M(R)$ has the localizable Iversen property. However if we remove a disk K from $R \in O_{HB}^{\infty}$ this is no longer true since $R - K \in U_{HB}$ (cf. IV.5 C). Nevertheless we can show (CONSTANTINESCU-CORNEA [3]):

THEOREM. *For every $f \in M(R)$ with $R \in U_{HB}$ the exceptional set $\{w \mid n_f(w) < \infty\}$ has capacity zero.*

Let $\{R_n\}_1^{\infty}$ be an exhaustion of R and set

$$E_n = \{|w| \leq \infty\} - f(R - \bar{R}_n).$$

Then clearly $E_1 \subset E_2 \subset \cdots$ and

$$E = \{w \mid n_f(w) < \infty\} = \bigcup_{n=1}^{\infty} E_n.$$

Suppose contrary to the assertion that E has positive capacity. Then some E_n has positive capacity. Let $R' = R - \bar{R}_n$. By IV.5 C, $R' \in U_{HB}$ and $f' = f \mid R' \in M(R')$. Because of Corollary 1 in IV.8 C, $f'(R') = \{|w| \leq \infty\} - E_n \in O_G$, a contradiction.

6 D. A Criterion for O_{MD^*}. Let $\tilde{R} \in O_{A^0B}$ and let (R, α) be a bordered subregion with compact border α such that $\tilde{R} - \bar{R}$ is a regular subregion of $\tilde{R}$ with $\alpha = \partial(\tilde{R} - \bar{R})$. The case $\alpha = \emptyset$ is not excluded. If $\alpha \neq \emptyset$ we can choose an exhaustion $\{\tilde{R}_n\}_0^{\infty}$ of $\tilde{R}$ with $\tilde{R}_0 = \tilde{R} - \bar{R}$. Then $\{R_n\}_1^{\infty}$ with $R_n = \tilde{R}_n - \bar{\tilde{R}}_0$ is an "exhaustion" of R with $\bar{R}_n \supset \alpha$.

By Theorem 3 D every nonconstant $f \in MD^*(R \cup \alpha)$ is continuous on $R_{S,\alpha}^*$. Moreover f covers $|w| \leq \infty$ at most a finite number of times, say N, and $f(\Gamma_S)$ is a totally disconnected closed set. From this we have the following test for O_{MD^*}:

THEOREM. *Let (R, α) be a bordered surface with compact border α (which may be empty) whose extension $\tilde{R}$ is in O_{A^0B}. Suppose there exists a Stoïlow ideal boundary point p whose every defining sequence $\{G_n\}$ has the property*

$$\sup_n \nu(n) = \infty \tag{65}$$

where $\nu(n)$ is the number of components of ∂G_n. Then $R \in O_{MD^*}$.

Let $w = f(p) \in f(\Gamma_S)$ where $f \in MD^*(R)$ is assumed nonconstant. Take a simply connected region W_n containing w such that $\partial W_n \cap f(\Gamma_S) = \emptyset$, ∂W_n is analytic, $W_1 \supset W_2 \supset \cdots$, and $\bigcap W_n = w$. Select a component G_n of

$f^{-1}(W_n)$ which is a neighborhood of p in the sense that $\bar{G}_n - \overline{\partial G}_n$ is an open set containing p. Clearly $\{G_n\}$ is a defining sequence of p. The relative boundary ∂G_n is compact and lies over ∂W_n. Hence

$$\nu(n) \leq N$$

which contradicts (65).

Obviously criterion (65) also applies to $R \in O_{AD^*}$.

6 E. Table of Strict Inclusion Relations. We summarize the main strict inclusion relations obtained in our book in the following table:

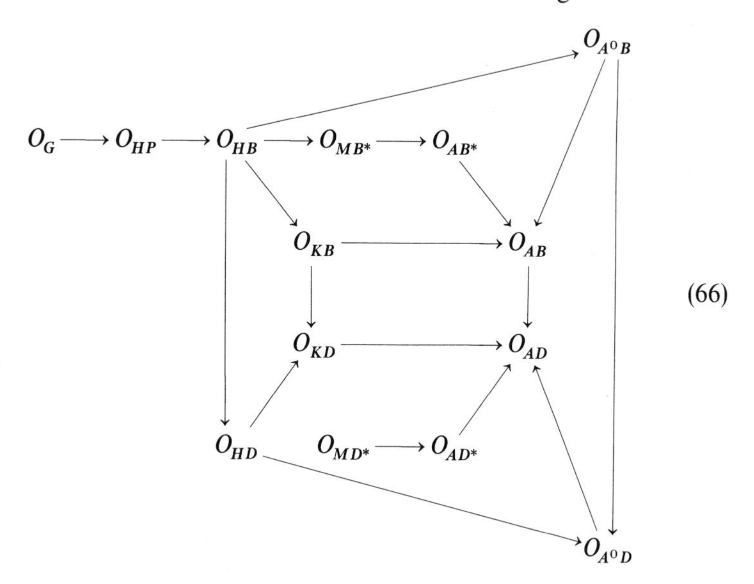

(66)

Here $O_X \to O_Y$ stands for $O_X < O_Y$.

Since closed surfaces do not belong to O_{MD^*} but do belong to every other class O_X in the above table we have $O_{MD^*} \not\supset O_X$. Similarly the finite plane serves to show that $O_{AD^*} \not\supset O_X$ where O_X is now any class in the table except O_{MD^*} and O_{AD^*}.

The surface R in IV.9 B belongs to O_G and has only one Stoïlow ideal boundary point p, which clearly satisfies (65). Remove a disk U from R and set $R' = R - \bar{U}$. Then $R' \in O_{MD^*}$. However there exists a nonconstant $u \in HD(R')$. Since R' has only one Stoïlow ideal boundary point and this has measure zero we obtain $u \in KD(R')$. Thus $O_{AD^*} \not\subset O_{KD}$. Next take the surface $R \in O_{HD} - O_G \subset O_{MD^*}$ (cf. 6 B) constructed in III.4 H. Since the projection f of R onto $|z| < 1$ is a nonconstant AB-function we conclude that $R \notin O_{AB}$ and consequently $O_{AD^*} \not\subset O_{AB}$. Therefore from the table we see that O_{MD^*} (resp. O_{AD^*}) $\not\subset O_X$ for any O_X except O_{MD^*}, O_{AD^*}, and O_{AD} (resp. O_{AD^*} and O_{AD}).

Appendix

Higher Dimensions

The class of Riemann surfaces is identical with the class of Riemannian manifolds of dimension 2. Therefore a classification theory of Riemannian manifolds based on the existence or nonexistence of harmonic functions with some specified property is a natural generalization of the classification theory of Riemann surfaces. If we replace harmonic functions by solutions of elliptic partial differential equations other than the Laplace-Beltrami equation we are led to further extensions. Moreover since the global theory of such solutions can be discussed in a more general setting, for example on harmonic spaces of Brelot, the classification theory of such spaces presents a challenging new problem.

Keeping these aspects in mind we shall however restrict our attention in the present book to Riemannian manifolds because of their concrete geometric significance. Most classification problems of Riemann surfaces can be extended to the general case of Riemannian manifolds of arbitrary dimension insofar as they concern merely the harmonic aspect and do not involve complex analytic structure. As for the classification theory of complex analytic manifolds of higher dimensions it falls outside of the scope of our book and presents itself as one of the most promising – and probably most difficult – topics for future research.

In this appendix we pay particular attention to those problems on Riemannian manifolds which bring forth new features because of the higher dimension. In 1 we briefly review fundamental definitions and properties of Riemannian manifolds, differential forms, and harmonic functions. The harmonic modulus of a higher dimensional annulus is then introduced in 2. This has no geometric interpretation as in the 2-dimensional case but it retains most of its analytic features.

In 3 we discuss the problem of parabolicity. The Euclidean ball can be endowed with a Riemannian structure conformally equivalent to the Euclidean one and such that the resulting manifold is parabolic. This illustrates the decisive distinction between dimension two and higher dimensions. If the Laplace-Beltrami equation is replaced by $\Delta u = Pu$ $(P \geq 0)$ then there are no parabolic manifolds unless $P \equiv 0$.

In 4 we establish the quasiconformal noninvariance and quasi-isometric invariance of the classes O_G, O_{HD}, and O_{KD}. This phenomenon also reveals a striking contrast between dimension two and higher dimensions.

No attempt at completeness has been made in this appendix. The main object is to call the attention of the reader to the variety and richness that is in store for further development of the classification problem.

1. Fundamentals

1 A. Riemannian Manifolds. By a *Riemannian manifold* $R=R^n$ we mean a connected, orientable, C^∞-manifold of dimension n with a smooth metric tensor g_{ij} corresponding to a positive definite symmetric form $g_{ij}\xi^i\xi^j$. Here we adopt the Einstein convention that an index in both lower and upper positions implies summation. We shall often use the same letter x for a generic point in R and a local parametric system $x=(x^1,\ldots,x^n)$ at the generic point x. If we can always choose $x=(x^1,\ldots,x^n)$ so that $g_{ij}(x)\equiv\delta_{ij}$ (the Kronecker delta) then we say that R is *locally flat*, or *locally Euclidean.*

We denote by (g^{ij}) the inverse matrix of (g_{ij}) and by g its determinant, i.e.

$$(g_{ij})(g^{ij})=(g^{ij})(g_{ij})=(\delta_{ij}),\qquad g=\det(g_{ij}). \tag{1}$$

The metric tensor g_{ij} determines the *arc element* ds by

$$ds^2=g_{ij}\,dx^i\,dx^j \tag{2}$$

and the *volume element* dV by

$$dV=\sqrt{g}\,dx=\sqrt{g}\,dx^1\wedge\cdots\wedge dx^n \tag{3}$$

in terms of the local parameter $x=(x^1,\ldots,x^n)$.

An open set G of R will be called *smooth* if the relative boundary ∂G of G has the following property: for every $x\in\partial G$ there exists a parametric ball V about x such that $V\cap\partial G$ is an open subset of a hyperplane in terms of some local parameters of V. If moreover G is relatively compact in R then we call G *regular.*

Since R is a metric space equipped with the natural distance

$$d(x_1,x_2)=\inf_\gamma\int ds \tag{4}$$

where the infimum is taken with respect to all arcs γ which join x_1 and x_2, R satisfies the second countability axiom and thus admits an *exhaustion* $\{R_m\}_1^\infty$, i.e. a sequence of regular subregions R_m of R with

$$\bar{R}_m\subset R_{m+1},\qquad R=\bigcup_1^\infty R_m. \tag{5}$$

In the special case $n=2$ we can always find an isothermal parametric system, i.e. one with $g_{11}=g_{22}$ and $g_{12}=g_{21}=0$ over the entire parametric ball (LICHTENSTEIN [1], KORN [1]). Therefore $R=R^2$ is a Riemann surface. Conversely for a Riemann surface R there always exists a positive density ρ such that $\rho|dz|$ is invariant. This gives isothermal coordinates with $g_{ij}=\rho^2\delta_{ij}$ for the parameter $z=x^1+ix^2$ and R becomes a Riemannian manifold. In short every 2-dimensional Riemannian manifold is a Riemann surface and conversely:

The conformal structure and the Riemannian structure are equivalent in the 2-dimensional case.

For higher dimensions the Riemannian structure provides us with more generality than the conformal structure and this naturally brings forth a new aspect in the classification problem, one in which we are particularly interested in 4.

1 B. Differential Forms. We denote by $\mathfrak{A}^p$ $(0\le p\le n)$ the space of *differential p-forms* or simply p-forms

$$\varphi=\varphi_{\alpha_1\ldots\alpha_p}dx^{\alpha_1}\wedge\cdots\wedge dx^{\alpha_p} \tag{6}$$

where the $\alpha=(\alpha_1,\ldots,\alpha_p)$ are p-tuples of integers with $1\le\alpha_1<\cdots<\alpha_p\le n$ and the $\varphi_\alpha=\varphi_{\alpha_1\ldots\alpha_p}$ are covariant tensors of rank p. Here 0-forms are nothing but functions.

The *Hodge star operator* $*$ is the operator $\varphi\to*\varphi$ of $\mathfrak{A}^p$ into $\mathfrak{A}^{n-p}$ given by

$$\begin{aligned}*\varphi&=(*\varphi)_{\beta_1\ldots\beta_{n-p}}dx^{\beta_1}\wedge\cdots\wedge dx^{\beta_{n-p}},\\(*\varphi)_{\beta_1\ldots\beta_{n-p}}&=\operatorname{sgn}\begin{pmatrix}1\,\ldots\ldots\ldots\ldots\,n\\\gamma_1\ldots\gamma_p\;\;\beta_1\ldots\beta_{n-p}\end{pmatrix}\sqrt{g}\,\varphi_{\alpha_1\ldots\alpha_p}g^{\gamma_1\alpha_1}\ldots g^{\gamma_p\alpha_p}\end{aligned} \tag{7}$$

where $\alpha=(\alpha_1,\ldots,\alpha_p)$, $\beta=(\beta_1,\ldots,\beta_{n-p})$, and $\gamma=(\gamma_1,\ldots,\gamma_p)$ are as in (6) and γ is disjoint from β.

The *exterior derivative* $d\varphi$ of the p-form φ in (6) is the $(p+1)$-form

$$\begin{aligned}d\varphi&=d\varphi_{\alpha_1\ldots\alpha_p}\wedge dx^{\alpha_1}\wedge\cdots\wedge dx^{\alpha_p},\\d\varphi_{\alpha_1\ldots\alpha_p}&=\frac{\partial\varphi_{\alpha_1\ldots\alpha_p}}{\partial x^i}dx^i.\end{aligned} \tag{8}$$

Here and in the sequel we postulate that all functions under consideration are sufficiently smooth. The *coderivative* $\delta\varphi$ of the p-form φ is given by

$$\delta\varphi=(-1)^{np+n+1}*d*\varphi. \tag{9}$$

This is a $(p-1)$-form for $p\ge1$ whereas $\delta\varphi=0$ for 0-forms.

The *Laplace-Beltrami operator* Δ is defined as

$$\Delta = \delta d + d\delta. \tag{10}$$

Hence $\Delta\varphi$ is a p-form together with φ. A p-form φ with $\Delta\varphi = 0$ is said to be *harmonic*.

We are primarily interested in 0-forms, i.e. functions. We have

$$du = \frac{\partial u}{\partial x^i} dx^i, \tag{11}$$

$$*du = \sum_{j=1}^{n} (-1)^{j-1} \sqrt{g}\, g^{ij} \frac{\partial u}{\partial x^i} dx^1 \wedge \cdots \wedge dx^{j-1} \wedge dx^{j+1} \wedge \cdots \wedge dx^n, \tag{12}$$

$$\Delta u = -\frac{1}{\sqrt{g}} \frac{\partial}{\partial x^i} \left(\sqrt{g}\, g^{ij} \frac{\partial u}{\partial x^j} \right). \tag{13}$$

Given a regular region G and an $(n-1)$-form φ on $\bar{G}$, *Stokes' formula* reads

$$\int_G d\varphi = \int_{\partial G} \varphi. \tag{14}$$

For functions u, v on an open set G the *Dirichlet inner product* $D_G(u, v)$ is by definition

$$D_G(u, v) = \int_G du \wedge * dv = \int_G g^{ij} \frac{\partial u}{\partial x^i} \frac{\partial v}{\partial x^j} dV \tag{15}$$

whenever the integral is meaningful. The *Dirichlet integral* $D_G(u)$ of u is

$$D_G(u) = D_G(u, u) = \int_G du \wedge * du = \int_G g^{ij} \frac{\partial u}{\partial x^i} \frac{\partial u}{\partial x^j} dV. \tag{16}$$

From (14) it follows that

$$D_G(u, v) = \int_{\partial G} u * dv + \int_G u \,\triangle\, v \, dV \tag{17}$$

for a regular G.

Taking coordinates $x = (x^1, \ldots, x^n)$ such that $x^n \equiv 0$ for $x \in \partial G$ we obtain a Riemannian structure for each component of ∂G. The *surface element* dS on ∂G is by definition the volume element of ∂G considered as an $(n-1)$-dimensional Riemannian manifold in the above sense. If v is defined on an open set containing $\bar{G}$ then the *normal derivative* $\partial v / \partial n$ of v is defined by

$$\frac{\partial v}{\partial n} dS = * dv \tag{18}$$

on ∂G. In particular if $v | \partial G$ is constant then by (12)

$$\frac{\partial v}{\partial n} = (-1)^{n-1} \sqrt{g^{nn}} \frac{\partial v}{\partial x^n} \tag{19}$$

and we can write

$$D_G(u,v)=\int_{\partial G} u\frac{\partial v}{\partial n}\,dS+\int_G u\,\Delta\, v\,dV. \tag{20}$$

Similarly

$$D_G(u)=\int_G |\mathrm{grad}\,u|^2\,dV=\int_{\partial G} u\frac{\partial u}{\partial n}\,dS+\int_G u\,\Delta\, v\,dV \tag{21}$$

where we agree to set

$$|\mathrm{grad}\,u|^2=g^{ij}\frac{\partial u}{\partial x^i}\,\frac{\partial u}{\partial x^j}. \tag{22}$$

For a detailed account of the topics in 1 A and 1 B we refer the reader to e.g. DE RHAM [1] and HODGE [1].

1 C. Harmonic Functions. A harmonic function u is a harmonic 0-form (cf. (10)) and therefore by (13) a solution of the second order strongly elliptic partial differential equation

$$\frac{\partial}{\partial x^i}\left(\sqrt{g}\,g^{ij}\frac{\partial u}{\partial x^j}\right)=0. \tag{23}$$

We continue denoting by $H(R)$ the class of harmonic functions on $R=R^n$ and by $HP(R)$, $HB(R)$, and $HD(R)$ the subclasses of functions in $H(R)$ which are positive, bounded, and Dirichlet finite respectively. We also retain the notation $K(R)$ for the class of functions u in $H(R)$ which have vanishing flux $\int_\gamma *du=0$ across every dividing $(n-1)$-cycle γ.

The existence of the fundamental solution for (23) leads to the existence of a Green's function $q_V(\cdot,a)$ for each parametric ball V. Here $q_V(\cdot,a)$ with pole $a\in V$ is the unique function on $\bar{V}$ with the following properties:

$$\begin{aligned}
&\text{(q.1)}\ \ q_V(\cdot,a)\in HP(V-a),\\
&\text{(q.2)}\ \ q_V(\cdot,a)\in C(\bar{V})\ \text{and}\ q_V(\cdot,a)|\partial V=0,\\
&\text{(q.3)}\ \ q_V(x,a)=O(d(x,a)^{2-n})\ \text{for}\ n>2,\ O(-\log d(x,a))\ \text{for}\\
&\qquad\ \ n=2\ \text{as}\ d(x,a)\to 0,\\
&\text{(q.4)}\ \ \textstyle\int_{\partial V} *dq_V(\cdot,a)=-1.
\end{aligned} \tag{24}$$

The last normalization is immaterial; for Riemann surfaces we chose the constant -2π but in this appendix we take -1.

For $u\in B(\bar{V})\cap H(V)$ we obtain by (14)

$$u(a)=-\int_{\partial V} u*dq_V(\cdot,a). \tag{25}$$

This property actually characterizes harmonicity of functions u in $B(\bar{V})$.

A *superharmonic function* v on R is a lower semicontinuous function on R with $-\infty < v \leq \infty$, $v \not\equiv \infty$, such that for every parametric ball V

$$v(a) \geq -\int_{\partial V} v * dq(\cdot, a). \tag{26}$$

An upper semicontinuous function v is called *subharmonic* if $-v$ is superharmonic.

From the existence of a local Green's function for (23) the following properties follow easily:

(I) *Harnack's inequality:* Harnack's function

$$k(x, y) = \inf\{c \mid c^{-1} u(x) \leq u(y) \leq c\, u(x) \text{ for every } u \in HP(R)\}$$

is finite, where $x, y \in R$, and $\log k(x, y)$ is a pseudometric on R.

(II) *Maximum principle:* $u \in H(R)$ does not take on its maximum on R unless it is constant.

(III) *Completeness:* if $\{u_n\} \subset H(R)$ converges to a finite function u on compacta of R then $u \in H(R)$.

(IV) *Monotone compactness:* if $\{u_n\} \subset H(R)$ is a montone sequence bounded at a point of R then it converges to a $u \in H(R)$.

(V) *Bounded compactness:* $HB(R)$ is sequentially compact.

It follows from these properties that the Perron-Brelot method applies to $H(R)$ and therefore the *Dirichlet problem* is solvable for regular regions. Moreover we can take for V in (24) an arbitrary regular region and obtain the *Green's function* for such a V.

Finally we remark that a sequence $\{u_n\} \subset HD(R)$ which converges at a point and converges in Dirichlet norm converges uniformly on every compact subset of R.

For a detailed account of subjects in 1 C we refer the reader to FELLER [1] and to the monographs of MIRANDA [1], DUFF [1], HÖRMANDER [1], and BRELOT [5].

1 D. Compactifications. The concept of *Tonelli function* generalizes to a Riemannian manifold R in analogy with III.1 A. Therefore we can define *Royden's algebra* $\mathbf{M}(R)$ and *Royden's compactification* $R^*_{\mathbf{M}}$ for R. Most of the fundamental properties can be shown to remain valid.

Since the Perron-Brelot method is applicable as mentioned in 1 C we can also consider *Wiener's functions* on a Riemannian manifold R (see IV.1), *Wiener's algebra* $\mathbf{N}(R)$, and *Wiener's compactification* $R^*_{\mathbf{N}}$. Much of the theory can be developed parallel to IV.§ 1.

Normal operators can also be considered on Riemannian manifolds and the *fundamental theorem on the existence of principal functions* proved in the same way as in I.7 A. By virtue of the solvability of the Dirichlet and Neumann problems for Eq. (23) the normal operators L_0

and L_1 introduced in I.7 C and II.12 B can be generalized to Riemannian manifolds (cf. SARIO-SCHIFFER-GLASNER [25], RODIN-SARIO [3, Chapter VI]).

Thus the basic tools for the classification theory, i.e. normal operators, Royden's compactification, and Wiener's compactification are available for Riemannian manifolds. As a consequence it can be shown that those aspects of the classification theory of Riemann surfaces which do not make essential use of the conformal structure carry over to Riemannian manifolds. However in the case of Riemann surfaces one naturally tends to make use of the conformal structure for simplicity and geometric clarity of the reasoning even if this structure is in reality not needed to reach the desired conclusions. For this reason it is difficult to distinguish what is simple reproduction and what is not, without actually carrying out the argument; often it reveals unforeseen new aspects (cf. the doctoral dissertations of LARSEN [1], SMITH [1], GLASNER [1], BREAZEAL [1], OW [1], KATZ [1], COUNCILMAN [1], and CHANG [1]).

Here we shall discuss some of the facets of the theory which are essentially different from the case of Riemann surfaces, i.e. whose counterparts for Riemann surfaces rely on the conformal structure.

2. Moduli of Annuli

2 A. Modulus. In a Riemannian manifold R consider a bordered compact region E. Suppose the border of E consists of several components divided into two classes α and β. For short we shall refer to the configuration (E, α, β) as an *annulus*. Let h be a harmonic function on E with continuous boundary values 0 on α and $\log \mu > 0$ on β such that

$$\int_\alpha * dh = 2\pi. \tag{27}$$

The number $\mu > 1$ will be called the *modulus* of the annulus (E, α, β),

$$\mu = \operatorname{mod}(E, \alpha, \beta).$$

Let w be the *harmonic measure* of β with respect to E, i.e. the harmonic function on E with continuous boundary values 0 on α and 1 on β. As in I.1 A we have

$$\log \mu = \frac{2\pi}{D_E(w)}. \tag{28}$$

2 B. Bisection into Small Annuli. Let (E_j, α_j, β_j) $(j=1, \ldots, m)$ be disjoint annuli. Set $E = \bigcup_{j=1}^m E_j$, $\alpha = \bigcup_{j=1}^m \alpha_j$, $\beta = \bigcup_{j=1}^m \beta_j$ and call (E, α, β)

again an annulus. Its modulus μ is defined as before and identity (28) remains valid. In terms of the moduli μ_j of the (E_j, α_j, β_j) $(j=1, \ldots, m)$ we have

$$\frac{1}{\log \mu} = \sum_{j=1}^{m} \frac{1}{\log \mu_j}. \tag{29}$$

Let γ_j be a hypersurface in E_j such that $E_j - \gamma_j = E_j' \cup E_j''$, $E_j' \cap E_j'' = \emptyset$, and $(E_j', \alpha_j, \gamma_j)$ and $(E_j'', \gamma_j, \beta_j)$ are annuli. We call $\gamma = \bigcup_{j=1}^m \gamma_j$ a bisecting surface of (E, α, β) and set $E' = \bigcup_{j=1}^m E_j'$, $E'' = \bigcup_{j=1}^m E_j''$. We shall prove the following counterpart of Theorem I.4 B (GLASNER-KATZ-NAKAI [2]):

THEOREM. *For every annulus (E, α, β) and an arbitrarily small $\varepsilon > 0$ there exists a bisecting surface γ of (E, α, β) such that*

$$\operatorname{mod}(E', \alpha, \gamma) < 1 + \varepsilon, \qquad \operatorname{mod}(E'', \gamma, \beta) < 1 + \varepsilon. \tag{30}$$

Proofs of this theorem in the 2-dimensional case given in I.4 B and I.4 C made essential use of the conformal structure and therefore are not applicable to the present general situation. Moreover they were restricted to doubly connected (planar) annuli. In the proof given below any need of conformal structure is of course dispensed with.

In view of (29) we may assume that (E, α, β) is connected.

2 C. An Estimate. Denote by $C(a, b) = C_{x_0}(a, b)$ the Euclidean cylinder

$$\sum_{i=1}^{n-1} (x^i - x_0^i)^2 < a^2, \qquad x_0^n < x^n < x_0^n + b \tag{31}$$

where $a, b > 0$ and $x_0 = (x_0^1, \ldots, x_0^n)$ is a fixed point. Let $\mathscr{F}(a, b)$ be the class of C^1-functions f on $C(a, b)$ with continuous boundary values 0 on $\overline{C(a, b)} \cap \{x^n = x_0^n\}$ and 1 on $\overline{C(a, b)} \cap \{x^n = x_0^n + b\}$. Designate by D^e the Dirichlet integral with respect to the Euclidean metric and by s the surface area of $\sum_{i=1}^{n-1} (x^i)^2 < 1$, $x^n = 0$.

LEMMA. *For every $f \in \mathscr{F}(a, b)$*

$$D^e_{C(a, b)}(f) \geq s \cdot \frac{a^{n-1}}{b} \tag{32}$$

with equality for $f_0(x) = b^{-1}(x^n - x_0^n)$, $x = (x^1, \ldots, x^n)$.

It is clear that $D^e_{C(a, b)}(f_0) = s \cdot (a^{n-1}/b)$. To prove (32) we may assume that $f \in C^1$ in a neighborhood of $\overline{C(a, b)}$. By Stokes' formula

$$D^e_{C(a, b)}(f - f_0, f_0) = \int_{\partial C(a, b)} (f - f_0) \frac{\partial f_0}{\partial n} \, dS = 0$$

since $f-f_0=0$ on the top and on the bottom of the cylinder and $\partial f_0/\partial n=0$ on its side S. By Schwarz's inequality

$$D^e_{C(a,b)}(f)\,D^e_{C(a,b)}(f_0)\geq(D^e_{C(a,b)}(f,f_0))^2=(D^e_{C(a,b)}(f_0))^2$$

from which (32) follows.

2 D. Proof of the Theorem. Take a point x_0 in α and a point y_0 in β. Let $x^1,\ldots,x^n$ be a local parametric system at $x_0=(x_0^1,\ldots,x_0^n)$ valid in a neighborhood U of x_0 such that $U\cap\alpha$ is given by $x^n=x_0^n$, with x^n increasing toward E. Similarly let $y^1,\ldots,y^n$ be a local parametric system at $y_0=(y_0^1,\ldots,y_0^n)$ valid in a neighborhood V of y_0 such that $V\cap\beta$ is given by $y^n=y_0^n$, with y^n increasing toward E. Choose a constant $c>0$ such that

$$\sqrt{g}|U\cup V\geq\sqrt{c} \tag{33}$$

and

$$(g^{ij}|U\cup V)\,\xi_i\,\xi_j\geq\sqrt{c}\sum_{i=1}^{n}\xi_i^2 \tag{34}$$

for every vector $(\xi_1,\ldots,\xi_n)$.

First take $a>0$ so small that $\sum_{i=1}^{n-1}(x^i-x_0^i)^2<a^2$ with $x^n=x_0^n$ and $\sum_{i=1}^{n-1}(y^i-y_0^i)^2<a^2$ with $y^n=y_0^n$ are contained in $U\cap\alpha$ and $V\cap\beta$ respectively. Next select $b>0$ satisfying the conditions

$$0<b<\frac{sca^{n-1}\log(1+\varepsilon)}{2\pi}, \tag{35}$$

$\overline{C_{x_0}(a,b)}-\{x^n=x_0^n\}\subset E$, $\overline{C_{y_0}(a,b)}-\{y^n=y_0^n\}\subset E$, and $\overline{C_{x_0}(a,b)}\cap\overline{C_{y_0}(a,b)}=\emptyset$. Then choose a bisecting surface γ of (E,α,β) such that $\gamma\cap(C_{x_0}(a,b)\cup C_{y_0}(a,b))=\emptyset$ and

$$\gamma\supset[\overline{C_{x_0}(a,b)}\cap\{x^n=x_0^n+b\}]\cup[\overline{C_{y_0}(a,b)}\cap\{y^n=y_0^n+b\}].$$

Let w' be the harmonic measure of γ with respect to E'. Since $E'\supset C_{x_0}(a,b)$ we obtain by (3), (21), (22), (33), and (34)

$$D_{E'}(w')>D_{C_{x_0}(a,b)}(w')\geq c\,D^e_{C_{x_0}(a,b)}(w').$$

We infer by (28), (32), and (35) that

$$\operatorname{mod}(E',\alpha,\gamma)<1+\varepsilon.$$

A similar consideration for E'' then gives (30).

3. Parabolicity

3 A. Equivalences and Inclusions. Parabolicity of a Riemannian manifold R can be defined in terms of vanishing of the harmonic measure of the ideal boundary as in III.2 F. We retain the symbol O_G for the class of parabolic Riemannian manifolds. Theorem III.2F is also valid in the present case. In particular $R \in O_G$ if and only if

$$1 \in \mathrm{M}_\Delta(R). \tag{36}$$

Another characterization of parabolicity is by the nonexistence of Green's functions (cf. V.8 A) and also by the vanishing of the capacity (cf. V.7 F). Particularly important is the counterpart of V.7 B.(56):

$$O_G = O_{\mathscr{S}} \tag{37}$$

where $\mathscr{S}(R)$ is the class of nonnegative superharmonic functions on R and $O_{\mathscr{S}}$ is the corresponding null class of Riemannian manifolds. We can deduce from (37) the inclusions

$$O_G < O_{HP} < O_{HB} \tag{38}$$

where the strictness is obvious in the present case $n \geq 3$. In fact Euclidean n-space E^n gives the strictness of $O_G < O_{HP}$ and $E^n - (0, \ldots, 0)$ that of $O_{HP} < O_{HB}$. Here we have employed the Kelvin transformation and the HB-removability of one point in E^n.

Using Newtonian potentials instead of logarithmic potentials we can develop an analogue of the argument in Chapter V to prove that $R \in O_G$ is characterized by the existence of an Evans-Selberg potential.

3 B. Parabolic Riemannian Ball. As seen above the general features of parabolicity for Riemannian manifolds are the same as in the 2-dimensional case. However as soon as the conformal structure is involved higher dimension brings forth entirely new aspects.

Let

$$V\colon\ |x|^2 = \sum_{i=1}^{n} (x^i)^2 < 1 \tag{39}$$

in Euclidean n-space E^n ($n \geq 3$). We assert:

THEOREM. *There exists a positive C^∞-function λ on the unit ball V such that the Riemannian manifold $(V, \lambda\delta_{ij})$ is parabolic.*

Observe that $(V, \lambda\delta_{ij})$ is conformally equivalent to the Euclidean ball (V, δ_{ij}). Since O_G is conformally invariant for dimension 2 the assumption $n \geq 3$ is essential in the above theorem.

We shall show that a required function λ is given by

$$\lambda(x) = |x|^{\frac{2-2n}{n-2}} (1 - |x|)^{\frac{4}{n-2}} \tag{40}$$

for $\frac{1}{2} \leq |x| < 1$. We can extend λ to all of $|x| < 1$ such that $\lambda \in C^\infty(V)$ and $\lambda > 0$ on V. Since the metric tensor is now given by

$$g_{ij}(x) = \lambda(x)\, \delta_{ij} \tag{41}$$

we have

$$g^{ij}(x)=\frac{1}{\lambda(x)}\delta_{ij},\qquad g(x)=\lambda^n(x) \tag{42}$$

on V. Therefore 1 B.(13) takes the form

$$\Delta u=-\lambda^{-\frac{n}{2}}\sum_{i=1}^{n}\frac{\partial}{\partial x^i}\left(\lambda^{\frac{n-2}{2}}\frac{\partial u}{\partial x^i}\right). \tag{43}$$

We exhaust V by balls with boundaries

$$\beta_m\colon |x|=1-\frac{1}{2+m},\qquad m=0,1,\ldots. \tag{44}$$

Between β_0 and β_m $(m\geq 1)$ let u_m be the harmonic measure of β_m, i.e. the harmonic function with continuous boundary values 0 on β_0 and 1 on β_m. By definition $R\in O_G$ if $\lim_m u_m=0$.

Since $u=1/(1-|x|)$ makes (43) vanish on $\frac{1}{2}\leq|x|<1$ we infer that

$$u_m(x)=\frac{1}{m(1-|x|)}-\frac{2}{m} \tag{45}$$

which clearly converges to zero as m tends to infinity.

3 C. Hyperbolic Punctured Torus. Consider a locally flat torus T^n $(n\geq 3)$, that is an n-dimensional cube $[-2,2]\times\cdots\times[-2,2]$ in E^n with the opposite faces identified by pairs. Remove from it a point, the center of the unit ball $V\colon |x|<1$, and denote by T_0^n the resulting punctured torus. It is also a locally flat Riemannian manifold and clearly $T_0^n\in O_G$. At the other extreme from Theorem 3 B we state:

THEOREM. *There exists a Riemannian structure on the punctured torus T_0^n which is conformally equivalent to the locally flat structure and yet makes T_0^n hyperbolic.*

Here the term "hyperbolic" is used, as in the 2-dimensional case, to mean "not parabolic." Again by the conformal invariance of O_G for $n=2$ the assumption $n\geq 3$ is essential for the validity of the above theorem.

Exhaust T_0^n by relatively compact subregions with boundaries

$$\beta_m\colon |x|=\frac{1}{1+m}. \tag{46}$$

Thus β_0 is the unit sphere and the β_m close in on the point boundary.

Take a C^∞-function λ on T_0^n with

$$\lambda(x)=|x|^{\frac{2-2n}{n-2}} \tag{47}$$

on $\overline{V}$ and endow T_0^n with the new Riemannian structure $\lambda\,\delta_{ij}$. Again Δu is given by (43). In the present case

$$u_m = \frac{1-|x|}{1-1/(1+m)} \tag{48}$$

is seen to be the harmonic measure of β_m between β_0 and β_m $(m=1, 2, \ldots)$. Since

$$u(x) = \lim_{m\to\infty} u_m(x) = 1-|x| \neq 0$$

we conclude that $(T_0^n, \lambda\,\delta_{ij}) \notin O_G$.

3 D. The Equation $\Delta u = Pu$ $(P \geq 0)$. A solution u of the equation

$$\Delta u \equiv -\frac{1}{\sqrt{g}}\frac{\partial}{\partial x^i}\left(\sqrt{g}\,g^{ij}\frac{\partial u}{\partial x^j}\right) = Pu \tag{49}$$

may be called P-harmonic. The usual harmonic functions are then simply 0-harmonic. Here P is a nonnegative C^1-function on a given Riemannian manifold $R = R^n$. We shall suppose henceforth that

$$P \not\equiv 0 \tag{50}$$

on R. Observe that nonnegative P-harmonic functions are subharmonic.

The local existence of a Green's function for equation (49) is also assured. Hence properties (I), (III), (IV), and (V) in 1 C remain valid for (49) (cf. e.g. MIRANDA [1]). However constants $\neq 0$ are no longer P-harmonic and thus the maximum principle in its original form does not hold. Nevertheless the following weaker version is valid:

(II′) *Maximum principle:* a nonnegative P-harmonic function $(\not\equiv 0)$ on R does not take on its maximum on R; a P-harmonic function which changes sign on R assumes neither its maximum nor minimum on R.

A classification theory based on (49) can be developed largely in the same fashion as for harmonic functions. It is actually more diversified than the harmonic case. The most significant new feature is the following:

THEOREM. *A Green's function exists for every compact or noncompact R.*

This result was obtained by L. MYRBERG [4] for Riemann surfaces. In the present case the Green's function g on R with pole $x_0 \in R$ is the smallest positive P-harmonic function on $R - x_0$ with a positive singularity at x_0. Thus there are no parabolic manifolds.

The proof will be given in 3 H after a general theorem has been established in 3 E.

For fundamentals concerning Eq. (49) we refer the reader to DUFF [1]. In the 2-dimensional case the classification theory based on (49) was initiated by OZAWA [2] and continued by L. MYRBERG [4–6], OZAWA

[4, 6], ROYDEN [9], HAYASHI [1], and others. LOEB [2] and WALSH [1] discussed the problem in the axiomatic setting of BRELOT [4].

3 E. Linear Operators. Let R_0 be a regular subregion of R with $\alpha=\partial R_0$. We shall consider operators L from the class $B(\alpha)$ of bounded continuous functions f on α into the class of solutions of (49) on $R-\bar{R}_0$ with continuous boundary values at α. The operators are to satisfy

(L.1) $Lf|\alpha=f$,
(L.2) $(\min_\alpha f)\cap 0\leq Lf\leq(\max_\alpha f)\cup 0$,
(L.3) $L(cf)=c(Lf)$,
(L.4) $L(f_1+f_2)=Lf_1+Lf_2$.

An important example of such an operator L is the *Dirichlet operator D* defined as follows. If R is compact then D is simply the operator which gives the solution Df of the Dirichlet problem in $R-\bar{R}_0$. If R is noncompact let $\{R_n\}_1^\infty$ be an exhaustion of R with $\bar{R}_0\subset R_1$. Denote by $D_n f$ the solution of the Dirichlet problem on $R_n-\bar{R}_0$ with boundary values f at α and 0 at ∂R_n. Clearly $\{D_n f\}_1^\infty$ is convergent on compacta of $R-R_0$; we set $Df=\lim_n D_n f$. It is easily verified that D satisfies (L.1)–(L.4).

Given a solution s of (49) on $R-\bar{R}_0$ which is continuous on $R-R_0$ consider the equation

$$L((u-s)|\alpha)=(u-s)|R-\bar{R}_0 \tag{51}$$

where the solution u is to be found in the class of solutions of (49) on R. Such a P-harmonic function u may be called the *L-principal P-harmonic function* with respect to s. We shall establish the following main existence theorem:

THEOREM. *For any s and L there always exists a unique L-principal P-harmonic function u on R with respect to s.*

Here condition (50) is of course essential. Again observe the striking contrast with the case of harmonic functions (cf. I.7 A).

The uniqueness is obvious since for two principal functions u_1 and u_2 Eq. (51) gives $L(v|\alpha)=v$ where $v=u_1-u_2$. By maximum principle (II′) and (L.2), v must vanish identically.

The existence proof will be given in 3 F, 3 G.

3 F. Reduction to Fredholm's Equation. On replacing s by $s-Ls$ we may assume that $s|\alpha=0$. Take a regular subregion R_1 of R such that $R_1\supset\bar{R}_0$ and $P\not\equiv 0$ on R_1. Denote by δ the relative boundary ∂R_1 of R_1. For $f\in B(\delta)$ let Kf be the P-harmonic function on R_1 with continuous boundary values f at δ. Finding a u on R which satisfies (51) is equivalent to determining an $f \in B(\alpha)$ and a $\varphi \in B(\delta)$ with the properties

$$K\varphi|\alpha = f, \quad \varphi - s|\delta = Lf|\delta. \tag{52}$$

In fact if we can find a u then $f = u|\alpha$ and $\varphi = u|\delta$ satisfy (52). Conversely if we have f and φ then u is obtained by

$$u|R_1 = K\varphi, \quad u|R - \bar{R}_0 = s + Lf. \tag{53}$$

Since (52) implies that $K\varphi = s + Lf$ on $\partial(\bar{R}_1 - R_0)$ and consequently on $\bar{R}_1 - R_0$, u is well defined and P-harmonic on R. Moreover (52) and (53) give $L((u-s)|\alpha) = L(u|\alpha) = L(K\varphi|\alpha) = Lf = u - s$ on $R - \bar{R}_0$, i.e. u satisfies (51). Thus in order to solve (51) it is sufficient to solve the simultaneous Eqs. (52).

It is readily seen that (52) is equivalent to the single equation

$$\varphi - L(K\varphi|\alpha)|\delta = s|\delta. \tag{54}$$

Let $\Lambda\varphi = L(K\varphi|\alpha)|\delta$. Clearly Λ is a linear operator from $B(\delta)$ into itself. Consider $B(\delta)$ as a Banach space with norm $\|\varphi\|_\infty = \sup_\delta |\varphi|$. Then by (L.2) and (II′) the operator norm

$$\|\Lambda\|_\infty = \sup_{\varphi \in B(\delta),\, \varphi \not\equiv 0} \frac{\|\Lambda\varphi\|_\infty}{\|\varphi\|_\infty} \tag{55}$$

is bounded and actually dominated by 1. We set $s|\delta = \lambda \in B(\delta)$ and write (54) as the Fredholm equation

$$(I - \Lambda)\varphi = \lambda \tag{56}$$

on the Banach space $B(\delta)$, with I the identity operator on $B(\delta)$.

It is elementary that (56) is solved by the Neumann series

$$\varphi = \sum_{n=0}^{\infty} \Lambda^n \lambda \tag{57}$$

if Λ satisfies

$$\|\Lambda\|_\infty < 1. \tag{58}$$

3 G. Proof of $\|\Lambda\|_\infty < 1$. Let w be the P-harmonic function on R_1 with continuous boundary values 1 at δ:

$$w = K1. \tag{59}$$

Since $P \not\equiv 0$ on R_1, $w \not\equiv 1$ on R_1. Therefore (II′) implies

$$0 < q = \max_\alpha w < 1. \tag{60}$$

From the inequality $\pm\varphi \le \|\varphi\|_\infty$ it follows that

$$\pm K\varphi \le \|\varphi\|_\infty K1 \tag{61}$$

and this with (60) gives

$$\pm K\varphi|\alpha \leq q\|\varphi\|_{\infty}. \tag{62}$$

On applying L to both sides we obtain

$$\pm L(K\varphi|\alpha) \leq q\|\varphi\|_{\infty} L1. \tag{63}$$

Since $0 < L1 < 1$ on $R - \bar{R}_0$, (63) implies

$$|L(K\varphi|\alpha)|\delta| \leq q\|\varphi\|_{\infty}$$

or equivalently

$$\|\Lambda\varphi\|_{\infty} \leq q\|\varphi\|_{\infty}. \tag{64}$$

From (55) and (64) it follows that

$$\|\Lambda\|_{\infty} \leq q. \tag{65}$$

This together with (60) gives assertion (58).

The proof of Theorem 3 E is herewith complete.

Remark. Theorem 3 E can easily be extended to harmonic spaces of Brelot with the property that the constant function 1 is superharmonic but not harmonic. For harmonic spaces see Loeb [2], Loeb-Walsh [1], Walsh [1], among others.

3 H. Proof of Hyperbolicity. We are now able to complete the proof of Theorem 3 D. Remove from R a point x_0, let B be a ball with center x_0, and B' be a larger concentric ball. The set $R_0 = B' - \bar{B}$ is a regular subregion of $R - x_0$. Take the Dirichlet operator D of 3 E with respect to $R - R_0$ and let $g(\cdot, x_0)$ be the Green's function for (49) on B. Set $s|\bar{B} - x_0 = g(\cdot, x_0)|\bar{B} - x_0$ and $s|R - B' = 0$. By Theorem 3 E there exists a unique D-principal P-harmonic function u on $R - x_0$ with respect to s. It is clearly the required Green's function for (49) on R.

Remark. If R is noncompact then there always exists a nonconstant positive P-harmonic function on R for $P \not\equiv 0$ (L. Myrberg [5]). In fact let $\{x_n\}$ be a sequence of points in R tending to the ideal boundary and set $k_n(x) = g(x, x_n)/g(x_0, x_n)$ with $g(x, x_n)$ the Green's function for (49) on R. Since $k_n(x_0) = 1$, $\{k_n\}$ forms a normal family and a subsequence converges to a P-harmonic function u. Clearly $u(x_0) = 1$ and $u > 0$.

4. Invariance under Deformation

4 A. Inclusion Relations. An analogue of any one of the proofs of the inclusion $O_{HB} \subset O_{HD}$ and of the equality $O_{HD} = O_{HBD}$ given in II.12 B, III.4 G, IV.6 M for Riemann surfaces qualifies as the demonstration of $O_{HB} \subset O_{HD} = O_{HBD}$ for Riemannian manifolds R^n (cf. 1 D).

We can also introduce the classes K and R as follows: $K=K(R^n)$ (resp. $R=R(R^n)$) is the subclass of H consisting of all $u\in H=H(R^n)$ such that

$$\int_\alpha *du=0 \tag{66}$$

for every dividing $(n-1)$-cycle (resp. every $(n-1)$-cycle) α in R^n. Thus K is a direct generalization from the 2-dimensional case. In contrast R for $n=2$ does not coincide with A but rather with $\{\mathrm{Re}\, f \mid f\in A\}$. However the latter consists of constants if and only if the former does. Therefore the null class O_{RD} actually coincides for $n=2$ with O_{AD} and is its natural generalization to higher dimensions. Clearly $O_{HB}\subset O_{KB}\subset O_{RB}$ and $O_{HD}\subset O_{KD}\subset O_{RD}$. The same proof as in II. 12 A gives $O_{KB}\subset O_{KD}=O_{KBD}$.

From the above observations and from (38) we have

$$O_G<O_{HP}<O_{HB}\begin{array}{l} \subset\; O_{KB}\subset O_{RB} \\ \qquad\;\; \cap \\ \subset\; O_{HD}\subset O_{KD}\subset O_{RD}. \end{array} \tag{67}$$

Whether or not examples can be constructed to prove the strictness for $n\geq 3$ of inclusions from O_{HB} on, in particular of $O_{HB}\subset O_{HD}$, is an important unsolved problem. The classes $O_{H\Phi}$, O_{KP}, $O_{K\Phi}$, O_{RP}, $O_{R\Phi}$ may also offer some interest.

4 B. Wiener's and Royden's Mappings. Consider two Riemannian manifolds R_1 and R_2 and a topological mapping T of R_1 onto R_2. We define T as a *Wiener mapping* in the same way as in IV. 10 A and obtain the counterparts of IV. 10 B – IV. 11 C. In particular:

The classes O_G and O_{HB} are invariant under Wiener's mappings.

We also define T as a *Royden mapping* in the same manner as in III. 8 B. Theorems III. 8 C – 8 H remain valid in the present general situation. We have in particular:

The classes O_G and O_{HD} are invariant under Royden's mappings.

As a special subclass of the class of Royden's mappings we also consider *Dirichlet mappings T* defined by the following two properties. First, $f\circ T\in\mathscr{D}(R_1)$ if and only if $f\in\mathscr{D}(R_2)$, where $\mathscr{D}$ is the class of Tonelli functions with finite Dirichlet integrals. Second, there exists a constant K such that

$$K^{-1}D_{R_2}(f)\leq D_{R_1}(f\circ T)\leq KD_{R_2}(f) \tag{68}$$

for every $f\in\mathscr{D}(R_2)$. Dirichlet mappings are quasiconformal mappings and conversely for $n=2$ (cf. II. 14 A and III. 7). Therefore the theory in III. 7 can be generalized to the present case if we replace quasiconformal mappings by Dirichlet mappings. Actually the process is quite trivial since the delicate analytical considerations are absorbed in the tailor-made assumption (68).

Since Dirichlet mappings are Royden's mappings, O_G and O_{HD} are invariant under Dirichlet mappings. In addition the characterization of O_{KD} given in II. 13 can be easily seen to hold in the present situation, and thus the same argument as in II. 14 enables us to obtain the invariance of O_{KD} under Dirichlet mappings. We conclude:

The classes O_G, O_{HD}, and O_{KD} are invariant under Dirichlet mappings.

4 C. Quasiconformal Mappings. Wiener's, Royden's and Dirichlet mappings are natural types of topological mappings related to the classification problem. In particular Dirichlet mappings in the 2-dimensional case are nothing but quasiconformal mappings and these are interesting in their own right because of their geometric significance. In this connection let us consider how to generalize the concept of quasiconformality to topological mappings T between two Riemannian manifolds R_1 and R_2.

Some straightforward extensions are obtained by taking (68) or III. 7 A. (104) as the definition, both being characteristic properties of quasiconformality in the 2-dimensional case. These would certainly be appropriate for the classification theory since they preserve every 2-dimensional aspect. However they have the drawback of lacking clear geometric meaning.

The most geometric, natural, and also most widely accepted definition is as follows. For a point $x \in R_1$ and a positive number $r > 0$ set

$$\begin{aligned} l(x, r) &= \inf_{d_1(x, y) = r} d_2(T(x), T(y)), \\ L(x, r) &= \sup_{d_1(x, y) = r} d_2(T(x), T(y)), \end{aligned} \tag{69}$$

where d_1 and d_2 are metrics on R_1 and R_2 given by 1 A. (4). If there exists a constant $K \geq 1$ such that

$$\limsup_{r \to 0} \frac{L(x, r)}{l(x, r)} \leq K \tag{70}$$

for every $x \in R_1$ then we call T a *quasiconformal mapping*. This condition is also characteristic of quasiconformal mappings for $n = 2$.

Since the theory of quasiconformal mappings between Euclidean regions obviously carries over to Riemannian manifolds we obtain the following properties (see GEHRING [4]).

A quasiconformal mapping T is a Tonelli function in terms of local parameters of R_1 and R_2, and it is totally differentiable a.e., the measures being Euclidean ones considered locally. Moreover at any point $x \in R_1$

$$I(x)^n \leq K^{n-1} J(x) \tag{71}$$

with

$$I(x) = \limsup_{r \to 0} \frac{L(x, r)}{r} \tag{72}$$

and

$$J(x) = \limsup_{r \to 0} \frac{V_2(TB_r)}{V_1(B_r)}. \tag{73}$$

Here B_r is the geodesic r-ball at x, and $V_2(TB_r)$ and $V_1(B_r)$ are the volumes of TB_r and B_r. At a point x where T is totally differentiable $J(x)$ is the Jacobian and we have

$$J(x) > 0 \tag{74}$$

a.e. If $E \subset R_1$ is measurable then so is TE, and its volume is given by

$$V_2(TE) = \int_E J \, dV_1. \tag{75}$$

Corresponding properties, with the same K, are possessed by the inverse mapping T^{-1} of T.

4 D. Conformal Noninvariance. In the special case where K in (70) can be taken as 1 we call the mapping T conformal. For $n=2$ we then actually have a univalent analytic mapping.

Let us return to the example $(V, \lambda \delta_{ij})$ given in 3 B. The identity mapping T of the ball V onto itself gives rise to a topological mapping T of $R_1 = (V, \delta_{ij})$ onto $R_2 = (V, \lambda \delta_{ij})$. We claim that T is conformal. In fact

$$d_2(T(x), T(y)) = \lambda(x) r + \varepsilon(x, y)$$

where $r = d_1(x, y)$ and $\varepsilon(x, y) \to 0$ as $r \to 0$ locally uniformly. In view of

$$L(x, r) = \lambda(x) r + \sup_{d_1(x, y) = r} \varepsilon(x, y),$$

$$l(x, r) = \lambda(x) r + \inf_{d_1(x, y) = r} \varepsilon(x, y)$$

we conclude that

$$\lim_{r \to 0} \frac{L(x, r)}{l(x, r)} = 1$$

as claimed.

Since for example $u(x) = x^1$ belongs to both $RB(R_1)$ and $RD(R_1)$, $R_1 = (V, \delta_{ij})$ does not belong to any of the null classes in table (67). On the other hand we know from 3 B that $R_2=(V,\lambda\delta_{ij})$ belongs to O_G and consequently all null classes in (67). We conclude:

The classes O_G, O_{HP}, O_{HB}, O_{HD}, O_{KB}, O_{KD}, O_{RB}, and O_{RD} for $n\geq 3$ fail to be invariant under not only quasiconformal but also conformal mappings.

This distinguishes higher dimensional cases from the 2-dimensional one. The reason for quasiconformal invariance of the classes O_G, O_{HD}, and O_{KD} of 2-manifolds lies in the validity of (68) for their quasiconformal mappings. For the present quasiconformal mappings (68) is also true if we replace the Dirichlet integral D by

$$D_R^{(n)}(f)=\int_R |\operatorname{grad} f|^n\, dV. \tag{76}$$

A function which satisfies the Dirichlet principle for $D^{(n)}$ is no longer harmonic. This breaks up any parallel argument for $n=2$ and $n\geq 3$. On the other hand one is led to a classification theory based on functions minimizing $D^{(n)}$.

4 E. Quasi-Isometric Invariance. The rather surprising phenomenon encountered in 4 D stems from the fact that conformal structures determine equivalence classes of Riemannian 2-manifolds whereas those of higher dimensional manifolds are characterized by isometric, not conformal, structures. In other words for $n\geq 3$ isometry is to replace conformal mapping.

Similarly a counterpart of a quasiconformal mapping is what we shall call a *quasi-isometry*. It is a topological mapping T of a Riemannian manifold R_1 onto another R_2 such that there exists a constant $K\geq 1$ with

$$K^{-1}r\leq l(x,r)\leq L(x,r)\leq Kr \tag{77}$$

for all $x\in R_1$. In particular if $K=1$ then we have an *isometry*. Clearly (77) implies (70) with K replaced by K^2. Therefore a quasi-isometry is a quasiconformal mapping and all analytical properties described in 4 C for the latter are shared by quasi-isometries.

We are now able to state the following result:

THEOREM. *The classes O_G, O_{HD}, and O_{KD} are preserved under quasi-isometries.*

In view of the last statement in 4 B it is sufficient to show that *a quasi-isometry is a Dirichlet mapping*. The proof will be given in 4 G.

4 F. Preliminaries. By way of preparation we insert here for the convenience of the reader two elementary properties of nonnegative matrices. We recall that a matrix A is nonnegative (resp. positive) if the form (Ax,x) is positive semidefinite (resp. definite).

Let A and B be symmetric $n \times n$ matrices. The inequality $A \leq B$ ($A < B$) means that $B - A$ is a nonnegative (resp. positive) matrix. The standard notation A^{-1}, A^t, and $|A|$ is used for the inverse matrix, the transposed matrix, and the determinant of A.

(I) *Let $A \geq 0$ and $B > 0$. Then the inequality $A \leq B$ implies that $|A| \leq |B|$.*

To see this take the orthogonal matrix P such that $P^t BP$ is a diagonal matrix with elements $\lambda_1, \ldots, \lambda_n$. Since $B > 0$ implies $\lambda_i > 0$ ($i = 1, \ldots, n$) we can consider the diagonal matrix C with elements $\lambda_1^{-\frac{1}{2}}, \ldots, \lambda_n^{-\frac{1}{2}}$. Let Q be the orthogonal matrix such that $Q^t (PC)^t A (PC) Q$ is diagonal with elements $\mu_1, \ldots, \mu_n$. Then $A \leq B$ implies that for $S = PCQ$

$$S^t AS \leq S^t BS. \tag{78}$$

In terms of the unit $n \times n$ matrix E and the $n \times n$ diagonal matrix with diagonal $(\mu_1, \ldots, \mu_n)$ we have

$$S^t AS = \mu, \qquad S^t BS = E, \tag{79}$$

and therefore

$$\mu \leq E. \tag{80}$$

This in turn implies that $\mu_i \leq 1$. Since $A \geq 0$ we obtain $S^t AS = \mu \geq 0$ and thus $\mu_i \geq 0$. Therefore $\prod_1^n \mu_i \leq 1$, i.e. $|\mu| \leq |E|$. Substitution of (79) gives $|A||S|^2 \leq |B||S|^2$, and the assertion $|A| \leq |B|$ follows.

A similar reasoning gives for positive matrices:

(II) *The inequality $A \leq B$ with $A > 0$ and $B > 0$ implies that $B^{-1} \leq A^{-1}$.*

In fact using the same notation as above we observe that $A > 0$ yields $S^t AS > 0$ and a fortiori $\mu_i > 0$. This together with (80) gives $0 < \mu_i \leq 1$ and consequently $1 \leq \mu_i^{-1}$. Therefore $E^{-1} \leq (\mu)^{-1}$, or in terms of (79), $(S^t BS)^{-1} \leq (S^t AS)^{-1}$. Hence $S(S^t BS)^{-1} S^t \leq S(S^t AS)^{-1} S^t$ which is $B^{-1} \leq A^{-1}$.

4 G. Proof of Theorem 4 E. To show that quasi-isometries $T\colon R_1 \to R_2$ are Dirichlet mappings we are to compare Dirichlet integrals over R_1 and R_2. To this end we first compare volume elements, then gradients.

Let (g_{ij}) and $(\tilde{g}_{ij})$ be the metric tensors for R_1 and R_2 respectively. The arc elements will be denoted by ds_1 and ds_2. Choose a set $E \subset R_1$ such that (i) meas $E = 0$, (ii) T is totally differentiable on $R_1 - E$, (iii) T^{-1} is totally differentiable on $R_2 - T(E)$, and (iv) $J > 0$ on $R_1 - E$. The existence of such an E is assured by the fact that T is a quasiconformal mapping (cf. 4 E). However (77) also implies the existence of E directly by virtue of the well-known Rademacher-Stepanoff theorem (see e.g. Saks [1, p. 310]). All considerations below will be in $R_1 - E$ and $R_2 - T(E)$.

Let $y = T(x)$. Then by (77)

$$K^{-1} ds_1 \leq ds_2 \leq K ds_1$$

or equivalently

$$K^{-2} g_{ij}\, dx^i\, dx^j \le \tilde{g}_{ij} \frac{\partial y^i}{\partial x^k} \frac{\partial y^j}{\partial x^l}\, dx^k\, dx^l \le K^2 g_{ij}\, dx^i\, dx^j. \tag{81}$$

In terms of the matrix $M=(\partial y^i/\partial x^j)$, with i indicating the row, j the column, this can be written

$$K^{-2}(g_{ij}) \le M^t(\tilde{g}_{ij})\, M \le K^2(g_{ij}). \tag{82}$$

For the corresponding determinants we have by 4 F.(I)

$$K^{-2n} g \le \tilde{g}\, |M|^2 \le K^{2n} g.$$

For volume elements $dV_1 = \sqrt{g}\, dx^1 \wedge \cdots \wedge dx^n$ of R_1 and $dV_2 = \sqrt{\tilde{g}}\, J\, dx^1 \wedge \cdots \wedge dx^n$ of R_2 this takes the form

$$K^{-n} dV_1 \le dV_2 \le K^n dV_1. \tag{83}$$

To compare gradients on R_1 and R_2 we obtain from 4 F.(II) and (82)

$$K^{-2}(g^{ij}) \le M^{-1}(\tilde{g}^{ij})(M^t)^{-1} \le K^2(g^{ij}).$$

A left and right multiplication by M and M^t respectively gives

$$K^{-2}(\tilde{g}^{ij}) \le M(g^{ij})\, M^t \le K^2(\tilde{g}^{ij}). \tag{84}$$

First consider functions $f \in \mathscr{D}^\infty(R_2) = \mathscr{D}(R_2) \cap C^\infty(R_2)$. Then clearly $f \circ T \in \mathscr{D}(R_1)$. On applying the $n \times 1$ matrix $(\partial f/\partial y^i)$ on the right and its transpose on the left to (84) we see that

$$K^{-2}\, \tilde{g}^{ij} \frac{\partial f}{\partial y^i} \frac{\partial f}{\partial y^j} \le g^{kl} \frac{\partial y^i}{\partial x^k} \frac{\partial y^j}{\partial x^l} \frac{\partial f}{\partial y^i} \frac{\partial f}{\partial y^j} \le K^2\, \tilde{g}^{ij} \frac{\partial f}{\partial y^i} \frac{\partial f}{\partial y^j}$$

a.e. on $R_1 - E$. The expression in the middle is

$$g^{kl} \frac{\partial f}{\partial x^k} \frac{\partial f}{\partial x^l} = |\text{grad}\, f \circ T|^2$$

and it follows that

$$K^{-2}\, |\text{grad}\, f|^2 \le |\text{grad}\, f \circ T|^2 \le K^2\, |\text{grad}\, f|^2 \tag{85}$$

a.e. on $R_1 - E$.

On combining (83) and (85) we obtain the desired relations

$$K_1^{-1} D_{R_2}(f) \le D_{R_1}(f \circ T) \le K_1\, D_{R_2}(f) \tag{86}$$

with $K_1 = K^{n+2}$.

In the general case of $f \in \mathscr{D}(R_2)$ the D-denseness of $\mathscr{D}^\infty(R_2)$ in $\mathscr{D}(R_2)$ implies that $f \circ T \in \mathscr{D}(R_1)$ and that f satisfies (86). Therefore T is a Dirichlet mapping.

The proof of Theorem 4 E is herewith complete.

Bibliography

Accola, R.

[1] On semi-parabolic Riemann surfaces. Trans. Amer. Math. Soc. **108** (1963), 437–448.

[2] On a class of Riemann surfaces. Proc. Amer. Math. Soc. **15** (1964), 607–611.

[3] Some classical theorems on open Riemann surfaces. Bull. Amer. Math. Soc. **73** (1967), 13–26.

Ahlfors, L.

[1] Zur Bestimmung des Typus einer Riemannschen Fläche. Comment. Math. Helv. **3** (1931), 173–177.

[2] Sur le type d'une surface de Riemann. C. R. Acad. Sci. Paris **201** (1935), 30–32.

[3] Bounded analytic functions. Duke Math. J. **14** (1947), 1–11.

[4] Open Riemann surfaces and extremal problems on compact subregions. Comment. Math. Helv. **24** (1950), 100–134.

[5] (– Beurling, A.) Conformal invariants and function-theoretic null-sets. Acta Math. **83** (1950), 101–129.

[6] Remarks on the classification of open Riemann surfaces. Ann. Acad. Sci. Fenn. Ser. A. I. No. 87 (1951), 8 pp.

[7] (– Royden, H. L.) A counterexample in the classification of open Riemann surfaces. Ibid. No. 120 (1952), 5 pp.

[8] On the characterization of hyperbolic Riemann surfaces. Ibid. No. 125 (1952), 5 pp.

[9] Remarks on Riemann surfaces. Lectures on functions of a complex variable, pp. 45–48. Ann Arbor: Univ. of Michigan Press 1955.

[10] Extremalprobleme in der Funktionentheorie. Ann. Acad. Sci. Fenn. Ser. A. I. No. 249/1 (1958), 9 pp.

[11] The method of orthogonal decomposition for differentials on open Riemann surfaces. Ibid. No. 249/7 (1958), 15 pp.

[12] (– Sario, L.) Riemann surfaces. Princeton Mathematical Series, No. 26. Princeton, N.J.: Princeton Univ. Press 1960. 382 pp.

[13] Lectures on quasiconformal mappings. Princeton, N.J.: D.Van Nostrand Co., Inc. 1966. 146 pp.

[14] Cf. Beurling, A.; Ahlfors, L. [3].

Ahlfors, L., et al. (edited by)

[1] Contributions to the theory of Riemann surfaces. Princeton, N.J.: Princeton Univ. Press 1953. 264 pp.

Akaza, T.

[1] On the weakness of some boundary component. Nagoya Math. J. **17** (1960), 219–223.

[2] On the sufficient conditions for some boundary component of a domain bounded by an infinite number of circles to be parabolic. Sci. Rep. Kanazawa Univ. **7** (1961), 41 – 54.

[3] (– Kuroda, T.) Module of annulus. Nagoya Math. J. **18** (1961), 37 – 41.

[4] (– Oikawa, K.) Examples of weak boundary components. Ibid. **18** (1961), 165 – 170.

[5] Length of the singular set of Schottky group. Kōdai Math. Sem. Rep. **15** (1962), 62 – 66.

Andreian-Cazacu, C.

[1] Sur le problème du type. [Romanian. Russian and French summaries.] An. Univ. "C.I. Parhon" Bucureşti. Ser. Şti. Nat. No. 22 (1959), 23 – 37.

Atchison, T. A.

[1] A class of Riemann surfaces. Proc. Amer. Math. Soc. **16** (1965), 731 – 738.

Bader, R.

[1] (– Parreau, M.) Domaines non compacts et classification des surfaces de Riemann. C.R. Acad. Sci. Paris **232** (1951), 138 – 139.

Behnke, H.

[1] (– Sommer, F.) Theorie der analytischen Funktionen einer komplexen Veränderlichen. Berlin-Göttingen-Heidelberg: Springer 1962. 603 pp.

Bers, L.

[1] On a theorem of Mori and the definition of quasi-conformality. Trans. Amer. Math. Soc. **84** (1957), 78 – 84.

[2] The equivalence of two definitions of quasiconformal mappings. Comment. Math. Helv. **37** (1962/63), 148 – 154.

Beurling, A.

[1] Études sur un problème de majoration. Thèse, Upsala, 1935. 109 pp.

[2] Ensembles exceptionnels. Acta Math. **72** (1939), 1 – 13.

[3] (– Ahlfors, L.) The boundary correspondence under quasiconformal mappings. Ibid. **96** (1956), 125 – 142.

[4] Cf. Ahlfors, L.; Beurling, A. [5].

Blanc, C.

[1] Le type des surfaces de Riemann simplement connexes. C.R. Acad. Sci. Paris **202** (1936), 623 – 625.

[2] Une décomposition du problème du type des surfaces de Riemann. Ibid. **206** (1938), 1078 – 1080.

[3] Les demi-surfaces de Riemann. Application au problème du type. Comment. Math. Helv. **11** (1938), 130 – 150.

[4] (– Fiala, F.) Le type d'une surface et sa courbure totale. Ibid. **14** (1942), 230 – 233.

Blatter, C.

[1] Über ein Typenproblem. Comment. Math. Helv. **37** (1962/63), 198 – 213.

BOBOC, N.

[1] (– MOCANU, G.) Sur la notion de métrique harmonique sur une surface riemannienne hyperbolique. Bull. Math. Soc. Sci. Math. Phys. R. P. Roumaine (N.S.) **4** (52) (1961), No. 1–2, 3–21 (1963).

BREAZEAL, N.

[1] The class O_{AD} of Riemannian 2-spaces. Doctoral dissertation, Univ. of Calif., Los Angeles, Calif., 1966. 43 pp.

BRELOT, M.

[1] (– CHOQUET, G.) Espaces et lignes de Green. Ann. Inst. Fourier **3** (1951), 199–263.

[2] Principe et problème de Dirichlet dans les espaces de Green. C.R. Acad. Sci. Paris **235** (1952), 598–600.

[3] Lignes de Green et problème de Dirichlet. Ibid. **235** (1952), 1595–1597.

[4] Lectures on potential theory. Tata Inst. Fund. Res., Bombay, 1960. 153 pp.

[5] Éléments de la théorie classique du potentiel. 3. ed., Centre de Documentation Universitaire, Paris, 1965. 209 pp.

[6] Norbert Wiener and potential theory. Bull. Amer. Math. Soc. **72** (1966), No. 1, pt. 2, 39–41.

BRUCKNER, J.

[1] Triangulations of bounded distortion in the classification theory of Riemann surfaces. Doctoral dissertation, Univ. of Calif., Los Angeles, Calif., 1960. 64 pp.

CARATHÉODORY, C.

[1] Funktionentheorie. I. Basel: Birkhäuser 1950. 288 pp.

[2] Funktionentheorie. II. Ibid. 1950. 194 pp.

[3] Conformal representation. 2. ed., Cambridge: Cambridge Univ. Press 1952. 115 pp.

CARLESON, L.

[1] On the connection between Hausdorff measures and capacity. Ark. Mat. **3** (1958), 403–406.

[2] Maximal functions and capacities. Ann. Inst. Fourier (Grenoble) **15** (1965), 59–64.

CHANG, S.

[1] Royden's compactification of Riemannian spaces. Doctoral dissertation, Univ. of Calif., Los Angeles, Calif., 1968. 74 pp.

CHOQUET, G.

[1] Cf. BRELOT, M.; CHOQUET, G. [1].

CONSTANTINESCU, C.

[1] On the behavior of analytic functions at boundary elements on Riemann surfaces. Rev. Math. Pures Appl. **2** (1957), 269–276.

[2] Sur le comportement d'une fonction analytique à la frontière idéale d'une surface de Riemann. C.R. Acad. Sci. Paris **245** (1957), 1995–1997.

[3] (– CORNEA, A.) Über den idealen Rand und einige seiner Anwendungen bei der Klassifikation der Riemannschen Flächen. Nagoya Math. J. **13** (1958), 169–233.

[4] (– CORNEA, A.) Comportement des transformations analytiques des surfaces de Riemann sur la frontière de Martin. C.R. Acad. Sci. Paris **249** (1959), 355–357.

[5] (– CORNEA, A.) Über einige Probleme von M. Heins. Rev. Math. Pures Appl. **4** (1959), 277–281.

[6] Über die Klassifikation der Riemannschen Flächen. Acta Math. **102** (1959), 47–78.

[7] Ideale Randkomponenten einer Riemannschen Fläche. Rev. Math. Pures Appl. **4** (1959), 43–76.

[8] (– CORNEA, A.) Über das Verhalten der analytischen Abbildungen Riemannscher Flächen auf dem idealen Rand von Martin. Nagoya Math. J. **17** (1960), 1–87.

[9] (– CORNEA, A.) Über den Martinschen idealen Rand einer Riemannschen Fläche. Rev. Math. Pures Appl. **5** (1960), 21–25.

[10] (– CORNEA, A.) Sur la frontière Martin d'une surface riemannienne. Acad. R. P. Romîne Fil. Cluj Stud. Cerc. Mat. **11** (1960), 261–265.

[11] (– CORNEA, A.) Spaces of harmonic functions on Riemann surfaces. Rev. Math. Pures Appl. **6** (1961), 373–390.

[12] (– CORNEA, A.) Le théorème de Beurling et la frontière idéale de Kuramochi. C.R. Acad. Sci. Paris **254** (1962), 1732–1734.

[13] Dirichletsche Abbildungen. Nagoya Math. J. **20** (1962), 75–89.

[14] Ein neues Dirichletsches Problem auf Riemannschen Flächen. Rev. Math. Pures Appl. **7** (1962), 127–133.

[15] (– CORNEA, A.) Analytische Abbildungen Riemannscher Flächen. Ibid. **8** (1963), 67–72.

[16] (– CORNEA, A.) Normale Kompaktifizierungen Riemannscher Flächen. Ibid. **8** (1963), 73–75.

[17] (– CORNEA, A.) Ideale Ränder Riemannscher Flächen. Berlin-Göttingen-Heidelberg: Springer 1963. 244 pp.

[18] (– CORNEA, A.) Frontières idéales des surfaces de Riemann. Rev. Math. Pures Appl. (Bucarest) **8** (1963), 227–242.

CORNEA, A.

[1] On the behaviour of analytic functions in the neighborhood of the boundary of a Riemann surface. Nagoya Math. J. **12** (1957), 55–58.

[2] Au sujet de parties singulières de la frontière idéale. [Romanian. Russian and French summaries.] Com. Acad. R. P. Romîne **8** (1958), 639–642.

[3–14] Cf. CONSTANTINESCU, C.; CORNEA, A. [3–5], [8–12], [15–18].

COUNCILMAN, S.

[1] The class of K-functions on Riemannian spaces. Doctoral dissertation, Univ. of Calif., Los Angeles, Calif., 1968. 79 pp.

DOOB, J.

[1] Boundary properties of functions with finite Dirichlet integrals. Ann. Inst. Fourier (Grenoble) **12** (1962), 573–621.

DUFF, G.

[1] Partial differential equations. Toronto: Univ. of Toronto Press 1956. 248 pp.

ENDL, K.

[1] Zum Typenproblem Riemannscher Flächen. Mitt. Mathem. Sem. Univ. Giessen No. 49 (1954), 35 pp.

[2] Sur des problèmes du type de Dirichlet utilisant les lignes de Green. C.R. Acad. Sci. Paris **244** (1957), 1705–1707.

EVANS, G.

[1] Potentials and positively infinite singularities of harmonic functions. Monatsh. Math. Phys. **43** (1936), 419–424.

FÉKETE, M.

[1] Über die Verteilung der Wurzeln bei gewissen algebraischen Gleichungen mit ganzzahligen Koeffizienten. Math. Z. **17** (1923), 228–249.

FELLER, W.

[1] Über die Lösungen der linearen partiellen Differentialgleichungen zweiter Ordnung vom elliptischen Typus. Math. Ann. **102** (1930), 633–649.

FIALA, F.

[1] Cf. BLANC, C.; FIALA, F. [4].

FLORACK, H.

[1] Reguläre und meromorphe Funktionen auf nicht geschlossenen Riemannschen Flächen. Schr. Math. Inst. Univ. Münster No. 1 (1948), 34 pp.

FORD, L.

[1] Automorphic functions. New York: McGraw-Hill Book Co., Inc. 1929. 333 pp.

FOURÈS, L.

[1] Décomposition en feuillets des surfaces de Riemann de type parabolique. C.R. Acad. Sci. Paris **228** (1949), 644–646.

FROSTMAN, O.

[1] Potentiel d'équilibre et capacité des ensembles avec quelques applications à la théorie des fonctions. Medd. Lunds Univ. Mat. Sem. **3** (1935), 115 pp.

FUJIIE, T.

[1] On weak boundary components of a Riemann surface. J. Math. Soc. Japan **15** (1963), 396–403.

[2] Extremal length and Kuramochi boundary. J. Math. Kyoto Univ. **4** (1964), 149–159.

[3] Applications of extremal length to classification of Riemann surfaces. Nagoya Math. J. **24** (1964), 159–166.

FUKUDA, N.

[1] Cf. SARIO, L.; FUKUDA, N. [23].

GARNETT, J.

[1] Positive length but zero analytic capacity. Proc. Amer. Math. Soc. (To appear.)

GARWICK, J. V.

[1] Über das Typenproblem. Arch. Math. Naturvid. **43** (1940), 33–46.

GEHRING, F.

[1] Rings and quasi-conformal mappings in space. Proc. Nat. Acad. Sci. U.S.A. **47** (1961), 98–105.

[2] (– VÄISÄLÄ, J.) On the geometric definition for quasiconformal mappings. Comment. Math. Helv. **36** (1961), 19–32.

[3] Rings and quasiconformal mappings in space. Trans. Amer. Math. Soc. **103** (1962), 353–393.

[4] Quasiconformal mappings in space. Bull. Amer. Math. Soc. **69** (1963), 146–164.

GELFAND, I.

[1] Normierte Ringe. Rec. Math. N.S. **9** (51) (1941), 3–24.

GLASNER, M.

[1] Harmonic functions with prescribed boundary behavior in Riemannian spaces. Doctoral dissertation, Univ. of Calif., Los Angeles, Calif., 1966. 57 pp.

[2] (– KATZ, R.; NAKAI, M.) Bisection into small annuli. Pacific J. Math. **24** (1968), 457–461.

[3] Cf. SARIO, L.; SCHIFFER, M.; GLASNER, M. [26].

GODEFROID, M.

[1] Une propriété des fonctions B.L.D. dans un espace de Green. Ann. Inst. Fourier (Grenoble) **9** (1959), 301–304.

GOL'DBERG, A. A.

[1] A class of Riemann surfaces. [Russian.] Mat. Sb. (N.S.) **49** (91) (1959), 447–458.

GOLDSTEIN, M.

[1] K- and L-kernels on an arbitrary Riemann surface. Pacific J. Math. **19** (1966), 449–459.

GOLUSIN, G.

[1] Sur la représentation conforme. Ruc. Math. Moscu (2) **1** (1936), 272–282.

GRUNSKY, H.

[1] Neue Abschätzungen zur konformen Abbildung ein- und mehrfach zusammenhängender Bereiche. Schriften Sem. Univ. Berlin **1** (1932), 95–140.

GUNNING, R. C.

[1] Lectures on Riemann surfaces. Princeton, N.J.: Princeton Univ. Press 1966. 254 pp.

AF HÄLLSTRÖM, G.

[1] Zur Beziehung zwischen den Automorphiefunktionen und dem Flächentypus. Acta Acad. Abo. **20** No. 10 (1956), 12 pp.

HARDY, G.

[1] (– LITTLEWOOD, J.; PÓLYA, G.) Inequalities. Cambridge: Cambridge Univ. Press 1959. 324 pp.

HARMON, S.

[1] Regular covering surfaces of Riemann surfaces. Pacific J. Math. **10** (1960), 1263–1289.

HAYASHI, K.

[1] Les solutions positives de l'équation $\Delta u = Pu$ sur une surface de Riemann. Kōdai Math. Sem. Rep. **13** (1961), 20–24.

[2] Sur une frontière des surfaces de Riemann. Proc. Japan Acad. **37** (1961), 469–472.

[3] Une frontière des surfaces de Riemann ouvertes et applications conformes. Kōdai Math. Sem. Rep. **14** (1962), 169–188.

HAYMAN, W. K.

[1] Meromorphic functions. Oxford: Clarendon Press 1964. 191 pp.

HEINS, M.

[1] A lemma on positive harmonic functions. Ann. of Math. (2) **52** (1950), 568–573.

[2] Riemann surfaces of infinite genus. Ibid. (2) **55** (1952), 296–317.

[3] On the Lindelöf principle. Ibid. (2) **61** (1955), 440–473.

[4] Lindelöfian maps. Ibid. (2) **62** (1955), 418–446.

[5] Some remarks concerning parabolic Riemann surfaces. J. Math. Pures Appl. (9) **36** (1957), 305–312.

[6] Asymptotic spots of entire and meromorphic functions. Ann. of Math. (2) **66** (1957), 430–439.

[7] On certain meromorphic functions of bounded valence. Rev. Math. Pures Appl. **2** (1957), 263–267.

[8] Functions of bounded characteristic and Lindelöfian maps. Proc. Internat. Congr. Math. 1958, pp. 376–388. New York: Cambridge Univ. Press 1960.

[9] A property of the asymptotic spots of a meromorphic function or an interior transformation whose domain is the open unit disk. J. Indian Math. Soc. (N.S.) **24** (1960), 265–268 (1961).

HILLE, E.

[1] Remarks on transfinite diameters. General topology and its relations to modern analysis and algebra (Proc. Sympos., Prague, 1961), pp. 211–220. New York: Academic Press; Prague: Publ. House Czech. Acad. Sci. 1962.

[2] Analytic function theory. Vol. II. Boston: Ginn and Co. 1962. 496 pp.

HODGE, W.

[1] The theory and applications of harmonic integrals. 2. ed., Cambridge: Cambridge Univ. Press 1952. 282 pp.

HÖRMANDER, L.

[1] Linear partial differential operators. Berlin-Göttingen-Heidelberg: Springer 1963. 287 pp.

HUBER, H.

[1] Zur analytischen Theorie hyperbolischer Raumformen und Bewegungsgruppen. Math. Ann. **138** (1959), 1–26.

[2] Riemannsche Flächen von hyperbolischem Typus im euklidischen Raum. Ibid. **139** (1959), 140–146.

[3] Riemann surfaces of hyperbolic type in the Euclidean 3-space. Contributions to function theory (Internat. Colloq. Function Theory, Bombay, 1960), pp. 219–221. Bombay: Tata Institute of Fundamental Research 1960.

[4] Zur analytischen Theorie hyperbolischer Raumformen und Bewegungsgruppen. II. Math. Ann. **143** (1961), 463–464.

HUCKEMANN, F.

[1] Typusänderung bei Riemannschen Flächen durch Verschiebung von Windungspunkten. Math. Z. **59** (1954), 385–387.

[2] Zur Darstellung von Riemannschen Flächen durch Streckenkomplexe. Ibid. **65** (1956), 215–239.

IKEGAMI, T.

[1] On the theorems of Constantinescu-Cornea. Proc. Japan Acad. **40** (1964), 196–199.

[2] On the non-minimal Martin boundary points. Nagoya Math. J. **29** (1967), 287–290.

IKOMA, K.

[1] On a property of the boundary correspondence under quasiconformal mappings. Nagoya Math. J. **16** (1960), 185–188.

[2] On the distortion and correspondence under quasiconformal mappings in space. Ibid. **25** (1965), 175–203.

IVERSEN, F.

[1] Recherches sur les fonctions inverses des fonctions méromorphes. Thèse, Helsinki, 1914.

JENKINS, J. A.

[1] On quasiconformal mappings. J. Rational Mech. Anal. **5** (1956), 343–352.

[2] Univalent functions and conformal mapping. Berlin-Göttingen-Heidelberg: Springer 1958. 169 pp.

[3] On a type problem. Canad. J. Math. **11** (1959), 427–431.

[4] On hyperbolic surfaces in three-dimensional Euclidean space. Michigan Math. J. **8** (1961), 1–5.

JOHNSON, W.

[1] Harmonic functions of bounded characteristic in locally Euclidean spaces. Doctoral dissertation, Univ. of Calif., Los Angeles, Calif., 1964. 52 pp.

JURCHESCU, M.

[1] L'invariance K-quasi conforme de la parabolicité d'un élément frontière. C.R. Acad. Sci. Paris **246** (1958), 2997–2999.

[2] Modulus of a boundary component. Pacific J. Math. **8** (1958), 791–804.

[3] Ensembles AB-enlevables dans des espaces complexes. [Romanian. Russian and French summaries.] Acad. R. P. Romîne. Fil. Iaşi. Stud. Cerc. Şti. Mat. **10** (1959), 13–26.

[4] On AB-removability in complex spaces. Rev. Math. Pures Appl. **4** (1959), 357–368.

KAKUTANI, S.

[1] Einführung einer Metrik auf die Riemannsche Fläche und der Typus der Riemannschen Fläche. Proc. Imp. Acad. Tokyo **13** (1937), 89–92.

[2] Applications of the theory of pseudo-regular functions to the type problem of Riemann surfaces. Japan. J. Math. **13** (1937), 375–392.

[3] Two-dimensional Brownian motion and the type problem of Riemann surfaces. Proc. Japan Acad. **21** (1949), 138–140.

[4] Random walk and the type problem of Riemann surfaces. Contributions to the theory of Riemann surfaces, pp. 95–101. Princeton, N.J.: Princeton Univ. Press 1953.

KAMETANI, S.

[1] The exceptional values of functions with the set of linear measure zero of essential singularities. I, II. Proc. Imp. Acad. Tokyo **17** (1941), 117–120; **19** (1943), 438–443.

KAPLAN, W.

[1] Construction of parabolic Riemann surfaces by the general reflection principle. Contributions to the theory of Riemann surfaces, pp. 103–106. Princeton, N.J.: Princeton Univ. Press 1953.

KATZ, R.

[1] Some results in the classification theory of Riemannian manifolds. Doctoral dissertation, Univ. of Calif., Los Angeles, Calif., 1967. 44 pp.

[2] Cf. GLASNER, M.; KATZ, R.; NAKAI, M. [2].

KAWAMURA, M.

[1] On completeness of Royden's algebra. Proc. Japan Acad. **40** (1964), 166–170.

KELLEY, J.

[1] General topology. Princeton, N.J.: D. Van Nostrand Co., Inc. 1955. 298 pp.

KIMURA, T.

[1] Sur la propriété d'Iversen et l'équation différentielle ordinaire du second ordre. Comment. Math. Univ. St. Paul **8** (1960), 63–70.

KOBAYASHI, Z.

[1] On the type of Riemann surfaces. Sci. Rep. Tokyo Bunrika Daigaku **2** (1935), 217–233.

[2] Ein Satz über ein Problem von Herrn Speiser. Ibid. **3** (1936), 29–32.

[3] A remark on the type of Riemann surfaces. Ibid. **3** (1937), 185–193.

[4] On Kakutani's theory of the type of Riemann surfaces. Ibid. **4** (1940), 9–44.

KOEBE, P.

[1] Abhandlungen zur Theorie der konformen Abbildung. VI. Abbildung mehrfach zusammenhängender schlichter Bereiche auf Kreisbereiche. Uniformisierung hyperelliptischer Kurven. (Iterationsmethode.) Math. Z. **7** (1920), 235–301.

[2] Allgemeine Theorie der Riemannschen Mannigfaltigkeiten. (Konforme Abbildung und Uniformisierung.) Acta Math. **50** (1927), 27–157.

KORN, A.

[1] Zwei Anwendungen der Methode der sukzessiven Anwendungen. Schwarz Festschrift, pp. 215–229. 1914.

KÜNZI, H. P.

[1] Quasikonforme Abbildungen. Berlin-Göttingen-Heidelberg: Springer 1960. 182 pp.

KURAMOCHI, Z.

[1] On covering surfaces. Osaka Math. J. **5** (1953), 155–201.

[2] Relations between harmonic dimensions. Proc. Japan Acad. **30** (1954), 576–580.

[3] Dirichlet problem on Riemann surfaces. I–IV. Ibid. **30** (1954), 731–735; 825–830; 831–836; 946–950.

[4] Harmonic measures and capacity of sets of the ideal boundary. I. Ibid. **30** (1954), 951–956.

[5] An example of a null-boundary Riemann surface. Osaka Math. J. **6** (1954), 83–91.

[6] (– KURODA, T.) A note on the set of logarithmic capacity zero. Proc. Japan Acad. **30** (1954), 566–569.

[7] On the existence of harmonic functions on Riemann surfaces. Osaka Math. J. **7** (1955), 23–28.

[8] Dirichlet problem on Riemann surfaces. V. On covering surfaces. Proc. Japan Acad. **31** (1955), 20–24.

[9] Harmonic measures and capacity of sets of the ideal boundary. II. Ibid. **31** (1955), 25–30.

[10] On the behavior of analytic functions on abstract Riemann surfaces. Osaka Math. J. **7** (1955), 109–127.

[11] Evans's theorem on abstract Riemann surfaces with null-boundaries. I, II. Proc. Japan Acad. **32** (1956), 1–6; 7–9.

[12] Capacity of subsets of the ideal boundary. Ibid. **32** (1956), 111–116.

[13] Evans-Selberg's theorem on abstract Riemann surfaces with positive boundaries. I, II. Ibid. **32** (1956), 228–233; 234–236.

[14] Mass distributions on the ideal boundaries of abstract Riemann surfaces. I, II. Osaka Math. J. **8** (1956), 119–137; 145–186.

[15] Analytic functions in the neighborhood of the ideal boundary. Proc. Japan Acad. **33** (1957), 84–86.

[16] On the ideal boundaries of abstract Riemann surfaces. Osaka Math. J. **10** (1958), 83–102.

[17] On harmonic functions representable by Poisson's integral. Ibid. **10** (1958), 103–117.

[18] Mass distributions on the ideal boundaries of abstract Riemann surfaces. III. Ibid. **10** (1958), 119–136.

[19] Representation of Riemann surfaces. Ibid. **11** (1959), 71–82.

[20] Correspondence of sets on the boundaries of Riemann surfaces. Proc. Japan Acad. **36** (1960), 112–117.

[21] Mass distributions on the ideal boundaries. Ibid. **36** (1960), 118–122.

[22] Relations among topologies on Riemann surfaces. I–IV. Ibid. **38** (1962), 310–315, 457–462, 463–467, 468–472.

[23] Potentials on Riemann surfaces. J. Fac. Sci. Hokkaido Univ. Ser. I **16** (1962), 5–79.

[24] On the behaviour of analytic functions on the ideal boundary. I. Proc. Japan Acad. **38** (1962), 150–155.

[25] On the behaviour of analytic functions on the ideal boundary. II. Ibid. **38** (1962), 188–193.

[26] On the behaviour of analytic functions on the ideal boundary. III. Ibid. **38** (1962), 194–198.

[27] Correction to the paper "On the behaviour of analytic functions". Ibid. **39** (1963), 27–32.

[28] Correspondence of boundaries of Riemann surfaces. J. Fac. Sci. Hokkaido Univ. Ser. I **17** (1963), 96–122.

[29] Examples of non minimal points on Riemann surfaces of planar character. Ibid. **19** (1965), 28–48.

[30] Relations between two Martin topologies on a Riemann surface. Ibid. **19** (1966), 146–153.

Kuroda, T.

[1] On the type of an open Riemann surface. Proc. Japan Acad. **27** (1951), 57–60.

[2] Some remarks on an open Riemann surface with null boundary. Tôhoku Math. J. (2) **3** (1951), 182–186.

[3] A property of some open Riemann surfaces and its application. Nagoya Math. J. **6** (1953), 77–84.

[4] On analytic functions on some Riemann surfaces. Ibid. **10** (1956), 27–50.

[5] Remarks on some covering surfaces. Rev. Math. Pures Appl. **2** (1957), 239–244.

[6] On some theorems of Sario. Bull. Math. Soc. Sci. Math. Phys. R. P. Roumaine (N.S.) **2** (50) (1958), 411–417.

[7] Cf. AKAZA, T.; KURODA, T. [3].

[8] Cf. KURAMOCHI, Z.; KURODA, T. [6].

[9] Cf. OZAWA, M.; KURODA, T. [8].

KUSUNOKI, Y.

[1] Über die hinreichenden Bedingungen dafür, daß eine Riemannsche Fläche nullberandet ist. Mem. Coll. Sci. Univ. Kyoto Ser. A. Math. **27** (1952), 99–108.

[2] Some classes of Riemann surfaces characterized by the extremal length. Proc. Japan Acad. **32** (1956), 406–408.

[3] Contributions to Riemann-Roch's theorem. Mem. Coll. Sci. Univ. Kyoto Ser. A. Math. **31** (1958), 161–180.

[4] (– MORI, S.) On the harmonic boundary of an open Riemann surface. I. Japan. J. Math. **29** (1959), 52–56.

[5] (– MORI, S.) On the harmonic boundary of an open Riemann surface. II. Mem. Coll. Sci. Univ. Kyoto Ser. A. Math. **33** (1960/61), 209–223.

[6] Supplements and corrections to my former papers. Ibid. **33** (1960/61), 429–433.

[7] On a compactification of Green spaces. Dirichlet problem and theorems of Riesz type. J. Math. Kyoto Univ. **1** (1961/62), 385–402.

LAASONEN, P.

[1] Zum Typenproblem der Riemannschen Flächen. Ann. Acad. Sci. Fenn. Ser. A.I. No. 11 (1942), 7 pp.

[2] Beiträge zur Theorie der Fuchsoiden Gruppen und zum Typenproblem der Riemannschen Flächen. Ibid. No. 25 (1944), 87 pp.

LARSEN, K.

[1] Extremal harmonic functions in locally Euclidean n-spaces. Doctoral dissertation, Univ. of Calif., Los Angeles, Calif., 1964. 49 pp.

LEHTO, O.

[1] On the existence of analytic functions with a finite Dirichlet integral. Ann. Acad. Sci. Fenn. Ser. A.I. No. 67 (1949), 7 pp.

LEHTO, O.; NEVANLINNA, R. (edited by)

[1] Proceedings of the international colloquium on the theory of functions, Helsinki, 1957. Suomalainen Tiedeakatemia, Helsinki, 1958.

LE-VAN, T.

[1] Beitrag zum Typenproblem der Riemannschen Flächen. Comment. Math. Helv. **20** (1947), 270–287.

[2] Un problème de type généralisé. C. R. Acad. Sci. Paris **228** (1949), 1270–1272.

[3] On the type of Riemann surfaces defined by the linear substitution group. [Russian.] Sibirsk. Mat. Ž. **5** (1964), 853–857.

LICHTENSTEIN, L.

[1] Zur Theorie der konformen Abbildung nichtanalytischer, singularitätenfreier Flächenstücke auf ebene Gebiete. Bull. Internat. Acad. Sci. Gracovie, Cl. Sci. Math. Nat. Ser. A. (1916), 192–217.

LITTLEWOOD, J.

[1] Cf. HARDY, G.; LITTLEWOOD, J.; PÓLYA, G. [1].

LOEB, P.

[1] (– WALSH, B.) The equivalence of Harnack's principle and Harnack's inequality in the axiomatic system of Brelot. Ann. Inst. Fourier (Grenoble) **15** (1965), 597–600.

[2] An axiomatic treatment of pairs of elliptic differential equations. Ibid. **16** (1966), 167–208.

LOKKI, O.

[1] Beiträge zur Theorie der analytischen und harmonischen Funktionen mit endlichem Dirichletintegral. Ann. Acad. Sci. Fenn. Ser. A.I. No. 92 (1951), 11 pp.

[2] Über eindeutige analytische Funktionen mit endlichem Dirichletintegral. Ibid. No. 105 (1951), 13 pp.

[3] Über harmonische Funktionen mit endlichem Dirichletintegral. 11. Scand. Congr. Math. Trondheim 1949, pp. 239–242.

LOOMIS, L.

[1] An introduction to abstract harmonic analysis. Princeton, N.J.: D. Van Nostrand Co., Inc. 1953. 190 pp.

MAEDA, F.

[1] Notes on Green lines and Kuramochi boundary of a Green space. J. Sci. Hiroshima Univ. Ser. A-I Math. **28** (1964), 59–66.

[2] Normal derivatives on an ideal boundary. Ibid. **28** (1964), 113–131.

MARDEN, A.

[1] (– RODIN, B.) Extremal and conjugate extremal distance on open Riemann surfaces with applications to circular-radial slit mappings. Acta Math. **115** (1966), 237–269.

MARIA, A.

[1] The potential of a positive mass and the weight function of Wiener. Proc. Nat. Acad. Sci. U.S.A. **20** (1934), 485–489.

MARTIN, R.

[1] Minimal positive harmonic functions. Trans. Amer. Math. Soc. **49** (1941), 137–172.

MATSUMOTO, K.

[1] Remarks on some Riemann surfaces. Proc. Japan Acad. **34** (1958), 672–675.

[2] On subsurfaces of some Riemann surfaces. Nagoya Math. J. **15** (1959), 261–274.

[3] An extension of a theorem of Mori. Japan J. Math. **29** (1959), 57–59.

[4] Analytic functions on some Riemann surfaces. II. Nagoya Math. J. **23** (1963), 153–164.

[5] Cf. TODA, N.; MATSUMOTO, K. [1].

MEEHAN, H.

[1] Capacity problems in locally Euclidean spaces. Doctoral dissertation, Univ. of Calif., Los Angeles, Calif., 1964. 80 pp.

MESCHKOWSKI, H.

[1] Über die konforme Abbildung gewisser Bereiche von unendlich hohem Zusammenhang auf Vollkreisbereiche. I, II. Math. Ann. **123** (1951), 392–405; **124** (1952), 178–181.

MINKOWSKI, H.

[1] Diophantische Approximationen. New York: Chelsea Publishing Co. 1957. 235 pp.

MIRANDA, C.

[1] Equazioni alle derivate parziali di tipo ellittico. Berlin-Göttingen-Heidelberg: Springer 1955. 222 pp.

MIZUMOTO, H.

[1] On Riemann surfaces with finite spherical area. Kōdai Math. Sem. Rep. **7** (1957), 87–96.

[2] Cf. OZAWA, M.; MIZUMOTO, H. [5].

MOCANU, G.

[1] Cf. BOBOC, N.; MOCANU, G. [1].

MOPPERT, K. F.

[1] Über eine gewisse Klasse von elliptischen Riemannschen Flächen. Comment. Math. Helv. **23** (1949), 174–176.

MORI, A.

[1] On Riemann surfaces, on which no bounded harmonic function exists. J. Math. Soc. Japan **3** (1951), 285–289.

[2] On the existence of harmonic functions on a Riemann surface. J. Fac. Sci. Univ. Tokyo Sect. I. **6** (1951), 247–257.

[3] A remark on the class O_{HD} of Riemann surfaces. Kōdai Math. Sem. Rep. (1952), 57–58.

[4] A remark on the prolongation of Riemann surfaces of finite genus. J. Math. Soc. Japan **4** (1952), 27–30.

[5] A note on unramified abelian covering surfaces of a closed Riemann surface. Ibid. **6** (1954), 162–176.

[6] On quasi-conformality and pseudo-analyticity. Trans. Amer. Math. Soc. **84** (1957), 56–77.

MORI, S.

[1] (– ÔTA, M.) A remark on the ideal boundary of a Riemann surface. Proc. Japan Acad. **32** (1956), 409–411.

[2] A remark on a subdomain of a Riemann surface of the class O_{HD}. Ibid. **34** (1958), 251–254.

[3] On a compactification of an open Riemann surface and its application. J. Math. Kyoto Univ. **1** (1961/62), 21–42.

[4] On a ring of bounded continuous functions on an open Riemann surface (supplements and corrections to my former paper). Ibid. **2** (1962), 25–42.

[5] Remarks on the harmonic boundary of a plane domain. Ibid. **4** (1964), 99–115.

[6–7] Cf. KUSUNOKI, Y.; MORI, S. [4–5].

MORREY, C.

[1] On the solutions of quasi-linear elliptic partial differential equations. Trans. Amer. Math. Soc. **43** (1938), 126–166.

MYRBERG, L.

[1] Über die Existenz von positiven harmonischen Funktionen auf Riemannschen Flächen. Ann. Acad. Sci. Fenn. Ser. A.I. No. 146 (1953), 6 pp.

[2] Über die Integration der Poissonschen Gleichung auf offenen Riemannschen Flächen. Ibid. No. 161 (1953), 10 pp.

[3] Über die Existenz von positiven harmonischen Funktionen auf offenen Riemannschen Flächen. 12. Scand. Congr. Math. Lund 1953, pp. 214–216.

[4] Über die Existenz der Greenschen Funktion der Gleichung $\Delta u = c(P) \cdot u$ auf Riemannschen Flächen. Ann. Acad. Sci. Fenn. Ser. A.I. No. 170 (1954), 8 pp.

[5] Über die Integration der Differentialgleichung $\Delta u = c(P) \cdot u$ auf offenen Riemannschen Flächen. Math. Scand. **2** (1954), 142–152.

[6] Über subelliptische Funktionen. Ann. Acad. Sci. Fenn. Ser. A.I. No. 290 (1960), 9 pp.

[7] Über meromorphe Funktionen auf nullberandeten Riemannschen Flächen. Ibid. No. 312 (1962), 11 pp.

MYRBERG, P. J.

[1] L'existence de la fonction de Green pour un domaine plan donné. C. R. Acad. Sci. Paris **190** (1930), 1372–1374.

[2] Über die Existenz der Greenschen Funktionen auf einer gegebenen Riemannschen Fläche. Acta Math. **61** (1933), 39–79.

[3] Sur la détermination du type d'une surface riemannienne simplement connexe. C. R. Acad. Sci. Paris **200** (1935), 1818–1820.

[4] Über die Bestimmung des Typus einer Riemannschen Fläche. Ann. Acad. Sci. Fenn. Ser. A. 45 No. 3 (1935), 30 pp.

[5] Die Kapazität der singulären Menge der linearen Gruppen. Ibid. No. 10 (1941), 19 pp.

[6] Über die analytische Fortsetzung von beschränkten Funktionen. Ibid. No. 58 (1949), 7 pp.

NAKAI, M.

[1] On a ring isomorphism induced by quasiconformal mappings. Nagoya Math. J. **14** (1959), 201–221.

[2] A function algebra on Riemann surfaces. Ibid. **15** (1959), 1–7.

[3] Purely algebraic characterization of quasiconformality. Proc. Japan Acad. **35** (1959), 440–443.

[4] On a problem of Royden on quasiconformal equivalence of Riemann surfaces. Ibid. **36** (1960), 33–37.

[5] Algebraic criterion on quasiconformal equivalence of Riemann surfaces. Nagoya Math. J. **16** (1960), 157–184.

[6] The space of bounded solutions of the equation $\Delta u = pu$ on a Riemann surface. Proc. Japan Acad. **36** (1960), 267–272.

[7] Some topological properties on Royden's compactification of a Riemann surface. Ibid. **36** (1960), 555–559.

[8] The space of non-negative solutions of the equation $\Delta u = pu$ on a Riemann surface. Kōdai Math. Sem. Rep. **12** (1960), 151–178.

[9] A measure on the harmonic boundary of a Riemann surface. Nagoya Math. J. **17** (1960), 181–218.

[10] The space of Dirichlet-finite solutions of the equation $\Delta u = Pu$ on a Riemann surface. Ibid. **18** (1961), 111–131.

[11] Bordered Riemann surface with parabolic double. Proc. Japan Acad. **37** (1961), 553 – 555.

[12] On the function-ring method in Riemann surfaces. [Japanese.] Sûgaku **13** (1961/62), 129 – 140.

[13] Genus and classification of Riemann surfaces. Osaka Math. J. **14** (1962), 153 – 180.

[14] On Evans potential. Proc. Japan Acad. **38** (1962), 624 – 629.

[15] On rings of analytic functions on Riemann surfaces. Ibid. **39** (1963), 79 – 84.

[16] Evans' harmonic functions on Riemann surfaces. Ibid. **39** (1963), 74 – 78.

[17] On Evans' solution of the equation $\Delta u = Pu$ on Riemann surfaces. Kōdai Math. Sem. Rep. **15** (1963), 79 – 93.

[18] On Wiener homeomorphism between Riemann surfaces. Proc. Japan Acad. **40** (1964), 468 – 473.

[19] A property of Green's star domain. Ibid. **40** (1964), 161 – 165.

[20] Behaviour of Green lines at Royden's boundary of Riemann surfaces. Nagoya Math. J. **24** (1964), 1 – 27.

[21] Green potential of Evans type on Royden's compactification of a Riemann surface. Ibid. **24** (1964), 205 – 239.

[22] On a criterion of quasi-boundedness of positive harmonic functions. Proc. Japan Acad. **41** (1965), 215 – 217.

[23] Φ-bounded harmonic functions and classification of Riemann surfaces. Pacific J. Math. **15** (1965), 1329 – 1335.

[24] Finite interpolation for analytic functions with finite Dirichlet integrals. Proc. Amer. Math. Soc. **17** (1966), 362 – 364.

[25] Existence of positive harmonic functions. Ibid. **17** (1966), 365 – 367.

[26] Royden's map between Riemann surfaces. Bull. Amer. Math. Soc. **72** (1966), 1003 – 1005.

[27] Radon-Nikodým densities between harmonic measures on the ideal boundary of an open Riemann surface. Nagoya Math. J. **27** (1966), 71 – 76.

[28] On Φ-bounded harmonic functions. Ann. Inst. Fourier **16** (1966), 145 – 157.

[29] Potentials of Sario's kernel. J. Analyse Math. **17** (1966), 225-240.

[30] (– Sario, L.) Construction of principal functions by orthogonal projection. Canad. J. Math. **18** (1966), 887 – 896.

[31] (– Sario, L.) Normal operators, linear liftings and the Wiener compactification. Bull. Amer. Math. Soc. **72** (1966), 947 – 949.

[32] On Evans' kernel. Pacific J. Math. **22** (1967), 125 – 137.

[33] Sario's potentials and analytic mappings. Nagoya Math. J. **29** (1967), 93 – 101.

[34] (– Sario, L.) Completeness and function-theoretic degeneracy of Riemannian spaces. Proc. Nat. Acad. Sci. U.S.A. **57** (1967), 29 – 31.

[35] (– Sario, L.) Classification and deformation of Riemannian spaces. Math. Scand. **20** (1967), 193 – 208.

[36] (– Sario, L.) A parabolic Riemannian ball. Proceedings of the 1966 Amer. Math. Soc. Summer Institute, pp. 341 – 349. Amer. Math. Soc., Providence, R.I., 1968.

[37] Cf. Glasner, M.; Katz, R.; Nakai, M. [2].

Narasimhan, M. S.

[1] The type and the Green's kernel of an open Riemann surface. Ann. Inst. Fourier (Grenoble) **10** (1960), 285 – 296.

NEVANLINNA, R.

[1] Das harmonische Maß von Punktmengen und seine Anwendung in der Funktionentheorie. 8. Scand. Congr. Math. Stockholm 1934, pp. 116–133.

[2] Ein Satz über offene Riemannsche Flächen. Ann. Acad. Sci. Fenn. Ser. A. 54 No. 3 (1940), 16 pp.

[3] Quadratisch integrierbare Differentiale auf einer Riemannschen Mannigfaltigkeit. Ibid. No. 1 (1941), 34 pp.

[4] Eindeutigkeitsfragen in der Theorie der konformen Abbildung. 10. Scand. Congr. Math. Copenhagen 1946, pp. 225–240.

[5] Sur l'existence de certaines classes de différentielles analytiques. C.R. Acad. Sci. Paris **228** (1949), 2002–2004.

[6] Über die Anwendung einer Klasse von Integralgleichungen für Existenzbeweise in der Potentialtheorie. Acta Sci. Math. Szeged **12** (1950), 146–160.

[7] Über die Existenz von beschränkten Potentialfunktionen auf Flächen von unendlichem Geschlecht. Math. Z. **52** (1950), 599–604.

[8] Über die Polygondarstellung einer Riemannschen Fläche. Ann. Acad. Sci. Fenn. Ser. A.I. No. 122 (1952), 9 pp.

[9] Uniformisierung. Berlin-Göttingen-Heidelberg: Springer 1953. 391 pp.

[10] Eindeutige analytische Funktionen. 2. Aufl. Berlin-Göttingen-Heidelberg: Springer 1953. 379 pp.

[11] Polygonal representation of Riemann surfaces. Lectures on functions of a complex variable, pp. 65–70. Ann Arbor: Univ. Michigan Press 1955.

NOSHIRO, K.

[1] Open Riemann surface with null boundary. Nagoya Math. J. **3** (1951), 73–79.

[2] The modern theory of functions. Tokyo: Iwanami 1954. 428 pp.

[3] Cluster sets. Berlin-Göttingen-Heidelberg: Springer 1960. 135 pp.

[4] Cf. SARIO, L.; NOSHIRO, K. [25].

OĞUZTÖRELI, M. N.

[1] Sur les fonctions à type borné. [Turkish summary.] Rev. Fac. Sci. Univ. Istanbul. Sér. A **22** (1957), 141–149.

[2] Sur les cas d'épuisement des surfaces de Riemann à type hyperbolique. [Turkish summary.] Ibid. A **22** (1957), 123–126.

OHTSUKA, M.

[1] Dirichlet problems on Riemann surfaces and conformal mappings. Nagoya Math. J. **3** (1951), 91–137.

[2] Note on the harmonic measure of the accessible boundary of a covering Riemann surface. Ibid. **5** (1953), 35–38.

[3] Boundary components of Riemann surfaces. Ibid. **7** (1954), 65–83.

[4] Boundary components of abstract Riemann surfaces. Lectures on functions of a complex variable, pp. 303–307. Ann Arbor: Univ. Michigan Press 1955.

[5] Capacité d'ensembles de Cantor généralisés. Nagoya Math. J. **11** (1957), 151–160.

[6] Some examples in potential theory. Japan J. Math. **29** (1959), 101–110.

[7] On weak and unstable components. J. Sci. Hiroshima Univ. Ser. A-I **28** (1964), 53–58.

[8] An elementary introduction of Kuramochi boundary. Ibid. **28** (1964), 271–299.

OIKAWA, K.

[1] On the prolongation of an open Riemann surface of finite genus. Kōdai Math. Sem. Rep. **9** (1957), 34 – 41.

[2] Some properties of quasi-conformal mapping. [Japanese.] Sûgaku **9** (1957/58), 13 – 14.

[3] On the stability of boundary components. Doctoral dissertation, Univ. of Calif., Los Angeles, Calif., 1958. 128 pp.

[4] On a criterion for the weakness of an ideal boundary component. Pacific J. Math. **9** (1959), 1233 – 1238.

[5] A constant related to harmonic functions. Japan. J. Math. **29** (1959), 111 – 113.

[6] Sario's lemma on harmonic functions. Proc. Amer. Math. Soc. **11** (1960), 425 – 428.

[7] On the uniqueness of the prolongation of an open Riemann surface of finite genus. Ibid. **11** (1960), 785 – 787.

[8] On the stability of boundary components. Pacific J. Math. **10** (1960), 263 – 294.

[9] On the type problem of Riemann surfaces constructed by conformal sewing. [Japanese.] Sûgaku **12** (1960/61), 160 – 164.

[10] Welding of polygons and the type of Riemann surfaces. Kōdai Math. Sem. Rep. **13** (1961), 37 – 52.

[11] A remark to the construction of Riemann surfaces by welding. J. Sci. Hiroshima Univ. Ser. A-I **27** (1963), 213 – 216.

[12] Minimal slit regions and linear operator method. Kōdai Math. Sem. Rep. **17** (1965), 187 – 190.

[13] Remarks to conformal mappings onto radially slit disks. Sci. Papers Coll. Gen. Ed. Univ. Tokyo **15** (1965), 99 – 109.

[14] (– SUITA, N.) Circular slit disk with infinite radius. Nagoya Math. J. **30** (1967), 57 – 70.

[15] Cf. AKAZA, T.; OIKAWA, K. [4].

[16] Cf. SARIO, L.; OIKAWA, K. [27].

[17] Cf. TAMURA, J.; OIKAWA, K. [1].

[18] Cf. TAMURA, J.; OIKAWA, K.; YAMAZAKI, K. [2].

OSGOOD, W. F.

[1] On the existence of the Green's function for the most general simply connected plane region. Trans. Amer. Math. Soc. **1** (1900), 310 – 314.

OSSERMAN, R.

[1] A hyperbolic surface in 3-space. Proc. Amer. Math. Soc. **7** (1956), 54 – 58.

[2] Riemann surfaces of class A. Trans. Amer. Math. Soc. **82** (1956), 217 – 245.

ÔTA, M.

[1] Cf. MORI, S.; ÔTA, M. [1].

OW, W.

[1] Capacity functions in Riemannian spaces. Doctoral dissertation, Univ. of Calif., Los Angeles, Calif., 1966. 63 pp.

OZAWA, M.

[1] On classification of the function-theoretic null-sets on Riemann surfaces of infinite genus. Kōdai Math. Sem. Rep. **3** (1951), 43 – 44.

[2] Classification of Riemann surfaces. Ibid. **4** (1952), 63–76.

[3] On harmonic dimension. I, II. Ibid. **6** (1954), 33–37; 55–58 and 70.

[4] Some classes of positive solutions of $\Delta u = Pu$ on Riemann surfaces. I, II. Ibid. **6** (1954), 121–126; **7** (1955), 15–20.

[5] (– Mizumoto, H.) On rings of analytic functions. Japan. J. Math. **29** (1959), 114–117.

[6] A set of capacity zero and the equation $\Delta u = Pu$. Kōdai Math. Sem. Rep. **12** (1960), 76–81.

[7] Positive harmonic functions on an end. Ibid. **12** (1960), 143–150.

[8] (– Kuroda, T.) On Pfluger's sufficient condition for a set to be of class N_B. Ibid. **13** (1961), 113–117.

[9] A supplement to "On Pfluger's sufficient condition for a set to be of class N_B". Ibid. **13** (1961), 118–122.

[10] Remarks on unramified abelian covering surfaces of a closed Riemann surface. Ibid. **16** (1964), 101–104.

Parreau, M.

[1] Sur les moyennes des fonctions harmoniques et la classification des surfaces de Riemann. C. R. Acad. Sci. Paris **231** (1950), 679–681.

[2] Sur certaines classes de fonctions analytiques uniformes sur les surfaces de Riemann. Ibid. **231** (1950), 751–753.

[3] Fonctions harmoniques et classification des surfaces de Riemann. Ibid. **234** (1952), 286–288.

[4] Sur les moyennes des fonctions harmoniques et analytiques et la classification des surfaces de Riemann. Ann. Inst. Fourier (Grenoble) **3** (1951), (1952), 103–197.

[5] Cf. Bader, R.; Parreau, M. [1].

Perron, O.

[1] Eine neue Behandlung der ersten Randwertaufgabe für $\Delta u = 0$. Math. Z. **18** (1923), 42–54.

Pfluger, A.

[1] Über das Anwachsen eindeutiger analytischer Funktionen auf offenen Riemannschen Flächen. Ann. Acad. Sci. Fenn. Ser. A.I. No. 64 (1949), 18 pp.

[2] La croissance des fonctions analytiques et uniformes sur une surface de Riemann ouverte. C. R. Acad. Sci. Paris **229** (1949), 505–507.

[3] Sur l'existence des fonctions non constantes, analytiques, uniformes et bornées sur une surface de Riemann ouverte. Ibid. **230** (1950), 166–168.

[4] Über das Typenproblem Riemannscher Flächen. Comment. Math. Helv. **27** (1953), (1954), 346–356.

[5] Extremallängen und Kapazität. Ibid. **29** (1955), 120–131.

[6] Theorie der Riemannschen Flächen. Berlin-Göttingen-Heidelberg: Springer 1957. 248 pp.

[7] Über die Äquivalenz der geometrischen und der analytischen Definition quasikonformer Abbildungen. Comment. Math. Helv. **33** (1959), 23–33.

[8] Über die Konstruktion Riemannscher Flächen durch Verheftung. J. Indian Math. Soc. (N.S.) **24** (1960), 401–412 (1961).

[9] (– Sutter, J.) Riemannsche Flächen vom hyperbolischen Typus, erzeugt durch Asymmetrien. Contemporary problems in theory anal. functions (Internat. Conf., Erevan, 1965), pp. 253–257. [Russian.] Moscow: Izdat. "Nauka" 1966.

Pólya, G.

[1] Cf. Hardy, G.; Littlewood, J.; Pólya, G. [1].

Potyagaĭlo, D. B.

[1] Condition of hyperbolicity of a class of Riemannian surfaces. [Russian.] Ukrain. Mat. Ž. **5** (1953), 459–463.

[2] The type of a conformal sewing of a strip. [Russian.] Dokl. Akad. Nauk SSSR **138** (1961), 1025–1028.

[3] On the type problem for a simply connected Riemann surface. [Russian.] Sibirsk. Mat. Ž. **2** (1961), 895–903.

[4] On a class of Riemann surfaces. [Russian.] Mat. Sb. (N.S.) **56** (98) (1962), 393–402.

Radojčić, M.

[1] Remarque sur le problème des types des surfaces de Riemann. Acad. Serbe Sci. Publ. Inst. Math. **1** (1947), 97–100.

[2] Certains critères concernant le type des surfaces de Riemann à points de ramification algébriques. Ibid. **3** (1950), 25–52, 305–306.

[3] Sur le discernement des types des surfaces de Riemann. [Serbo-Croatian. French summary.] 1. Congr. Math. Phys. R.P.F.Y., 1949, 2. Communications et Exposés Scient., pp. 163–167. Nauîna Knjiga, Belgrade, 1951.

[4] Sur le problème des types des surfaces de Riemann. [Serbo-Croatian. French summary.] Srpska Akad. Nauka. Zbornik Radova 35. Nat. Inst. **3** (1953), 15–28.

Rao, K. V. R.

[1] Lindelöfian maps and positive harmonic functions. Doctoral dissertation, Univ. of Calif., Los Angeles, Calif., 1962. 48 pp.

[2] Lindelöfian meromorphic functions. Proc. Amer. Math. Soc. **15** (1964), 109–113.

[3] Remarks on the classification of Riemann surfaces. Ibid. **15** (1964), 632–634.

[4] On certain algebras of analytic functions. Michigan Math. J. **11** (1964), 231–235.

Reich, E.

[1] (– Warschawski, S.) On canonical conformal maps of regions of arbitrary connectivity. Pacific J. Math. **10** (1960), 965–985.

[2] On a characterization of quasiconformal mappings. Comment. Math. Helv. **37** (1962/63), 44–48.

de Rham, G.

[1] Variétés différentiables. Paris: Hermann 1960. 196 pp.

Richards, I.

[1] On the classification of noncompact surfaces. Trans. Amer. Math. Soc. **106** (1963), 259–269.

Rickart, C.

[1] General theory of Banach algebras. Princeton, N.J.: D. Van Nostrand Co., Inc. 1960. 394 pp.

Riemann, B.

[1] Gesammelte mathematische Werke und wissenschaftlicher Nachlaß. Herausgegeben unter Mitwirkung von R. Dedekind und H. Weber. Leipzig: Teubner 1876. 526 pp.

RODIN, B.

[1] Reproducing kernels and principal functions. Proc. Amer. Math. Soc. **13** (1962), 982–992.

[2] Extremal length and removable boundaries of Riemann surfaces. Bull. Amer. Math. Soc. **72** (1966), 274–276.

[3] (– SARIO, L.) Principal functions. Princeton, N.J.: D. Van Nostrand Co., Inc. 1968. 347 pp.

[4] Cf. MARDEN, A.; RODIN, B. [1].

ROYDEN, H. L.

[1] Some remarks on open Riemann surfaces. Ann. Acad. Sci. Fenn. Ser. A. I. No. 85 (1951), 8 pp.

[2] Harmonic functions on open Riemann surfaces. Trans. Amer. Math. Soc. **73** (1952), 40–94.

[3] On the ideal boundary of a Riemann surface. Contributions to the theory of Riemann surfaces, pp. 107–109. Princeton, N.J.: Princeton Univ. Press. 1953.

[4] Some counterexamples in the classification of open Riemann surfaces. Proc. Amer. Math. Soc. **4** (1953), 363–370.

[5] A property of quasi-conformal mapping. Ibid. **5** (1954), 266–269.

[6] Rings of analytic and meromorphic functions. Trans. Amer. Math. Soc. **83** (1956), 269–276.

[7] Open Riemann surfaces. Ann. Acad. Sci. Fenn. Ser. A.I. No. 249/5 (1958), 13 pp.

[8] Rings of meromorphic functions. Proc. Amer. Math. Soc. **9** (1958), 959–965.

[9] The equation $\Delta u = Pu$, and the classification of open Riemann surfaces. Ann. Acad. Sci. Fenn. Ser. A.I. No. 271 (1959), 27 pp.

[10] The Riemann-Roch theorem. Comment. Math. Helv. **34** (1960), 37–51.

[11] A class of null-bounded Riemann surfaces. Ibid. **34** (1960), 52–66.

[12] The boundary values of analytic and harmonic functions. Math. Z. **78** (1962), 1–24.

[13] Riemann surfaces with the AB-maximum principle. Ann. Acad. Sci. Fenn. Ser. A.I. No. 336/16 (1963), 7 pp.

[14] Real analysis. New York: The Macmillan Co. 1963. 284 pp.

[15] Algebras of bounded analytic functions on Riemann surfaces. Acta Math. **114** (1965), 113–142.

[16] Riemann surfaces with the absolute AB-maximum principle. Proc. Conf. Complex Analysis (Minneapolis, 1964), pp. 172–175. Berlin-Heidelberg-New York: Springer 1965.

[17] Cf. AHLFORS, L.; ROYDEN, H. L. [7].

RUDIN, W.

[1] Analytic functions of class H_p. Trans. Amer. Math. Soc. **78** (1955), 46–66.

[2] Essential boundary points. Bull. Amer. Math. Soc. **70** (1964), 321-324.

SAGAWA, A.

[1] A note on a Riemann surface with null boundary. Tôhoku Math. J. (2) **3** (1951), 273–276.

[2] On the existence of Green's function. Ibid. (2) **7** (1955), 136–139.

[3] On the existence of harmonic functions on a Riemann surface. Ibid. (2) **15** (1963), 265–272.

SAKS, S.

[1] Theory of the integral. Second revised edition. English translation by L. C. YOUNG. New York: Hafner Publishing Co. 1937. 347 pp.

SARIO, L.

[1] Über Riemannsche Flächen mit hebbarem Rand. Ann. Acad. Sci. Fenn. Ser. A.I. No. 50 (1948), 79 pp.

[2] Sur la classification des surfaces de Riemann. 11. Scand. Congr. Math. Trondheim 1949, pp. 229–238.

[3] Sur le problème du type des surfaces de Riemann. C. R. Acad. Sci. Paris **229** (1949), 1109–1111.

[4] Existence des fonctions d'allure donnée sur une surface de Riemann arbitraire. Ibid. **229** (1949), 1293–1295.

[5] Quelques propriétés à la frontière se rattachant à la classification des surfaces de Riemann. Ibid. **230** (1950), 42–44.

[6] Questions d'existence au voisinage de la frontière d'une surface de Riemann. Ibid. **230** (1950), 269–271.

[7] On open Riemann surfaces. Internat. Congr. Math. Cambridge 1950, pp. 398–399.

[8] A linear operator method on arbitrary Riemann surfaces. Trans. Amer. Math. Soc. **72** (1952), 281–295.

[9] An extremal method on arbitrary Riemann surfaces. Ibid. **73** (1952), 459–470.

[10] Construction of functions with prescribed properties on Riemann surfaces. Contributions to the theory of Riemann surfaces, pp. 63–76. Princeton, N.J.: Princeton Univ. Press 1953.

[11] Modular criteria on Riemann surfaces. Duke Math. J. **20** (1953), 279–286.

[12] Minimizing operators on subregions. Proc. Amer. Math. Soc. **4** (1953), 350–355.

[13] Capacity of the boundary and of a boundary component. Ann. of Math. (2) **59** (1954), 135–144.

[14] Functionals on Riemann surfaces. Lectures on functions of a complex variable, pp. 245–256. Ann Arbor: Univ. Michigan Press 1955.

[15] Positive harmonic functions. Ibid. pp. 257–263.

[16] Extremal problems and harmonic interpolation on open Riemann surfaces. Trans. Amer. Math. Soc. **79** (1955), 362–377.

[17] Strong and weak boundary components. J. Analyse Math. **5** (1956/57), 389–398.

[18] On univalent functions. 13. Scand. Congr. Math. Helsinki 1957, pp. 202–208.

[19] Stability problems on boundary components. Seminars on analytic functions. II, pp. 55–72. Princeton, N.J.: Institute for Advanced Study 1958.

[20] On locally meromorphic functions with single-valued moduli. Pacific J. Math. **13** (1963), 709–724.

[21] An integral equation and a general existence theorem for harmonic functions. Comment. Math. Helv. **38** (1964), 284–292.

[22] Classification of locally Euclidean spaces. Nagoya Math. J. **25** (1965), 87–111.

[23] (– FUKUDA, N.) Harmonic functions with given values and minimum norms in Riemannian spaces. Proc. Nat. Acad. Sci. U.S.A. **53** (1965), 270–273.

[24] (– WEILL, G.) Normal operators and uniformly elliptic self-adjoint partial differential equations. Trans. Amer. Math. Soc. **120** (1965), 225–235.

[25] (– SCHIFFER, M.; GLASNER, M.) The span and principal functions in Riemannian spaces. J. Analyse Math. **15** (1965), 115–134.

[26] (– NOSHIRO, K.) Value distribution theory. Princeton, N.J.: D. Van Nostrand Co., Inc. 1966. 236 pp.

[27] (– OIKAWA, K.) Capacity functions. Berlin-Heidelberg-New York: Springer 1969. 361 pp.

[28] Cf. AHLFORS, L.; SARIO, L. [12].

[29–33] Cf. NAKAI, M.; SARIO, L. [30–31], [34–36].

[34] Cf. RODIN, B.; SARIO, L. [3].

SAVAGE, N.

[1] Weak boundary components of an open Riemann surface. Duke Math. J. **24** (1957), 79–95.

[2] Ahlfors' conjecture concerning extreme Sario operators. Bull. Amer. Math. Soc. **72** (1966), 720–724.

SCHIFFER, M.

[1] The span of multiply connected domains. Duke Math. J. **10** (1943), 209–216.

[2] Cf. SARIO, L.; SCHIFFER, M.; GLASNER, M. [26].

SCHMIDT, E.

[1] Über die isoperimetrische Aufgabe im *n*-dimensionalen Raum konstanter negativer Krümmung. Math. Z. **46** (1940), 204–230.

[2] Die isoperimetrischen Ungleichungen auf der gewöhnlichen Kugel und für Rotationskörper im *n*-dimensionalen sphärischen Raum. Ibid. **46** (1940), 743–794.

SCHWARTZ, L.

[1] Théorie des distributions. Paris: Hermann 1966. 420 pp.

SCHWARZ, H.

[1] Gesammelte mathematische Abhandlungen. 2 Bände. Berlin: Springer 1890. 338 pp., 370 pp.

SEEWERKER, J.

[1] The extendability of a Riemann surface. Doctoral dissertation, Univ. of Calif., Los Angeles, Calif., 1957, 42 pp.

SEIBERT, P.

[1] Typus und topologische Randstruktur einfach-zusammenhängender Riemannscher Flächen. Ann. Acad. Sci. Fenn. Ser. A.I. No. 250/34 (1958), 11 pp.

SELBERG, H.

[1] Über die ebenen Punktmengen von der Kapazität Null. Avh. Norske Vid.-Akad. Oslo, No. 10 (1937), 10 pp.

SIBONY, D.

[1] Généralisation de la théorie de Constantinescu-Cornea-Doob sur les propriétés "à la frontière" des fonctions analytiques. C.R. Acad. Sci. Paris **260** (1965), 2686–2688.

SMITH, S.

[1] Classification of Riemannian spaces. Doctoral dissertation, Univ. of Calif., Los Angeles, Calif., 1965. 57 pp.

SOMMER, F.

[1] Cf. BEHNKE, H.; SOMMER, F. [1].

SPEISER, A.

[1] Über Riemannsche Flächen. Comment. Math. Helv. **2** (1930), 284–293.

[2] Riemannsche Flächen vom hyperbolischen Typus. Ibid. **10** (1938), 232–242.

SPRINGER, G.

[1] Introduction to Riemann surfaces. Reading, Mass.: Addison-Wesley Publ., 1957. 307 pp.

STOÏLOW, S.

[1] Les surfaces de Riemann à frontière nulle. VI Sjazed Mat., Warszawa, 1948, 36–37.

[2] Leçons sur les principes topologiques de la théorie des fonctions analytiques. 2. ed., Paris: Gauthier-Villars 1956. 194 pp.

[3] Sur quelques points de la théorie moderne des surfaces de Riemann. Rend. Matem. e sue appl. S. V. **16** (1957), 170–196.

[4] Sur la théorie topologique des recouvrements Riemanniens. Ann. Acad. Sci. Fenn. Ser. A.I. No. 250/35 (1958), 7 pp.

STREBEL, K.

[1] Eine Bemerkung zur Hebbarkeit des Randes einer Riemannschen Fläche. Comment. Math. Helv. **23** (1949), 350–352.

[2] A remark on the extremal distance of two boundary components. Proc. Nat. Acad. Sci. U.S.A. **40** (1954), 842–844.

[3] Die extremale Distanz zweier Enden einer Riemannschen Fläche. Ann. Acad. Sci. Fenn. Ser. A.I. No. 179 (1955), 21 pp.

[4] On the maximal dilation of quasiconformal mappings. Proc. Amer. Math. Soc. **6** (1955), 903–909.

SUITA, N.

[1] On certain criteria for a set to be of class $N_{\mathscr{B}}$. Nagoya Math. J. **19** (1961), 189–194.

[2] Minimal slit domains and minimal sets. Kōdai Math. Sem. Rep. **17** (1965), 166–186.

[3] On radial slit disc mappings. Ibid. **18** (1966), 219–228.

[4] Cf. OIKAWA, K.; SUITA, N. [14].

SUMITA, Y.

[1] Minimal harmonic functions on a Riemann surface. Kōdai Math. Sem. Rep. **18** (1966), 51–60.

SUTTER, J.

[1] Cf. PFLUGER, A.; SUTTER, J. [9].

SZEGÖ, G.

[1] Bemerkungen zu einer Arbeit von Herrn M. FEKETE: Über die Verteilung der Wurzeln bei gewissen algebraischen Gleichungen mit ganzzahligen Koeffizienten. Math. Z. **21** (1924), 203–208.

TAMURA, J.

[1] (– OIKAWA, K.) Examples of minimal slit domains and $N_{\mathscr{D}}$-sets. [Japanese.] Sûgaku **17** (1965), 99–101.

[2] (– OIKAWA, K.; YAMAZAKI, K.) Examples of minimal parallel slit domains. Proc. Amer. Math. Soc. **17** (1966), 283–284.

TEICHMÜLLER, O.

[1] Eine Anwendung quasikonformer Abbildungen auf das Typenproblem. Deutsche Math. **2** (1937), 321–327.

TODA, N.

[1] (– MATSUMOTO, K.) Analytic functions on some Riemann surfaces. Nagoya Math. J. **22** (1963), 211–217.

TÔKI, Y.

[1] On the classification of open Riemann surfaces. Osaka Math. J. **4** (1952), 191–201.

[2] On the examples in the classification of open Riemann surfaces I. Ibid. **5** (1953), 267–280.

TSUJI, M.

[1] On the capacity of general Cantor sets. J. Math. Soc. Japan **5** (1953), 235–252.

[2] A metrical theorem on the singular set of a linear group of Schottky type. Ibid. **6** (1954), 115–121.

[3] On a Riemann surface which is conformally equivalent to a Riemann surface with a finite spherical area. Comment. Math. Univ. St. Paul **6** (1957), 1–7.

[4] On Abelian and Schottkyan covering surfaces of a closed Riemann surface. Ibid. **6** (1957), 8–28.

[5] Potential theory in modern function theory. Tokyo: Maruzen. Co. 1959. 590 pp.

ULRICH, F. E.

[1] The problem of type for a certain class of Riemann surfaces. Duke Math. J. **5** (1939), 567–589.

USKILA, L.

[1] Über die Existenz der beschränkten automorphen Funktionen. Ark. Mat. **1** (1949), 1–11.

VÄISÄLÄ, J.

[1] Cf. GEHRING, F.; VÄISÄLÄ, J. [2].

VIRTANEN, K. I.

[1] Über die Existenz von beschränkten harmonischen Funktionen auf offenen Riemannschen Flächen. Ann. Acad. Sci. Fenn. Ser. A.I. No. 75 (1950), 8 pp.

VITUŠKIN, A. G.

[1] Example of a set of positive length but zero analytic capacity. [Russian.] Dokl. Acad. Nauk SSSR **127** (1959), 246–249.

VOLKOVYSKIĬ, L. I.

[1] On the problem of type of simply connected Riemann surfaces. [Russian.] Rec. Math. (Mat. Sbornik) (N.S.) **18** (60) (1946), 185–212.

[2] Investigations on the problem of type for a simply connected Riemann surface. [Russian.] Uspehi Matem. Nauk (N.S.) **3** (25) (1948), 215–216.

[3] The determination of the type of certain classes of simply-connected Riemann surfaces. [Russian.] Mat. Sbornik (N.S.) **23** (65) (1948), 229–258.

[4] The influence of the closeness of the branch points on the type of a simply connected Riemann surface. [Russian.] Ibid. **25** (67) (1949), 415–450.

[5] On the type problem of a simply connected Riemann surface. [Russian.] Ukrain. Mat. Ž. **1** (1949), 39–48.

[6] An example of a simply connected Riemann surface of hyperbolic type. [Russian.] Ibid. **1** (1949), 60–67.

[7] Investigation of the type problem for a simply connected Riemann surface. [Russian.] Trudy Mat. Inst. Steklov. **34** (1950), 171 pp.

[8] Contemporary investigations on the theory of Riemann surfaces. Uspehi Mat. Nauk **6** (1956), 101–105.

WALSH, B.

[1] Flux in axiomatic potential theory. I; Cohomology. Invent. Math. **8** (1969), 175–221.

[2] Cf. LOEB, P.; WALSH, B. [1].

WARSCHAWSKI, S.

[1] Cf. REICH, E.; WARSCHAWSKI, S. [1].

WEILL, G.

[1] Reproducing kernels and orthogonal kernels for analytic differentials on Riemann Surfaces. Pacific J. Math. **12** (1962), 729–767.

[2] Cf. SARIO, L.; WEILL, G. [24].

WEYL, H.

[1] Die Idee der Riemannschen Fläche. Dritte vollständig umgearbeitete Auflage. Stuttgart: Teubner 1958. 162 pp.

WIRTH, E. M.

[1] Über die Bestimmung des Typus einer Riemannschen Fläche. Comment. Math. Helv. **31** (1956), 90–107.

WITTICH, H.

[1] Ein Kriterium zur Typenbestimmung von Riemannschen Flächen. M. Math. Phys. **44** (1936), 85–96.

[2] Bemerkung zum Typenproblem. Comment. Math. Helv. **26** (1952), 180–183.

YAMASHITA, S.

[1] On some families of analytic functions on Riemann surfaces. Nagoya Math. J. **31** (1968), 57–68.

YAMAZAKI, K.

[1] Cf. TAMURA, J.; OIKAWA, K.; YAMAZAKI, K. [2].

YOSHIDA, M.

[1] The method of orthogonal decomposition for differentials on open Riemann surfaces. J. Sci. Hiroshima Univ. Ser. A-I **32** (1968), 181–210.

YOSIDA, K.

[1] Functional analysis. Berlin-Heidelberg-New York: Springer 1965. 458 pp.

YOSIDA, T.

[1] On a sufficient condition for a given Riemann surface to be of hyperbolic type. Sci. Rep. Tokyo Bunrika Daigaku, Sec. A. **4** (1941), 89–92.

YÛJÔBÔ, Z.

[1] On the Riemann surfaces, no Green function of which exists. Math. Japonicae **2** (1951), 61–68.

Author Index

Subject and Notation Index

Italicized page numbers refer to definitions

Universitätsdruckerei H. Stürtz AG Würzburg

Die Grundlehren der mathematischen Wissenschaften in Einzeldarstellungen mit besonderer Berücksichtigung der Anwendungsgebiete

---

76. Tricomi: Vorlesungen über Orthogonalreihen. DM 68,– ; US $ 18.70
77. Behnke/Sommer: Theorie der analytischen Funktionen einer komplexen Veränderlichen. DM 79,– ; US $ 21.80
78. Lorenzen: Einführung in die operative Logik und Mathematik. DM 54,– ; US $ 14.90
80. Pickert: Projektive Ebenen. DM 48,60; US $ 13.40
81. Schneider: Einführung in die transzendenten Zahlen. DM 24,80; US $ 6.90
82. Specht: Gruppentheorie. DM 69,60; US $ 19.20
84. Conforto: Abelsche Funktionen und algebraische Geometrie. DM 41,80; US $ 11.50
86. Richter: Wahrscheinlichkeitstheorie. DM 68,– ; US $ 18,70
88. Müller: Grundprobleme der mathematischen Theorie elektromagnetischer Schwingungen. DM 52,80; US $ 14.60
89. Pfluger: Theorie der Riemannschen Flächen. DM 39,20; US $ 10.80
90. Oberhettinger: Tabellen zur Fourier Transformation. DM 39,50; US $ 10.90
91. Prachar: Primzahlverteilung. DM 58,– ; US $ 16.00
93. Hadwiger: Vorlesungen über Inhalt, Oberfläche und Isoperimetrie. DM 49,80; US $ 13.70
94. Funk: Variationsrechnung und ihre Anwendung in Physik und Technik. DM 120,– ; US $ 33.00
95. Maeda: Kontinuierliche Geometrien. DM 39,– ; US $ 10.80
97. Greub: Linear Algebra. DM 39,20; US $ 9,80
98. Saxer: Versicherungsmathematik. 2. Teil. DM 48,60; US $ 13.40
99. Cassels: An Introduction to the Geometry of Numbers. DM 69,– ; US $ 19.00
100. Koppenfels/Stallmann: Praxis der konformen Abbildung. DM 69,– ; US $ 19.00
101. Rund: The Differential Geometry of Finsler Spaces. DM 59,60; US $ 16.40
103. Schütte: Beweistheorie. DM 48,– ; US $ 13.20
104. Chung: Markov Chains with Stationary Transition Probabilities. DM 56,– ; US $ 14.00
105. Rinow: Die innere Geometrie der metrischen Räume. DM 83,– ; US $ 22.90
106. Scholz/Hasenjaeger: Grundzüge der mathematischen Logik. DM 98,– ; US $ 27.00
107. Köthe: Topologische lineare Räume I. DM 78,– ; US $ 21.50
108. Dynkin: Die Grundlagen der Theorie der Markoffschen Prozesse. DM 33,80; US $ 9.30
110. Dinghas: Vorlesungen über Funktionentheorie. DM 69,– ; US $ 19.00
111. Lions: Equations différentielles opérationnelles et problèmes aux limits. DM 64,– ; US $ 17.60
112. Morgenstern/Szabó: Vorlesungen über theoretische Mechanik. DM 69,– ; US $ 19.00
113. Meschkowski: Hilbertsche Räume mit Kernfunktion. DM 58,– ; US $ 16.00
114. MacLane: Homology. DM 62,– ; US $ 15.50
115. Hewitt/Ross: Abstract Harmonic Analysis. Vol. 1: Structure of Topological Groups, Integration Theory, Group Representations. DM 76,– ; US $ 20.90
116. Hörmander: Linear Partial Differential Operators. DM 42,– ; US $ 10.50
117. O'Meara: Introduction to Quadratic Forms. DM 48,– ; US $ 13.20
118. Schäfke: Einführung in die Theorie der speziellen Funktionen der mathematischen Physik. DM 49,40; US $ 13.60
119. Harris: The Theory of Branching Processes. DM 36,– ; US $ 9.90
120. Collatz: Funktionalanalysis und numerische Mathematik. DM 58,– ; US $ 16.00
121./122. Dynkin: Markov Processes. DM 96,– ; US $ 26.40
123. Yosida: Functional Analysis. DM 66,– ; US $ 16.50
124. Morgenstern: Einführung in die Wahrscheinlichkeitsrechnung und mathematische Statistik. DM 38,– ; US $ 10.50

125. Itô/McKean: Diffusion Processes and Their Sample Paths. DM 58,–; US $ 16.00
126. Letho/Virtanen: Quasikonforme Abbildungen. DM 38,–; US $ 10.50
127. Hermes: Enumerability, Decidability, Computability. DM 39,–; US $ 10.80
128. Braun/Koecher: Jordan-Algebren. DM 48,–; US $ 13.20
129. Nikodým: The Mathematical Apparatus for Quantum-Theories. DM 144,–; US $ 36.00
130. Morrey: Multiple Integrals in the Calculus of Variations. DM 78,–; US $ 19.50
131. Hirzebruch: Topological Methods in Algebraic Geometry. DM 38,–; US $ 9.50
132. Kato: Perturbation Theory for Linear Operators. DM 79,20; US $ 19.80
133. Haupt/Künneth: Geometrische Ordnungen. DM 68,–; US $ 18.70
134. Huppert: Endliche Gruppen I. DM 156,–; US $ 42.90
135. Handbook for Automatic Computation. Vol. 1/Part a: Rutishauser: Description of ALOGL 60. DM 58,–; US $ 14.50
136. Greub: Multilinear Algebra. DM 32,–; US $ 8.00
137. Handbook for Automatic Computation. Vol. 1/Part b: Grau/Hill/Langmaack: Translation of ALOGL 60. DM 64,–; US $ 16.00
138. Hahn: Stability of Motion. DM 72,–; US $ 19.80
139. Mathematische Hilfsmittel des Ingenieurs. Herausgeber: Sauer/Szabó. 1. Teil. DM 88,–; US $ 24.20
140. Mathematische Hilfsmittel des Ingenieurs. Herausgeber: Sauer/Szabó. 2. Teil. DM 136,–; US $ 37.40
141. Mathematische Hilfsmittel des Ingenieurs. Herausgeber: Sauer/Szabó. 3. Teil. DM 98,–; US $ 27.00
142. Mathematische Hilfsmittel des Ingenieurs. Herausgeber: Sauer/Szabó. 4. Teil. In Vorbereitung.
143. Schur/Grunsky: Vorlesungen über Invariantentheorie. DM 32,–; US $ 8.80
144. Weil: Basic Number Theory. DM 48,–; US $ 12.00
145. Butzer/Berens: Semi-Groups of Operators and Approximation. DM 56,–; US $ 14.00
146. Treves: Locally Convex Spaces and Linear Partial Differential Equations. DM 36,–; US $ 9.90
147. Lamotke: Semisimpliziale algebraische Topologie. DM 48,–; US $ 13.20
148. Chandrasekharan: Introduction to Analytic Number Theory. DM 28,–; US $ 7.00
149. Sario/Oikawa: Capacity Functions. DM 96,–; US $ 24.00
150. Iosifescu/Theodorescu: Random Processes and Learning. DM 68,–; US $ 18.70
151. Mandl: Analytical Treatment of One-dimensional Markov Processes. DM 36,–; US $ 9.80
152. Hewitt/Ross: Abstract Harmonic Analysis. Vol. II. DM 140,–; US $ 38.50
153. Federer: Geometric Measure Theory. DM 118,–; US $ 29.50
154. Singer: Bases in Banach Spaces I. DM 112,–; US $ 30.80
155. Müller: Foundations of the Mathematical Theory of Electromagnetic Waves. DM 58,–; US $ 16.00
156. van der Waerden: Mathematical Statistics. DM 68,–; US $ 18.70
157. Prohorov/Rozanov: Probability Theory. DM 68,–; US $ 18.70
159. Köthe: Topological Vector Spaces I. DM 78,–; US $ 21.50
160. Agrest/Maksimov: Theory of Incomplete Cylindrical Functions and their Applications. In preparation
161. Bhatia/Szegö: Stability Theory of Dynamical Systems. In preparation
162. Nevanlinna: Analytical Functions. DM 76,–; US $ 20.90
163. Stoer/Witzgall: Convexity and Optimization in Finite Dimensions I. DM 54,–; US $ 14.90
164. Sario/Nakai: Classification Theory of Riemann Surfaces. DM 98,–; US $ 27.00
165. Mitrinović: Analytic Inequalities. In preparation

166. Grothendieck/Dieudonné: Eléments de Géometrie Algébrique I. In preparation
167. Chandrasekharan: Arithmetical Functions. DM 58,– ; US $ 16.00
168. Palamodov: Linear Differential Operators with Constant Coefficients. In preparation
169. Rademacher: Topics in Analytic Number Theory. In preparation
170. Lions: Optimal Control Systems Governed by Partial Differential Equations. In preparation
171. Singer: Best Approximation in Normed Linear Spaces by Elements of Linear Subspaces. DM 60,– ; US $ 16.50